Achieving sustainable cultivation of grain legumes

Volume 2: Improving cultivation of particular grain legumes

It is widely recognised that agriculture is a significant contributor to global warming and climate change. Agriculture needs to reduce its environmental impact and adapt to current climate change whilst still feeding a growing population, i.e. become more 'climate-smart'. Burleigh Dodds Science Publishing is playing its part in achieving this by bringing together key research on making the production of the world's most important crops and livestock products more sustainable.

Based on extensive research, our publications specifically target the challenge of climate-smart agriculture. In this way we are using 'smart publishing' to help achieve climate-smart agriculture.

Burleigh Dodds Science Publishing is an independent and innovative publisher delivering high quality customer-focused agricultural science content in both print and online formats for the academic and research communities. Our aim is to build a foundation of knowledge on which researchers can build to meet the challenge of climate-smart agriculture.

For more information about Burleigh Dodds Science Publishing simply call us on +44 (0) 1223 839365, email info@bdspublishing.com or alternatively please visit our website at www.bdspublishing.com.

Related titles:

Achieving sustainable cultivation of grain legumes Volume 1:
Advances in breeding and cultivation techniques
Print (ISBN 978-1-78676-136-1); Online (ISBN 978-1-78676-138-5, 978-1-78676-139-2)

Achieving sustainable cultivation of cassava Volume 1: Cultivation techniques
Print (ISBN 978-1-78676-000-5); Online (ISBN 978-1-78676-002-9, 978-1-78676-003-6)

Achieving sustainable cultivation of cassava Volume 2: Genetics, breeding, pests and diseases
Print (ISBN 978-1-78676-004-3); Online (ISBN 978-1-78676-006-7, 978-1-78676-007-4)

Managing soil health for sustainable agriculture Volume 1: Fundamentals
Print (ISBN 978-1-78676-188-0); Online (ISBN 978-1-78676-190-3, 978-1-78676-191-0)

Managing soil health for sustainable agriculture Volume 2: Monitoring and management
Print (ISBN 978-1-78676-192-7); Online (ISBN 978-1-78676-194-1, 978-1-78676-195-8)

Chapters are available individually from our online bookshop: https://shop.bdspublishing.com

climate•SMART•publishing
IN AGRICULTURAL SCIENCE

BURLEIGH DODDS SERIES IN AGRICULTURAL SCIENCE

NUMBER 36

Achieving sustainable cultivation of grain legumes

Volume 2: Improving cultivation of particular grain legumes

Edited by Dr Shoba Sivasankar, formerly International Crops Research Institute for the Semi-Arid Tropics (ICRISAT), India; Dr David Bergvinson and Dr Pooran Gaur, ICRISAT, India; Dr Shiv Kumar Agrawal, International Center for Agricultural Research in the Dry Areas (ICARDA), Morocco; Dr Steve Beebe, International Center for Tropical Agriculture (CIAT), Colombia; and Dr Manuele Tamò, International Institute of Tropical Agriculture (IITA), Benin

burleigh dodds
SCIENCE PUBLISHING

Published by Burleigh Dodds Science Publishing Limited
82 High Street, Sawston, Cambridge CB22 3HJ, UK
www.bdspublishing.com

Burleigh Dodds Science Publishing, 1518 Walnut Street, Suite 900, Philadelphia, PA 19102-3406, USA

First published 2018 by Burleigh Dodds Science Publishing Limited

Library of Congress Control Number: 2017938497

British Library Cataloguing in Publication Data
A catalogue record for this book is available from the British Library

ISBN 978-1-78676-140-8 (print)
ISBN 978-1-78676-142-2 (online)
ISBN 978-1-78676-143-9 (online)
ISSN 2059-6936 (print)
ISSN 2059-6944 (online)

Typeset by Deanta Global Publishing Services, Chennai, India
Printed by Lightning Source

Contents

Series list xi

Foreword xv

Acknowledgements xvii

Introduction xviii

Part 1 Cultivation of common beans, lentils, soybeans and groundnuts

1 Developing improved varieties of common bean 3
James D. Kelly, Michigan State University, USA
1 Introduction 3
2 Yield and yield gap in common beans 4
3 Factors affecting bean yields 5
4 Quantitative trait loci (QTL) analysis 8
5 Genomic analysis 12
6 Conclusion 13
7 Future trends 13
8 Where to look for further information 14
9 References 14

2 Improving cultivation practices for common beans 19
John O. Ojiem, Kenya Agricultural and Livestock Research Organization, Kenya
1 Introduction to the common bean: origin, domestication and economic importance 19
2 World production trends 20
3 Production constraints 23
4 Cropping system 27
5 Integrated pest management 30
6 Management of diseases 31
7 Water management 33
8 Nutrient management: phosphorus (P) and nitrogen (N) 36
9 Nutrient management: further issues 40
10 Conclusion 43
11 Where to look for further information 43
12 Abbreviations and acronyms 44
13 References 44

3 Developing improved varieties of lentil 51
William Erskine, University of Western Australia, Australia; Ashutosh Sarker, International Center for Agricultural Research in the Dry Areas (ICARDA), India; and Shiv Kumar, International Center for Agricultural Research in the Dry Areas (ICARDA), Morocco
1 Introduction 51
2 Production regions and their agro-ecologies 52

3 Phylogeny, domestication/spread and genetic resources 54
4 Problems addressable by breeding 56
5 Breeding methods 58
6 New technologies 59
7 Success stories 62
8 Future trends 63
9 Where to look for further information 64
10 References 64

4 Improving cultivation of lentil 71
Fred J. Muehlbauer, Washington State University, USA; and Ashutosh Sarker, International Center for Agricultural Research in the Dry Areas (ICARDA), India
1 Introduction 71
2 Lentil description 73
3 Climate requirements for lentil cultivation 74
4 Land requirements for lentil cultivation 74
5 Nutrient requirements for lentil cultivation 74
6 Seed quality and treatments 75
7 Land preparation and planting 75
8 Innovations in lentil cropping systems 77
9 Conclusion and future trends 80
10 Where to look for further information 80
11 References 80

5 The use of marker-assisted selection in developing improved varieties of soybean 83
Y.-C. Lee, R. Lemes Hamawaki, V. Colantonio, M. J. Iqbal and D. A. Lightfoot, Southern Illinois University, USA
1 Introduction 83
2 Genomes as intellectual property 84
3 Methods for MAS 85
4 Marker development 85
5 Choice of markers 88
6 Identification of polymorphism 89
7 Genetic and association map development 91
8 Marker-assisted recovery of recurrent parent genome 92
9 MAS in recurrent cross populations 92
10 Scoring of phenotypes 93
11 Use of molecular markers to select desirable traits in soybean 93
12 Future trends and conclusion 98
13 Where to look for further information 98
14 References 99

6 Improving cultivation practices for soybeans in sub-Saharan Africa 105
Frederick P. Baijukya and Harun M. Murithi, International Institute of Tropical Agriculture (IITA), Tanzania; and Fred Kanampiu, International Institute of Tropical Agriculture (IITA), Kenya
1 Introduction 105
2 Trends in soybean production and consumption in SSA 107

3 Best practices to enhance soybean yields: inoculation with rhizobia and use of improved varieties 108
4 Best practices to enhance soybean yields: improved crop management 113
5 Other considerations to improve soybean productivity in SSA 116
6 Conclusion 119
7 Where to look for further information 120
8 References 120

7 Developing improved varieties of groundnut 123
C. Michael Deom, University of Georgia, USA; and David Kalule Okello, National Semi-Arid Resources Research Institute, Uganda
1 Introduction 123
2 Genetic resources for trait mining 126
3 Breeding targets 126
4 Groundnut oil quality and content 136
5 Case study: groundnut improvement in Uganda 137
6 Conclusion and future trends 143
7 Where to look for further information 144
8 Acknowledgements 145
9 References 145

8 Improving cultivation of groundnuts 155
S. N. Nigam, International Crops Research Institute for the Semi-Arid Tropics (ICRISAT), India; D. L. Jordan, North Carolina State University, USA; and P. Janila, International Crops Research Institute for the Semi-Arid Tropics (ICRISAT), India
1 Introduction 155
2 Limitations of present agronomic recommendations 157
3 Choice of variety/cultivar 158
4 Field preparation and soil resources management 159
5 Seed preparation, planting and weed and water management 162
6 Plant protection practices 167
7 Harvesting, drying, curing and storage 170
8 Precision cultivation 172
9 Seed systems 174
10 Conclusion 174
11 Where to look for further information 175
12 References 175

9 Preventing mycotoxin contamination in groundnut cultivation 181
David Jordan, Rick Brandenburg and Gary Payne, North Carolina State University, USA; David Hoisington, Nick Magnan and James Rhoads, The University of Georgia, USA; Mumuni Abudulai, Savanna Agricultural Research Institute, Ghana; Koushik Adhikari and Jinru Chen, The University of Georgia, USA; Richard Akromah, William Appaw and William Ellis, Kwame Nkrumah University of Science and Technology, Ghana; Maria Balota and Kumar Mallikarjunan, Virginia Polytechnic Institute and State University, USA; Kenneth Boote and Greg MacDonald, University of Florida, USA; Kira Bowen, Auburn University, USA; Boris

Bravo-Ureta and Jeremy Jelliffe, University of Connecticut, USA; Agnes Budu, University of Ghana, Ghana; Hendrix Chalwe, Alice Mweetwa and Munsanda Ngulube, University of Zambia, Zambia; Awere Dankyi and Brandford Mochia, Crops Research Institute, Ghana; Vivian Hoffmann, International Food Policy Research Institute, USA; Amade Muitia, Mozambique Institute of Agricultural Research, Mozambique; Agnes Mwangwela, Lilongwe University of Agriculture and Natural Resources, Malawi; Sam Njoroge, International Crops Research Institute for the Semi-Arid Tropics (ICRISAT), Malawi; David Okello, National Semi-Arid Resources Research Institute (NaSARRI), Uganda; and Nelson Opoku, University for Development Studies, Ghana

1 Introduction 181
2 Factors impacting aflatoxin development 182
3 Prediction models for aflatoxin development 184
4 Minimizing aflatoxin contamination 185
5 Aflatoxin contamination during drying, storage and processing 190
6 Challenges in mycotoxin research 192
7 Value chain projects 193
8 Conclusion 203
9 Where to look for further information 204
10 References 204

Part 2 Cultivation of cowpea, faba beans and pigeonpea

10 Breeding improved varieties of cowpea 215
B. B. Singh, G.B. Pant University of Agriculture and Technology, India

1 Introduction 215
2 Production constraints 217
3 Cowpea breeding programmes and past challenges 217
4 Cowpea breeding at IITA 218
5 Highlights of progress made 220
6 Cowpea international trials 227
7 Future trends and conclusion 227
8 Where to look for further information 230
9 References 230

11 Improving cultivation of cowpea in West Africa 235
Alpha Y. Kamara, Lucky O. Omoigui and Nkeki Kamai, International Institute of Tropical Agriculture (IITA), Nigeria; Sylvester U. Ewansiha, University of Benin, Nigeria; and Hakeem A. Ajeigbe, International Crops Research Institute for the Semi-Arid Tropics (ICRISAT), Nigeria

1 Introduction 235
2 Optimal plant population 236
3 Plant configuration in intercropping systems in West Africa 237
4 Manipulating planting dates to improve cowpea productivity 241
5 Nutrient management for increased cowpea productivity 244
6 Integrated pest management (IPM) in cowpea production 245

7 Future trends and conclusion 247
8 Where to look for further information 247
9 References 248

12 Developing improved varieties of faba bean 253
Fouad Maalouf, International Center for Agricultural Research in the Dry Areas (ICARDA), Lebanon; and Seid Ahmed and Somanagouda Patil, International Center for Agricultural Research in the Dry Areas (ICARDA), Morocco
1 Introduction 253
2 Heat tolerance 255
3 Drought tolerance 255
4 Herbicide tolerance 256
5 Breeding for resistance to foliar diseases 257
6 Breeding for broomrape resistance 258
7 Biological nitrogen fixation 258
8 Major achievements 259
9 Future trends and conclusion 262
10 References 263

13 Efficient and sustainable production of faba bean 269
R. Redden, RJR Agricultural Consultants, Australia; X. Zong, Chinese Academy of Agricultural Sciences (CAAS), China; R. M. Norton, International Plant Nutrition Institute and The University of Melbourne, Australia; F. L. Stoddard, University of Helsinki, Finland; F. Maalouf, International Center for Agricultural Research in the Dry Areas (ICARDA), Lebanon; S. Ahmed and M. El Bouhssini, International Center for Agricultural Research in the Dry Areas (ICARDA), Morocco; Y. Tao and L. Rong, Chinese Academy of Agricultural Sciences (CAAS), China; and L. Ling, Liaoning Academy of Agricultural Science, China
1 Introduction 269
2 Faba bean production in China 270
3 Faba bean production in Central and West Asia, North Africa regions (CWANA) 274
4 Faba bean breeding 276
5 Faba bean genetic and genomic resources 278
6 Faba bean management with limited water resources 280
7 Faba bean BNF 282
8 Faba bean mineral nutrient requirements 283
9 Conclusion 287
10 Where to look for further information 288
11 References 288

14 Developing improved varieties of pigeonpea 297
K. B. Saxena, United Arab Emirates; Y. S. Chauhan, Department of Agriculture and Fisheries, Australia; C. V. S. Kumar, A. J. Hingane, R. V. Kumar, R. K. Saxena and G. V. R. Rao, International Crops Research Institute for the Semi-Arid Tropics (ICRISAT), India
1 Introduction 297
2 Pigeonpea for nutritional security 298

3 Factors affecting stability of pigeonpea production 299
4 Genetic factors influencing sustainability of pigeonpea production 303
5 Enhancing pigeonpea sustainability through crop modelling 306
6 Enhancing sustainability through an efficient seed system 307
7 Enhancing sustainability through plant breeding 308
8 Pigeonpea hybrids for greater productivity and sustainability 316
9 Future trends and conclusion 320
10 Where to look for further information 321
11 References 321

15 Improving the cultivation of pigeonpea 327
K. R. Latha and L. Vimalendran, Tamil Nadu Agricultural University, India
1 Introduction 327
2 Seeds and sowing 329
3 Intercropping/mixed cropping 334
4 Nutrient management 336
5 Water management 341
6 Efficient use of fertilizer and water 343
7 Weed management 347
8 Agronomic management of pigeonpea hybrids 349
9 Early-maturing pigeonpea varieties 351
10 Rabi and summer pigeonpea cultivation 355
11 Harvesting, threshing and storage 356
12 Conclusion 356
13 Guidelines for future research 357
14 Where to look for further information 358
15 References 359

Index 367

Series list

Title	Series number
Achieving sustainable cultivation of maize - Vol 1 From improved varieties to local applications Edited by: Dr Dave Watson, CGIAR Maize Research Program Manager, CIMMYT, Mexico	001
Achieving sustainable cultivation of maize - Vol 2 Cultivation techniques, pest and disease control Edited by: Dr Dave Watson, CGIAR Maize Research Program Manager, CIMMYT, Mexico	002
Achieving sustainable cultivation of rice - Vol 1 Breeding for higher yield and quality Edited by: Prof. Takuji Sasaki, Tokyo University of Agriculture, Japan	003
Achieving sustainable cultivation of rice - Vol 2 Cultivation, pest and disease management Edited by: Prof. Takuji Sasaki, Tokyo University of Agriculture, Japan	004
Achieving sustainable cultivation of wheat - Vol 1 Breeding, quality traits, pests and diseases Edited by: Prof. Peter Langridge, The University of Adelaide, Australia	005
Achieving sustainable cultivation of wheat - Vol 2 Cultivation techniques Edited by: Prof. Peter Langridge, The University of Adelaide, Australia	006
Achieving sustainable cultivation of tomatoes Edited by: Dr Autar Mattoo, USDA-ARS, USA & Prof. Avtar Handa, Purdue University, USA	007
Achieving sustainable production of milk - Vol 1 Milk composition, genetics and breeding Edited by: Dr Nico van Belzen, International Dairy Federation (IDF), Belgium	008
Achieving sustainable production of milk - Vol 2 Safety, quality and sustainability Edited by: Dr Nico van Belzen, International Dairy Federation (IDF), Belgium	009
Achieving sustainable production of milk - Vol 3 Dairy herd management and welfare Edited by: Prof. John Webster, University of Bristol, UK	010
Ensuring safety and quality in the production of beef - Vol 1 Safety Edited by: Prof. Gary Acuff, Texas A&M University, USA & Prof.James Dickson, Iowa State University, USA	011
Ensuring safety and quality in the production of beef - Vol 2 Quality Edited by: Prof. Michael Dikeman, Kansas State University, USA	012
Achieving sustainable production of poultry meat - Vol 1 Safety, quality and sustainability Edited by: Prof. Steven C. Ricke, University of Arkansas, USA	013
Achieving sustainable production of poultry meat - Vol 2 Breeding and nutrition Edited by: Prof. Todd Applegate, University of Georgia, USA	014
Achieving sustainable production of poultry meat - Vol 3 Health and welfare Edited by: Prof. Todd Applegate, University of Georgia, USA	015

Achieving sustainable production of eggs - Vol 1 016
Safety and quality
Edited by: Prof. Julie Roberts, University of New England, Australia

Achieving sustainable production of eggs - Vol 2 017
Animal welfare and sustainability
Edited by: Prof. Julie Roberts, University of New England, Australia

Achieving sustainable cultivation of apples 018
Edited by: Dr Kate Evans, Washington State University, USA

Integrated disease management of wheat and barley 019
Edited by: Prof. Richard Oliver, Curtin University, Australia

Achieving sustainable cultivation of cassava - Vol 1 020
Cultivation techniques
Edited by: Dr Clair Hershey, formerly International Center for Tropical Agriculture (CIAT), Colombia

Achieving sustainable cultivation of cassava - Vol 2 021
Genetics, breeding, pests and diseases
Edited by: Dr Clair Hershey, formerly International Center for Tropical Agriculture (CIAT), Colombia

Achieving sustainable production of sheep 022
Edited by: Prof. Johan Greyling, University of the Free State, South Africa

Achieving sustainable production of pig meat - Vol 1 023
Safety, quality and sustainability
Edited by: Prof. Alan Mathew, Purdue University, USA

Achieving sustainable production of pig meat - Vol 2 024
Animal breeding and nutrition
Edited by: Prof. Julian Wiseman, University of Nottingham, UK

Achieving sustainable production of pig meat - Vol 3 025
Animal health and welfare
Edited by: Prof. Julian Wiseman, University of Nottingham, UK

Achieving sustainable cultivation of potatoes - Vol 1 026
Breeding, nutritional and sensory quality
Edited by: Prof. Gefu Wang-Pruski, Dalhousie University, Canada

Achieving sustainable cultivation of oil palm - Vol 1 027
Introduction, breeding and cultivation techniques
Edited by: Prof. Alain Rival, Center for International Cooperation in Agricultural Research for Development (CIRAD), France

Achieving sustainable cultivation of oil palm - Vol 2 028
Diseases, pests, quality and sustainability
Edited by: Prof. Alain Rival, Center for International Cooperation in Agricultural Research for Development (CIRAD), France

Achieving sustainable cultivation of soybeans - Vol 1 029
Breeding and cultivation techniques
Edited by: Prof. Henry Nguyen, University of Missouri, USA

Achieving sustainable cultivation of soybeans - Vol 2 030
Diseases, pests, food and non-food uses
Edited by: Prof. Henry Nguyen, University of Missouri, USA

Achieving sustainable cultivation of sorghum - Vol 1 031
Genetics, breeding and production techniques
Edited by: Prof. Bill Rooney, Texas A&M University, USA

Achieving sustainable cultivation of sorghum - Vol 2 032
Sorghum utilisation around the world
Edited by: Prof. Bill Rooney, Texas A&M University, USA

Achieving sustainable cultivation of potatoes - Vol 2 033
Production and storage, crop protection and sustainability
Edited by: Dr Stuart Wale, Potato Dynamics Ltd, UK

Achieving sustainable cultivation of mangoes 034
Edited by: Professor Víctor Galán Saúco, Instituto Canario de Investigaciones Agrarias (ICIA), Spain & Dr Ping Lu, Charles Darwin University, Australia

Achieving sustainable cultivation of grain legumes - Vol 1 035
Advances in breeding and cultivation techniques
Edited by: Dr Shoba Sivasankar et al., CGIAR Research Program on Grain Legumes, ICRISAT, India

Achieving sustainable cultivation of grain legumes - Vol 2 036
Improving cultivation of particular grain legumes
Edited by: Dr Shoba Sivasankar et al., CGIAR Research Program on Grain Legumes, ICRISAT, India

Achieving sustainable cultivation of sugarcane - Vol 1 037
Cultivation techniques, quality and sustainability
Edited by: Prof. Philippe Rott, University of Florida, USA

Achieving sustainable cultivation of sugarcane - Vol 2 038
Breeding, pests and diseases
Edited by: Prof. Philippe Rott, University of Florida, USA

Achieving sustainable cultivation of coffee 039
Edited by: Dr Philippe Lashermes, Institut de Recherche pour le Développement (IRD), France

Achieving sustainable cultivation of bananas - Vol 1 040
Cultivation techniques
Edited by: Prof. Gert Kema, Wageningen University, The Netherlands & Prof. André Drenth, University of Queensland, Australia

Global Tea Science 041
Current status and future needs
Edited by: Dr V. S. Sharma, Formerly UPASI Tea Research Institute, India & Dr M. T. Kumudini Gunasekare, Coordinating Secretariat for Science Technology and Innovation (COSTI), Sri Lanka

Integrated weed management 042
Edited by: Emeritus Prof. Rob Zimdahl, Colorado State University, USA

Achieving sustainable cultivation of cocoa - Vol 1 043
Genetics, breeding, cultivation and quality
Edited by: Prof. Pathmanathan Umaharan, Cocoa Research Centre – The University of the West Indies, Trinidad and Tobago

Achieving sustainable cultivation of cocoa - Vol 2 044
Diseases, pests and sustainability
Edited by: Prof. Pathmanathan Umaharan, Cocoa Research Centre – The University of the West Indies, Trinidad and Tobago

Water management for sustainable agriculture 045
Edited by: Prof. Theib Oweis, Formerly ICARDA, Lebanon

Improving organic animal farming 046
Edited by: Dr Mette Vaarst, Aarhus University, Denmark & Dr Stephen Roderick, Duchy College, Cornwall, UK

Improving organic crop cultivation 047
Edited by: Prof. Ulrich Köpke, University of Bonn, Germany

Managing soil health for sustainable agriculture - Vol 1 048
Fundamentals
Edited by: Dr Don Reicosky, USDA-ARS, USA

Managing soil health for sustainable agriculture - Vol 2 049
Monitoring and management
Edited by: Dr Don Reicosky, USDA-ARS, USA

Rice insect pests and their management 050
E. A. Heinrichs, Francis E. Nwilene, Michael J. Stout, Buyung A. R. Hadi & Thais Freitas

Improving grassland and pasture management in temperate agriculture 051
Edited by: Prof. Athole Marshall & Dr Rosemary Collins, University of Aberystwyth, UK

Precision agriculture for sustainability 052
Edited by: Dr John Stafford, Silsoe Solutions, UK

Achieving sustainable cultivation of temperate zone tree fruit and berries – Vol 1 053
Physiology, genetics and cultivation
Edited by: Prof. Gregory Lang, Michigan State University, USA

Achieving sustainable cultivation of temperate zone tree fruit and berries – Vol 2 054
Case studies
Edited by: Prof. Gregory Lang, Michigan State University, USA

Agroforestry for sustainable agriculture 055
Edited by: Prof. María Mosquera-Losada, University of Santiago de Compostela, Spain & Dr Ravi Prabhu, World Agroforestry Centre (ICRAF), Kenya

Achieving sustainable cultivation of tree nuts 056
Edited by: Prof. Ümit Serdar, Ondokuz Mayis University, Turkey & Emeritus Prof. Dennis Fulbright, Michigan State University, USA

Assessing the environmental impact of sustainable agriculture 057
Edited by: Prof. Bo P. Weidema, Aalborg University/2.-0 LCA Consultants, Denmark

Critical issues in plant health: 50 years of research in African agriculture 058
Edited by: Dr. Peter Neuenschwander, IITA & Dr. Manuele Tamò, IITA

Achieving sustainable cultivation of vegetables – Vol 1 059
Physiology, breeding, cultivation and quality
Edited by: Emeritus Prof. George Hochmuth, University of Florida, USA

Achieving sustainable cultivation of vegetables – Vol 2 060
Case studies
Edited by: Emeritus Prof. George Hochmuth, University of Florida, USA

Foreword

The world continues to face serious challenges of hunger and malnutrition as well as the challenges of increasing food production in a sustainable way that protects the environment and maintains the productivity of agricultural land. According to the FAO, despite all the efforts to address these challenges over many years, about 795 million people of the 7.3 billion people in the world, or one in nine, suffered from chronic undernourishment in 2014–2016. Despite their vital importance to address these global challenges, pulses have not received sufficient world attention for the potential role they can play. Pulses are at least 3 to 4 times richer in plant-based protein compared to major cereals crops such as rice and wheat. They have a complementary amino acid profile with cereals and make a balanced diet when combined with cereals. This is particularly important for the low-income strata of the population. They are also high in micro-nutrients, especially iron and zinc, which address the problem of hidden hunger and anemia. Pulses are also high in dietary fiber and a good source of carbohydrates which makes them a great functional food.

In addition to their unique nutritional qualities, pulses, as leguminous plants, fix atmospheric nitrogen which enhances soil fertility and maintains soil productivity, protecting the environment. They have a relatively low carbon footprint compared to other crops since they do not require chemical nitrogen fertilizer. Including pulses in crop rotations reduces the overall N fertilizer requirement of crop production systems which has great benefits to the subsequent crops in the rotation. This makes them important climate change crops. Pulses are certainly the best crops to solve the soil problems caused by the cereals monoculture that many developing countries follow due to the subsidies for cereals.

Only very recently the world community recognized the importance of pulses through the declaration of 2016 as the International Year of Pulses (IYP 2016). This was announced by the United Nations General Assembly upon the recommendation of FAO General Conference. Although the recognition was belated, it has been an excellent opportunity 'to raise public awareness of the nutritional benefits of pulses as part of sustainable food production aimed towards food security and nutrition'. The IYP 2016 was successful in achieving its objectives considering the many activities held throughout the world, as was well documented by The IYP 2016 Legacy developed by FAO with other stakeholders on pulses as 'nutritious seeds for sustainable future'.

However, pulses still face major challenges that require immediate attention so that they can play their potential role to help achieve food and nutritional security as well as sustainable agricultural production systems. The areas that require immediate attention include the following:

- Pulses should be considered as major crops and not as secondary crops mostly grown under rainfed marginal conditions. Thus, they deserve much more investment in both research and development (R & D) which are currently negligible compared to the R & D investment in cereals.
- More research efforts should focus on ensuring high yield potential by utilizing a wider range of genetic diversity to mine desirable genes from both conserved land races and wild relatives.
- More work needs to be done on yield instability through breeding for tolerance and durable resistance to a range of abiotic and biotic stresses.

- More attention needs to be paid to emerging new constraints and stresses that are caused by climate change, including excessive drought and heat as well as emerging diseases and insect pests.
- Overcoming inappropriate policies favouring cereal production at the expense of pulses as a result of subsidies. This could involve either lifting subsidies on cereals or providing similar subsidies to pulses as India rightly did recently.
- Last but not least, overcoming the limited access of farmers to high quality seed from improved varieties and other important inputs for high pulses productivity.

The two volumes of *Achieving sustainable cultivation of grain legumes*, the first on advances in breeding and cultivation and the second on improving cultivation of particular grain legumes, are very timely. They bring more visibility to these important crops as well as updated knowledge on both the science and practical advances in cultivation addressing the challenges facing specific pulses so that they can play their role in achieving food and nutritional security as well as sustainable agricultural production.

Mahmoud El Solh, Ph.D.
Former Director General
International Center for Agricultural Research in the Dry Areas (ICARDA)

Acknowledgements

We wish to acknowledge the following for their help in reviewing particular chapters:

- Chapter 7: Dr Emmanuel Monyo, ICRISAT, Kenya
- Chapter 8: Dr Jamie Rhoads, University of Georgia, USA

Introduction

Grain legumes are widely cultivated, particularly for their dry seeds (known as pulses). The FAO defines pulses as crop plant members of the *Leguminosae* family (commonly known as the pea family) that produce edible seeds. In this definition, only legumes harvested for dry grain are classified as pulses, excluding species such as soybean and groundnut. However, broader definitions of grain legumes include groundnut and soybean as well as common bean and lentils which are grown globally. Other more regionally-specific types of pulses include cowpea, faba beans and pigeonpea.

Grain legumes are important in the developing world for a number of reasons. They are a rich source of protein and fibre, minerals and vitamins. In addition, their rapid growth and ability to fix nitrogen and improve soil health makes them a key rotation crop in the sustainable intensification and diversification of smallholder farming. This makes grain legumes a key food security crop. However, yields in developing countries are low as a result of such factors as the need for improved varieties of seed, poor seed distribution, the impact of pests and diseases, as well as vulnerability to poor soils, drought and other effects of climate change. There is now a rich body of research addressing these challenges.

The challenges facing grain legumes are addressed in the two volumes of *Achieving sustainable cultivation of grain legumes*:

- Volume 1 Advances in breeding and cultivation techniques
- Volume 2 Improving cultivation of particular grain legumes

This volume reviews key research on particular types of grain legumes with chapters on developing improved varieties as well as improvements in cultivation techniques. Part 1 covers common beans, lentils, soybeans and groundnuts. Part 2 discusses cowpea, faba beans and pigeonpea.

Part 1 Cultivation of common beans, lentils, soybeans and groundnuts

There is a long history of genetic improvement of the common bean. Chapter 1 describes breeding programs focusing on a wide range of biotic and abiotic production constraints, traits for local adaptation and consumer quality, with yield being the overriding challenge for most bean breeding programs. The chapter assesses the wide range of breeding procedures used to improve bean yields, from ideotype to molecular breeding. It discusses the impact of these techniques on yield gains and future trends in genomic analysis research on yield improvement.

Chapter 2 looks at the origin, domestication and economic importance of common bean, including world production trends. It also assesses production constraints, such as soil fertility and acidity, pests and diseases, drought and heat stresses. It goes on to describe ways of addressing these constraints through cropping systems, integrated pest management, and management of diseases such as angular leaf spot (ALS), anthracnose, common bacterial blight and halo blight. Water and nutrient management, including phosphorus and nitrogen, are discussed, as well as sections on liming and foliar fertilization.

Lentil is a popular pulse consumed primarily in Asia. It has a high protein content and also contains high amounts of macro- and micro-nutrients. Lentil is an important food legume in the semi-arid regions of the world where it can be grown successfully on limited soil moisture and in relatively poor soils. The inclusion of lentils in rotations also benefits succeeding crops as a result of biological nitrogen fixation. Chapter 3 reviews the global production of lentils, including production regions and their agro-ecologies, and shows how breeding methods and technology, and the use of new varieties with higher yield potential and improved disease resistance, have led to increased productivity in many countries. It discusses successful attempts to broaden the genetic base of lentil in South Asia and to cross domestic varieties with wild relatives to access new disease-resistance genes. Finally, it considers the scope for breeding new climate-smart varieties of lentil in response to emerging climate changes and variability.

Chapter 4 describes the climate, land and nutrient requirements for lentil cultivation, before going on to address agronomic practices used in lentil-producing countries, including methods of land preparation, sowing, harvesting, threshing and cleaning. Procedures used in developed countries where the crop is entirely mechanized are also included, as well as innovations in lentil cropping systems. The chapter suggests improvements to seed supply systems, seed varieties, planting methods, weed control and harvesting methods.

After discussing research on the soybean genome, Chapter 5 describes methods for developing and using DNA markers derived from genomic sequences for monogeneic, oligogeneic and polygeneic traits, together with examples of successful mapping, fine mapping and gene isolation. The chapter discusses identification of polymorphism, genetic and association map development, and marker-assisted selection (MAS) of the recurrent parent genome. It also looks at MAS in recurrent cross populations, and the scoring of phenotypes, before examining ways in which marker-assisted selection can be used to isolate and select desirable traits in soybeans.

Soybean production in sub-Saharan Africa (SSA) has increased significantly in response to demand arising from increasing populations and improved incomes, though many countries still reply on imports There is therefore an urgent need to further improve the production of this important legume. Chapter 6 discusses trends in soybean production and consumption across the region, and reviews best practices for soybean cultivation, together with evidence for their effectiveness in improving soybean yields. These best practices include use of improved varieties, inoculation, improved crop management, as well as nutrient, disease and pest management. The authors discuss the importance of supporting these practices with good input delivery and financing systems, agricultural advisory services and functioning output markets.

Groundnut is an important nutrient-dense crop grown in over 100 countries. Breeding for improved varieties is critical for increasing yields and enhancing quality. Chapter 7 describes the genetic resources of groundnut and their potential for mining desirable traits, potential breeding targets and ways to maximise groundnut oil quantity and quality. The chapter provides a detailed case study of groundnut production in Uganda, and outlines the potential benefits of improved groundnut varieties, including disease resistance, as well as suggesting future directions for groundnut research.

Groundnut is largely grown under rainfed conditions in Asia and Africa by resource-poor small-holder farmers, requiring low-cost technologies based on locally-available resources. The diverse growing conditions of the crop necessitate development of crop management techniques to meet specific requirements. Genetic and management techniques can realize optimal groundnut pod yield, meet the needs of processors and consumers, and ensure food safety standards. Chapter 8 discusses the limitations of current agronomic guidelines, as well as best practice in field preparation, soil resource management, seed preparation, planting, weed and water management, as well as plant protection. The chapter goes on to describe harvesting, drying, curing and storage processes, as well as trends in seed production.

Aflatoxin contamination of crops and food poses a substantial threat to humans and livestock worldwide. Preventing various *Aspergillus* species from becoming established and growing on peanuts (groundnut, *Arachis hypogaea* L.) can reduce aflatoxin contamination. Chapter 9 describes factors that affect the growth of *A. flavus* and *A. parasiticus* on groundnuts. The chapter also discusses the use of models to predict contamination, as well as cultural and biological control measures designed to minimize contamination. The chapter also features three case studies from current value chain projects in Ghana, Haiti, and Malawi. The chapter argues the need for a more focused and concerted effort to address the issue of aflatoxin contamination in groundnuts.

Part 2 Cultivation of cowpea, faba beans and pigeonpea

Cowpea is a legume crop of vital importance to the livelihoods of millions of people, providing a nutritious grain and an inexpensive source of protein for both rural poor and urban consumers. It is grown for both food and animal feed and is an integral component of various cropping systems in the semi-arid tropics and sub-tropics covering over 65 countries. Chapter 10 describes production constraints, existing cowpea breeding programs and past challenges, with a particular focus on cowpea breeding at the International Institute for Tropical Agriculture (IITA). The chapter gives an account of progress in breeding improved varieties made to date, including cowpea international trials, and suggests lines of research for the future.

Chapter 11 examines what constitutes an optimal cowpea plant population and explains how to integrate the crop in intercropping systems in West Africa. The chapter explores how planting dates can be manipulated to improve cowpea productivity, and how nutrient management can be used to increase cowpea yields. Finally, the chapter examines the application of Integrated Pest Management (IPM) in cowpea production and looks ahead to future trends in this area.

The faba bean is an important cool-season food legume crop grown under different cropping systems for food, animal feed and as a green manure. Chapter 12 discusses key advances in producing new varieties of faba bean that can achieve high nitrogen fixation, are tolerant of heat, drought and herbicides, as well as varieties resistant to broomrape, and disease. The chapter looks ahead to future research trends in this area.

Faba bean is mainly grown under rain-fed conditions, although irrigated production is important in Egypt, parts of China and Central Asia. The main focus of Chapter 13 is faba bean cultivation under rain-fed growing conditions, covering production in China, West Asia, North and East Africa. The chapter discusses the diseases, weeds and pests affecting this crop and ways they can be managed. It also reviews faba bean breeding, genetic resources and markers for breeding, water deficit management, tillage systems, biological nitrogen fixation (BNF) and mineral nutrient requirements.

Pigeonpea is a high protein pulse crop that grows well under biotic and abiotic stress situations. It has the potential to play a significant role in meeting the challenges of food security in the tropics and sub-tropics. Chapter 14 examines the role of pigeonpea in global nutritional security for humans and animals, and addresses the physical, environmental and genetic factors that may affect the sustainability of pigeonpea production. The chapter examines four ways of enhancing pigeonpea production: through crop modelling, an efficient seed system, plant breeding and hybridization. Finally, the chapter considers the latest trends in pigeonpea breeding and production.

Production of pigeonpea is constrained by poor agronomic practices such as improper methods of sowing, incorrect sowing time, inadequate seed rate, insufficient weeding, inappropriate fertilization and insufficient irrigation. Low plant densities, low soil fertility, insufficient weeding and insufficient/inappropriate use of fungicides and herbicides may also be constraints. Chapter 15 discusses seeds and sowing, and analyses integrated crop management techniques for pigeonpea with an emphasis on efficient fertilizer, nutrient management and water use to maximize sustainable yields. The chapter examines the role of weed management and the potential contributions of pigeonpea hybrid cultivars, early maturing pigeonpea varieties, as well as rabi and summer pigeonpea. It concludes by looking at harvesting, threshing and storage techniques, before looking ahead to future trends in this area.

Part 1

Cultivation of common beans, lentils, soybeans and groundnuts

Chapter 1

Developing improved varieties of common bean

James D. Kelly, Michigan State University, USA

1 Introduction
2 Yield and yield gap in common beans
3 Factors affecting bean yields
4 Quantitative trait loci (QTL) analysis
5 Genomic analysis
6 Conclusion
7 Future trends
8 Where to look for further information
9 References

1 Introduction

Common bean (*Phaseolus vulgaris* L.) is the most widely grown and consumed pulse crop in the American and the sub-Saharan regions of the African continent (Akibode and Maredia, 2012). The crop is well adapted to a wide range of geoecological zones ranging from 200 m in temperate zones to 3000 m in tropical regions. Adaptation is associated with growth-habit differences where type IV climbing beans are grown at higher elevations and the three non-climbing habits are produced in shorter-season regions compromised by limited rainfall or frost-free dates. The non-climbing types are further separated into three growth habits: type I is a determinate bush; type II is an upright indeterminate short vine; and type III possesses a long prostrate vine (Singh, 1982). Breeding for higher yield has largely focused on the non-climbing types, as climbing beans, by nature, are more productive based on their higher node and pod numbers, and longer pod-filling periods assuming the inputs are adequate (Checa and Blair, 2012). Yields of climbing beans have been reported as high as 6 t ha^{-1} in Rwanda (Musoni et al., 2014) but yields are generally reduced in half when beans are grown in association with maize (*Zea mays*; Isaacs et al., 2016). One of the challenges for bean breeders has been to increase yield of short-season, bush beans, not just to enhance disease resistance, plant architecture or quality traits but to focus on actual yield improvement in the less favourable environments where beans are typically grown. Lower consumer demand compared to cereals fuelled

http://dx.doi.org/10.19103/AS.2017.0023.18

by government policies and higher commodity costs have forced bean production to marginal production areas where yields are compromised by shorter seasons, variable and limited rainfall patterns, poor soil fertility and marginal production systems. The need to provide a hungry world with a better balanced and nutritious diet makes it imperative to produce beans in a competitive and profitable manner, and one approach is to increase yield through breeding within the constraints imposed by the marketplace and production system (Beaver and Osorno, 2009).

2 Yield and yield gap in common beans

2.1 Defining the yield gap

In a recent review article, Vandemark et al. (2014) detailed the basis for yield improvement in dry beans in the United States. Yield gains over the past 30 years have varied across seed classes and have ranged from 13.9 kg ha^{-1} yr^{-1} (0.77% yr^{-1}) for navy beans to 17.4 kg ha^{-1} yr^{-1} (0.65% yr^{-1}) for pinto beans. The authors also detailed the significant yield gap as the difference between on-farm yield and potential yield and recommended continued improvement in management practices to allow for higher production efficiency (Vandemark et al., 2014). The yield gap is much greater in developing nations where bean production has been pushed to marginal areas. In Mexico 86% of the harvested acreage is grown under semiarid rain-fed conditions and accounts for only 69% of the total production (Zahniser et al., 2010). While yields in the United States averaged 1980 kg ha^{-1} during 2007–9, yields in Mexico averaged 760 kg ha^{-1} for the same period (Zahniser et al., 2010). On closer observation the data show rain-fed yields in Mexico averaged around 500 kg ha^{-1} as compared to 1500 kg ha^{-1} for irrigated acreage. Similar situations exist in other countries in Central America where beans are grown under the short rainy season known as the '*Postera*', and the crop confronts terminal drought during the critical reproductive period. As such, bean breeders are forced to breed for yield under these yield-limiting conditions, and advances are correspondingly lower as a result of the limited genetic potential of materials grown under these environmental conditions (Beebe et al., 2013). In addition to recognizing the constraints of growth habit, maturity, seed type and stress factors that must be considered when working to improve yield, breeders should also consider a number of common factors when breeding to improve yield in their local environment. These include the following concepts and approaches that researchers have studied and applied towards yield improvement in the past.

2.2 Defining yield

Yield as a complex trait was first described in *P. vulgaris* by Adams and Duarte (1961) in a study of the two components of total leaf area in an intergene pool cross of the bean varieties Algarrobo and Michelite. The larger-seeded varieties like Algarrobo belong to the Andean gene pool, whereas the small-seeded Michelite variety belongs to the Middle American gene pool (Gepts, 1998). Leaflet number showed dominance in the F1 generation, was equivalent in number to that of Michelite, whereas leaf size was intermediate resulting in total leaf area that significantly exceeded both parents. They explained the result as epistatic gene action between gene(s) acting additively and gene(s) acting dominantly. They concluded that the observed heterosis was a case of both additive × additive and additive × dominance gene action not at the nuclear level but at the level of morphological integration, a kind of somatic interaction which was the result of component interaction at the level of integrated morphogenetic systems (Adams and Duarte, 1961). However, the autogamous pollination nature of beans has prevented the exploitation of heterosis that could be applied to yield improvement in beans.

2.3 Yield component compensation

The concept of yield component compensation as it is applied to all grain crops was first demonstrated in beans by Adams (1967). He described an increase in one of the three yield components (X, Y, Z) as resulting in a corresponding decrease in the other components. In beans, breeders are usually locked into a single seed size (Z-component) due to market constraints, so improvements in yield are restricted to adjustments in the other two components, either pods per plant (X) or seeds per pod (Y). The X-component has the greatest impact on yield, but also fluctuates more based on environmental conditions. This view was reinforced by Duarte and Adams (1972), who state that 'the direct effect of a particular component upon yield is in nearly every case partially counterbalanced by negative indirect effects through the other two components'. The same authors go on to state that pods per plant (X-component) exerts a predominant effect upon yield and that leaf number is highly associated with pod number and leaf size is highly associated with seed size (Duarte and Adams, 1972). These findings have since been confirmed in many studies where high negative correlations between seed size and yield have resulted in overall low yield potential of most large-seeded Andean beans (White and Gonzales, 1990).

3 Factors affecting bean yields

3.1 Ideotype

The concept of a high-yielding ideotype in beans was first proposed by Adams (1973) and was later refined by Denis and Adams (1978) and Coyne (1980). High-yielding cultivars must be based on relatively large plants bearing numerous nodes, leaves and reproductive structures with an architectural display of phytomeric units that facilitate more uniform light interception through the plant canopy (Denis and Adams, 1978). The stacking of phytomeric units to achieve more efficient partitioning and higher yields in an upright bean ideotype was illustrated by Adams (1982) and the concept was implemented by Kelly and Adams (1987) to develop Durango race genotypes with enhanced performance. The concept has had particular application in the highly mechanized production systems in the northern temperate regions, where beans are now directly harvested (Kelly, 2001; Kelly and Cichy, 2013).

3.2 Partitioning

Wallace et al. (1993a) proposed that high yield is achieved through the accumulation of biomass, partitioning the biomass and the time the genotype–environment uses to accumulate the biomass and to partition the biomass to the final economic yield. Optimizing these processes has resulted in the development of the Redkloud kidney bean that accumulates the same yield as does the older landrace Redkote variety in 20 days less time (Wallace et al., 1993b). Redkote possesses the dominant allele of the *Ppd* photoperiod gene, whereas Redkloud has the recessive *ppd* allele. Efficient partitioning conditioned by the *ppd* gene appears to be more critical to Andean kidney beans as most breeders of Middle American genotypes find that higher yields are generally associated with later maturity (Kelly et al., 1987), unless foreshortened in certain production areas by terminal drought (Beebe et al., 2008). Despite the importance of partitioning, the plant must first produce sufficient biomass to sustain a pod load that will result in higher yields.

3.3 Breeding methods

Numerous breeding methods have been proposed as strategies to improve yield in dry bean (Singh et al., 2007). These include intergene pool crosses (Singh et al., 1989), early generation testing (Singh et al., 1990), gamete selection (Singh et al., 1998) with or without the aid of markers, all necessitating a determination of yield in earlier generations. The cost of these approaches may not justify the outcome as breeders are actively selecting other critical phenotypic (agronomic, phenological, seed) traits in earlier generations where progress can be assured as heritability levels are higher for these traits. Selection for yield is better delayed until additive genetic variance is fixed and can be determined with higher accuracy in replicated trials. The driving force behind these studies based on early generation selection is the concern over the level of genetic variability associated with yield that is being retained in bean-breeding nurseries when breeders are fixing genetic variability for other key traits. Interestingly soybean breeders have faced a similar situation and have moved away from early generation yield testing first proposed by Cooper (1985) to the exclusive use of the single seed descent (SSD) method. In this system, mechanization of yield evaluation is possible, and other characteristics for seed type, agronomic and maturity are fixed in the parents, so SSD allows for selection mainly for yield. A similar approach could be applied to beans if crosses were restricted to single seed sizes or types with similar agronomic and phenological traits. Off-season nurseries would shorten the inbreeding process as selection for yield is delayed until later generations when it is more effective. A possible explanation for the underutilization of the SSD method in bean breeding is the diversity of seed classes that confront breeders and the need to fix seed market traits in earlier selection generations. The benefits of mechanization should not be overlooked as it provides an opportunity to increase the number of lines being tested and has assured access to adequate genetic variability for yield improvement in commercial class seeds.

3.4 Yield pyramid

Given the wide diversity of bean seed types, growth habits and maturities, Kelly et al. (1998) proposed an integrated approach to yield breeding based on a three-tiered yield pyramid. The pyramid was designed not only to meet constraints for specific seed and agronomic types faced by breeders, but also to face the genetic constraints imposed by the gene pool structure of *P. vulgaris* (Gepts, 1998). Such constraints on free combining ability resulting from masked traits and hybrid weakness have prevented breeders from accessing the genetic potential of Mesoamerican beans for the yield improvement of Andean beans (Beaver and Kelly, 1994). Different suggested breeding methods to exploit these limitations were outlined at each level of the pyramid and improved recombinant materials would move upwards in the pyramid (Kelly et al., 1998). In addition, the pyramid provided an overview that allowed individual breeders to determine where they best fit in the overall improvement scheme based on different institutional goals imposed by the same institutions. Private breeders tend to work at the apex of the pyramid and rely on germplasm coming from public scientists working at the lower tiers of the pyramid. This overview should allow for a better distribution of credit, workloads and collaboration among public, private and international bean breeders as few breeders can effectively work at all three levels of the pyramid.

3.5 Wild germplasm

One area of specific interest outlined as a strategy at the base of the pyramid is the use of wild bean germplasm for yield improvement despite the obvious lack of phenotypic performance in wild germplasm (Acosta et al., 2007; Porch et al., 2013). The interest was stimulated by the study of the domestication syndrome in beans (Koinange et al., 1996) where the genetic control of the syndrome involves genes that have a major effect and account for most of the variation observed (>60%) in traits for growth habit, phenology, seed dispersal and seed size. Inbred-backcross IBC2 populations between cultivated and wild germplasm made at Centro Internacional de Agricultura Tropical (CIAT) were evaluated in Michigan and Colombia (Blair et al., 2006; Kelly, 2004). Major yield gains in IBC line – 115M of 27% – over the cultivated recurrent parent were observed in Michigan (Kelly, 2004). However, genomic regions associated with yield gain do not appear to come from the wild parent (Beebe, 2012). The challenge for breeders interested in pursuing this avenue of research is in determining which wild germplasm to introgress with the cultivated species as no phenotypic traits exist in the wild germplasm that identify it as valuable for yield improvement. In other studies, using 115M as a parent, Wright and Kelly (2011) found a major QTL for yield that resided on chromosome Pv10. In a follow-up study with single nucleotide polymorphism (SNP) markers, the same 115M parental line retained the highest levels of heterozygosity (36%) when compared to 95 genotypes in a Mesoamerican diversity panel (Hoyos-Villegas et al., 2016a). The residual heterozygosity resulting from a lack of inbreeding in this IBC line may account for the higher yield due to residual heterotic effects that are absent in more highly inbred pure lines.

3.6 Yield under stress

The extensive literature on improving bean yields under drought-stress conditions needs to be considered as 60% of bean production is under some level of drought stress (Beebe et al., 2013). The focus of these studies tends to be on the best-adapted germplasm, and the approaches are quite variable as drought stress varies greatly across locations, but some common themes arise and would be applicable to breeding for yield improvement. The use of the geometric mean yield to identify the highest-yielding lines under contrasting water stress environments (Schneider et al., 1997) has broad appeal. This parameter could have broader use in yield breeding across contrasting locations that differ for other stress factors, such as low fertility, temperature extremes, water stress or water logging. The other theme focuses on improving partitioning first described by Wallace et al. (1993a). The concept has been refined by Beebe et al. (2013) to measure pod harvest index, as some genotypes need to better partition resources from pod into seed rather than the more general partitioning of resources from vegetative plant organs into the seed. This partitioning trait is receiving broad use in breeding for yield under drought and would have application in any breeding programme focused on improving economic bean yields. The evaluation of this trait, however, relies heavily on hand labour and will restrict the volume of germplasm that can be effectively analysed.

4 Quantitative trait loci (QTL) analysis

Given that yield is a quantitative trait influenced by many genes of minor effect, with a large environmental component, research in recent years has focused on QTL analysis of a wide range of recombinant inbred line (RIL) populations contrasting in genetic origins and evaluated for yield across a wide range of environmental conditions. Genetic maps were built with limited genomic coverage using the only available SSR markers, yet researchers found and reported on yield QTL. Unfortunately, many of these QTL could not be verified across populations because the maps were relatively crude as the physical position of these markers was not known. Many of these studies examined agronomic components of yield (Beattie et al., 2003; Blair et al., 2006, 2012; Tar'an et al., 2002; Wright and Kelly, 2011) and although similar QTL were identified, many were unique to specific populations or environments, and impossible to compare in the absence of a physical position for the QTL. The development of the first physical SNP-based BARCBean3_6K Beadchip courtesy of colleagues at USDA-ARS (Song et al., 2015) allowed breeders to identify specific genomic regions associated with yield traits. Knowledge of where specific traits are located in the genome allows for comparisons with other studies where yield traits were mapped in contrasting populations for performance under drought, low-N and optimum conditions. Since 2010, a number of studies have reported on yield-related traits mapped using SNP markers (Table 1).

The first study to integrate QTL analysis with SNP markers was one by Mukeshimana et al. (2014) who mapped yield under drought stress and non-stress conditions in an intergene pool RIL population in Rwanda. Yield QTL were identified on chromosomes Pv03 and Pv09. The yield QTL (SY3.3SC) on Pv03 was associated with yield under non-stress conditions in Karama, Rwanda, in 2011 and 2012, and was linked to a QTL associated with seed weight. Various seed yield QTL had been previously identified on Pv03 (Blair et al., 2006; Wright and Kelly, 2011), but none of these could be validated as the physical positions of the QTL/markers were not known. Following this study, a genome-wide association study (GWAS) was conducted on a panel of 237 Andean beans to measure symbiotic nitrogen fixation (SNF) under low-N conditions in Michigan (Kamfwa et al., 2015). The authors identified the same SY3.3SC QTL for seed yield in a marker interval that contained the same three SNPs (Table 2). Analysis of linkage disequilibrium (LD) between these SNPs, and the significant SNP ss715649410 for seed yield in the GWAS study, indicated that two of three SNPs were in LD with ss715649410. One of these three SNPs in SY3.3SC interval was in strong LD with the most significant SNP ss715648538 for seed yield in the GWAS study (Kamfwa et al., 2015). Another QTL for seed yield identified by Mukeshimana et al. (2014) on Pv03 was in the marker interval ss715646941-ss715648035 containing 19 SNPs. Eight of these 19 SNPs were in LD with the significant SNP ss715649410 in the GWAS study (Kamfwa et al., 2015). These results suggest that the gene(s) underlying the QTL for seed yield identified by Mukeshimana et al. (2014) are the same ones in LD with significant SNP ss715649410 identified by Kamfwa et al. (2015). A third study conducted by Hoyos-Villegas et al. (2015) determined seed yield

Table 1 Author, location, year, population, size, gene pool and mapping study, environmental conditions and traits associated with the most significant SNPs for seed yield on *Phaseolus vulgaris* chromosome Pv03, alias B3, determined over eight studies from 2004 to 2014

Lead Author	Location	Years	Population(s)	Size	Gene Pool	Study	Conditions	Traits &Yield
Heilig, 2017	Puerto Rico	2012–2013	RIL[†]	132	Meso[‡]	QTL	Low-N	SNF[§]
Kamfwa, 2015	Michigan	2012–2013	Panel	237	Andean	GWAS	Irrigated +N	Agronomic
Farid, 2015	Ontario	2011–2013	RIL	140	Meso	QTL	Optimum	SNF
Hoyos, 2016	Michigan	2013–2014	3-RIL	160	Meso	QTL	Rain-fed	Agronomic
Hoyos, 2014	Michigan	2007–2010	RIL	96	Meso	QTL	Irrigated +N	White Mould
Mukeshimana, 2014	Rwanda	2010–2012	RIL	125	Meso/Andean	QTL	Stress/Non-Stress	Drought
Blair, 2012	Colombia	2005–2007	RIL	117	Meso	QTL	Irrigated	Drought
Wright, 2011	Michigan	2004–2007	RIL	96	Meso	QTL	Optimum	Yield
Blair, 2006	Colombia	2 seasons	BC2F5	157	A/M wild	QTL	Optimum	Agronomic

[†]RIL= Recombinant Inbred Line; [‡]Meso = Mesoamerican; [§]SNF = Symbiotic Nitrogen Fixation.

Table 2 Author, location, year, SNP, position, p-values, proportion of phenotypic variation explained (R^2), additive effect and contributing parent for the most significant SNPs for seed yield on *Phaseolus vulgaris* chromosome Pv03, alias B3, measured over eight studies from 2004 to 2014

Lead Author	Location	Year	SNP [SSR][†]	SNP Position (Mb)	P-value (LOD)	$R^{2§}$	Additive Effect kg/ha	Parent
Heilig, 2017	PR	2012	ss715639345	39.59	3.2	14.6	139	Zorro
Heilig, 2017	PR	2013	ss715646621	39.44	3.1	12.1	193	Zorro
Kamfwa, 2015	Michigan	2012	ss715648538	38.27	6.3	0.14		237 lines
Farid, 2015	Ontario	2011, 2013	NA	38.27	3.3	0.11	110	Mist
Hoyos, 2016	Michigan	2014	ss715647671-ss715639244	37.06–45.59	4.1	0.308	25–435	Merlot
Hoyos, 2014	Michigan	2007–2010	ss715647849	33.55	2.3	12.2	0.7	P02630
Mukeshimana, 2014	Karama, Rwanda	2010	ss715647636-	39.64-	3.1	8.7	9.6	CAL96
		2011	ss715649325	36.70	3.2	9.2	16.5	CAL96
Blair, 2012	Palmira, Colombia	2005 2006 2006 2006	[AE103, Q1001, BM181, AD1801]	NA NA NA NA	NA NA NA NA	0.16 0.19 0.14 0.14	98 170 156 150	DOR364 DOR364 DOR364 DOR364
Wright, 2011	Michigan	2004, 2006	[ME1/6.550, PVat008]	3.2cM	2.7	0.11	168	Jaguar
Blair, 2006	Popayan, Colombia	A season, B season	[BM172, BM98]	NA		0.10, 0.11	219 99	ICA Cerinza G24404

[†]SNP = Single Nucleotide Polymorphic code, SSR = Simple Sequence Repeat; [‡]P = significance level and E = exponential; [§]R^2 is phenotypic variation explained by the SNP.

in a pinto bean RIL population evaluated for four years under white mould infection. Three SNPs within the yield QTL interval on Pv03 identified in this study were in strong LD with the most significant SNP for yield in the previous GWAS of the Andean diversity panel (Kamfwa et al., 2015). The data suggests that the QTL for seed yield in these two studies co-localizes on Pv03 and the QTL is stable across a broad range of genetic backgrounds and environments. A meta-QTL study of three RIL populations grown under contrasting rain-fed regimes in Michigan detected the largest QTL for seed yield on Pv03 (Hoyos-Villegas et al. 2016b). This QTL had the largest R^2 of all the QTL for seed yield detected with a value of 30.8% and the largest additive effect among all three RIL populations. The QTL was located between markers ss715647671 (37.0 Mb) and ss715639244 (45.5 Mb) in the same region as SY3.3SC QTL (Table 2). In another RIL population developed from a cross of two navy bean varieties, Farid (2015) detected a QTL for yield in two growing seasons. The experiment was conducted over five location-years in Ontario from 2011 to 2013 and was focused on improving SNF. The QTL for yield was located at 38.27Mb on Pv03 (Table 2) in the exact same position as the QTL detected for yield in the GWAS by Kamfwa et al. (2015). The code names for the SNPs in this region differed from those of the BARCBean3_6K Beadchip and could not be compared directly, other than both studies focused on improving yield under N-fixation, but contrasted in the gene pools and populations evaluated. The final study on SNF was conducted on a black bean RIL population grown for two years in low-N soils in Puerto Rico (Heilig et al., 2017). A QTL for yield was identified on Pv03 in the same interval (39.44 to 39.59Mb) as yield QTL were detected by Mukeshimana et al. (2014), Hoyos-Villegas et al. (2016b) and others.

4.1 Seed yield QTL

Seed yield QTL have been identified on the same genomic region of Pv03 in six independent studies with very diverse mapping populations. If these QTL are stable and expressed in diverse genetic backgrounds they could be used as potential candidates for marker-assisted breeding for seed yield. All studies point to a genomic region on Pv03 that is a critical component of high-yielding genotypes. If this region is lacking or absent from ones' germplasm, breeders should consider it essential to fix this region in future breeding materials. Breeders are often accused of reducing genetic variability in new cultivars when contrasted with landrace or non-cultivated species. Recent studies clearly show that breeders have increased the genetic variability for economic traits in new varieties. During the breeding for type II Durango race beans, Soltani et al. (2016) showed that the genetic variance increased for five traits, and the two architectural traits that exhibited increased genetic variances were canopy height and stem diameter. The data on QTL for yield that co-localized on Pv03 detected in six distinct studies suggest that breeders have been indirectly selecting for this region as they focus on yield improvement. As breeders build on past genetic gains, key genomic areas become fixed in future varieties. These studies also suggest that major genes play a role in yield performance in beans and contradict the traditionally held belief that yield is controlled largely by minor genes. Similar findings of major genomic regions controlling yield are now emerging in other crops such as wheat, *Triticum aestivum* (Zhang et al., 2010), as more detailed mapping studies become available. However, bean breeders are only finding major QTL at present because of small population sizes and limited genomic coverage (6K chip) so detecting minor QTL that may enhance yield improvement further will most likely require better genomic coverage and more extensive phenotypic data on larger mapping populations.

5 Genomic analysis

5.1 Other genomic regions

Other genomic regions that have been consistently identified as contributing to yield have been mapped to Pv07 (Asfaw et al., 2012; Hoyos-Villegas et al., 2016b), Pv09 (Blair et al., 2006; Mukeshimana et al., 2014; Kamfwa et al., 2015; Tar'an et al., 2002) and Pv10 (Hoyos-Villegas et al., 2016b; Trapp et al., 2015; Wright and Kelly, 2011). The QTL on Pv07 was identified by Asfaw et al. (2012) in a three-location study in Africa and Colombia, which was evaluated under drought and non-stressed conditions. The same genomic region was identified as contributing to yield by Hoyos-Villegas et al. (2016b) in Michigan. The QTL on Pv09 identified by Mukeshimana et al. (2014) under drought conditions in Rwanda appears to be in the same region identified by Kamfwa et al. (2015) in the GWAS in Michigan. Confirmation that this is the same region identified in earlier studies (Blair et al., 2006; Tar'an et al., 2002) was not possible as genomic locations were not known in these studies. The QTL SY10.1 identified by Trapp et al. (2015) at 39.9 Mb on Pv10 is in proximity to the QTL identified by Hoyos-Villegas et al. (2016b) at 40.1 Mb. This may be the SY10.2 QTL named by Wright and Kelly (2011), but the location has not yet been confirmed due to different markers used. As new data validate the contribution of these genomic regions to yield, breeders will be better assured of which genomic regions to combine to further enhance yield of common bean.

5.2 Genome-wide association studies (GWAS)

One of the interesting findings to date with the current GWAS is that these studies may still be limited to only detecting QTL already identified in prior QTL/RIL populations, even those studies with low population numbers and SSR markers and relatively crude mapping tools. This may change as population sizes are increased and as genomic coverage is improved or when GWAS are focused on traits not previously mapped. However, GWAS is a conservative breeding approach as it only identifies and isolates genetic variance that currently exists in the germplasm under study. Breeders need to expand genetic variability and enhance it through crossing and selection and GWAS may provide an avenue to identify regions of interest, but these regions still need to be introgressed into new varieties. Searching out the genes controlling economic traits through candidate gene discovery with tools like GWAS is becoming popular. This approach is based on localizing the trait of interest to a specific genomic region when popular genetic theory supports multigenic control of economic traits such as yield. It would be interesting to fine-map and study the genomic basis behind the region controlling yield on Pv03. Hopefully this might provide insights into the control or structure of this region and ways to more effectively utilize it. If we agree with the theory of gene duplication and the resulting changes that have contributed to genomic regions that control disease resistance, do similar regions that control yield exist and have these regions been selected for over time from ancient farming communities to modern-day scientific breeding? As breeders expand the genetic variability in regions controlling economic traits, and if independent events have occurred in different genetic backgrounds, it behoves breeders to identify and combine these regions in an effort to improve yields of future bean varieties.

6 Conclusion

The success of yield breeding in beans appears modest when compared to yield gains in small grain cereals, maize and soybean (*Glycine max*). The reasons for the disparity are numerous, but focus largely on the financial investment in breeding research. The disparity in investment in breeding cereals compared to beans is huge as is the private sector investment in breeding maize or soya bean. Most bean-breeding programmes are modestly funded and even public sector funding of basic research on beans is minimum. Other contributing factors are the nature of the seed product produced by beans. Beans are a nutrient-dense food and the cost of producing that food source is greater for the plant than producing a lower-grade carbohydrate food source. Yield gains in cereals have resulted from improvements in partitioning nutrients from the plant to the grain. Beans are a short-season crop that already are relatively efficient in partitioning nutrients to the seed with values approaching 50% in most cultivars, so future gains in the area will be limited. Most non-climbing beans mature within 100 days and many mature in 70 days. Given the short growing season, the expectations to fix more carbon and/or nitrogen in this time period is again limiting when compared to soya bean that takes 30 additional days to mature. Bean breeders do not have the luxury to focus on a single seed type but must focus on a range of seed types that differ in quality and consumer traits. Gains made in one seed type are not directly transferable to another seed type, requiring breeders to run concurrent programmes as opposed to breeding a single seed or grain type that most soya bean or cereal breeders focus on. The final challenge rests with the environments where beans are produced. In most countries beans are relegated to less favourable environments, prone to stress, limiting in nutrients, with growing seasons shortened by elevation or latitude. In this regard the yield gap looms large as yield gains made under the ideal conditions of experiment stations cannot be directly transferred to the low-input fields of smallholder farmers. Rectifying this disparity will be the biggest challenge for bean breeders in the future.

7 Future trends

The BARCBean3_6K Beadchip has been a very useful tool in mapping quantitative traits, particularly yield-related traits. The chip needs to be expanded well beyond the current 5398 SNP markers to provide better coverage of the bean genome. This is particularly critical for its application to beans from the Andean gene pool as the coverage of Andean genome is inadequate with the 6K chip. The first genome sequence of an Andean bean was published by Schmutz et al. (2014) and two other international consortiums have recently published genomic sequences of two other bean genomes, one Middle American in origin (Vlasova et al., 2016) and the other has introgressions from the related tepary bean (http://www.beangenomics.ca/). With the low cost of sequencing, bean researchers are now conducting genome by sequencing (GBS) analyses of a wider array of bean germplasm and this sequence information is public and should permit the development of a more comprehensive SNP chip in the range of 50K SNPs. This would permit more in-depth analysis of specific regions such as on Pv03 and other regions where genes of economic importance are located. Better genomic coverage would permit the detection of minor QTL for yield as these will undoubtedly contribute to further gains in yield of common bean. In the future as more genomic information is acquired, bean breeders will look to new selection tools such as genomic selection to further advance the science of bean improvement that contributes positively to the wide array of bean germplasm grown and consumed worldwide.

8 Where to look for further information

Dry bean research

Bean Improvement Cooperative – http://bic.css.msu.edu/
BeanCAP – http://www.beancap.org/
USDA – http://www.ers.usda.gov/topics/crops/vegetables-pulses/dry-beans.aspx
CIAT – https://ciat.cgiar.org/category/crops/beans

US public breeding programmes

Michigan State University – http://bean.css.msu.edu/
North Dakota State University – https://www.ag.ndsu.edu/crops/dry-bean
University of Nebraska-Lincoln – http://cropwatch.unl.edu/drybeans
Colorado State University – http://beans.agsci.colostate.edu/
University of Idaho – https://www.uidaho.edu/cals/kimberly-research-and-extension-center/dry-beans

Other publications

Gepts, P. (ed.). 1988. *Genetic Resources of Phaseolus beans. Their maintenance, domestication, evolution and utilization.* Kluwer Acad. Pub. Dordrecht, Holland. 613p. https://books.google.com/books?hl=enandlr=&id=SZ_6CAAAQBAJ&oi=fnd&pg=PR7&dq=Genetic+Resources+of+Phaseolus+beans&ots=_X7wa6R37z&sig=-hV1XZIBREmVm5hGGhK9dHIYxgo#v=onepage&q=Genetic%20Resources%20of%20Phaseolus%20beans&f=false

Kandel, H (ed.). 2013. *Dry Bean Production Guide.* NDSU Extension Service. https://www.ag.ndsu.edu/publications/landing-pages/crops/dry-bean-production-guide-a-1133

Siddiq, M., and M. A. Uebersax (eds), 2012. *Dry Beans and Pulses Production, Processing and Nutrition.* Wiley-Blackwell, 408p. http://onlinelibrary.wiley.com/book/10.1002/9781118448298

van Schoonhoven, A., and O. Voysest (eds). 1991. *Common beans: Research for crop improvement.* CAB International; Cali, CO: Centro Internacional de Agricultura Tropical (CIAT), Oxon, GB, 980p. https://books.google.com.co/books?id=CvcbUopfa54C&printsec=frontcover&source=gbs_atb&hl=es-419#v=onepage&q&f=false

9 References

Acosta-Gallegos, J. A., J. D. Kelly and P. Gepts. 2007. Prebreeding in common bean and use of genetic diversity from wild germplasm. *Crop Sci.* 47(S3): S44–59.

Adams, M. W. 1967. Basis of yield component compensation in crop plants with special reference to the field bean (Phaseolus vulgaris). *Crop Sci.* 7: 505–10.

Adams, M. W. 1973. Plant architecture and physiological efficiency in the field bean. In D. Wall (ed.), *Potentials of Field Beans and other Food Legumes in Latin America*, Series Seminars No. 2E, CIAT. Cali, Colombia, pp. 226–78.

Adams, M. W. 1982. Plant architecture and yield breeding in Phaseolus vulgaris L. *Iowa State J. Res.* 56: 225–54.

Adams, M. W., and R. Duarte. 1961. The nature of heterosis for a complex trait in a field bean cross. *Crop Sci.* 1: 380.

Akibode, C. S., and M. Maredia. 2012. Global and regional trends in production, trade and consumption of food legume crops. Staff Paper 2012–10. Department of Agricultural, Food and Resource Economics, Michigan State University.

Asfaw, A., M. W. Blair and P. C. Struik. 2012. Multienvironment quantitative trait loci analysis for photosynthate acquisition, accumulation, and remobilization traits in common bean under drought stress. *G3-Genes Genome Genet.* 2: 579–95. doi:10.1534/g3.112.002303.

Beattie, A. D., J. Larsen, T. E. Michaels and K. P. Pauls. 2003. Mapping quantitative trait loci for a common bean (Phaseolus vulgaris L.) ideotype. *Genome* 46: 411–22.

Beaver, J. S., and Kelly, J. D. 1994. Comparison of selection methods for dry bean populations derived from crosses between gene pools. *Crop Sci.* 34: 34–7.

Beaver, J. S., and J. M. Osorno. 2009. Achievements and limitations of contemporary common bean breeding using conventional and molecular approaches. *Euphytica* 168: 145–75.

Beebe, S. E. 2012. Common bean breeding in the tropics. *Plant Breed. Rev.* 36: 357–426.

Beebe, S. E., I. M. Rao, C. Cajiao and M. Grajales. 2008. Selection for drought resistance in common bean also improves yield in phosphorus limited and favourable environments. *Crop Sci.* 48: 582–92.

Beebe, S. E., I. M. Rao, M. W. Blair and J. A. Acosta-Gallegos. 2013. Phenotyping common beans for adaptation to drought. *Front. Physiol.* 4. doi:10.3389/fphys.2013.00035.

Blair, M. W., G. Iriarte and S. E. Beebe. 2006. QTL analysis of yield traits in an advanced backcross population derived from a cultivated Andean × wild common bean (Phaseolus vulgaris L) cross. *Theor. Appl. Genet.* 112: 1149–63.

Blair, M. W., C. H. Galeano, E. Tovar, M. C. Munoz Torres, A. Velasco Castrillon, S. E. Beebe and I. M. Rao. 2012. Development of a Mesoamerican intra-genepool genetic map for quantitative trait loci detection in a drought tolerant susceptible common bean (Phaseolus vulgaris L.) cross. *Mol. Breed.* 29: 71–88. doi:10.1007/s11032-010-9527-9.

Checa, O. E., and W. M. Blair. 2012. Inheritance of yield-related traits in climbing beans (Phaseolus vulgaris L.). *Crop Sci.* 52: 1998–2013.

Cooper, R. L. 1985. Breeding semidwarf soybeans. *Plant Breed. Rev.* 3: 289–311.

Coyne, D. P. 1980. Modification of plant architecture and crop yield by breeding. *HortScience* 15: 244–7.

Denis, J. C., and M. W. Adams. 1978. A factor analysis of plant variables related to yield in dry beans. I. Morphological traits. *Crop Sci.* 18: 74–8.

Duarte, R. A., and M. W. Adams. 1972. A path coefficient analysis of some yield component interrelations in field beans (Phaseolus vulgaris L.). *Crop Sci.* 12: 579–82.

Farid, M. 2015. Symbiotic nitrogen fixation in common bean. Doctoral Dissertation, University of Guelph, Ontario, 271p.

Gepts, P. 1998. Origin and evolution of common bean: Past events and recent trends. *HortScience* 33: 1124–30.

Heilig, J. A., J. S. Beaver, E. M. Wright, Q. Song, and J. D. Kelly. 2017. QTL analysis of symbiotic nitrogen fixation in a black bean population. *Crop Sci.* 57: 118–129. doi:10.2135/cropsci2016.05.0348.

Hoyos-Villegas, V., W. Mkwaila, P. B. Cregan and J. D. Kelly. 2015. QTL analysis of white mold avoidance in pinto bean (Phaseolus vulgaris). *Crop Sci.* 55: 2116–29.

Hoyos-Villegas, V., Q. Song, and J. D. Kelly. 2016a. Genomewide association analysis for drought tolerance and associated traits in common bean. *The Plant Genome* 9. doi:10.3835/plantgenome2015.12.0122.

Hoyos-Villegas, V., Q. Song, E. M. Wright, S. E. Beebe and J. D. Kelly. 2016b. Joint linkage QTL mapping for yield and agronomic traits in a composite map of three common bean RIL populations. *Crop Sci.* 56: 2546–63. doi:10.2135/cropsci2016.01.0063.

Isaacs, K. B., S. S. Snapp, J. D. Kelly and K. R. Chung. 2016. Farmer knowledge identifies competitive bean ideotype for maize-bean intercrop systems in Rwanda. *Agriculture & Food Security* 5: 15. doi:10.1186/s40066-016-0062-8.

Kamfwa, K., K. A. Cichy and J. D. Kelly. 2015. Genome-wide association study of agronomic traits in common bean. *The Plant Genome* 8. doi:10.3835/plantgenome2014.09.0059.

Kelly, J. D. 2001. Remaking bean plant architecture for efficient production. *Adv. Agron.* 71: 109–43.

Kelly, J. D. 2004. Advances in common bean improvement: some case histories with broader applications. *Acta Hort.* (ISHS) 637: 99–122. doi:10.17660/ActaHortic.2004.637.11.

Kelly, J. D., and M. W. Adams. 1987. Phenotypic recurrent selection in ideotype breeding of pinto beans. *Euphytica* 36: 69–80.

Kelly, J. D., and K. A. Cichy. 2013. Dry bean breeding and production technologies. In M. Siddiq and M. A. Uebersax (eds), *Dry Beans and Pulses: Production, Processing, and Nutrition.* Wiley-Blackwell Publishing Co. Oxford U.K, pp. 23–54. doi:10.1002/9781118448298.ch2.

Kelly, J. D., M. W. Adams and G. V. Varner. 1987. Yield stability of determinate and indeterminate dry bean cultivars. *Theor. Appl. Genet.* 74: 516–21.

Kelly, J. D., J. Kolkman and K. Schneider. 1998. Breeding for yield in dry bean (Phaseolus vulgaris L.). *Euphytica* 102: 343–56.

Koinange, E. M. K., S. P. Singh and P. Gepts. 1996. Genetic control of the domestication syndrome in common-bean. *Crop Sci.* 36: 1037–45.

Miklas, P. N., K. M. Larsen, K. A. Terpstra, D. C. Hauf, K. F. Grafton and J. D. Kelly. 2007. QTL analysis of ICA Bunsi-derived resistance to white mold in a pinto × navy bean cross. *Crop Sci.* 47: 174–9.

Mukeshimana, G., L. Butare, P. B. Cregan, M. W. Blair and J. D. Kelly. 2014. Quantitative trait loci associated with drought tolerance in following processing of black bean common bean. *Crop Sci.* 54: 923–38.

Musoni, A., J. Kayumba, L. Butare, F. Mukamuhirwa, E. Murwanashyaka, D. Mukankubana, J. D. Kelly, J. Ininda and D. Gahakwa. 2014. Innovations to overcome staking challenges to growing climbing beans by smallholders in Rwanda. In B. Vanlauwe et al. (eds), *Challenges and Opportunities for Agricultural Intensification of the Humid Highland Systems of Sub-Saharan Africa.* Springer International Publishing Switzerland, pp. 129–36. doi:10.1007/978-3-319-07662-1_11.

Porch, T. G., J. S. Beaver, D. G. Debouck, S. Jackson, J. D. Kelly and H. Dempewolf. 2013. Use of wild relatives and closely related species to adapt common bean to climate change. *Agronomy* 3: 433–61. doi:10.3390/agronomy3020433.

Schmutz, J., P. E. McClean, S. Mamidi, G. A. Wu, S. B. Cannon, J. Grimwood, et al. 2014. A reference genome for common bean and genome-wide analysis of dual domestications. Nat. Genet. 46: 707–713. doi:10.1038/ng.3008.

Schneider, K. A., R. Rosales-Serna, F. Ibarra-Perez, B. Cazares-Enriquez, J. A. Acosta-Gallegos, P. Ramirez-Vallejo, N. Wassimi and J. D. Kelly. 1997. Improving common bean performance under drought stress. *Crop Sci.* 37: 43–50.

Singh, S. P. 1982. A key for identification of different growth habits of Phaseolus vulgaris L. *Annu. Rept. Bean Improv. Coop.* 25: 92–4.

Singh, S. P., C. Cajiao, J. A. Gutiérrez, J. García, M. A. Pastor-Corrales and F. J. Morales. 1989. Selection for seed yield in intergene pool crosses of common bean. *Crop Sci.* 29: 1126–31.

Singh, S. P., R. Lépiz, J. A. Gutiérrez, C. Urrea, A. Molina and H. Terán. 1990. Yield testing of early generation populations of common bean. *Crop Sci.* 30: 874–8.

Singh, S. P., C. Cardona, F. J. Morales, M. A. Pastor-Corrales and O. Voysest. 1998. Gamete selection for upright carioca bean with resistance to five diseases and a leafhopper. *Crop Sci.* 38: 666–72.

Singh, S. P., H. Terán, M. Lema, D. M. Webster, C. A. Strausbaugh, P. N. Miklas, H. F. Schwartz and M. A. Brick. 2007. Seventy-five years of breeding dry bean of the western USA. *Crop Sci.* 47: 981–9. doi:10.2135/cropsci2006.05.0322.

Soltani, A., M. Bello, E. Mndolwa, S. Schroder, S. M. Moghaddam, J. Osorno, P. N. Miklas and P. E. McClean. 2016. Targeted analysis of dry bean growth habit: interrelationship among architectural, phenological and yield components. *Crop Sci.* 56: 3005–15. doi:10.2135/cropsci2016.02.0119.

Song, Q., G. Jia, D. L. Hyten, J. Jenkins, E.-Y. Hwang, S. Schroeder, J. Schmutz, S. A. Jackson, P. McClean and P. B. Cregan. 2015. SNP assay development for linkage map construction, anchoring whole genome sequence and other genetic and genomic applications in common bean. *G3: Genes| Genomes| Genetics*, 5: 2285–90. doi:10.1534/g3.115.020594.

Tar'an, B., T. E. Michaels and K. P. Pauls. 2002. Genetic mapping of agronomic traits in common bean. *Crop Sci.* 42: 544–56.

Trapp, J. J., C. A. Urrea, P. B. Cregan and P. N. Miklas. 2015. Quantitative trait loci for yield under multiple stress and drought conditions in a dry bean population. . *Crop Sci.* 55: 1596–1607. doi:10.2135/cropsci2014.11.0792.

Vandemark, G. J., M. A. Brick, J. M. Osorno, J. D. Kelly and C. A. Urrea. 2014. Edible grain legumes. In S. Smith, B. Diers, J. Specht and B. Carver (eds), *Yield Gains in Major U. S. Field Crops*. CSSA Spec. Publ. 33. CSSA, Madison, WI, pp. 87–123. doi:10.2135/cssaspecpub33.c5.

Vlasova, A., S. Capella-Gutiérrez, M. Rendon-Anaya, M. Hernandez-Onate, A. E. Minoche, I. Erb, et al. 2016. Genome and transcriptome analysis of the Mesoamerican common bean and the role of gene duplications in establishing tissue and temporal specialization of genes. *Genome Biol.* 17: 32. doi:10.1186/s13059-016-0883-6.

Wallace, D. H., J. P. Baudoin, J. S. Beaver, D. P. Coyne, D. E. Halseth, P. N. Masaya, H. M. Munger, J. R. Myers, M. Silbernagel, K. S. Yourstone and R. W. Zobel. 1993a. Improving efficiency of breeding for higher crop yield. *Theor. Appl. Genet.* 86: 27–40.

Wallace, D. H., K. S. Yourstone, P. N. Masaya and R. W. Zobel. 1993b. Photoperiod gene control over partitioning between reproductive and vegetative growth. *Theor. Appl. Genet.* 86: 6–16.

White, J. W., and A. Gonzalez. 1990. Characterization of the negative association between seed yield and seed size among genotypes of common bean. *Field Crops Res.* 23: 159–75.

Wright, E. M., and J. D. Kelly. 2011. Mapping QTL for seed yield and canning quality (Phaseolus vulgaris L.). *Euphytica* 179: 471–84.

Zahniser, S., M. V. Torres, J. A. Cuellar Alvarez, N. F. Lopez and R. Bhatta. 2010. The U. S. and Mexican dry bean sectors. VGS-341-01. USDA, Economic Research Service. http://www.ers.usda.gov/media/134896/vgs34101.pdf.

Zhang, L.-Y., D.-C. Liu, X.-L. Guo, W.-L. Yang, J.-Z. Sun, D.-W. Wang and A. Zhang. 2010. Genomic distribution of quantitative trait loci for yield and yield-related traits in common wheat. *J. Integr. Plant Biol.* 52: 996–1007.

Chapter 2

Improving cultivation practices for common beans

John O. Ojiem, Kenya Agricultural and Livestock Research Organization, Kenya

1 Introduction to the common bean: origin, domestication and economic importance
2 World production trends
3 Production constraints
4 Cropping system
5 Integrated pest management
6 Management of diseases
7 Water management
8 Nutrient management: phosphorus (P) and nitrogen (N)
9 Nutrient management: further issues
10 Conclusion
11 Where to look for further information
12 Abbreviations and acronyms
13 References

1 Introduction to the common bean: origin, domestication and economic importance

1.1 Origin and domestication

The common bean (*Phaseolus vulgaris* L.) is perhaps the most widely cultivated grain legume, having been domesticated several thousand years ago (Freytag and Debouck, 2002; Chacón et al., 2005). It is grown in all continents, except Antarctica (Gepts, 1998; Singh, 1999), for its seeds and pods. The seeds are harvested at physiological maturity, while pods are harvested before seed development phase commences. Common bean is an essential part of the diet in Tropical America, as well as in eastern and southern Africa, where it was introduced about four centuries ago (Greenway, 1945), and is also consumed in many other parts of the world.

http://dx.doi.org/10.19103/AS.2017.0023.19

Although the exact origin of cultivated common bean has not been clearly identified (Gentry, 1969), Tropical America, more specifically Mexico, Guatemala and Peru, are considered to be centres of origin, although there is also evidence of its multiple domestication within Central America (Kay, 1979). The current distribution of the wild common bean suggests that the origin encompasses a large geographical area between Mexico and Argentina (Toro et al., 1990). The cultivated common bean is generally well adapted and thrives under a wide range of agro-ecological conditions. In Latin America, the average temperature of the common bean growing regions ranges between 17.5 and 25°C (Gepts, 1998). However, common bean is also grown in regions where the average temperatures during the growing season reaches 27°C, and at relatively high altitudes, for example 60°N in northern Europe (Gepts, 1998). It is well adapted to areas that receive an annual average rainfall ranging from 500 to 1500 mm with optimum temperature range of 16–24°C, and a frost-free period of 105 to 120 days. Generally, common bean performs best on deep, friable and well-aerated soils with optimum pH range of 6.0–6.8 (Kay, 1979).

1.2 Economic importance

The economic significance of common bean cannot be overstated. It is perhaps the world's most important grain legume for direct human consumption. This nutritious food security crop plays a major dietary role, especially in sub-Saharan Africa (SSA) where it meets more than 50% of dietary protein requirements of households (Broughton et al., 2003; Wortmann et al., 1998). Although the *per capita* consumption of common bean varies between regions of the world, its consumption is greater among low-income people who cannot afford to buy nutritious food stuff, such as meats and fish (Broughton et al., 2003). Not only is common bean a food security crop, it is also an important source of protein, vitamins, minerals and fibre, especially for the poorer populations of Africa and Latin America (Brougthon et al., 2003). Additionally, common bean contributes significantly to household income in regions where it is grown (Ferris and Kaganzi, 2008), especially among the majority of the rural peasants in SSA (Wortmann et al., 1998). In eastern Africa, the *per capita* consumption of common bean is 50–60 kg year^{-1} in Rwanda, Kenya and Uganda. This is considerably higher than in Latin America where *per capita* consumption is 4 and 17 kg year^{-1} in Colombia and Brazil, respectively (Beebe et al., 2013; Broughton et al., 2003). Although common bean is cultivated primarily for its dry seeds and green pods, young tender leaves and flowers are used as fresh vegetables in some parts of central and eastern African, and in some Latin American countries (Kay, 1979; Singh, 1999). Due to its acknowledged role in management of human health, for example, reduction of blood cholesterol levels and combating heart diseases, cancers and diabetes (Singh, 1999), the economic importance of common bean is steadily growing.

2 World production trends

According to the statistics available on common bean production for the period 1995 to 2014 (FAO, 2016), Uganda, Tanzania, Kenya and Burundi lead in production in eastern Africa, Malawi and Mozambique in southern Africa, while in western Africa, Cameroon and Ghana are the major producers of common bean (Table 1a). In Latin America, where common bean is a significant traditional food, Brazil, Mexico and Argentina are the major producers (Table 1b), while the United States and Canada are the leading producers in North America and the

Table 1a Bean production trend in major producing countries in Africa (Figures are sum for production period)

Country	Production (million tonnes) 1995–2004	2005–2014
Burundi	2.51	2.11
Kenya	3.35	5.00
Rwanda	1.87	3.40
Tanzania	4.08	5.83
Uganda	4.83	8.45
Malawi	0.86	1.52
Mozambique	0.30*	1.74
Republic of South Africa	0.73	0.54
Zimbabwe	0.49	0.28
Cameroon	1.64	3.11
Ivory Coast	0.49*	0.32
Ghana	0.17*	1.88

*Ivory Coast (data not available for 1995 and 1996), Ghana (data only available for 2001, 2002, 2003 and 2004) and Mozambique (data only available for 2002, 2003 and 2004). Source: FAOSTAT

Table 1b Bean production trend in major producing countries of Latin America (Figures are sum for production period)

Country	Production (million tonnes) 1995–2004	2005–2014
El Salvador	0.68	0.88
Guatemala	0.89	1.76
Honduras	0.67	0.81
Mexico	11.98	10.74
Nicaragua	1.47	2.02
Argentina	2.57	2.81
Brazil	28.08	32.17
Colombia	1.29	1.46
Paraguay	0.54	0.54
Peru	0.64	0.88

Source: FAOSTAT

Table 1c Bean production trend in major producing countries in North America and the Caribbean (Figures are sum for production period)

	Production (million tonnes)	
Country	1995–2004	2005–2014
Canada	2.50	2.63
USA	12.16	12.00
Cuba	0.78	1.08
Dominican Republic	0.28	0.31
Haiti	0.37	0.79

Source: FAOSTAT

Table 1d Bean production trend in major producing countries in Europe (Figures are sum for production period)

	Production (million tonnes)	
Country	1995–2004	2005–2014
Belarus	1.42	1.69
Poland	0.42	0.32
Romania	0.41	0.24
Ukraine	0.53	0.34
Ireland	0.13	0.14
Latvia	0.01	0.06
Lithuania	0.04	0.12
Albania	0.23	0.25
Greece	0.22	0.20
Italy	0.20	0.13
Serbia and Montenegro	0.60	0.06

Source: FAOSTAT

Caribbean (Table 1c). Common bean production in Europe is relatively low in comparison with other regions of the world and Belarus appears to be the leading producer (Table 1d). According to the available FAO statistics, common bean production share by region is 35% for the Americas, 17.7% for Africa, 2.6% for Europe and 0.2% for Oceania (FAO, 2015).

Globally, common bean is cultivated on about 28 million ha, producing on average, approximately 715 kg ha^{-1}year^{-1} (Broughton et al., 2003). There has been a slight increase in average yields, although the increase is not significant in the resource-poor SSA, where the average yields are still below 600 kg ha^{-1} (Akibode and Maredia, 2012). Average yields of 500, 490, 280, 490 and 540 kg ha^{-1} have been reported in Uganda, Kenya, Angola, Malawi and Democratic Republic of Congo, respectively (Akibode and Maredia, 2012).

These values are far below the potential yield of 1500–3000 kg ha^{-1} when common bean production occurs under conditions that meet the requirements (Hillocks et al., 2006). The low level of production in SSA has been attributed partly to low levels of soil plant-available P and drought stress caused by climate change variability (Beebe et al., 2011, 2013; Wortmann et al., 1998). Comparatively, farm-level genetic diversity of common bean in Africa is usually broader than in Latin America, where bean was domesticated (Grisley and Mwesigwa, 1991), but where consumers are more specialized in preferences. In Africa, in addition to grain appearance, a complex of characteristics influence preference (Wortmann et al., 1998). These include culinary quality, taste and market value.

3 Production constraints

In the major common bean production regions of Latin America and Africa, the crop is primarily grown by smallholders. Production generally takes place with few purchased inputs and is besieged by an array of abiotic and biotic production problems. Consequently, yields are notoriously low, particularly when compared with the average yields obtained in temperate regions of North America and Europe (Schwartz and Pastor-Corrales, 1989). Important production constraints in tropical bean production regions include nitrogen (N) deficiency, phosphorus (P), aluminium (Al) and manganese (Mn) toxicities in acid soils, drought, heat stress, and diseases and insect pests (Table 2). Most of the landraces and improved varieties grown in these areas are vulnerable to one or more of these production constraints, preventing the realization of their full yield potential and causing production instability from one year to the next (Schwartz and Pastor-Corrales, 1989).

The yield of common bean is therefore poor in many growing regions and the improvement in production has generally not kept pace with demand. In Ethiopia, as in many countries in SSA, the observed low national mean yield is attributed to various constraints, such as low adoption of improved agricultural technologies, drought, limited use of improved varieties, poor cultural practices, diseases and environmental degradation (Legesse et al., 2006). Smallholder bean production takes place in systems that involve complex associations. For this reason, it is difficult to objectively evaluate the relative importance of the various production constraints. This is because some tropical agro-ecosystems are self-supporting subsistence systems (Igbozurike, 1971; Janzen, 1974) in which coevolved crops have achieved an equilibrium not only with one another and with their environment (Bunting, 1975) but also with their parasites (Schwartz and Pastor-Corrales, 1989). Some of the economically significant biophysical constraints to common bean production in different regions are given in Table 2. Table 2 shows that the major biotic and abiotic constraints to common bean production have a variable distribution in the bean growing regions.

3.1 Soil fertility and acidity

Soil fertility, particularly low soil N and P levels, and acid soil conditions, are important constraints to bean production in most of the areas where the crop is grown (Graham et al., 2003). In Latin America and Africa, bean is grown in different soil types which often limit their growth and yields due to nutritional deficiencies and toxicities. N and P deficiencies, and soil acidity, are encountered in Latin America and Caribbean, as well as in eastern, southern and western Africa (Table 2). However, P deficiency is of high relative importance in Latin America and Caribbean, while in eastern, southern and western Africa, both N and P deficiencies are of high relative importance. Soil acidity is a relatively more important production constraint in Latin America and Caribbean than in the common bean growing regions of Africa. However, soil acidity affects more than 1.5 billion ha worldwide, and its importance is likely to increase as a result of acid rain, long-term N fertilization and natural weathering (Graham and Vance, 2003). Soil acidity

Table 2 Major biotic and abiotic constraints to common bean production in the main growing regions of Latin America, Caribbean and Africa

	Constraint*			
Region	Abiotic	Relative importance	Biotic	Relative importance
Latin America and the Caribbean	N deficiency	Medium	BCMV/BCMNV	Low
	P deficiency	High	BGMV/BGYMV	High
	Acid soils (Al, Mn toxicities)	Medium	Angular leaf spot	Medium
	Drought	High	Anthracnose	Medium
	Heat	High	*Pythium*	Low
			Fusarium	Low
			Rust	Medium
			Common bacterial blight	Low
			Bruchids	Medium
			Bean stem maggot	–
Eastern, western and southern Africa	N deficiency	High	BCMV/BCMNV	Medium
	P deficiency	High	BGMV/BGYMV	–
	Soil acid (Al, Mn toxicities)	Low	Angular leaf spot	High
	Drought	High	Anthracnose	Medium
	Heat	Medium	*Pythium*	High
			Fusarium	Medium
			Rust	Medium
			Common bacterial blight	High#
			Bruchids	Medium
			Bean stem maggot	High

*Al=Aluminium, Mn=Manganese, BCMV= bean common mosaic virus, BCMNV=bean common mosaic necrotic virus, BGMV=bean golden mosaic virus, BGYMV=bean golden yellow mosaic virus.
#Common bacterial blight high priority especially in southern Africa.
Source: Adapted from Beebe et al. 2011, Schwartz and Pastor-Corrales, 1989.

affects plant growth by inducing Al and Mn toxicities, and P, Mo or Ca deficiency, all of which contribute to the problem of low legume productivity (Graham, 1992). Both soil acidity and P deficiency affect nodulation, survival of rhizobia and BNF. These twin soil constraints greatly exacerbate the production problem, given that symbiotic N fixation rates and tolerance to P deficiency are already low in common bean in comparison to other legumes (Broughton et al., 2003).

In general, nutrient depletion is an important production constraint, especially among smallholders in developing countries, where use of fertilizers is limited by cost. According to Sanchez (2002), average annual nutrient depletion rates across 37 African countries is 22 kg N ha^{-1}, 2.5 kg P ha^{-1} and 15 kg K ha^{-1}. The relatively lower capacity of smallholder farmers to replenish these nutrients is one of the factors responsible for low common bean productivity. Increasing population pressure in SSA has necessitated intense cultivation with limited opportunities for fallow periods or crop rotation, leading to declining soil fertility and a build-up of pathogen inoculum in the soil (Wortmann et al., 1998). Effect of diseases on plants is more pronounced in soils of low fertility status since root growth is inhibited and this reduces the potential for recovery after infection.

3.2 Pests and diseases

Similar to soil-related constraints, the major pests of common bean have a variable distribution in the bean growing regions (Table 2). For example, bean fly is extremely important in Africa but is not present in Latin America. However, other insect pests, such as bruchids, are present in both regions but are of medium importance. The major common bean diseases in the two regions include bean common mosaic virus (BCMV), bean common mosaic necrotic virus (BCMNV), angular leaf spot (ALS), anthracnose, *Pythium* and *Fusarium* root rots, rust, halo blight and common bacterial blight. While *Fusarium* and *Pythium* are relatively more important in Africa than in Latin America and the Caribbean, bean golden mosaic virus (BGMV) and bean golden yellow mosaic virus (BGYMV) are absent in the major bean growing regions in Africa.

3.3 Drought and heat stresses

Moisture stress is becoming an important production constraint due to climate change and expansion of bean growing into more marginal rainfall areas. Bean has a relatively short growing season and often subjected to intermittent or terminal drought, hence there is need for increased production of drought-tolerant varieties, alongside application of supplementary irrigation, and use of appropriate water conservation strategies to improve yields. Heat stress is a greater problem in lowland environments where terminal drought can be aggravated by high temperatures (Graham and Ranalli, 1997). High temperatures during the reproductive growth stage of common bean results in a reduction in pod and seed set due to enhanced abscission of flower buds, flowers and pods (Ahmed et al., 1992). While drought and heat are production constraints of high relative importance in Latin America, in Africa heat is ranked medium as a production constraint. Bean breeders in Central America and the Caribbean, where heat is an important factor in bean production, have focused on addressing heat as a constraint to expanding bean production in the lowland tropics (Graham and Ranalli, 1997; Beaver et al., 2003) and have made significant progress in developing bean cultivars with improved levels of heat tolerance.

3.4 Variety selection and seed quality considerations

Low access by farmers to high-yielding improved varieties, in conjunction with edaphic and biotic constraints, are the major causes for the low bean yields reported in many bean growing regions. Many factors inform the choice of bean varieties grown by farmers. There are significant yield differences between varieties, as well as differences in genotype by environment (GxE) stability. Farmers also have different production objectives in the sense that some farmers produce mainly for home consumption while others are market oriented. This situation leads to a wide range of common bean varieties being planted. Besides, it has long been appreciated that there is no premium on genetic uniformity in tropical subsistence farming (Leakey, 1970). In such systems involving complex associations, an intrinsic balance has been achieved between crops and their environment. In addition, resource-constrained farmers operating in riskier production environments are generally more concerned with yield stability and often plant varietal mixtures, including their local land races. This practice stems from the need to maximize stability of performance rather than productivity *per se* (CIAT 1989). Participatory development of bean varieties is therefore crucial, and this should be accompanied by seed systems to ensure seed is available to farmers.

High-quality, preferably certified seed with high germination percentage (at least 80%) is essential for successful production of common beans. Many smallholder farmers, particularly in SSA, rely on own recycled seed either due to cost considerations or unavailability of certified bean seed systems. Recycling of seed is a very significant production constraint considering that the cost of investing in certified seed is quite small compared to yield losses that would occur due to diseases or poor stand resulting from use of low-quality seed.

In a study in western Kenya supported by The McKnight Foundation through its Collaborative Crop Research Program (CCRP), Kadari et al. (2014) collected soil and farm-saved bean seed samples from farmers' fields and subjected them to physical purity, germination and seed health tests to determine levels of bean disease inoculum. The results showed high levels of disease inoculum in the soil and farm-saved bean seeds. Symptoms of seed infection included shrivelling and discolouration and mouldiness. The soils contained inoculum levels of up to 20 000 CFU/g, while the farm-saved bean seeds showed low purity, low germination percentage (less than 85%) and high levels of infection. The seed samples contained common bacterial blight (*Xanthomonas axonopodis* pv. *phaseoli* inoculum levels of up to 456 CFU/g). The inoculum level of halo blight causing pathogen (*Pseudomonas savastanoi* pv. *phaseolicola*) was up to 132 CFU/g, while the major root rot pathogens identified in the soil were *Rhizoctonia, Fusarium solani, F. oxysporum, Pythium* and *Macrophomina*. These results point towards the need for appropriate seed systems to supply high-quality seed to the smallholder farmers. In the absence of this, smallholder farmers should be encouraged and supported to treat their farm-saved seed before planting in order to raise bean productivity.

Using certified seed, or seed produced by farmers under supervision from seed experts, will reduce the incidence of seed-borne diseases such as BCMV, bacterial diseases (common blight, halo blight and bacterial brown spot) and fungal diseases, such as anthracnose. FAO, recognizing the crucial role appropriate seed systems can play in agricultural development, innovated the Quality Declared Seed (QDS) system, which is a system of seed quality control that is less expensive than regular certification procedures (FAO, 2006). Increasing the quality of seeds increases the yield potential of the crop and is considered one of the most economical and efficient inputs for improving crop productivity. Rigorous seed quality control has long been seen as a key tool to achieve the goal of delivering good quality seed of improved and new varieties to farmers. However, this has not been very successful due to many inherent problems. QDS is an attempt to reconcile the continuing need to improve seed supply to farmers with the desire to reflect and accommodate the diversity of farming systems, particularly in the more difficult areas where highly organized seed systems do not function well, without compromising basic standards of seed quality.

4 Cropping system

In many of the common bean growing regions, especially in the tropics, multiple cropping is a common practice. Multiple cropping includes growing of two or more crops at the same time (intercropping) or in a sequence (rotation). Multiple cropping allows exploitation of various kinds of crop-crop interactions, including symbiosis, commensalism and mutualism. Intercropping is an old and widespread practice and has numerous advantages over sole cropping. The mixing of crops of different rooting depth, height, canopy architecture and nutrient requirements leads to complementary utilization of growth resources. The understory legume increases soil and moisture conservation through greater ground cover than sole cropping. Intercropping also allows resource-poor smallholder farmers to minimize production risks, control weeds and balance household nutrition. Many advantages of intercropping have been reported, including optimum utilization of land (Searle et al., 1981), weed suppression (Haggard-Nelson et al., 2001), improvement in yield and yield stability (Nazir et al., 2002; Bhatti et al., 2006), and soil fertility improvement through biological nitrogen fixation (BNF) (Jensen, 1996).

Common bean is grown either as a sole crop or intercropped with other crops, mostly cereals. In Africa, it is commonly grown intercropped with maize and sorghum. Intercropping systems involving non-legume and legume combination have significant yield advantages (Morris and Garrity, 1993) over mono-cropping systems. Intercropping is generally more productive and sustainable due to its nutrient facilitation, and its effect of increasing N uptake for the intercropped cereal via BNF (Latati et al., 2013). Nitrogen fixing legume is an important resource for improving the grain yield of cereals (Shen and Chu, 2004; Betencourt et al., 2012). Due to these advantages, intercropping has been, and remains a popular practice among smallholder farmers in the tropics (Ofori and Stern, 1987; Fujita et al., 1990; Rusinamhodzi et al., 2012). Apart from maize-common bean associations, this positive interaction has been confirmed in other cereal-legume associations, for example, cowpea-maize, faba bean-wheat and wheat-soybean intercrops (Zhang and Li., 2003; Li et al., 2005; Dahmardeh et al., 2010). The characteristics of the two intercropped species determine the nature and degree of inter-specific competition. Inter-specific competition is an interaction between two species that reduces the fitness of one or both of the component crops. The species compete both above- and below-ground for growth resources. In nature, this plays a role in determining the structure and dynamics of plant communities (Aerts, 1999). In legume-cereal association, the cereal component crop is usually taller and has a faster growing or more extensive system of fine roots (Lehmann et al., 1998) and very competitive for soil nitrogen than the legumes which usually fix atmospheric N (Jensen,

1996). The morphological and physiological differences between the species benefit their mutual association (Akuda, 2001).

4.1 Nitrogen requirement and biological nitrogen fixation (BNF)

Even though common bean is often intercropped with cereals, particularly in Africa, little attention has been given to evaluation of soil fertility management technologies appropriate for these intercrops. As a result, maize and bean yields, and BNF remain perpetually low. A study by Snapp et al. (1998) researched these factors. Their study investigated the effect of phosphorus at 30 kg P ha^{-1} and improved bean cultivars on intercrop yields and bean BNF (measured by the N difference method) at two sites with nutrient-deficient soils in Malawi. Maize was the main crop and bean was planted at low density as the intercrop. According to their results, phosphorus enhanced maize and bean yields and BNF. The yields of the improved bean cultivars were about 150% higher than farmer cultivars at both 0 and 30 kg P ha^{-1}, while BNF was increased by about 90% in improved cultivars compared to farmer cultivars. In a second trial conducted on-farm at nine sites, fertilizer P (30 kg ha^{-1}) and N (30 kg ha^{-1}) were evaluated on a maize-bean intercrop versus sole maize. Yields and economic evaluation indicated higher net benefits of maize-bean intercrop (US$192 ha^{-1}) and fertilized sole maize (US$86 ha^{-1}) compared to unfertilized sole maize (US$62 ha^{-1}). This study concluded that fertilizing maize-bean intercrops is a promising technology for smallholder farmers in P-deficient soils, and in areas where bean sole crop is not sufficiently valued by subsistence farmers to warrant fertilization.

4.2 Plant population and spatial arrangement

The spatial distribution of plants in a crop community is an important determinant of yield (Egli, 1988). The many experiments that are conducted to establish spacing and plant population testifies to this. The seed yield of common beans is influenced by many variables, such as pods/plant, seeds/pod and unit weight of seed. The importance of row spacing, row orientation (e.g. NS or EW) and plant density in determination of yield can be explained by two general principles. First, production of sufficient leaf area for maximum light interception for photosynthesis (Gardiner and Craker, 1981; Tsubo and Walker, 2002), and secondly, distance between plants affects interplant competition (Pendleton and Hartwing, 1973) for nutrients and soil moisture. For determinate cultivars, a row spacing of 75 cm is generally recommended. However, many smallholder subsistence farmers use lower seed rates than recommended resulting in low plant population and poor yields, for example, smallholders in Ethiopia (Ali et al., 2003), and other parts of eastern Africa.

Various spatial arrangements have evolved for planting maize-bean intercrops. In Kenya, where smallholder farmers commonly plant maize intercropped with common beans, farmers use a variety of spatial arrangements. Whereas maize is usually planted in rows, row bean planting has not been fully adopted. Farmers may randomly plant beans in between the maize rows, or use the conventional method of planting one row of bean in between a pair of maize rows. An intercropping row arrangement devised by Woomer et al. (2004) provides greater opportunity to understorey legume by reducing competition with maize. In this system referred to as MBILI (meaning 'two' in Kiswahili), every other maize row is staggered by 25 cm and maize and beans are planted as paired rows (see Fig. 1). This results in the same plant population as the conventional intercropping

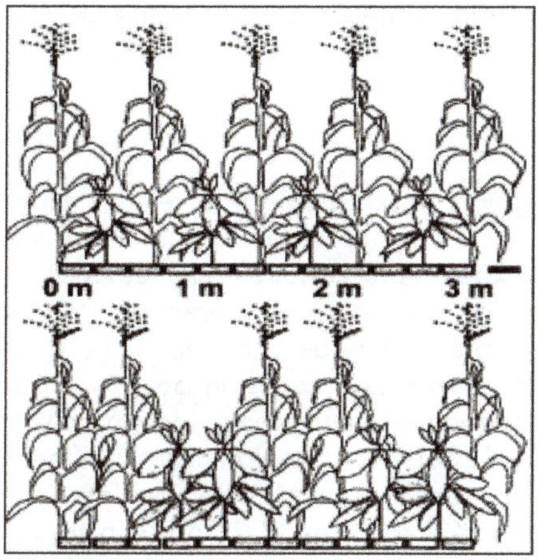

Conventional intercropping system with 0.75 cm rows of maize and legumes planted between every row

MBILI (2-by-2) arrangement with staggered 0.5 m and 1.0 m rows of maize and paired, alternating bean rows

Figure 1 Row arrangement for conventional and MBILI intercropping systems. Source: Woomer et al., 2004.

Table 3 Average daily light penetration and crop productivity in conventional and MBILI intercropping row arrangements

Row arrangement	Solar radiation above		Crop yield		Net return (KSh ha^{-1})
	Maize	Beans	Maize	Beans	
	--------LUX--------		--------(kg ha^{-1})--------		
Conventional	91,560	38,403	4,835	760	26,333
MBILI	91,823	59,327	4,900	1075	31,689
LSD $_{0.05}$	ns	20,448	ns	207	3,359

ns: not significant.
Source: Woomer et al., 2004.

system (44 444 and 88 888 plants ha^{-1} for maize and beans, respectively) while allowing bean to grow in a 100 cm gap between maize rows.

The MBILI intercropping system repeatedly outperformed the more common conventional alternate row intercropping system over three seasons of on-farm trials. The results of these on-farm trials, averaged over three seasons, are presented in Table 3. Maize yields were either unaffected or slightly improved. The MBILI system increased bean yield by 15%. Part of the bean yield response was attributed to improved availability of solar radiation. Solar radiation available to the understorey legume increased by 54% (Table 3), resulting in 41% increase in bean yield and an additional Ksh. 3359 return. Maize grain yields in the conventional and staggered MBILI row arrangements were not significantly different but tended to be higher within the MBILI staggered rows. These results suggested that MBILI's advantage is related to aboveground competition for light between maize and the understorey bean. Apart from competition for light, intercropped species compete for moisture and nutrients. Both plant population and spatial arrangement influence the use of these resources and this competition determines the performance and ultimate benefits.

4.3 Nutrient facilitation

Another important aspect of cereal-legume intercropping is related to its nutrient facilitation, and its effect to increase N uptake by the intercropped cereal via symbiotic nitrogen fixation. In multi-locational field experiments conducted in Setif region of Algeria, Latati et al. (2013) investigated the effect of maize-common bean intercropping system on the growth, nodulation, yield and N uptake at three sites representing major agro-ecological conditions where intercropping maize and common bean is practised. They used one cultivar of maize, and one of common bean commonly grown by farmers in the region. $CaCO_3$, total-P, Olsen-P and total-N contents were significantly different between the sites. Their results showed a large site x management interaction. Intercropped bean nodule biomass increased by 65, 23 and 34% for sites 1, 2 and 3, respectively, while intercropped bean shoot N concentration increased by 6.1% in site 2 and by 8.2 % in site 3. However, site 1 recorded a decrease in intercropped bean shoot N concentration of 28.5%. While the root N concentration of intercropped treatments decreased in sites 1 and 3, in site 2 there was an increase.

In maize, intercropping significantly increased shoot N concentration by 14.7, 18.6 and 20.8% for sites 1, 2 and 3, respectively. However, root N concentration was enhanced under intercropping treatment in sites 1 and 3, but decreased by 31.5% in site 2. Intercropping maize with beans increased maize grain yield by more than 12.5% for all sites, with the largest increase of 46% in site 1. According to the authors, this increase in maize grain yield was partly due to enhanced N concentration in rhizosphere of intercropped common bean. This led to increased N uptake in shoot and seed of intercropped maize, especially for the sites with low N concentration in indigenous soil. Common bean had a positive effect on inter-specific competition through N fixation. Consequently, maize grain yield was significantly increased by intercrop effect. These results demonstrate the advantage of intercropping maize with common bean over sole cropping.

5 Integrated pest management

Pests are among the most significant factors affecting the production of common bean in many countries, greatly contributing to low yields and poor quality of the produce. A large number of field and storage pests affect common bean in the major growing regions in Africa, Asia, Latin America and the Caribbean. In Africa, the most important field and storage pests of common bean are the bean fly or the stem maggot complex (*Ophiomyia* spp.) and bruchids, respectively (Abate and Ampofo, 1996). Bean fly is aggravated by environmental stresses, such as poor soil fertility and moisture deficit. Also, repeated planting of bean in the same plot leads to bean fly population build-up, worsening the damage (Buruchara et al., 2010). Foliage beetles, flower thrips, pollen beetles, pod borers and sap suckers, such as aphids, also cause significant damage. Bean fly and bean borer, *Etiella zinckenella* (Treitschke), are of considerable importance in Asia (Jackai, 1995). In the American continents, leafhoppers (*Empoasca kraemeri* and

E. fabae) are the most widely distributed insect pests in common bean fields, especially in relatively drier areas, bean pod weevil (*Apion godmani* and *A. aurichalceum* Wagner) causes severe damage to developing pods and seeds in the highlands of Mexico, in Guatemala, El Salvador, Honduras and Nicaragua, while Mexican bean beetles (*Epilachna varivestis* Mulsant) cause severe leaf damage in the highlands of Mexico and Western US (Singh et al., 2010).

The common bean field pests of economic importance often require more than one tactic for effective management. Integrated pest management (IPM) is an ecosystem-based control strategy that combines local knowledge and science, and emphasizes long-term prevention of pests and their damage through a combination of techniques. These techniques include biological control, judicious use of pesticides, application of appropriate cultural practices, use of host plant resistance and so on. The aim is to maintain pest populations below economic threshold levels and minimize risks to human health and the environment. Chemical control of pests generally provides more complete protection compared to other available methods, and even though it usually has favourable cost-to-benefit ratio, it is not a sustainable option (Jackai, 1995). In Africa, where crops are produced under extremely variable agro-ecological conditions, and two or more crops are grown in the same field at the same time, the resulting diversity of the production systems enhances natural enemy abundance and generally keeps pest numbers at low levels (Abate et al., 2000). This is quite important, given that pesticide use in Africa is quite low, accounting for only 2–4% of the global pesticide market (Williamson et al., 2008; Belmain et al., 2013). IPM strategies, particularly in the African smallholder setting, include appropriate sowing dates, optimum plant density, varietal mixtures and intercropping (Abate and Ampofo, 1995). Generally, good crop husbandry practices are essential in IPM. These include post-harvest tillage to expose the insects to heat, crop rotation with non-hosts, timing of sowing to avoid susceptible stages of the crop coinciding with peak pest population cycle and so on (Buruchara et al., 2010).

6 Management of diseases

6.1 Angular leaf spot (ALS)

ALS, caused by *Phaeoisariopsis griseola* (Sacc.) Ferraris, is widely distributed in the major common bean growing regions and can cause yield losses of up to 80% (Stenglein et al., 2003), particularly in eastern Africa. Humid conditions and moderate temperatures (20–25°C) favour infection and disease development (Buruchara et al., 2010). The disease may affect all parts of the bean plant above the ground. Symptoms include angular or polygonal lesions which usually appear as brown spots with a tan or silvery centre. Infected pods may bear poorly developed or shrivelled and discoloured seeds. The main sources of infection are plant debris or residue and infected volunteer plants. The disease can be controlled by a combination of strategies, including crop rotation with a non-host crop, removal or deep ploughing of crop debris, seed dressing with a suitable fungicide, planting resistant varieties, if available, and growing of bean variety mixtures, including resistant ones to reduce the spread of infection. Use of resistant varieties is an effective, economical and environmentally friendly strategy of managing ALS, especially for small-scale farmers in SSA.

6.2 Anthracnose

Bean anthracnose (*Colletotrichum lindemuthianum*) is widely distributed and is one of the most important diseases of common bean worldwide. Frequent rainfall favours disease development and damage is greater under cool to moderate temperatures with high relative humidity. The disease affects any part of the plant, depending upon time of infection and source of inocula. Initial symptoms include dark brown to black lesions along the veins on the underside of the leaves. Under severe infection, young pods may shrivel and dry prematurely. The fungus may penetrate through the pod causing discolouration and distortion of the seeds. The infected seed and crop debris serve as sources of infection to the succeeding crop. Control of the disease can be achieved through the use of a combination of strategies, including using disease-free seed, crop rotation with non-hosts, burying of plant debris from fields soon after harvest, planting resistant varieties, dressing seeds with appropriate fungicides, cleaning of seed storage facilities to avoid contamination and so on.

6.3 Common bacterial blight

Common bacterial blight (*Xanthomonas axonopodis* pv. *phaseoli*) is widespread and is favoured by warm to high temperatures and high humidity. The major source of infection is seed. Symptoms first appear as water-soaked spots on the underside of leaves then the spots enlarge and coalesce, and yellowing of leaves becomes more general. The stem may rot at the first node where cotyledons were attached and cause the plant to break. Bacteria also invade seed and may remain dormant until germination. Other sources of infection include dry bean debris or residue on the soil surface and infected volunteer plants. Control measures include use of disease-free seeds, chemical seed dressing, crop rotation, deep ploughing of infected bean debris and planting of resistant bean varieties, where available.

6.4 Halo blight

Halo blight (*Pseudomonas savastanoi* pv. *phaseolicola, Pseudomonas syringae* pv. *phaseolicola*) is more prevalent in mid- to high altitude areas. Its development is favoured by cool temperatures (16–20°C) and moist, cloudy conditions. The disease is therefore not common in regions or seasons with high temperatures (Buruchara et al., 2010). The most characteristic symptoms of halo blight occur on the bean leaves. First symptoms appear as small water-soaked pinprick spots on the undersides of leaves. The tissues surrounding the spots gradually become yellow green. Seeds may become shrivelled and discoloured or rotten. Control strategies include spraying with copper-based fungicides, crop rotation, deep ploughing of infected debris, planting resistant varieties and so on.

6.5 Bean common mosaic viruses

BCMV and BCMNV are important viral diseases of bean in Africa. In Latin America, BGMV and BGYMV are important (see Table 2). Symptoms are usually light green or yellow and dark green mosaic patterns on leaves. Other symptoms on susceptible hosts include mottling, curling and malformation of leaves, as well as general stunting of the plants. Planting of resistant varieties, use of disease-free seed, early planting to escape high aphid population pressure period and destroying of other plant hosts during the entire bean growing period are some of the recommended control strategies.

6.6 Bean root rots

Bean root rots are widespread in Africa and are caused by a complex of several different soil-borne fungi (*Pythium* spp., *Fusarium solani* f. sp. *phaseoli, Rhizoctonia solani, Sclerotium rolfsii* and *Macrophomina phaseolina*) and are particularly more serious in common bean crops grown under stress, for example, low soil fertility, high humidity, warm to high temperatures, high or low soil moisture, drought and so on. Root rot disease symptoms vary depending on causative organisms and on environmental conditions. In general, leaves turn yellow and plants wilt and die. *Pythium* root rot causes seedling and post-emergence damping-off. The pathogen destroys much of the hypocotyls and main root system. In the case of *Fusarium* root rot, the primary and lateral roots are often killed and detached, resulting in hollow lower stems. Infected soil is the major source of bean root rots. However, plant debris and infected or contaminated seed are other sources of infection. A combination of strategies are recommended for management of root rots. These include seed and soil treatments with appropriate fungicides; planting resistant varieties; improvement of soil fertility, for example, through application of farmyard manure (FYM) and inorganic fertilizers; planting in raised beds or ridges to facilitate rapid plant growth; hilling up soil around the stem before flowering to stimulate development of adventitious roots for *Rhizoctonia* and *Fusarium* infected plants; and deep ploughing of infected plant material during land preparation.

7 Water management

Drought is one of the major production challenges to improved bean productivity, especially in marginal rainfall environments with limited irrigation capacity. Drought can occur as early, intermittent or terminal stress in a cropping season. However, its effect is greater on common bean growth and productivity during early establishment, vegetative expansion, flowering and grain filling (Rao, 2001). Drought is of particular importance to common bean production in SSA, where its frequent occurrence is a critical constraint (Beebe et al., 2013; Emam et al., 2010) affecting more than 40% of the growing areas (Beebe et al., 2013; Broughton et al., 2003). In eastern and southern Africa, drought lowers bean yields to less than 400 kg ha^{-1} $year^{-1}$ (Beebe et al., 2013), which is significantly smaller compared with the optimum production of 1500 kg ha^{-1} $year^{-1}$ realized in less drought-constrained areas in eastern Africa (Hillocks et al., 2006). Due to climate change, the increased temperatures and associated higher evapotranspiration rates, combined with erratic and lower rainfall, is expected to exacerbate the problems of bean production (Beebe et al., 2013) in bean growing regions with limited irrigation capacity. Climate models predict that many drought-stressed regions will generally become drier over the next decades (Jones and Thornton, 2003; Williams et al., 2007).

Water is an important component of a growing plant and constitutes more than 80% of the growing tissues, and is essential for most plant functions (Shaker et al., 2013). The amount of water available, the timing, the quality and the prevailing micro-meteorological conditions are important for plant growth and yield. The inability of a plant root system to supply water as demanded is one of the principal constraint to plant productivity (Baker et al., 1992). Under rain-fed production, a minimum of 400–500 mm of rain is required during the growing season for bean production, although an annual total of

600–650 mm is considered ideal (Liebenberg, 2002). In arid environments characterized by low soil moisture, bean yield can be greatly increased by applying appropriate moisture conservation techniques. Generally, soil and water conservation practices can be divided into agronomic and mechanical measures. Agronomic measures are easier to implement because they are less labour intensive and may include no-tillage, minimum tillage, ridging, crop spacing, use of cover crops and mulching. Mechanical measures may include contour bunds, bench terraces and so on. Suitability of these practices depends on factors such as microclimate, soil type, slope characteristics and ability of farmers to effect specific management options.

Irrigation can greatly improve common bean productivity. In contrast to rain-fed production, where moisture availability is subject to the erratic timing of rainfall, irrigation allows farmers to apply water at the most beneficial times for the crop. Scheduling can be done for the critical moisture-sensitive growth states, that is, flowering and early pod set, to obtain optimum yield per unit of water. Although irrigation is essential for meeting the current world demand for food, it is going to be even more vital in the future, given that the world population is projected to increase by about 33%, from the current figure of 7.3 billion to 9.7 billion by 2050 (FAO, 2009). Most of this increase will take place in the major common bean growing regions of Africa, Latin America and Asia, given their high projections of population growth (Howell, 2001). Population increase poses the danger of extension of bean production to marginal rainfall areas, a situation likely to have far-reaching consequences on food security and household incomes, particularly in SSA, where irrigation capacity is limited. Different systems of irrigation are available for enhancing bean production, especially in arid and semi-arid environments. These include furrow and drip irrigation.

7.1 Furrow irrigation

Furrow irrigation is a method of applying irrigation water to the field through shallow, evenly spaced channels (furrows). Furrows are small, parallel channels, made to carry water in order to irrigate the crop (Fig. 2). The crop is usually grown on the ridges between the furrows. Channels convey the water either down or across the slope of the field to the vicinity of plants. The water infiltrates the soil both vertically and horizontally. Land grading may be necessary to provide uniform slopes to permit uniform water application and efficient irrigation. Different grading systems (e.g. level, graded straight and graded contour) are used depending on the crops and furrows needed to distribute the irrigation water to clean-tilled crops planted in rows with one furrow between crop rows. The furrow stream is applied until the desired application depth and lateral penetration are obtained.

Advantages of furrow irrigation include lower initial investment in equipment, minimized water loss inefficiencies and lower pumping costs. However, the method has disadvantages as well. These include accumulation of salinity between furrows, an increased level of tail water losses if retention ponds are not available to capture the run-off and allow recycling, the need to grade the land to allow uniform dispersion of water over the field and so on.

7.2 Drip irrigation

Drip irrigation is another method of irrigation that can be used, especially where water availability is limited. A schematic diagram of the layout of this method is presented in Fig. 3. The method relies on the concept of irrigating only the root zone of a crop and

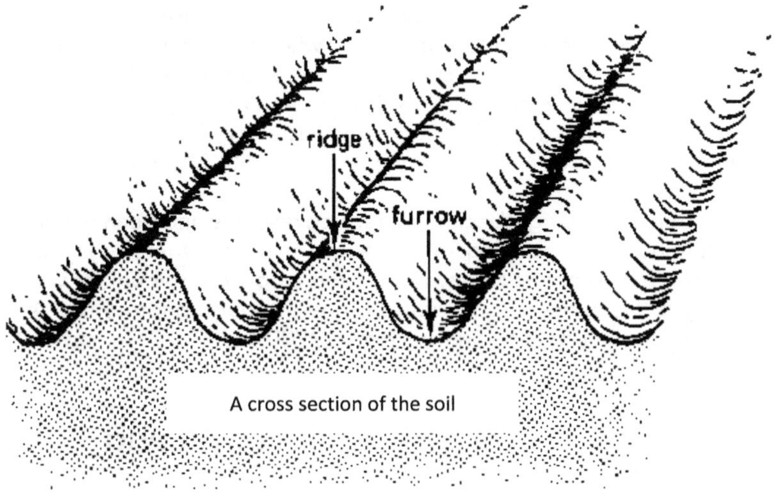

Figure 2 Diagram of furrow irrigation system.

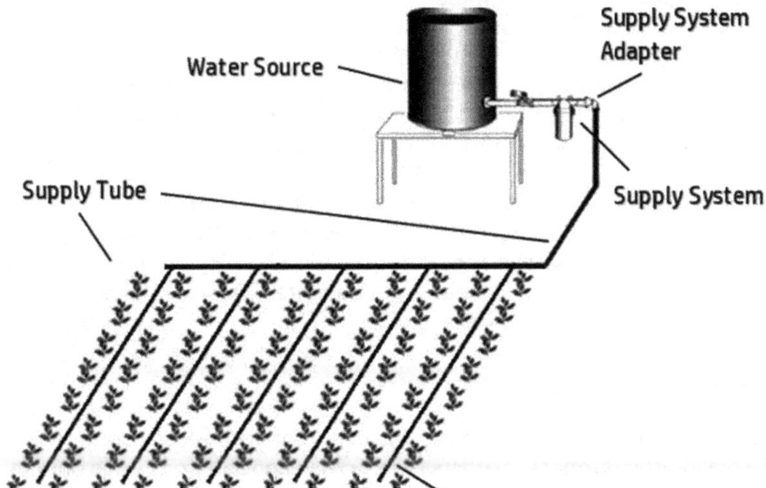

Figure 3 Diagram of gravity-fed layout for drip irrigation system.

maintaining moisture content of the root zone at nearly optimum level (Shaker et al., 2013). Apart from improving water use efficiency (WUE), drip irrigation has the effect of limiting evaporation, reducing weed growth and minimizing interruption of cultural operations. Manickasundaram et al. (2002) found that the WUE was 20–60% higher in drip irrigation compared to that of surface irrigation method. In an experiment comparing drip to surface irrigation on bean production in Sudan, Shaker et al. (2013) found that yield and yield components were very highly influenced by irrigation method. Bean yield was 6% higher with drip irrigation compared to surface irrigation, while drip irrigation used only 50% of the water used in surface irrigation. Greater yield differences have been reported in other grain legumes. For example, Raina et al. (1998) reported 49.5% higher pea grain yield with drip compared to surface irrigation. Irrigation efficiencies of 80–90% have been reported for drip irrigation (Al-Jamal et al., 2001) compared with 50–73% under furrow irrigation (Chimonides, 1995).

Worldwide, water is increasingly becoming a scarce resource and bean growing is rapidly expanding to marginal areas requiring irrigation. The solution is therefore to reduce losses and increase WUE. In its simplest terms, WUE is the crop yield per unit of water used, or on a more biological level, the carbohydrates formed through photosynthesis from CO_2, sunlight and water, per unit of transpiration (Howell, 2001).

8 Nutrient management: phosphorus (P) and nitrogen (N)

N is one of the most important elements limiting crop production in the tropics. This is more so in SSA, where crop yields are severely limited by N deficiency. However, low soil N levels also limit production of common bean in other regions where the crop is grown (Graham et al., 2003). In addition to N, P is another generally deficient nutrient for cultivation of common bean (Beebe et al., 2011; Cichy et al., 2009). In the resource-constrained farming systems of SSA, deficiency in plant-available P (Wortmann et al., 1998), coupled with soil moisture deficit (Beebe, 2012; Miller et al., 2003), constitute major limitations to common bean production. According to several sources (e.g. Broughton et al., 2003; Lunze et al., 2007; Wortmann et al., 1998), up to 80% of the bean producing areas in SSA are deficient in plant-available P. Soils with less than 40 mg kg^{-1} of plant-available P determined by Bray-1 method are considered P deficient (Mourice and Tryphone, 2012). Causes of deficiencies include prolonged degradation by erosion, repeated P removal in crop harvest without replacement (Henao and Baanante, 2006) and P fixation in acidic soils with high P-fixing capacities (Nziguheba, 2007).

8.1 P management

N and P deficiencies can be corrected by application of appropriate commercial fertilizers but many smallholder farmers, particular in SSA, cannot afford such inputs. Genetic manipulation of plants has been researched as a possible means of addressing certain production constraints, especially those related to nutrient and moisture stresses. With regard to low plant-available P and soil moisture deficit, specific architectural root traits that can improve bean productivity in soils with low levels of plant-available P and moisture deficit are needed. However, improvement in genetic aspects alone is not sufficient to address low levels of plant-available P and soil moisture deficit. Management practices, such as application of P fertilizers and supplementary irrigation, are needed to reduce the yield losses realized when production occurs under conditions of low plant-available P and soil moisture deficit (Namugwanya et al., 2014). However, farmers in resource-constrained SSA cannot easily afford commercial P fertilizers and irrigating their farms (Beebe et al., 2011) on top of concerns about the environmental problems as a result of such intensive agriculture (Namugwanya et al., 2014). A systematic approach integrating plant genetic manipulation, application of organic and inorganic fertilizers, and appropriate soil moisture conservation techniques is therefore necessary.

A number of bean genotypes with ability to thrive in P-deficient soils were screened by CIAT (Beebe, 2012; Lunze et al., 2007). These genotypes, designated as low P-tolerant bean materials, include MCM5001/K131, MCM 2001/NABE 3, AND932-A-1, AFR709-1, AFR708, AFR703-1, Nm126806-2A, FOT44-1, T842-6-9, AND1055-1, NR12634-13C-1, BZ12894-C-1, DRK137-1 and ARA8-B-1. They yield between 880 and 1352 kg ha^{-1} on soils deficient in plant-available P, which is quite a significant improvement, given that conventional common bean genotypes grown on P-deficient soils yield about 750 kg ha^{-1} (Lunze et al., 2007).

Common bean survival under low levels of plant-available soil P is dependent on two main mechanisms, namely P utilization efficiency (Atemkeng et al., 2011; Cichy et al., 2009) and high P acquisition efficiency. P utilization efficiency refers to response of common bean to low P availability by making more efficient use of absorbed P to produce biomass (Boutraa, 2009; Liao et al., 2004). P use efficient plants distribute more biomass to roots than shoots under low P levels and develop excellent ability to change the acquired P into plant biomass and yield (Cichy et al., 2009). In contrast, P acquisition efficiency, which confers tolerance to low soil P availability, is the ability of a plant to mobilize and absorb more P from the soil layers (Atemkeng et al., 2011) and is achieved by development of root architecture capable of foraging for the nutrient (Beebe et al., 2006; Cichy et al., 2009; Miller et al.,, 2003), whose content and availability are more in top than in sub-soil. Therefore, root architectural traits that allow the exploration and use of P from surface layers govern P acquisition (Beebe et al., 2006; Liao et al., 2004; Miller et al., 2003). Consequently, common bean genotypes which increase the below-ground biomass are able to acquire P in P-deficiency conditions (Liao et al., 2004; Namayanja et al., 2014) and root architecture that enable common bean genotypes to exploit the topsoil layers where soil P availability is more important for low P tolerance. Another important mechanism of P acquisition involves plant roots forming associations with mycorrhizal fungi that broadens the root absorption area, thereby improving P and moisture absorption by the plant (Song, 2005). Arbuscular mycorrhizal fungi (AMF) form symbiotic associations with majority of cultivated crops, including common bean, to enhance biological soil fertility and plant nutrition.

Several studies, for example Birhane et al. (2012), indicate that mycorrhizal association with plant roots improves P availability and moisture status. While the fungi obtains soluble organic compounds excreted by roots, the plant benefits from supply of P and moisture. The fungi hyphae has the capacity to explore the soil more widely, reaching the tiny soil pores that retain P and moisture as the soils dry (Birhane et al., 2012; Shen et al., 2011). The increased nutrient and water-absorbing surface created by the hyphae enhances the ability of plants to take up P and moisture from soils with low levels of P and moisture. AMF also increases nodulation and atmospheric nitrogen fixation potential in legumes (Turk et al., 2008). Thus, dual inoculation of plants with rhizobia and AMF would result in synergistic effects on nodulation and nitrogen fixation (Tajini et al., 2012). Colonization of plant roots with AMF makes them less likely to be colonized by other microbes, reducing their susceptibility to soil-borne pathogens, such as phytopathogenic fungi or nematodes (Selveraj and Chelleppan, 2006). AMF symbiosis is a vital option that should form part of integrated nutrient management practices in smallholder systems to reduce reliance on expensive chemical fertilizers. This can be made possible by availing affordable inocula to smallholder farmers. Oruru and Njeru (2016) recommend on-farm production of inocula. This would avoid most of the costs associated with production in greenhouses or laboratories, making AMF symbiosis and the associated environmental and economic benefits available to smallholder farmers. Also, on-farm methods have the potential of producing inocula containing the indigenous AMF already adapted to farm conditions.

8.2 N management and symbiotic N fixation

Generally, grain legumes have the ability to acquire N from the atmosphere through biological N fixation. However, common bean is considered to be a poor fixer of atmospheric N when compared with other crop legumes (Graham, 1981; Ojiem et al., 2007) and generally responds poorly to inoculation with rhizobia under field conditions (Buttery et al., 1987). Due to this, common bean rarely derives more than 50% of the N requirement from the atmosphere (Bliss, 1993). The crop has a shorter growth cycle compared to other grain legumes but with a high demand for N, which is the nutrient absorbed most and exported by the crop (Andraus et al., 2016). Nodule senescence often coincides with seed development, thus reducing yield (Giller, 2001). Common bean responds to N fertilizer even under conditions where it grows and fixes N well (Redden and Herridge, 1999) and is generally considered more responsive to N fertilization than other legumes (Graham, 1981).

Experiments conducted on-farm in western Kenya (Ojiem et al., 2007) evaluated BNF rates of five grain legumes: common bean, groundnut (*Arachis hypogaea* L.), soybean (*Glycine max* (L.) Merr.), lima (*Phaseolus lunatas* L.) and lablab (*Lablab purpureus* L.) were grown in three agro-ecologically diverse sites with application of P at 30 kg ha^{-1} at planting and no mineral N applied. The experiment was conducted in nine farms in each site. BNF was measured at mid-flowering stage using ^{15}N natural abundance method. The results (Table 4) indicated that common bean (KK8, medium maturing, type 1 growth habit) was

Table 4 Nitrogen fixation rates of some selected grain legumes in smallholder farms in western Kenya

	N fixed (kg ha^{-1} $season^{-1}$)		
Grain legume	Site 1	Site 2	Site 3
Common bean	31	14	8
Groundnut	124	43	41
Soybean	74	43	6
Lima	71	10	12
Lablab	172	86	42
$SED_{Legumes\ (P<0.05)}$	18	18	18
$SED_{sites(P<0.05)}$	5	5	5

Source: Modified from Ojiem et al., 2007.
Site 1: 1500 m asl, 2000 mm mean annual rainfall, 18°C.
Mean soil characteristics: Nitisols, pH 5.59, Total soil C 11.5 g kg^{-1}, Exchangeable P 15.8 mg kg^{-1}.
Site 2: 1385 m asl, 1600 mm mean annual rainfall, mean annual temperature 19°C.
Mean soil characteristics: Acrisols, pH 5.82, Total soil C 16.9 g kg^{-1}, Exchangeable P 2.5 mg kg^{-1}.
Site 3: 1270 m asl, 1200 mm mean annual rainfall, mean annual temperature 22°C.
Mean soil characteristics: Ferralsols, pH 5.88, Total soil C 15.0 g kg^{-1}, Exchangeable P 2.4 mg kg^{-1}.

relatively poor in N fixation compared with the other grain legumes, especially groundnut and lablab. In site 1, common bean fixed an average of 31 kg N ha^{-1}, which was only 25% of that of groundnut, and 18% of that of lablab. Common bean fixed an average of 14 kg N ha^{-1} in site 2. This was 32% of that fixed by groundnut, and only 16% of that fixed by lablab. Similarly, in site 3, common bean showed poorer BNF performance in comparison with groundnut and lablab, having fixed only 8 kg N ha^{-1}, which was 20% and 19% of that fixed by groundnut and lablab, respectively.

The poor N fixation exhibited by common bean has been attributed to the difficulty of establishing effective symbioses in the field, and plant genetic variability in the capacity to fix N (Graham, 1981), leading to a recommendation that N fertilization of field-grown plants should be undertaken (Graham, 1981; Westermann et al., 1981; Piha and Munns, 1987), especially in soils with low N levels due to the unreliability of N fixation. However, wide variations in the ability of the common bean to fix nitrogen have been reported by other studies, for example, 24–65 kg/ha (Ruschel et al., 1982), 40–125 kg/ha (Rennie and Kemp, 1983) and 20–115 kg/ha (Graham and Temple, 1984). Since enhancement of the N fixing capacity of common bean is an important agronomic goal for improving the productivity of smallholders operating under low resource endowment conditions, exploitation of the existing genetic diversity offers possibilities for realizing this goal.

The ability to fix N is variable among genotypes of common bean and ranges from 4 to 59% of plant N being derived from the atmosphere (Hardarson et al., 1993; Pena-Cabriales and Castellanos, 1993; Pena-Cabriales et al., 1993). In addition to genetic differences among cultivars in their ability to fix N, the field conditions under which they are grown also influence their N-fixation capacity. The rate of N fixation is initially slow during the early vegetative growth, when the soil mineral N is probably sufficient for crop requirement, but increases later with the crop growth rate, peaking during flowering and early pod fill (White, 1989). If flowering and pod fill processes are prolonged, N fixation may continue. Generally, indeterminate cultivars fix more N than those with a determinate growth habit (Piha and Munns, 1987). Research in the last three decades has demonstrated that climbing beans are superior to bush beans in nitrogen fixation under different agronomic conditions (Graham and Rosas, 1977; Graham and Temple, 1984; Kipe-Nolt and Giller, 1993). Therefore, unravelling the genetic and physiological basis for the superior BNF performance by climbing beans would immensely contribute to improved common bean production.

Bean cultivar and *Rhizobium* strain interactions, competition among efficient and inefficient bacterial strains, suppression of nodulation by residual soil N, high demand for P and photosynthates by *Rhizobium*, sensitivity to moisture stress, and interactions of these factors with environments complicate the development of bean cultivars capable of high N-fixation rates in the tropics and subtropics (Fageria et al., 2014). Supply of sufficient quantities of N is therefore necessary for improving its productivity. This calls for, among other things, addressing the factors that may be limiting nodule activity and efficiency, and the environmental factors likely to have negative impact on the plant-bacteria association (Neves et al., 1987). In a greenhouse experiment to determine the effects of N fertilization and inoculation with *Rhizobium* on yield and yield components of 15 dry bean genotypes, Fageria et al. (2014) found that N × genotypes interactions were significant for shoot dry weight, grain yield, number of pods per plant, number of grain per pod, 100-grain weight, grain harvest index (GHI), and maximum root length and root dry weight. Maximum grain yield was produced at 200 mg N kg^{-1}. Inoculation with rhizobial strains improved grain yield but did not reach the level of 200 mg N kg^{-1} applied with chemical fertilizer.

9 Nutrient management: further issues

9.1 Integrated soil fertility management (ISFM)

Integrated soil fertility management (ISFM) offers the best opportunity for improving common bean productivity. ISFM is an important principle that can be relied upon by resource-poor smallholders and involves efficient management of inorganic fertilizers, including combining organic and inorganic resources, and efficient recycling of nutrients through crop residues, FYM and compost. In addition to improving productivity, integrated management of these low-cost and eco-friendly nutrient sources can help in combating environmental degradation through reduction in the use of inorganic N fertilizers. In addition to use of inorganic fertilizers, FYM and compost, ISFM practices include the use of improved germplasm and combining the use of the inputs with the knowledge on how to adapt the practices to local conditions, aiming at maximizing agronomic use efficiency of the applied nutrients and improving crop productivity (Vanlauwe et al., 2010). Fig. 4 illustrates the management steps that are needed in moving towards full ISFM to improve agronomic efficiency of the applied inputs.

Knowledge is required for local adaptations to adjust the practice for variability in soil fertility conditions. Fig. 4 shows that improvement in agronomic efficiency is dependent on the status of the soil. While substantial improvements in agronomic efficiency can be expected on responsive soils (A), on poor, less-responsive soils, application of fertilizer alone does not result in improved agronomic efficiency (B). Better results are expected when fertilizers are applied in combination with organic inputs (C) (Alley and Vanlauwe, 2009).

Many studies have demonstrated the value of integrated management of organic and inorganic resources. Datt et al. (2013) assessed the impact of integrated nutrient management practices on symbiotic parameters and productivity of French beans in India. They evaluated eight treatments (see Table 5) on French beans over a period of four years.

Minimum nodule weight was recorded with T_7 chemical fertilizer treatment (Table 6). However, T_7 gave the maximum number of pods followed by integrated nutrients, organic

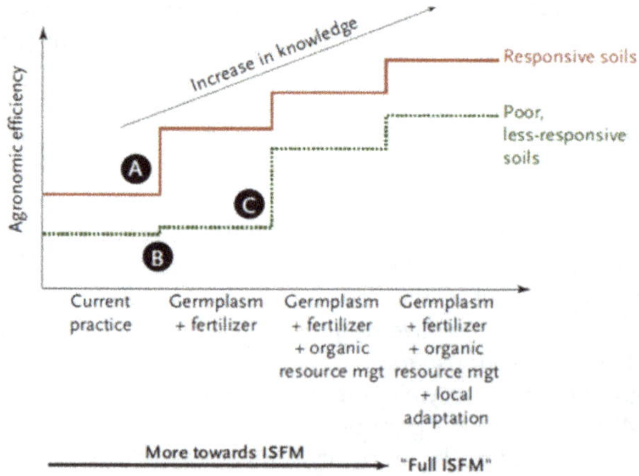

Figure 4 Steps towards full application of ISFM to increase agronomic efficiency of the inputs and improved productivity. Source: Vanlauwe et al., 2010; Alley and Vanlauwe, 2009.

Table 5 Details of treatments applied

Treatment	Details
T_1	Organic (10 t of farmyard manure per ha + nitrogen fixer-A, phosphate solubilizer and chopped crop residues of same plot)
T_2	Organic (10 t farmyard manure per ha + nitrogen fixer-B, phosphate solubilizers and chopped crop residues of same plot)
T_3	Integrated (5 t of farmyard manure per ha + nitrogen fixer-A, phosphate solubilizer + half dose of P_2O_5)
T_4	Integrated (5 t of farmyard manure per ha + nitrogen fixer-A + half nitrogen + phosphate solubilizer+half dose of P_2O_5)
T_5	Integrated (5 t of farmyard manure per ha + nitrogen fixer-B + phosphate solubilizer + half dose of P_2O_5)
T_6	Integrated (5 t of farmyard manure per ha + nitrogen fixer-B + half dose of nitrogen + phosphate solubilizer + half dose of P_2O_5)
T_7	Chemical fertilizer (recommended dose of N, P_2O_5 and K_2O)
T_8	Control (no fertilizer or farmyard manure)

Source: Datt et al. (2013).
The recommended dose for French bean is 50 kg nitrogen, 100 kg $P_2O_5$and 60 kg K_2O on hectare basis. On an average farmyard manure contained 0.60%N, 0.15% P and 0.54%K. The N, P and K were applied as urea, single super phosphate and muriate of potash, respectively. The full amount of farmyard manure, P and K, and whole of N was applied at sowing time. The *Rhizobium* culture was applied as seed treatment and phosphorus solublizing bacteria through farmyard manure at sowing.

Table 6 Effect of organic, inorganic and integrated use of nutrients on nodulation and yield of green bean pods (Data pooled over four years)

Treatment*	Nodule number pooled over four years	Nodule weight (mg plant^{-1}) pooled over four years	No. of pods plant^{-1} pooled over four years	Pooled yield over four years (t ha^{-1})
T_1	17	45.7	20	7.12
T_2	16	43.4	22	7.10
T_3	14	30.8	26	8.15
T_4	12	18.7	28	8.73
T_5	15	31.8	25	8.31
T_6	12	22.4	29	8.92
T_7	5	2.5	30	7.35
T_8	9	14.3	14	4.63
LSD (0.05)	2.6	4.7	5.7	0.692

* treatments are as detailed in Table 5.
Source: Modified from Datt et al. (2013).

treatments and control. This, according to the authors, indicates that application of organic fertilizers enhanced microbial activity while nitrogenous fertilizer enhanced nutrient mobility and availability but suppressed microbial activity. The ready-made supply of nitrogen from chemical fertilizer therefore suppressed nodule formation, resulting in poorer symbiosis between bean plants and rhizobia compared to the control. Mineral nitrogen is among the most important soil chemical factors that influence symbiotic nitrogen fixation in legume crops. Application of nitrogenous fertilizer in integrated treatments T_4 and T_6 resulted in more pods than T_5 but lower number of pods than chemical fertilizer treatment. Control treatment recorded the minimum and significantly less number of pods than other the treatments. Integrated treatments T_4 and T_6 gave significantly higher pod yields than organic treatments (T_1 and T_2) or chemical fertilizer treatment (T_7).

The study concluded that up to 25 and 20.5 kg ha^{-1} of nitrogen and phosphorus, respectively, can be saved in French bean production by use of 5 t of FYM and 50% of the nutrients can be supplied through the use of nitrogen and phosphate bio-fertilizers.

9.2 Liming for soil acidity management

Soil acidity, occurring either naturally or as a result of human activity, is a major constraint to sustainable production of annual crops, including common bean, in many regions of the world. Determining optimum soil acidity indices is an important strategy for achieving maximum economic crop yields on acid soils (Fageria, 2008). These indices include soil pH, base saturation, calcium (Ca), magnesium (Mg) and potassium (K) saturation. Reducing aluminium (Al) saturation is essential for achieving high productivity. Soils in the humid and sub-humid tropical regions, where significant production of bean occurs, are highly weathered and consequently low in basic cations and liming is the most effective way of replenishing the cations (Dierolf et al., 1997).

Soil acidity results in chemical imbalance and toxicities of Al, Mn and hydrogen (H) ions. This is often coupled with parallel deficiencies in available N, P, K, Ca, Mg, zinc (Zn) and Mo (Fageria, 2002). These problems can be addressed by liming to raise soil pH, Ca, and Mg concentrations and base saturation. Lime application neutralizes soil acidity; reduces the toxicity of Al, Fe and Mn; and improves the physical, chemical and biological properties of the soil (Kisinyo et al., 2005).

In liming experiments by Fageria (2008), common bean grain yield was significantly increased by improving soil pH; base saturation; calcium (Ca), Mg and K saturation; and reducing Al saturation. The optimum values for maximum bean yield were pH 6.5, base saturation 67%, Ca saturation 48% and Mg saturation 19%. Bean yield linearly increased with increasing K saturation in the range of 1.5 to 3% across two soil depths. Yield components, such as pod number, grains per pod and 100-grain weight were significantly improved with liming.

9.3 Foliar fertilization

Soil application is the most commonly used method to supply essential nutrients to plants. In this case, nutrients are absorbed by plant roots. However, higher plants can also absorb mineral nutrients when applied as foliar sprays in appropriate concentrations (Fageria et al., 2009). Foliar spraying has been advocated as a feasible and economic way to supplement the plant nutrients for more efficient fertilization (Girma et al., 2007). Successful foliar fertilization requires higher leaf area index for absorbing applied nutrient solution in sufficient amount, optimal nutrient concentration and optimal day temperature to avoid leaf burning (Fageria et al., 2009). Repeated application may be necessary, depending on severity of nutrient deficiency. Obviously, the nutrient source should be soluble in water for effective fertilization. However, caution should be taken to avoid leaf damage by spraying solutions of high nutrient concentration. With regard to foliar application of nitrogenous fertilizers, the risk of foliage burning is less likely when the N source is urea (Alkier et al., 1972). Since urea has a low salt index, desiccation of leaf cells through osmosis is reduced (Gooding and Davies, 1992).

The purpose of foliar feeding is not to replace soil fertilization. Soil application remains the most effective and economical method for supplying the major nutrients (N, P and K) needed for plant growth. However, foliar application has proven to be an excellent method of supplying plant requirements for secondary nutrients (Ca, Mg, S) and micronutrients, and has advantage over soil application in that it is a highly efficient and timely method of supplying needed plant nutrients at critical stages. Appropriate methods of application and use of the best suited nutrient materials are essential. In field experiments conducted at Mansehra in Pakistan, Rahman et al. (2014) evaluated three foliar applied fertilizers on productivity of common bean and concluded that foliar spray of NPK (20:20:20) was more suitable than poultry manure and DAP (di-ammonium phosphate) for the maximum yield of common bean, having significantly increased the number of pods per plant, seeds per pod, the number of seeds per plant, and biomass and grain yield.

10 Conclusion

Common bean is one of the most important grain legumes worldwide but its productivity is low in most regions where it is grown, particularly in sub-Saharan Africa. This is despite the huge potential that exists. Many biotic and abiotic factors constrain common bean production. These include soil nutrient deficiencies, low soil moisture, incidences of pests and diseases, limited access by smallholder farmers to certified seed of adapted germplasm and so on. For bean to play its role as an important food and cash crop, especially for the low-income populations in Africa and Latin America, these constraints must be addressed to ensure sustainable increases in productivity. Research conducted in the last two decades or so provides solutions to some of these constraints. For example, bean germplasm tolerant to important pests and diseases, some bean germplasm tolerant to low soil phosphorus levels, integrated pest management options, water management technologies, and integrated nutrient management strategies exist. However, the adoption of some of these technologies is generally poor in many of the major bean growing regions due to several factors, including limited access by smallholder farmers to markets and unfavourable policy environment. Sustainable improvements in bean productivity requires putting these management technologies together into an integrated crop management strategy, improving market access and providing supportive policy environment.

11 Where to look for further information

Further information on topics discussed in this chapter can be obtained from many sources, including Miklas et al. (2006), which provides a detailed treatment of common bean breeding for resistance against biotic and abiotic stresses. Further information on important pests and diseases of common bean can be obtained from a bean disease and pest identification and management handbook. CIAT publication no. 371. website: www.ciat.cgiar.org/work/Africa/Pages/Africa.aspx. Many publications, for example, Vanlauwe et al. (2015) deal with the need for intensification of smallholder production, particularly in sub-Saharan Africa and discuss in detail ISFM as a means of increasing the agronomic efficiency of inputs. Agroecological intensification (AEI), which refers to improving the performance of agriculture through integration of ecological principles into farm and system management, is an important concept related to ICM. More information about AEI concept can be obtained from the Collaborative Crops Research Program (CCRP) of The McKnight Foundation. Website: http://www.ccrp.org/program-essentials.

12 Abbreviations and acronyms

AEI	Agroecological intensification
Al	Aluminium
ALS	Angular leaf spot
AMF	Arbuscular mycorrhizal fungi
BCMNV	Bean common mosaic necrotic virus
BCMV	Bean common mosaic virus
BGMV	Bean golden mosaic virus
BGYMV	Bean golden yellow mosaic virus
BNF	Biological nitrogen fixation
Ca	Calcium
CFU	Colony-forming units
CIAT	International Centre for Tropical Agriculture
DAP	Di-ammonium phosphate
FAO	Food and Agriculture Organization
Fe	Iron
FYM	Farmyard manure
H	Hydrogen
ICM	Integrated crop management
IPM	Integrated pest management
ISFM	Integrated soil fertility management
K	Potassium
Mn	Manganese
Mo	Molybdenum
N	Nitrogen
P	Phosphorus
SSA	Sub-Saharan Africa
WUE	Water use efficiency
Zn	Zinc

13 References

Abate, T., van Huis, A. and Ampofo, J. K. O. 2000. Pest management strategies in traditional agriculture: an African perspective. *Annu. Rev. Entomol.*, 45(1), 631–659.

Abate, T. and Ampofo, J. K. O. 1996. Insect pests of beans in Africa: their ecology and management. *Annu. Rev. Entomol.*, 41, 45–73.

Aerts, R. 1999. Interspecific competition in natural plant communities: mechanisms, trade-offs and plant-soil feedbacks. *J. Exp. Bot.*, 50, 29–37.

Ahmed, F. E., Hall, A. E. and Demason, D. A. 1992. Heat injury during floral development in cowpea (*Vigna unguiculata*, Fabaceae). *Am. J. Bot.*, 79, 784–91.

Akibode, S. and Maredia, M. K. 2012. *Global and Regional Trends in Production, Trade and Consumption of Food Legume Crops.* Department of Agricultural, Food, and Resource Economics, Michigan State University.

Akuda, E. 2001. Intercropping and population density effects on yield component, seed quality and photosynthesis of sorghum and soybean Elijah. *J. Food Technol. Afr.*, 6, 96–100.

Ali, K., Keneni, G., Ahmed, H., Malhotra, R., Beniwal, S., Makkouk, K. and Halila, M. H. 2003. Food and forage legumes of Ethiopia: Progress and prospects. Proceedings of a Workshop on Food and Forage legumes. 22–6 September 2003, Addis Ababa, Ethiopia.

Al-Jamal, M. S., Ball, S. and Sammis, T. W. 2001. Comparison of sprinkler, trickle and furrow irrigation efficiencies for onion production. *Agric. Water Manag.*, 46, 253–66.

Alkier, A. C., Racz, G. J. and Soper, R. J. 1972. Effects of foliar and soil applied nitrogen and soil nitrogen level on the protein content of Neepawa wheat. *Can. J. Soil Sci.*, 52, 301–9.

Alley, M. M. and Vanlauwe, B. 2009. *The Role of Fertilizers in Integrated Plant Nutrient Management.* International Fertilizer Industry Association.

Andraus, M. P., Cardoso, A. A. and Ferreira, E. P. B. 2016. Differences in nodulation and grain yield on common bean cultivars with different growth cycles. *Commun. Soil Sci. Plant Anal.*, 47, 1148–61.

Atemkeng, F. M., Remans, R., Michiels, J., Tagne, A. and Ngonkeu, M. L. E. 2011. Inoculation with *Rhizobium etli* enhances organic acid exudation in common bean *(Phaseolus vulgaris* L.) subjected to phosphorus deficiency. *Afr. J. Agric. Res.*, 6, 2235–42.

Baker, J. T, Allen, Jr. L. H. and Boote, K. J. 1992. Temperature effects on rice at elevated CO_2 concentration. *J. Exp. Bot.*, 43, 959–64.

Beaver, J. S., Rosas, J. C., Myers, J., Acosta, J., Kelly, J. D., Nchimbi-Msolla, S., Misangu, R., Bokosi, J., Temple, S., Arnaud-Santana, E. and Coyne, D. P. 2003. Contributions of the Bean/Cowpea CRSP to cultivar and germplasm development in common bean. *Field Crops Res.*, 82, 87–102.

Beebe, S., Ramírez, J., Jarvis, A., Rao, I. M., Mosquera, G., Bueno, J. M. and Blair, M. W. 2011. Genetic improvement of common beans and the challenges of climate change. In Yadav, S. S., Redden, R. J., Hatfield, J. L., Lotze-Campen, H. and Hall, A. E. (Eds), *Crop Adaptation to Climate Change.* John Wiley & Sons, Ltd., published by Blackwell publishing Ltd: Richmond, Australia, pp. 356–69.

Beebe, S. E., Rao, I. M., Blair, M. W. and Acosta-Gallegos, J. A. 2013. Phenotyping common beans for adaptation to drought. *Front. Physiol.* 4, 35.

Beebe, S. E., Rojas-Pierce, M., Yan, X., Blair, M. W., Pedraza, F., Munoz, F., Tohme, J. and Lynch, J. P. 2006. Quantitative trait loci for root architecture traits correlated with phosphorus acquisition in common bean. *Crop Sci.* 46, 413–23.

Beebe, S. 2012. Common bean breeding in the tropics. *Plant Breed. Rev.* 36, 357–426.

Belmain, S. R., Haggar, J., Holt, J. and Stevenson, P. C. 2013. Managing legume pests in sub-Saharan Africa: Challenges and prospects for improving food security and nutrition through agro-ecological intensification. Chatham Maritime (United Kingdom): Natural Resources Institute, University of Greenwich, p. 34.

Betencourt, E., Duputel, M., Colomb, B., Desclaux, D. and Hinsinger, P. 2012. Intercropping promotes the ability of durum wheat and chickpea to increase rhizosphere phosphorus availability in a low P soil. *Soil Biol. Biochem.* 46, 21–33.

Bhatti, I. H., Ahmad, R., Jabbar, A., Nazir, M. S. and Mahmood, T. 2006. Competitive behavior of component crops in different sesame-legume intercropping systems. *Int. J. Agric. Biol.*, 8, 165–7.

Birhane, E., Sterck, J. F., Fetene, M., Bongers, F. and Kuyper, W. T. 2012. Arbuscular mycorrhizal fungi enhance photosynthesis, water use efficiency, and growth of frankincense seedlings under pulsed water availability conditions. *Oecologia*, 169, 895–904.

Bliss, F. A. 1993. Breeding common bean for improved biological nitrogen fixation. In *Enhancement of Biological Nitrogen Fixation of Common Bean in Latin America*, (pp. 71–9). Springer: Netherlands.

Boutraa, T. 2009. Growth and carbon partitioning of two genotypes of bean (*Phaseolus vulgaris*) grown with low phosphorus availability. *Eurasia. J. Biosci.*, 3, 17–24.

Broughton, W. J., Hernandez, G., Blair, M., Beebe, S., Gepts, P. and Vanderleyden, J. 2003. Beans (*Phaseolusspp*) - model food legumes. *Plant Soil*, 252, 55–128.

Bunting, A. H. 1975. Time, phenology and the yield of crops. *Weather*, 30, 312–25.

Buruchara, R., Mukaruziga, C. and Ampofo, K. O. 2010. *Bean Disease and Pest Identification and Management*. Handbook four. CIAT.

Buttery B. R., Park, S. J. and Findlay, W. I. 1987. Growth and yield of white bean (*phaseolus vulgaris* L.) in response to nitrogen, phosphorus and potassium fertilizer and to inoculation with rhizobium. *Can. J. Plant Sci.*, 67, 425–32.

Chacón, M. I., Pickersgill, S. B. and Debouck, D. G. 2005. Domestication patterns in common bean (*Phaseolus vulgaris* L.) and the origin of the Mesoamerican and Andean cultivated races. *Theor. Appl. Genet.*, 110, 432–44.

Chimonides, S. J., and Tsiourtis, N. X. 1995. Irrigation management under water shortage conditions. In *Water resources management under drought or water shortage conditions. Proceedings of the EWRA 95 Symposium, Nicosia, Cyprus*, 14–18 March 1995. (pp. 73–8). AA Balkema.

Cichy, A. K., Snapp, S. S. and Blair, W. M. 2009. Plant growth habit, root architecture traits and tolerance to low soil phosphorus in an Andean bean population. *Euphytica*, 165, 257–68. http://dx.doi.org/10.1007/s10681-008-9778-2.

Dahmardeh, M., Ghanbari, A., Syahsar, B. A. and Ramrodi, M. 2010. Effect of maize-common bean intercropping system on the growth, nodulation, yield and N uptake and deficiency. *Afr. J. Agric. Res.*, 6, 2235–43.

Datt, N., Dubey, Y. P. and Chaudhary, R. 2013. Studies on impact of organic, inorganic and integrated use of nutrients on symbiotic parameters, yield, and quality of French-bean (*Phaseolus vulgaris* L.) vis-à-vis soil properties of an acid alfisol. *Afr.J. Agric. Res.*, 8, 2645–54.

Dierolf, T. S., Arya, L. M. and Yost, R. S. 1997. Water and cation movement in an Indonesian Ultisol. *Agron. J.*, 89, 572–9.

Egli, D. B. 1988. Plant density and soybean yield. *Crop Sci.*, 28(6), 977–80.

Emam, Y., A. Shekoofa, F. Salehi and A. H. Jalali. 2010. Water stress effects on two common bean cultivars with contrasting growth habits. *Am.-Eurasian J. Agric. Environ. Sci.*, 9, 495–9.

Fageria, N. K., Melo, L. C., Ferreira, E. P. B., Oliveira, J. P. and Knupp, A. M. 2014. Dry matter, grain yield, and yield components of dry bean as influenced by nitrogen fertilization and Rhizobia. *Comm. Soil Sci. Plant Anal.*, 45, 111–25.

Fageria, N. K. 2002. Nutrient management for sustainable dry bean production in the tropics. *Comm. Soil Sci. Plant Anal.*, 29, 1219–28.

Fageria, N. K. 2008. Optimum soil acidity indices for dry bean production on an Oxisol in no-tillage system. *Comm. Soil Sci. Plant Anal.*, 39, 845–57.

Fageria, N. K., Filho, M. B., Moreira, A. and Guimaraes, C. M. 2009. Foliar fertilization of crop plants. *J. Plant Nutr.*, 32, 1044–64.

FAO, 2009. *Global Agriculture Towards 2050. High Level Expert Forum - How to Feed the World in 2050*. Office of the Director, Agricultural Development Economics Division, Economic and Social Development Department: Rome, Italy. http://www.un.org/en/development/desa/news/population/2015-report.html

FAO. 2006. *Quality Declared Seed System*. FAO Plant Production and Protection Paper 185. FAO: Rome, Italy.

FAO. 2006. FAOSTAT: FAO statistical databases Date Accessed: 28 November 2016. http:faostat.fao.org/faostat/. Last Updated 3 August 2016.

FAO. 2015. FAOSTAT: FAO statistical databases Date Accessed: 18 November 2016. http://faostat3.fao.org/browse/Q/QC/E.

Ferris, S. and Kaganzi, E. 2008. *Evaluating Marketing Opportunities for Haricot Beans in Ethiopia.* IPMS Working Paper 7. ILRI: Nairobi (Kenya).

Freytag, G. F. and Debouck, D. G. 2002. *Taxonomy, Distribution and Ecology of the Genus Phaseolus (Leguminosae–Papilionoideae) in North America, Mexico and Central America*. Sida, Botanical.

Fujita, K., Ogata, S., Matsumoto, K., Masuda, T., Ofosu-Budu, G. K. and Kuwata, K. 1990. Nitrogen transfer and dry matter production on in soybean and sorghum mixed cropping system at different population densities. *Soil Sci. Plant Nutr.*, 36, 233–41.

Gardiner, T. R. and Craker, L. E. 1981. Bean growth and light interception in a bean-maize intercrop. Field Crops Res., 4, 313–20.

Gentry, H. S. 1969. Origin of the common bean, Phaseolus vulgaris. *Econ. Bot.*, 23, pp.55–69.

Gepts, P. 1998. Origin and evolution of common bean: past events and recent trends. *HortScience*, 33, 1124–30.

Giller, K. E. 2001. *Nitrogen Fixation in Tropical Cropping Systems.* Cabi.

Girma, K., Martin, K. L., Freeman, K. W., Mosali, J., Teal, R. K., Raun, W. R., Moges, S. M. and Arnall, D. B. 2007. Determination of optimum rate and growth for foliar applied phosphorus in corn. *Comm. Soil Sci.Plant Anal.*, 38, 1137–54.

Gooding, M. J., and Davies, W. P. Foliar urea fertilization of cereals: a review. *Fert. Res.*, 32, 209–22.

Graham, P. H. and Ranalli, P. 1997. Common bean (Phaseolus vulgaris L.). *Field Crops Res.*, 53, 131–46.

Graham, P. H. 1992. Stress tolerance in Rhizobium and Bradyrhizobium, and nodulation under adverse soil conditions. *Can. J. Microbiol.*, 38, 475–84.

Graham, P. H. and Temple, S. R. 1984. Selection for improved N fixation in Glycine max. (L.) Merr. and Phaseolus vulgaris. *Plant Soil* 82, 315–27.

Graham, P. H. and Rosas, J. C. 1977. Growth and development of indeterminate bush and climbing cultivars of Phaseolus vulgaris L. inoculated with Rhizobium. *J. Agr. Sci.*, 88, 503–8.

Graham, P. H., Rosas, J. C., Estevez de Jensen, C., Peralta, E., Tlusty, B., Acosta-Gallegos J., and Pereira, P. A. 2003. Addressing edaphic constraints to bean production: the bean/cowpea CRSP project in perspective. *Field Crops Res.*, 82, 179–92.

Graham, P. H. 1981. Some problems of nodulation and symbiotic nitrogen fixation in Phaseolus vulgaris L.: a review. *Field Crops Res.*, 4, 93–112.

Graham, P. H. and Vance, C. P. 2003. Legumes: importance and constrains to greater use. *Plant Physiol.*, 131, 872–7.

Greenway, P. J. 1945. Origins of some East African food plants. *East Afr. Agric. J.*, 10, 34–9.

Grisley, W. and Mwesigwa, D. 1991. A report on the socioeconomics of bean production and marketing in Uganda: information for research planning. In: Grisley, W. (Ed.), Proceedings Workshop on National Research Planning for Bean Production in Uganda. CIAT, Kampala, Uganda, pp. 75–81. (Network on Bean Research in Africa Workshop Series, no. 9).

Haggard-Nelson, H., Ambus, P. and Jensen, E. S. 2001. Inter-specific competition, N use and interference with weeds in pea-barley intercropping. *Field Crops Res.*, 70, 101–9.

Hardarson, G. 1993. Methods for enhancing symbiotic nitrogen fixation. *Plant Soil* 152, 1–17.

Henao, J. and Baanante, C. 2006. *Agricultural Production and Soil Nutrient Mining in Africa: Implication for Resource Conservation and Policy Development.* IFDC-An International Centre for Soil Fertility and Agricultural Development.

Hillocks, R. J., Madata, S. C., Chirwa, R., Minja, M. E. and Msolla, S. 2006. Phaseolus bean improvement in Tanzania 1956–2005. *Euphytica*, 150, 215–31.

Howell, T. A. 2001. Enhancing water use efficiency in irrigated agriculture. *Agron. J.*, 93, 281–9.

Igbozurike, M. U. 1971. Ecological balance in tropical agriculture. *Geogr. Rev.*, 61, 521–9

Jackai, L. E. N. 1995. Integrated pest management of borers of cowpea and beans. *Int. J. Trop. Insect Sci.*, 16, 237–50.

Janzen, A. J. 1974. Agro-ecosystems in future society. *Agro-ecosystems*, 1, 69–80.

Jensen, E. S. 1996. Barley uptake of N deposited in the rhizosphere of associated field pea. *Soil Biol. Biochem.* 28, 159–68.

Jones, P. G. and Thornton, P. K. 2003. The potential impacts of climate change on maize production in Africa and Latin America in 2055. *Glob. Environ. Change*, 13, 51–9.

Kadari, A., Muthomi, J. W. and Ojiem, J. O. 2014. Bean diseases inoculum in soil and seeds and management of bean root by lablab residue, green manures, and seed dressing. Project Annual Report, Kenya Agricultural and Livestock Research Organization, Nairobi, Kenya.

Kay, D. E. 1979. Food Legumes. *TPI Crops and Product Digest*. Tropical Product Institutes, (3).

Kipe-Nolt, J. A. and Giller, K. E. 1993. A field evaluation using the 15N isotope dilution method of lines of Phaseolus vulgaris L. bred for increased nitrogen fixation. *Plant Soil*, 152: 107–14.

Latati, M., Pansu, M., Drevon, J. J. and Ounane, S. M. 2013. Advantage of intercropping maize (Zea mays L.) and common bean (Phaseolus vulgaris L.) on yield and nitrogen uptake in Northeast Algeria. *Int. J. Res. Appl. Sci.*, 1, 1–7.

Leakey, C. L. A. 1970. Need one grow pure lines in developing countries? Bean Improve. *Ann. Rep. Bean Improvement Co-operative*, 62.

Legesse, D. and Ayenew, T. 2006. Effect of improper water and land resource utilization on the central Main Ethiopian Rift lakes. *Quat. Int.*, 148, 8–18.

Lehmann, J., Peter, I., Steglich, C., Gebauer, G., Huwe, B. and Zech, W. 1998. Below-ground interactions in dryland agroforestry. *Forest Ecol. Manag.*, 111, 157–69.

Li, W., Li, L., Sun, J., Guo, T., Zhang, F., Baob, X., Peng, A. and Tang, C. 2005. Effects of intercropping and nitrogen application on nitrate present in the profile of an Orthic Anthrosol in Northwest China. *Agric. Eco. Environ.*, 105, 483–91.

Liao, H., Yan, X., Rubio, G., Beebe, S. E., Blair, M. W. and Lynch, J. P. 2004. Genetic mapping of basal root gravitropism and phosphorus acquisition efficiency in common bean. *Funct. Plant Biol.*, 31, 959–70.

Liebenberg, A. J. 2002. *Dry Bean Production*. Printed and published by Department of Agriculture, Resource Centre, Directorate Agricultural Information Services, Republic of South Africa, p. 27.

Lunze, L., Kimani, P. M., Ngatoluwa, R., Rabary, B., Rachier, G. O., Ugen, M. M. and Ruganza, V. 2007. Bean improvement for low soil fertility adaptation in Eastern and Central Africa. In *Advances in Integrated Soil Fertility Management in sub-Saharan Africa: Challenges and Opportunities* (pp. 325–32). Springer: Netherlands.

Manickasundaram, P., Selvaraj, P. K., Krishnamoorthi, V. V. and Gnanamurthy, P. 2002. Drip irrigation and fertigation studies in tapioca. *Madras Agric. J.*, 89, 466–8.

Miklas, P.N., Kelly, J. D., Beebe, S. E., and Blair, M.W. 1006. Common bean breeding for resistance against biotic and abiotic stresses:From classical to MAS breeding. *Euphytica* 147: 105–131.

Miller, C. R., Ochoa, I, Nielsen, KL, Beck, D. and Lynch, J. P. 2003. Genetic variation for adventitious rooting in response to low phosphorus availability: potential utility for phosphorus acquisition from stratified soils. *Funct. Plant Biol.* 30, 973–85.

Morris, R. A. and Garrity, D. P. 1993. Resource capture and utilization in intercropping: non-nitrogen nutrients. *Field Crops Res.*, 34, 319–34.

Mourice, S. K. and Tryphone, G. M. 2012. Evaluation of common bean (Phaseolus vulgaris L.) genotypes for adaptation to low phosphorus. *ISRN Agron.*, 1–9.

Namayanja, A., Semoka, J., Buruchara, R., Nchimbi, S. and Waswa, M. 2014. Genotypic variation for tolerance to low soil phosphorus in common bean under controlled screen house conditions. *Agric. Sci.*, 5, 270–85.

Namugwanya, M., Tenywa, J. S., Otabbong, E., Mubiru, D. N. and Basamba, T. A. 2014. Development of common bean (*Phaseolus Vulgaris* L.) production under low soil phosphorus and drought in sub-Saharan Africa: a review. *J. Sustain. Dev.*, 7, 128–37.

Nazir, M. S., Jabbar, A., Ahmad, I., Nawaz, S., and Bhatti, I. H. 2002. Production potential and economics of intercropping in autumn-planted sugarcane. *Int. J. Agric. Biol.* 4, 140–2.

Neves, M. C. P., and Hungria, M. 1987. The physiology of nitrogen fixation in tropical grain legumes. *Crit. Rev. Plant Sci.*, 6, 267–321.

Nziguheba, G. 2007. Overcoming phosphorus deficiency in soils of eastern Africa: Recent advances and challenges. In *Advances in Integrated Soil Fertility Management in Sub-Saharan Africa: Challenges and Opportunities* (pp. 149–60). Springer: Netherland.

Ofori, F. and Stern, W. 1987. Cereal-legume intercropping systems. *Adv. Agron.*, 41, 41–90.

Ojiem, J. O., Vanlauwe, B., de Ridder, N. and Giller, K. E. 2007. Niche-based assessment of contributions of legumes to the nitrogen economy of Western Kenya smallholder farms. *Plant Soil*, 292, 119–35.

Oruru, M. B., and Njeru, E. M. 2016. Upscaling arbuscular Mycorrhizal symbiosis and related agroecosystems services in smallholder farming systems. *BioMed Res. Int.*, 2016.

Pachico, D. 1989. Trends in world common bean production. *Bean Production Problems in the Tropics*, 10, 1.

Pena-Cabriales, J. J., Grageda-Cabrera, O. A., Kola, V. and Hardarson, G. 1993. Time course of N_2 fixation in common bean (*Phaseolus vulgaris* L.). *Plant Soil* 152, 115–21.

Pena-Cabriales, J. J. and Castellanos, J. Z. 1993. Effects of water stress on N_2 fixation and grain yield of *Phaseolus vulgaris* L. *Plant Soil* 152, 151–5.

Pendleton, J. W., Hartwig, E. E. 1973. Management, pp. 211–37. In Caldwell, B. E. (Ed.), Soybeans: improvement, production, and uses. American Society of Agronomy pub, *Madison, Wisconsin. Agron. Ser.* 16, p. 681.

Piha, M. I. and Munns, D. N. 1987. Nitrogen fixation capacity of field grown bean compared to other seed legumes. *Agron. J.*, 79, 690–6.

Rahman I. U., Afzal, A., Iqbal, Z., Ijaz, F., Manan, S., Ali, S. A., Khan, K., Karim, S. and Qadir, G. 2014. Growth and yield of Phaseolus vulgaris as influenced by different nutrient treatments in Mansehra. *Int. J. Agron. Agric. Res.*, 4, 20–6.

Raina, A. N., Thakur, B. C. and Bhournal, A. R. 1998. Effect of drip irrigation and plastic mulch on yield, water use efficiency and benefit cost ratio of pea cultivation. *J. Indian Soc. Soil Sci.*, 46, 562–7.

Rao, I. M. 2001. Role of physiology in improving crop adaptation to abiotic stresses in the tropics: the case of common bean and tropical forages. *Handbook of Plant and Crop Physiology*, pp. 583–613.

Redden, R. J., and Herridge, D. F. 1999. Evaluation of genotypes of navy and culinary bean (Phaseolus vulgaris L.) selected for superior growth and nitrogen fixation. *Anim. Prod. Sci.*, 39, 975–80.

Rennie, R. J. and Kemp, G. A. 1983. N2-fixation in field beans quantified by15N isotope dilution. II. Effect of cultivars of beans. *Agron. J.*, 75, 645–49.

Ruschel, A. P., Vose, P. B., Matsui, E., Victoria, R. L. and Tsai Saito, S. M. 1982. Field evaluation of N2-fixation and N-utilisation by Phaseolus bean varieties determined by N isotope dilution. *Plant Soil*, 65, 397–407.

Rusinamhodzi, L., Corbeels, M., Nyamangara, J. and Giller, K. E. 2012. Maize-grain legume intercropping is an attractive option for ecological intensification that reduces climatic risk for smallholder farmers in central Mozambique. *Field Crops Res.*, 136, 12–22. http://doi.org/10.1016/j.fcr.2012.07.014

Sanchez, P. A. 2002. Soil fertility and hunger in Africa. *Science*, 295, 2019–20.

Schwartz, H. F. and Pastor-Corrales, M. A. P. (Eds). 1989. *Bean Production Problems in the Tropics*. CIAT.

Searle, P. G. E., Comudom, Y., Sheddon, D. C. and Nance, R. A. 1981. Effects of maize legume intercropping systems and fertilizer -nitrogen on crop yields and residual nitrogen. *Field Crops Res.*, 4, 133–45.

Selveraj, T. and Chellepan, P. 2006. Arbuscular mycorrhizae: a diverse personality. *J. Cent. Eur. Agric.*, 2, 349–58.

Shaker B. A, Saeed, A. B., and Al-Khalifa, B. A. 2013. Effect of drip irrigation on Phaseolus Bean production under the open field condition of Sudan. *J. Agri-Food Appl. Sci.*, 1, 86–90.

Shen, J., Yuan, L., Zhang, J., Li, H., Bai, Z., Chen, X. and Zhang, F. 2011. Phosphorus dynamics: from soil to plant. *Plant Physiol.*, 156, 997–1005.

Shen, Q. R. and Chu, G. X. 2004. Bi-directional nitrogen transfer in an intercropping system of peanut with rice cultivated in aerobic soil. *Biol. Fert. Soils*, 40, 81–7.

Singh, S. 1999. *Common Bean Breeding in the Twenty- First Century. Developments in Plant Breeding.* Kluwer Academic Publishers: Dordrecht, Boston, London.

Singh, S. P. and Schwartz, H. F. 2010. Review: Breeding common bean for resistance to insect pests and nematodes. *Can. J. Plant Sci.*, 91, 239–50.

Snapp, S., Aggarwal, V. and Chirwa, R. 1998. Note on phosphorus and cultivar enhancement of biological nitrogen fixation and productivity of maize/bean intercrops in Malawi. *Field Crops Res.*, 58, 205–12.

Song, H. 2005. Effects of VAM on host plant in the condition of drought stress and its mechanisms. *Electron. J. Biol.*, 1, 44–8.

Stenglein, S. L. D., Ploper, O., Vizgarra, P. and Balatti, P. 2003. Angular leaf spot: a disease caused by the fungus Phaeoisariopsis griseola (Sacc.) Ferraris on Phaseolus vulgaris L. *Adv. Appl. Microbiol.*, 52, 209–43.

Tajini, F., Trabelsi, M. and Drevon, J. J. 2012. Combined inoculation with Glomus intraradices and Rhizobium tropici CIAT899 increases phosphorus use efficiency for symbiotic nitrogen fixation in common bean (Phaseolus vulgaris L.). *Saudi J. Biol. Sci.*, 19, 157–63.

Toro, O., Tohme, J., and Debouck, D. G. 1990. *Wild Bean (Phaseolus vulgaris L.): Description and Distribution* (Vol. 181). CIAT.

Tsubo, M. and Walker, S. 2002. A model of radiation interception and use by a maize–bean intercrop canopy. *Agric. Forest Meteorol.*, 110, 203–15.

Turk, M. A., Assaf, K. M. Hammad, and Al-Tawahi, A. M. 2008. Significance of Mycorrhizae. *World J. Agric. Sci.*, 2, 16–20.

Vanlauwe, B., Bationo, A., Chianu, J., Giller, K. E., Merckx, R., Mokwunye, U., Ohiokpehia, O., Pypers, P., Tabo, R., Shepherd, K. D., and Smaling, E. M. A. 2010. Integrated soil fertility management operational definition and consequences for implementation and dissemination. *Outlook Agric.*, 39, 17–24.

Vanlauwe, B., Descheemaeker, K., Giller, K. E., Huising, J. Merckx, R., Nziguheba, G., Wendt, J., and Zingore, S. 2015. Integrated soil fertility management in sub-Saharan Africa: unravelling local adaptation. *Soil*, 1, 491–508, www.soil-journal.net/1/491/2015/ doi:10.5194/soil-1-491-2015

Westermann, D. T., Kleinkopf, G. E., Porter, L. K. and Leggett, G. E. 1981. Nitrogen sources for bean seed production. *Agron. J.*, 73, 660–4.

White, J. G. H. (Ed.). 1989. Grain Legumes in Sustainable Cropping Systems: A Review. Agronomy Society of New Zealand, Lincoln, Special publication no 7, pp. 109–15.

Williams, J. W., Jackson, S. T. and Kutzbach, J. E. 2007. Projected distributions of novel and disappearing climates by 2100 AD. *Proc. Natl. Acad. Sci. U. S. A.*, 104, 5738–42.

Williamson, S., Ball, A. and Pretty, J. 2008. Trends in pesticide use and drivers for safer pest management in four African countries. *Crop Prot.*, 27, 1327–34.

Woomer, P. L., Lan'gat, M. and Tungani, J. O. 2004. Innovative maize-bean intercropping results in above- and below-ground competitive advantages for understorey legumes. *West Afr. J. Appl. Ecol.*, 6, 85–94.

Wortmann C. S., Kirkby, R. A., Aledu, C. A. and Allen, D. J. 1998. *Atlas of Common Bean (Phaseolus vulgaris L) Production in Africa.* CIAT, p. 131.

Zhang, F. and Li, L. 2003. Using competitive and facilitative interactions in intercropping systems enhances crop productivity and nutrient-use efficiency. *Plant Soil* 248, 305–12.

Chapter 3

Developing improved varieties of lentil

William Erskine, University of Western Australia, Australia; Ashutosh Sarker, International Center for Agricultural Research in the Dry Areas (ICARDA), India; and Shiv Kumar, International Center for Agricultural Research in the Dry Areas (ICARDA), Morocco

1 Introduction
2 Production regions and their agro-ecologies
3 Phylogeny, domestication/spread and genetic resources
4 Problems addressable by breeding
5 Breeding methods
6 New technologies
7 Success stories
8 Future trends
9 Where to look for further information
10 References

1 Introduction

The lentil (*Lens culinaris* Medik. ssp. *culinaris*) is a short, slender annual leguminous plant that produces a popular pulse that is widely consumed, especially in Asia. Its grain is a vital source of protein, with a mean content of 28.3% ranging from 15.9 to 31.4% (Grusak, 2009), especially for the poor, who cannot afford animal products. Lentil seed also contains high amounts of macro- and micro-nutrients such as Ca, Fe, K, P and Zn; the vitamins – ascorbic acid, inositol, niacin and vitamin A; fibre; and also carbohydrates for balanced nutrition (Bhatty, 1988; Savage, 1988; Kumar et al., 2016b; Thavarajah et al., 2016).

Lentil is part of the staple diet in many developing countries, being eaten in various ways. In South Asia it is predominantly eaten as boiled or fried *dal* – a stew-like preparation. Rice and lentil *dal* is a popular daily dish of the poor of Bangladesh, Nepal, Sri Lanka and India (Sarker et al., 2009). *Khichri* is another common food among poor people in South Asia, made from a mixture of split/dehulled lentil and cracked wheat or rice. In West Asia *mujaddarah*, made from whole lentils and either immature wheat seed or rice, is a popular dish. Lentil soup is also widely consumed in West Asia and North Africa.

Lentil seeds are classified into two types: large-seeded/Green (greater than 50 g per 1000 seeds) and small-seeded/Red (50 g or less per 1000 seeds). Most of world production and trade is in red lentil, which is marketed as a whole seed or in de-hulled and split form.

http://dx.doi.org/10.19103/AS.2017.0023.22

Dal and lentil soup are made from red lentils, whereas *mujaddarah* is made from green lentils.

Lentils are predominantly cropped under rainfed conditions and on residual/conserved soil moisture in different cropping systems. The inclusion of lentil, like other pulse crops, typically benefits succeeding crops by improving soil health as a result of biological nitrogen fixation and other rotational effects. The rhizobial bacterium involved in the symbiotic relationship of the lentil crop is *Rhizobium leguminosarum* bv. *viciae*. This is the symbiont of all *Lens*, and it also forms associations with other members of the Vicieae (i.e. *Vicia*, *Pisum* and *Lathyrus*). The amount of N fixed by the rhizobia with lentil crop varies from 0 to 192 kg total N/ha around a mean of 80 kg total N/ha (Quinn, 2009). This is similar to quantities fixed by chickpea and dry bean, but less than the 327–450 kg N/ha usually fixed by lupin, faba bean and soybean (Quinn, 2009). Most fixed N is removed as grain, with only a small percentage remaining in the soil. The carry-over of N from lentil is low at 23 to 45 kg N/ha but, in comparison to other legumes such as common bean and chickpea, it is generally positive. Increased yields and protein contents of cereals grown in rotation with lentil and other grain legumes, in comparison to continuous cereals have been reported (Quinn, 2009). Lentil helps not only to improve crop yield but also to enhance net carbon sequestration and lower the carbon footprint of cereal production. One study showed that the lentil-wheat system among tested rotations had the lowest per area carbon footprint at −552 kg CO_2 eq/ha (Gan et al., 2014). The study showed that wheat can convert more CO_2 from the atmosphere into soil carbon than is actually emitted during its production if key farming practices such as a lentil-wheat system are used. Such factors make lentil a key rotation crop in the sustainable intensification and diversification of smallholder farming.

Lentil is grown as a mono-crop in rotation with cereals in South Asia, North America and West Asia. Double cropping in South Asia is popular as a means of crop diversification, for example, lentil with rice, maize, cotton and so on (Ali et al., 2009). Lentil is also intercropped with other crops mainly wheat, linseed and sugarcane to provide dietary diversity and insurance of production from individual crop failures.

Additionally, lentil straw is a valued animal feed, which in dry seasons when feed prices rise can give a greater economic return to growers than the grain (Erskine et al., 1990).

2 Production regions and their agro-ecologies

2.1 Global production

Globally lentil production is dominated by two giants: Canada and India. From 2011 to 2013 Canada produced 1.65 million t – representing 35% of global production – from an area of 0.99 million ha giving an average yield of 1674 kg/ha, whereas India which had 41% of global harvested area (1.75 million ha) produced 1.04 million t giving the substantially lower yield levels of 597 kg/ha (FAOSTAT, 2016). Additionally in South Asia, Bangladesh and Nepal are also substantial lentil producing countries. In North America, the United States is the second major producer after Canada. The third region of lentil production after South Asia and North America is West Asia comprising Iran, Syria and Turkey, which collectively produced 13% of global lentils in the period 2011–13 (FAOSTAT, 2016). Outside these three regions the only other substantial producing nations are Australia (producing for export), China (relative newcomer to the lentil FAO statistics) and Ethiopia.

Australia, Canada and the United States dominate the export trade in lentils which focuses primarily on Asia and North Africa (Erskine, 2009).

World lentil production has climbed steadily from 1961 to 1963 – when FAO data collection started – by five-fold by 2011–13 from an average of 0.92 million tonnes to 4.69 million tonnes (FAOSTAT, 2016). This spectacular growth comprises a lift of ×2.5 in harvested area from 1.64 million ha in 1961–3 to 4.29 million ha in 2011–13, and importantly an almost doubling of productivity from an average yield of 560 kg/ha in 1961–3 to 1094 kg/ha by 2011–13 (FAOSTAT, 2016). Dis-aggregating the yield data for the last 15 years against the background of a 1% increase in yield/ha/year shows a patchy landscape with some of the poorest producing countries – Nepal and Bangladesh – posting above-average yield gains along with China and Turkey. However, yield gains have been below average in India, Iran, Syria and the United States (Erskine et al., 2009).

2.2 Production regions and their agro-ecologies

In South Asia the winter crop grows on residual moisture stored from monsoonal rain. During spring the crop is faced with a sudden rise in temperature and depleting soil moisture at the grain filling stage, causing forced maturity. The diseases *Fusarium* wilt caused by *F. oxysporum* f. sp. *lentis* W.L. Gordon, root rots caused by *Sclerotium rolfsii* Sacc., *Rhizoctonia solani* Kuhn and rust caused by *Uromyces viciae-fabae* (Pers.) J. Schröt inflict heavy crop losses. *Ascochyta* blight (AB) and *Stemphylium* blight (SB) also cause economic losses in damper areas. The only insect pests of concern are pod borer (*Etiella zinckenella* Tr.) and the aphids *Aphis craccivora* Koch and *Acyrthosiphon pisum* Harris. Collectively because of these issues and the low yield potential of landraces, yields in India barely increased from 450 to 646 kg/ha from the decade 1960–70 to 2004–13 in contrast to the rest of the world (FAOSTAT, 2016).

In West Asia both winter- and spring-sown lentils are grown with the winter crop dominating (~65%). The amount and distribution of rainfall received within the cropping season determines the seed yield, and the winter crop is typically grown in areas where the annual rainfall is 200–500 mm (Erskine and El Ashkar, 1993). Low temperature at seedling and flowering stages, intermittent drought throughout the cropping season, and terminal drought and heat are common events (Wery et al., 1994). Spring-sown lentils at high elevations experience a high incidence of root diseases, whereas the winter-planted crop has a high incidence of AB. Parasitic weeds such as broomrape (*Orobanche* sp.) cause very high crop losses, particularly in the winter-sown crop. The lentil in West Asia is also affected by the weevil *Sitona crinitus* Herbst which attacks root nodules reducing nitrogen fixing capacity. Within the region productivity is particularly low in Iran – 617 kg/ha in 2011–13 (FAOSTAT, 2016).

In North America lentil is sown – usually with rhizobial inoculants – in the spring, and late frosts can be a hazard at the seedling stage. Lentil production in Western Canada is most successful when grown in rotation with cereals, such as spring or durum wheat. With lentil rooting depth shallower than that of the following wheat crop, this allows the cereal to extract water from a greater depth and produce high yield and protein content (Anon., 2000). Lentil is a notoriously poor competitor to weeds, so careful weed management is required and should be considered strategically in concert with the entire crop rotation. Lentil is susceptible to soil residues of some herbicides used in previous years. Despite protection measures, yield losses have been considerable due to diseases such as AB caused by *Ascochyta fabae* Speg. f.sp. *lentis* and anthracnose from

Colletotrichum truncatum Andrus et W.D. Moore. Botrytis grey mould (*Botrytis cinerea* Pers.), *Sclerotinia* stem and stem/pod rot (*Sclerotinia sclerotiorum* (Lib.) de Bary), SB (*Stemphylium botryosum* Wallr.) and powdery mildew (*Erysiphe* spp.) appeared in the late 1990s and are potential threats to future lentil production in North America (Chongo et al., 2002).

3 Phylogeny, domestication/spread and genetic resources

3.1 Phylogeny

The genus of the cultivated lentil *Lens* has received many classifications. Cubero et al. (2009) provided a useful review of lentil phylogeny, origin, domestication and spread. Currently the most widely used classification (e.g. by the National Center for Biotechnology Information) is that of Ferguson et al. (2000), who developed this classification using morphology and molecular markers. As *L. orientalis* is clearly the wild ancestor of the cultigen, its status is sub-specific (i.e. *L. culinaris* ssp. *orientalis*; Harlan and De Wet, 1971) to the cultivated lentil (*L. culinaris* Medikus). Ferguson et al. (2000) considered both *odemensis* and *tomentosus* as subspecies of *L. culinaris*. The classification of the genus *Lens* as proposed by Ferguson et al. (2000) comprises *L. culinaris*, with four subspecies, namely, *culinaris, orientalis, tomentosus* and *odemensis*; *L. ervoides*; *L. nigricans*; and *L. lamottei*. The genus now consists of seven taxa split into four species:

1 *L. culinaris* Medikus

- ssp. *culinaris*
- ssp. *orientalis* (Boiss.) Ponert
- ssp. *tomentosus* (Ladiz.) M.E. Ferguson et al.
- ssp. *odemensis* (Ladiz.) M. E. Ferguson et al.

2 *L. ervoides* (Brign.) Grande
3 *L. nigricans* (M. Bieb.) Godr.
4 *L. lamottei* Czefr.

Despite the morphological similarity of the taxa – often leading to mis-classifications, of crucial importance to the breeder is access to valuable variation, and for this the gene pool concept of Harlan and De Wet (1971) is useful. Cubero et al. (2009) indicated that hybridization barriers support the concept of seven differentiated taxa. *L. c. orientalis* is clearly within the primary gene pool of the domesticated lentil *L. culinaris*. *L. odemensis* belongs to the secondary gene pool because it can be crossed with the cultivated lentil with difficulty, although success (needing embryo rescue or not) depends on the specific parents involved. *L. nigricans* and *L. ervoides* belong to the tertiary gene pool, but can become part of the secondary gene pool by means of embryo rescue (Tullu et al., 2013) – see Section 7.2. Further hybridization is needed to position *L. tomentosus* and *L. lamottei* in the secondary or tertiary gene pool.

3.2 Domestication and spread

The region from western Turkey to northern Iraq contains all the wild lentil taxa. Within this archaeological and botanical data point to southern Turkey – northern Syria as the most likely place of lentil domestication (Cubero et al., 2009). Here populations of *orientalis* were unconsciously subjected to automatic selection, leading to a new crop, *Lens culinaris*.

From this centre of origin, lentil spread as a component of the Near East crop complex in Neolithic agriculture (Harlan, 1992), being first cultivated in the Fertile Crescent, then moving to Greece, Central and Western Europe along the Danube, to the Nile Delta, and eastwards to India (Cubero et al., 2009). The above domestication event represents the first genetic bottleneck of variability in *L. culinaris*. A second genetic bottleneck of significance to breeders occurred in the spread of the cultivated lentil from highland Afghanistan into the Indo-Gangetic Plain, where it is of major importance today (Erskine et al., 2011a). Breaking this bottleneck is a lentil breeding success story (Erskine et al., 1998 – see Section 7.1).

3.3 Genetic resources

Extensive collections of genetic resources of lentil are conserved at ICARDA under FAO auspices and elsewhere at the National Bureau of Plant Genetic Resources, New Delhi, India; Vavilov Institute of Plant Industry, St. Petersburg, Russia; and the United States Department of Agriculture, the United States. Despite the war in Syria, the ICARDA collection is well backed-up and by far the largest comprising 13 907 cultivated and 602 wild relatives (Table 1).

The Genetic Resources Unit of the ICARDA has characterized 11 165 accessions for various morphological and phenological traits to date. Marked variation among the characters for use in breeding and selection programmes has been reported for various morphological characters, response in flowering to temperature and photoperiod, winter hardiness, iron deficiency chlorosis and boron imbalances, drought and heat tolerance, herbicide tolerance and resistance to fungal diseases and viruses. Wild relatives also have shown considerable variability for morphological traits, winter hardiness, and tolerance to drought, *Fusarium* wilt, AB, anthracnose, *Sitona* weevils and *Orobanche*.

Table 1 Genetic resources conserved at ICARDA

Name of taxon	Number of accessions	No. of countries of collection
L. culinaris ssp. *culinaris*		
Landraces	11 197	78
Breeding lines (ICARDA)	2710	–
L. culinaris ssp. *orientalis*	263	15
L. culinaris ssp. *tomentosus*	21	2
L. culinaris ssp. *odemensis*	66	5
L. ervoides	174	16
L. nigricans	68	8
L. lamottei	10	8

Within the cultivated lentil adaptation among landraces is primarily driven by crop phenology (time to flowering and maturity) (Erskine et al., 1989). In another study of the variability among world landrace accessions of lentil – this time using single nucleotide polymorphisms (SNPs) – three major groups that prominently reflected geographical origin (world's agro-ecological zones) were found. The three clusters complemented the origins, pedigrees and breeding histories of the germplasm. The three groups were (a) South Asia (subtropical savannah), (b) Mediterranean and (c) northern temperate (Khazaei et al., 2016) (see Section 2.2).

4 Problems addressable by breeding

4.1 Yield enhancement (including yield and yield components)

The global genetic resources of domesticated and wild lentils at the ICARDA genebank provide wide variability and scope for crop improvement. Overall lentil yields are generally low (global mean 1094 kg/ha – see Section 2.1) because of the limited yield potential of landraces, which are also vulnerable to a range of stresses – both biotic and abiotic, and also poor agronomic management (Section 2.2). Regarding yield potential, national lentil yields are highest (2.5 t/ha) in New Zealand (FAOSTAT, 2016), where farm yields of more than 3 t/ha are common under dryland Canterbury conditions because of the prevailing mild, extended winter conditions (McKenzie et al., 1985).

Wild species are valuable sources of novel genes and alleles for yield-enhancing traits and resistance to biotic (see Section 7.2) and abiotic stresses (4.2). These have largely remained under-utilized due to crossability barriers and linkage drag, but there are some examples of successful introgression of genes into the cultivated species from their wild relatives in lentil. For example, *L. ervoides* has also been identified as a good source of such agronomic traits as growth habit, biomass production and seed traits (Fiala et al., 2009; Tullu et al., 2011). Useful genetic variability for crop duration, secondary branches, number of pods/plant, biological yield/plant, grain yield/plant and 100-seed weight have been reported in other wild relatives of lentil (Singh et al., 2013, 2014; Kumar et al., 2011, 2014). Evaluation of recombinant inbred lines between *L. culinaris* and *L. ervoides* has resulted in transgressive segregants for various agronomic traits including 8% increase in seed size (Tullu et al., 2011). Identification of new sources of extra earliness in lentil from a wild accession, ILWL118, having less than 90 days maturity has resulted in the development of new breeding lines (Kumar et al., 2016). Further information on recent successes with wide hybridization in *Lens* is elaborated in Section 7.2.

Further agronomic problems include pod dehiscence, pod drop and lodging. Adequate variability for all these traits is known to exist within the genetic resources of the crop (wild and cultivated), allowing manipulation through plant breeding.

Seed traits such as seed shape are becoming important in developing new varieties to address market demand and consumer preference. For example, a roundish seed shape is more desirable than a flat shape in red lentils to reduce processing loss and to meet consumer demand. 'Football'-type red lentil after dehulling, where cotyledons remain intact, is more favoured in international markets than split lentils. In contrast in large-seeded green lentils a flat seed shape is demanded by consumers in North Africa, Europe and Latin America. Considerable variability exists among cultivated and wild lentils to address such seed traits.

4.2 Yield stability (abiotic and biotic stresses)

Regarding abiotic stresses, drought and heat stress are of prime importance in both South Asia and the Mediterranean region where the crop experiences a sudden rise in temperature and depleting soil moisture at the grain filling stage, causing forced maturity. Terminal drought and heat stress are common events in both spring- and winter-sown crops (Wery et al., 1994). Progress has been made in drought stress screening (see Section 7). However, to increase adaptation to potentially more extreme weather conditions associated with climate change, more research on heat stress tolerance is needed. A significant correlation was observed between deep root system and shoot growth and seed yield (Sarker et al., 2005) and between root and shoot traits such as dry shoot biomass and chlorophyll content (Kumar et al., 2012). A recent study suggested that markers for QTLs related to root and shoot traits associated with drought tolerance, such as dry root biomass, lateral root number, root-shoot ratio and specific root length, could be used in marker-assisted breeding after validation (Idrissi et al., 2015). A molecular marker associated with *Sdt* locus controlling seedling survival/drought tolerance (Singh et al., 2016a,b) and a QTL hot spot genomic region controlling a number of root and shoot characteristics (Idrissi et al., 2016) have been reported for improving drought tolerance in lentil. Similarly, the lentil crop sometimes encounters a few days of exposure to high temperatures (30–35°C) at reproductive stage, causing heavy yield losses due to flower drop and pod abortion. Recent studies have successfully identified heat tolerant lines in lentil (Kumar et al., 2016a; Singh et al., 2016b; Kumari et al., 2017), which will help develop climate-smart cultivars to counter challenges of climate variability and change.

The most important disease of lentil globally is vascular wilt. Resistance to vascular wilt has been widely reported and its inheritance studied (Bayaa et al., 1997; Eujayl et al., 1998; Khare, 1981). In order to identify resistant sources, screening of ~20 000 lines in a well-established wilt sick plot has resulted in identification of 325 donors (Kumar et al., 2010). Recently, Pouralibaba et al. (2016) identified pathotypes of *F. oxysporum* f.sp. *lentis* for the first time. In the light of this new information on pathotypes, approaches to breeding for resistance to vascular wilt need reassessment. Among foliar diseases, rust caused by *Uromyces vicia-fabae* (Pers.) Schroet is a serious yield constraint in Africa (Ethiopia and Morocco) and both South Asia and South America. Rust resistance is widely reported (Bejiga et al., 1995; Negussie and Pretorius, 2012; Singh and Sandhu, 1988). Pathogenic variability in the fungus is also known (Singh et al., 1995). The relative regional importance of other diseases is discussed in Section 2.1 but host plant resistance is the approach of choice for control of such fungal diseases of lentils such as AB and anthracnose, which is becoming a major disease in temperate lentil growing regions, for example, Canada (see Section 7.2). SB is also of growing importance in Western Canada and South Asia and is recently spreading in India and Nepal – previously it was only important in Bangladesh. Parasitic weeds such as broomrape (*Orobanche* sp.) cause very high crop losses, particularly to the winter-sown crop in North Africa and West Asia. The lentil in West Asia is also affected by the weevil *Sitona crinitus* Herbst. New developments in variation in resistance to anthracnose, *Stemphylium*, *Sitona* weevil and *Orobanche* are discussed in Section 7.2.

Weeds are a major production constraint in lentil. The crop is a very poor competitor with weeds and non-parasitic weeds compete for light and nutrients, often leading to yield losses up to 40% (Tepe et al., 2005). The species of weeds commonly infesting lentil fields vary with changing production environments. Soil characteristics, moisture amount, precipitation pattern, crop rotation, temperature, latitude, altitude, fertility, weed control technology and other factors interact to determine weed flora and intensity. Generally, weeds causing significant losses in lentils are categorized as annual monocots, annual dicots, perennial monocots, perennial dicots and parasitic (Yenish et al., 2009). Imidazolinone herbicide-tolerant lentil varieties have been released to Australian and Canadian growers and proven extremely popular in giving farmers more flexibility in herbicide control options, particularly around managing herbicide residues in soils (Slinkard et al., 2007; Muehlbauer et al., 2009). This has led to an increasing frequency of imidazolinone resistance in weed populations prompting the search for tolerance to other herbicides such as the broad-spectrum herbicide metribuzin. Lentil tolerance to metribuzin has been identified in both Australia and Canada (Meier et al., 2014). Tolerance to other herbicides is also required. Screening of lentil at ICARDA has resulted in the identification of ILL8112, ILL5988, ILL8009 and ILL4994 as tolerant to Imazethapyr and ILL1005, ILL0462, ILL5531, ILL6434, ILL0195, GCP10 and 06S 53110-02 to Metribuzin.

Besides common weeds, parasitic weeds like *Orobanche* have also emerged as a threat to lentil production in some Mediterranean countries (Rubiales et al., 2006). The evaluation of lentil germplasm accessions in infested field did not show resistance to *Orobanche* (Erskine et al., 1994). However, potential resistant sources with significant reduction in infestation density have been reported in Spanish germplasm (Fernandez-Aparicio et al., 2008) and wild species including *Lens ervoides*, *L. odemensis* and *L. orientalis* (Fernandez-Aparicio et al., 2009).

5 Breeding methods

The choice of breeding method for plant improvement depends on the objective and the feasibility with which a trait under consideration is amenable besides of course the available resources and progress to date. Pure line selection of lentil was the preferred method up to 1980s when it was replaced by the pedigree method of selection following hybridization. Though many variants of pedigree method such as single-seed descent, single-pod descent and so on have been suggested and applied, the classical bulk-pedigree method remains the most widely applied method to select better recombinants. In addition to bulk-pedigree method, back-cross and mutation breeding have also been employed occasionally to improve specific traits in an otherwise already improved variety and break repulsion linkages that limit simultaneous improvement in lentil.

Mutation breeding to induce mutations is useful to create variability, when a desired trait is unavailable in the existing germplasm. Seven cultivars, three each in India (Ranjan, Arun and HUL57) and Bangladesh (Bina Masur 1, Bina Masur 2 and Bina Masur 3) and one in Pakistan (NIAB Massor 2006) have been developed through mutation breeding. Induced mutation was also used to develop a lentil line with tolerance to imidazolinone herbicides in Canada. The trait was patented (US Patent 7232942) and licensed for use in Clearfield® lentil varieties (Muehlbauer et al., 2009). Mutant lentil lines with resistance to imidazolinone have also been developed in Australia (Materne et al., 2011).

Lentil improvement at ICARDA is built upon the foundation of its vast germplasm collection and its use to breed new varieties better adapted to different agro-ecological conditions. Except for a few traits (e.g. tolerance to aphids, bruchids and *Orobanche*), sufficient variability for important economic traits is reported in the existing germplasm. To increase the use of germplasm in breeding programmes, the Focused Identification of Germplasm Strategy (FIGS) has recently been pursued in pulses such as faba bean and lentil (Khazaei et al., 2013). The strategy has proven successful for various adaptive traits such as tolerance to heat, drought, cold and salt, besides resistance to diseases. Such FIGS sets in lentil are now available to NARS partners to discover and deploy useful genes into desired agronomic background. ICARDA breeding programme generally uses parents of diverse origin with an aim to combine traits contributing to yield, appropriate phenology, adaptation to major biotic and abiotic stresses, and market preferred traits. Following a selection-hybridization-selection cycle, ICARDA constructs new breeding lines to deliver to the NARS partners in the form of international nurseries. These nurseries comprise a range of genetically fixed materials and segregating populations to provide opportunities to NARS partners for selection. On the basis of phenological adaptation, agronomically desirable traits, resistance to prevailing stresses, quality aspects, and farmer and consumer preference, NARS partners identify and select promising lines/single plants for eventual release as variety for commercial cultivation. Over 137 varieties of lentil have been released for cultivation in target countries using ICARDA germplasm directly or indirectly. During the last ten years, NARS partners have released 41 varieties of lentil with various traits using ICARDA material. Focused programmes on breeding have shown profound effect on production and productivity. These gains become more spectacular when viewed with markedly reduced crop duration of these varieties. This has increased the per day productivity comparable to any rainfed crops besides making them suitable for introduction in new niches and diversification of the existing cropping systems (see Section 7.1).

6 New technologies

Accelerating the pace of genetic gain is a major objective of plant breeding. Recent advances in genomics reveal the wealth of genetic diversity available within crops, while new and more cost-effective genotyping and phenotyping technologies allow breeding advancements with greater precision and speed. Specifically in lentil significant progress has been made in the development of genetic and genomic resources during past two decades. Many molecular markers such as RAPD, SSR, AFLP, RFLP and SNP have been developed and are now available for marker-trait association studies and marker-assisted selection (MAS; Havey and Muehlbauer, 1989; Weeden et al., 1992; Tahir and Muehlbauer, 1994; Eujayl et al., 1998; Rubeena et al., 2003; Hamwieh et al., 2009; Kaur et al., 2011; Kumar et al., 2015). In recent years, next-generation sequencing platforms have accelerated development of SNP markers. For example, 44 879 SNP markers have been identified in lentil using Illumina Genome Analyzer (Sharpe et al., 2013). The existing resources of transcriptome, EST and whole-genome sequences have further enabled the discovery and validation of existing and new molecular markers. Recently, next-generation sequencing has accelerated development of ESTs and now in lentil about 10 163 ESTs are available in the public domain. The lentil genome assembly v1.0, based on the Canadian variety CDC Redberry, was released in January 2016 (Bett et al., 2016). The assembly consists of seven pseudomolecules anchored through the use of 6 high-density genetic linkage maps, with the total assembled bases representing approximately half of the 4.3 Gb lentil genome.

Recent studies have also been able to establish trait-marker association through linkage and association mapping, and identified genes/QTLs controlling some important traits

Table 2 Marker-trait association studies conducted in lentil

Trait	Gene/ QTL	Marker type	Phenotypic variation explained (%)	Reference
Ascochyta blight resistance	*QTL*	RAPD	90	Ford et al., 1999
	Ra/2	RAPD, SCAR	–	Chowdhury et al., 2001; Tar'an et al., 2003
	QTLs	AFLP	Up to 50	Rubeena et al., 2006
	QTLs	SNPs	52 and 69	Sudheesh et al., 2016
	QTL	–	~50	Taylor et al., 2006
Anthracnose resistance	*LCt-2*	AFLP, RAPD	–	Tullu et al., 2003; Tar'an et al., 2003
Fusarium wilt resistance	*Fw*	RAPD, SSR	–	Eujayl et al., 1998b; Hamwieh et al., 2005
Stemphylium blight resistance	QTLs	SSRs, SRAPs, RAPDs	46	Saha et al., 2010a
Rust resistance	*R*	STS, SSRs, RFPLs, RAPDs, CAPS and dCAPS		Saha et al., 2010b
Rust	*R*	SSR	–	Mekonnen et al., 2014
Rust	*R*	SSR	–	Dikshit et al., 2016
Earliness and plant height	QTL	RAPD, SSR, AFLP	31–46	Tullu et al., 2008
Plant structure, growth habit and yield, seed diameter, seed weight	QTL	RAPDs, ISSRs, AFLPs, SSRs	18.2–90.4	Fratini et al., 2007
Flowering time	*QTL*	SSR	57	Kahriman et al., 2015
Seed weight and size	*QTL*	SNP	27.5–48.4	Verma et al., 2015
Seed thickness, diameter, seed plumpness, days to 50% flowering	*QTL*	SSR and seed colour loci	8.4–60	Fedoruk et al., 2013
Seed weight, plant height, seed diameter	*QTL*	SSR, SRAP, RAPD	15.3–32.6	Saha et al., 2013
Boron tolerance	*QTL*	SNP	71	Kaur et al., 2014
Selenium content	*QTL*	SNP	6.3–16.9	Ates et al., 2016
Drought tolerance	*QTL*	SSR	69.7	Singh et al., 2016a,b

Trait	Gene/ QTL	Marker type	Phenotypic variation explained (%)	Reference
Root and shoot traits in relation to drought tolerance	*QTL*	SNP, SRAP	27.6–28.9%	Idrissi et al., 2016
Frost tolerance	*Frt*	RAPD, SSR	–	Eujayl et al., 1999
Winter hardiness	QTL	RAPD, SSR, AFLP	33.4	Kahraman et al., 2004
	QTL	RAPD, ISSR and AFLP m	20.5	Kahraman et al., 2010

Source: Modified from Kumar et al. (2012, 2015).

(Table 2). Evaluation of markers associated with AB resistance across a diverse panel of lentil germplasm revealed that the identity of alleles associated with AB_IH1 predicted the phenotypic responses with high level of accuracy (Sudheesh et al., 2016). The tightly linked molecular markers for AB resistance will enable MAS for AB resistant cultivars, based on introgression of QTL-containing genomic regions from donor to recipient germplasm. Fedoruk et al. (2013) used association mapping to identify QTL for seed size and seed shape in lentil. This opens up scope for genome-wide association selection in lentil as a properly designed association mapping panel has a greater frequency of alleles and covers more genetic variation in the crop, and may facilitate time and cost saving as compared to MAS. Access to high-throughput genotyping and sequencing technologies is expected to accelerate genetic gain across target environments in lentil. These developments ultimately will increase the utilization of genomic resources in genetic improvement of lentil and will lead to fast-track development of improved cultivars. Further, increasing number of re-sequencing database in coming days will allow identification of more SNPs and consequently, HTP cost-effective genotyping assays using only informative SNPs would become available for the development of high-density linkages for MAS. These advances also simultaneously encourage lentil breeders to develop specialized mapping populations such as nested association mapping and multi-parent advanced generation inter-cross populations to generate genome-wide allelic and haplotype data. Likewise, other techniques such as target-induced local lesion in genomes and RNA interference (RNAi) also have demonstrated potential scope for lentil improvement. At ICARDA, mutagenic lentil populations have been recently developed using ethyl methane sulphonate in order to identify point and knock-out mutations for traits such as pod shattering, herbicide tolerance and *Orobanche* tolerance. Likewise, other approaches including Gene Editing, RNAi technology and Virus-Induced Gene Silencing will help understand the molecular mechanisms of various different biological processes in lentil. The future should provide more opportunities to integrate genomic-assisted breeding tools into conventional breeding programmes. This coupled with rapid generation advancement technologies (Croser et al., 2016; Bermejo et al., 2016), crop simulation model (Ghanem et al., 2015) and appropriate experimental design and statistical models (Sarker and Singh, 2015) will contribute to more efficient breeding programmes.

7 Success stories

7.1 Widening the genetic base in South Asia

The narrow genetic base of indigenous *microsperma* germplasm (i.e. *pilosae* type) in South Asia has been recognized phenotypically despite the large area under cultivation (see Section 2.1) (Erskine et al., 1998). Molecular diversity studies of ecotypes confirmed both the particularity and lack of diversity of South Asian lentil germplasm (Ferguson et al., 1998; Kumar et al., 2014a). The realization of a narrow genetic base in South Asia then led to deliberate attempts by ICARDA with the relevant national programmes to break the genetic bottleneck and broaden the variation available. This involved the direct introduction of early flowering lines from other parts of the world and a massive hybridization effort of South Asian material with material from other areas with disease resistance genes and yield potential. Crosses were made at ICARDA, Syria and segregating populations were sent to South Asia for selection *in situ*. Initially, cross incompatibility of *macrosperma*-type germplasm due to their late flowering compared to Indian germplasm restricted their use in the Indian lentil breeding programme. However, identification of an early flowering exotic germplasm of *macrosperma* type 'Precoz' led to its introduction in India and involved in Indian lentil breeding programme to improve earliness, seed size and rust resistance (Asghar et al., 2010; Erskine et al., 1998; Kumar et al., 2004). As a result, breeding lines having Precoz genes in their derivatives have been developed. Recently in India a diverse sample of 21 frequently used local and exotic lentil lines in crossing revealed that among these, 16 genotypes either had 'Precoz' gene, an Argentine line in their pedigree or genes from introduced lines from ICARDA (Kumar et al., 2014a).

In Bangladesh the adoption rate of lentil cultivars from this base broadening effort has now reached 100% among 480 growers interviewed in 2015 (Shafiqul Islam et al., 2015). Recently developed varieties such as Barimasur-6, Barimasur-7, Barimasur-8, Binamasur-7 and Binamasur-10 with diverse exotic parentage have broadened the genetic base with respect to resistance to SB, rust and root diseases, seed traits, crop duration, drought tolerance and so on. Some of these varieties (Barimasur-4, Barimasur-6, Barimasur-8) are also high in iron (>75 ppm), thus large-scale production and their consumption contribute to mitigating anaemia among pre-school children and pregnant women in rural Bangladesh.

In Nepal varietal adoption from this base broadening thrust is slower with local lentil still used by 43% of growers followed by the new cultivars Khajura 2 (31%) and Khajura 1 (16%) (Thapa Magar et al., 2014). With recent releases of Maheswar-Bharati, Sagun, ILL 7723, Nepal is moving fast in spreading of improved varieties in Terai and mid-hill regions. These varieties have robust root systems and fast growing habits which are also being promoted in rice fallows in western and eastern districts of Nepal.

7.2 Introgression of genes from wild species

The most exciting recent development in lentil improvement has been the widening of the genetic base available to breeders following the successful hybridization of the cultivated lentil with *L. ervoides* using embryo rescue to obtain F_1 hybrids (Tullu et al., 2013). Despite being identified as a treasure trove of useful variability (see Section 4), *L. ervoides* was considered to lie within the tertiary gene pool of the cultivated lentil because of its lack of crossability with the cultigen (see Section 3.1). Through these important wide crosses, the interspecies transfer of resistance genes has been realized as the examples in the following paragraphs show. The introgression approach may also be the way to address other traits of potential importance to sustainable agriculture, where the cultivated lentil gene pool is lacking in variability, such as heat stress tolerance (Singh et al., 2016b), extreme earliness (Kumar, 2015) and various nutritional traits (Kumar et al., 2016b).

Anthracnose is a major disease of lentil in North America. Resistance to the more virulent race Ct0 of the pathogen is extremely rare within the cultivated gene pool and is limited to partial resistance. Transfer of resistance to race Ct0 from *L. ervoides* to the cultivated lentil was reported from glasshouse screening (Tullu et al., 2013) and confirmed in the field (Vail et al., 2012). Fiala et al. (2009) reported the inheritance of introgressed resistance to race Ct0 of anthracnose as conditioned by two recessive genes in an interspecific cross of cv. Eston × L-01-827A (*L. ervoides*).

Only partial resistance to the parasitic weed *Orobanche* sp. has been identified in the cultivated lentil (Fernandez-Aparicio et al., 2008), whilst in 2009 complete resistance was found among wild *Lens* accessions (Fernandez-Aparicio et al., 2009). Importantly Bucak et al. (2014) recently reported the interspecific transfer of resistance to the broomrape *Orobanche crenata* Forsk. from the wild *L. ervoides* into the domesticated lentil.

Using the wide hybridization approach several other previously intractable biotic stress problems are now potentially solvable by breeders (see Section 4). This includes the interspecific transfer of useful variability for resistance to *Sitona* weevil and SB expressed within the secondary gene pool. Useful variability for resistance to the *Sitona* weevil, which damages N-fixing root nodules, has not yet been found within cultivated lentil germplasm and only identified in wild *Lens* (El-Bouhssini et al., 2008). Among resistant wild lentil accessions, one accession (ILWL 245) crossable with the cultigen is currently in use at ICARDA to introgress resistance genes into agronomically superior cultivars. Wide crossing with resistant *L. ervoides* accessions also brings *Sitona*-resistant variation within the secondary gene pool into the breeding frame.

Few sources for resistance to SB exist among commercial lentil cultivars. Using a highly aggressive isolate of SB as inoculum to screen both wild and cultivated lentil accessions, most cultivated accessions were susceptible to SB, whereas more than 70% of wild lentil accessions screened had disease severity scores equal to or significantly lower than that of the SB-resistant check (Podder et al., 2013). Some wild species accessions previously identified with resistance to anthracnose and AB were also highly resistant to SB. The highest frequency of resistance to SB was found in *L. lamottei* followed by *L. ervoides* of the secondary gene pool. Following the above approach these sources can potentially be used to develop cultivars with single or multiple disease resistance. The significance of multiple disease resistance is clear in Bangladesh, eastern Indian states and Nepal where devastating attacks of foliar diseases: SB and rust in particular have recently been experienced. Lentil was about to disappear from cultivation due to these diseases. Introduction and use of exotic germplasm has resulted in the development of new crop varieties with combined resistance to these foliar diseases. For example, Maheswar Bharati, Sagun in Nepal; Moitree in eastern India; and Barimasur-8, Binamasur-10 and Barimasur-7 in Bangladesh showed high to moderate levels of resistance to these diseases.

8 Future trends

With a rising global population and limits to natural resource availability such as cultivated land and water on which food grain production relies, together with competition from other sectors, there is enormous pressure on agricultural production. This will be further challenged by changing climates which may manifest itself in the form of shifting rainfall patterns, untimely and erratic rains, extreme temperatures and so on which may also change cropping patterns. To develop climate-smart varieties, landraces and wild relatives, which are rich reservoir of useful alien genes, can play an important role in lentil. There is a strong need to formulate strategic planning to achieve the goal focusing on broadening the genetic base of lentil to narrow the yield gap and achieve new yield gains. In the past it may have been sufficient to develop a variety well adapted to a particular agro-ecological region, taking into account the well-understood abiotic and biotic constraints and end product quality. However, with climate and farming system changes becoming a reality, a dramatic shift in the production base is expected to take place. Wide genetic base of cultivated varieties provides a type of insurance against the epidemics of diseases and insect pests besides, of course, making the cultivated germplasm more amenable to breeding advances through conventional and biotechnological approaches. For efficient utilization of large germplasm, FIGS, core and mini-core sets representing maximum variability of the germplasm collection need to be phenotyped precisely and efficiently followed by their utilization in breeding programme and multi-location evaluation.

9 Where to look for further information

The two largest breeding programmes for lentil are at the International Center for Agricultural Research in the Dry Areas (ICARDA), Morocco (see www.icarda.cgiar.org), focused on the developing world and the Crop Development Centre at the University of Saskatoon, Canada (who have an important portal at http://knowpulse.usask.ca). There are many vibrant national improvement efforts – notably that in India coordinated by the Indian Institute of Pulses Research (http://www.iipr.res.in/) in Kanpur.

A key conference for lentil breeding improvement is the International Food Legume Research Conference, while for legume genomics the most important forum is the International Conference on Legume Genetics and Genomics, the former is held approximately every four years and the latter every alternate year.

An introduction to lentil improvement is to be found at Erskine et al. (2011b), 'Crops that feed the World 3. Investing in lentil improvement toward a food secure world', *Food Sec.*, 3, 127–139.

10 References

Ali, M., Singh, K. K., Pramanik, S. C. and Ali, O. (2009), 'Cropping systems and production agronomy', pp. 213–28. In: W. Erskine, F. J. Muehlbauer, A. Sarker and B. Sharma (Eds), *The Lentil–Botany, Production and Uses*. Wallingford, UK: Comm. Agric. Bureaux.

Anon. (2000), '*Lentils in Saskatchewan*', Fact Sheet – Production. Canada: Saskatchewan Ministry of Agriculture.

Asghar, M. J., Abbas. G. and Shah. T. M. (2010), 'Study of genetic diversity in some local and exotic lentil (*Lens culinaris* Medik) genotypes', *Pak. J. Bot.*, 42, 2681–90.

Ates, D., Sever, T., Aldemir, S., Yagmur, B., Temel, H. Y., Kaya, H. B., Alsaleh, A., Kahraman, A., Ozkan, H., Vandenberg, A. and Tanyolac, B. (2016), 'Identification of QTLs controlling genes for Se uptake in Lentil seeds', *PLoS ONE*, 11(3), e0149210.

Bayaa, B., Erskine, W. and Singh, M. (1997), 'Screening lentil for resistance to fusarium wilt: Methodology and sources of resistance', *Euphytica*, 98, 69–74.

Bejiga, G., Abebe, T., Seifu, T., Seid, A. and Yadeta, A. (1995), 'Evaluation of lentil genotypes for resistance to rust', *Petria*, 5, 189–94.

Bermejo, C., Gatti, I. and Cointry, E. (2016), '*In vitro* embryo culture to shorten the breeding cycle in lentil (*Lens culinaris* Medik)', *Plant Cell Tiss. Org. Cult.*, 127, 585–90.

Bett, K. and et al. (2016), *Lens culinaris* Genome v1.2. Available via KnowPulse BLAST and JBrowse. http://knowpulse2.usask.ca/portal/

Bhatty, R. S. (1988), 'Composition and quality of lentil (*Lens culinaris* Medik.): A review', *Can. Inst. Food Sci. Tech.*, 21, 144–60.

Bucak, B., Bett, K. E., Banniza, S. and Vandenberg, A. (2014), 'Transfer of resistance to broomrape (*Orobanche crenata*) from *Lens ervoides* to cultivated lentil', p. 62. In: *Abstracts of 6th International Food Legume Research Conference (IFLRCVI) and 7th International Conference on Legume Genetics and Genomics (ICLGGVII)*, 7–11 July 2014, Saskatoon, Canada.

Chongo, G., Banniza, S. and Morrall, R. A. A. (2002), 'Diseases of lentil in Saskatchewan in 2002', *Can. Plant Dis. Survey*, 83, 119.

Chowdhury, M. A., Andhrahannadi, C. P., Slinkard, A. E. and Vandenberg, A. (2001), 'RAPD and SCAR markers for resistance to Ascochyta blight in lentil', *Euphytica*, 118, 331–7.

Croser, J., Pazos-Navarro, M., Bennett, R., Tschirren, S., Edwards, K., Erskine, W., Creasy, R. and Ribalta, F. (2016) 'Time to flower of temperate pulses *in vivo* and generation turnover *in vivo – in vitro* of narrow-leaf lupin (are) accelerated by low red to far-red ratio and high intensity in the far-red region', *Plant Cell, Tiss. Org. Cult.*, 127, 591–9.

Cubero, J. I., de la Vega, M. P. and R. Fratini, R. (2009), 'Origin, phylogeny, domestication and spread', pp. 13–33. In: W. Erskine, F. J. Muehlbauer, A. Sarker and B. Sharma (Eds), *The Lentil–Botany, Production and Uses*. Wallingford, UK: Comm. Agric. Bureaux.

Dikshit, H. K., Singh, A., Singh, D., Aski, M., Jain, N., Hegde, V. S., Basandrai, A. K., Basandrai, D. and Sharma, T. R. (2016), 'Tagging and mapping of SSR marker for rust resistance gene in lentil (*Lens culinaris* Medikus ssp. *culinaris*)', *Indian J. Exp. Biol.*, 54, 394–9.

El-Bouhssini, M., Sarker, A., Erskine, W. and Joubi, A. (2008), 'First sources of resistance to *Sitona* weevil (*Sitona crinitus* Herbst) in wild *Lens* species', *Gen. Res. Crop Evol.*, 55, 1–4.

Erskine, W. (2009), 'Global production, supply and demand', pp. 4–12. In W. Erskine, F. J. Muehlbauer, A. Sarker and B. Sharma (Eds), *The Lentil–Botany, Production and Uses*. Wallingford, UK: Comm. Agric. Bureaux.

Erskine, W., Adham, Y. and Holly, L. (1989), 'Geographic distribution of variation in quantitative characters in a world lentil collection', *Euphytica*, 43, 97–103.

Erskine, W., Chandra, S., Chaudhry, M., Malik, I. A., Sarker, A., Sharma, B., Tufail, M. and Tyagi, M. C. (1998), 'A bottleneck in lentil: Widening the genetic base in South Asia', *Euphytica*, 101, 207–11.

Erskine, W. and El Ashkar, F. (1993), 'Rainfall and temperature effects on lentil seed yield in Mediterranean environments', *J. Agric. Sci.*, 121, 347–54.

Erskine, W., Rihawe, S. and Capper, B. S. (1990), 'Variation in lentil straw quality', *Animal Feed Sci. Tech.*, 28, 61–9.

Erskine, W., Sarker, A. and Ashraf, M. (2011a), 'Re-constructing an ancient bottleneck of the movement of the lentil (*Lens culinaris* ssp. *culinaris*) into South Asia', *Gen. Res. Crop Evol.*, 58, 373–81.

Erskine, W., Sarker, A. and Kumar, S. (2011b), 'Crops that feed the World 3. Investing in lentil improvement toward a food secure world', *Food Sec.*, 3, 127–39.

Erskine, W., Tufail, M., Russell, A., Tyagi, M. C., Rahman, M. M. and Saxena, M. C. (1994), 'Current and future strategies in breeding lentil for resistance to biotic and abiotic stresses', *Euphytica*, 73, 127–35.

Eujayl, I., Baum, M., Powell, W., Erskine, W. and Pehu, E. (1998), 'A genetic linkage map of lentil (Lens sp.*)* based on RAPD and AFLP markers using recombinant inbred lines', *Theor. Appl. Genet.*, 97, 83–9.

Eujayl, I., Erskine, W., Baum, M. and Pehu, E. (1999), 'Inheritance and linkage analysis of frost injury in lentil', *Crop Sci.*, 39, 639–42.

Eujayl, I., Erskine, W., Bayaa, B., Baum, M. and Pehu, E. (1998b), 'Fusarium vascular wilt in lentil: Inheritance and identification of DNA markers for resistance', *Plant Breed.*, 117, 497–9.

FAOSTAT (2016), *FAOSTAT statistical database of the United Nations Food and Agriculture Organization (FAO)*. Rome, Italy: FAO. http://faostat.fao.org/site/567/default.aspx#ancor

Fedoruk, M. J., Vandenberg, A. and Bett, K. E. (2013), 'Quantitative trait loci analysis of seed quality characteristics in lentil using single nucleotide polymorphism markers', *The Plant Genome*, 6(3), 1–10.

Ferguson, M. E., Maxted, N., van Slageren, M. and Robertson, L. D. (2000), 'A reassessment of the taxonomy of *Lens* Miller (*Leguminosae, Papilionoideae, Vicieae*)', *Bot. J. Linn. Soc.*, 133, 41–59.

Ferguson, M. E., Robertson, L. D., Ford-Lloyd, B. V., Newbury, H. J. and Maxted, N. (1998), 'Contrasting genetic variation amongst lentil landraces from different geographical origins', *Euphytica*, 102, 265–73.

Fernandez-Aparicio, M., Sillero, J. C., Pérez-de-Luque, A. and Rubiales, D. (2008), 'Identification of sources of resistance to crenate broomrape (*Orobanche crenata*) in Spanish lentil (*Lens culinaris*) germplasm', *Weed Res.*, 48, 85–94.

Fernandez-Aparicio, M., Sillero, J. C., Pérez-de-Luque, A. and Rubiales, D. (2009), 'Resistance to broomrape in wild lentils (*Lens* spp.)', *Plant Breed.*, 128, 266–70.

Fiala, J. V., Tullu, A., Banniza, S., Séguin-Swartz, G. and Vandenberg, A. (2009), 'Interspecies Transfer of Resistance to Anthracnose in Lentil (*Lens culinaris* Medic.)', *Crop Sci.*, 49, 825–30.

Ford, R., Pang, E. C. K. and Taylor, P. W. J. (1999), 'Genetics of resistance to *Ascochyta* blight (*Ascochyta lentis*) of lentil and the identification of closely linked RAPD markers', *Theor. Appl. Genet.*, 98, 93–8.

Fratini, R., Duran, Y., Garcia, P. and De La Vega, P. M. (2007), 'Identification of quantitative trait loci (QTL) for plant structure E: Growth habit and yield in lentil', *Spanish J. Agri. Res.*, 5, 348–56.

Gan, Y., Liang, C., Chai, Q., Lemke, R. L., Campbell, C. A. and Zentner, R. P. (2014), 'Improving farming practices reduces the carbon footprint of spring wheat production', *Nature Comm.*, 5, 5012.

Ghanem, M. E., Marrou, H., Soltani, A., Kumar, S. and Sinclair, T. H. (2015), 'Lentil variation in phenology and yield evaluated with a model', *Agron. J.*, 107, 1967–77.

Grusak, M. A. (2009), 'Nutritional and health-beneficial quality', pp. 368–90. In: W. Erskine, F. J. Muehlbauer, A. Sarker and B. Sharma (Eds), *The Lentil–Botany, Production and Uses*. Wallingford, UK: Comm. Agric. Bureaux.

Hamwieh, A., Udupa, S. M., Choumane, W., Sarker, A., Dreyer, F., Jung, C. and Baum, M. (2005) 'A genetic linkage map of *Lens* sp. based on microsatellite and AFLP markers and the localization of *Fusarium* vascular wilt resistance', *Theor. Appl. Genet.*, 77, 839–43.

Hamwieh, A., Udupa, S. M., Sarker, A., Jung, C. and Baum, M. (2009), 'Development of new microsatellite markers and their application in the analysis of genetic diversity in lentils', *Breed Sci.*, 59, 77–86.

Havey, M. J. and Muehlbauer, F. J. (1989), 'Linkages between restriction fragment length, isozyme, and morphological markers in lentil', *Theor. Appl. Genet.* 77, 395–401.

Harlan, J. (1992), *Crops and Man*. Madison, Wisconsin, USA: American Society of Agronomy.

Harlan, J. R. and De Wet, J. M. J. (1971), 'Towards a rational classification of cultivated plants', *Taxon*, 20, 509–17.

Idrissi, O., Udupa, S. M., Keyser, E. D McGee, R. J., Coyne, C. J., Saha, G. C., Muehlbauer, F. J., Damme, P. V. and Riek, J. D. (2016), 'Identification of quantitative trait loci controlling root and shoot traits associated with drought tolerance in a lentil (*Lens culinaris* Medik.) recombinant inbred line population', *Front. Plant Sci.*, 7, 1174.

Idrissi, O., Houasli, C., Udupa, S. M., De Keyser, E., Van Damme, P. and De Riek, J. (2015), 'Genetic variability for root and shoot traits in a lentil (*Lens culinaris* Medik.) recombinant inbred line population and their association with drought tolerance', *Euphytica*, 204, 693–709.

Kahraman, A., Demirel, U., Ozden, M. and Muehlbauer, F. J. (2010), 'Mapping of QTLs for leaf area and the association with winter hardiness in fall-sown lentil', *Afr. J. Biotechnol.*, 9, 8515–19.

Kahraman, A., Kusmenoglu, I., Aydin, N., Aydogan, A., Erskine, W. and Muehlbauer, F. J. (2004), 'QTL mapping of winter hardiness genes in lentil', *Crop Sci.*, 44, 13–22.

Kahriman, A., Temel, H. Y., Aydoğan, A. and Tanyolac, M. B. (2015), 'Major quantitative trait loci for flowering time in lentil', *Turkish J. Agric. Forest.*, 39, 588–95.

Kaur, S., Cogan, N. I., Stephens, A., Noy, D., Butsch, M. and Forster, J. (2014), 'EST-SNP discovery and dense genetic mapping in lentil (*Lens culinaris* Medik.) enable candidate gene selection for boron tolerance', *Theor. Appl. Genet.*, 127, 703–13.

Kaur, S., Cogan, N. O., Pembleton, L. W., Shinozuka, M., Savin, K. W., Materne, M. and Forster, J. W. (2011), 'Transcriptome sequencing of lentil based on second-generation technology permits large-scale unigene assembly and SSR marker discovery', *BMC Genomics*, 12, 265–75.

Khare, M. N. (1981), 'Diseases of lentils', pp. 163–72. In C. Webb and G. C. Hawtin (Eds), *Lentils*. Wallingford: Comm. Agric. Bureaux.

Khazaei, H., Street, K., Bari, A., Mackay, M. and Stoddard, F. L. (2013), 'The FIGS (Focused Identification of Germplasm Strategy) approach identifies traits related to drought adaptation in *Vicia faba* genetic resources', *PLoS ONE*, 8(5), e63107.

Khazaei, H., Caron, C. T., Fedoruk, M., Diapari, M., Vandenberg, A., Coyne, C. J., McGee, R. and Bett. K. E. (2016), 'Genetic diversity of cultivated lentil (*Lens culinaris* Medik.) and its relation to the world's agro-ecological zones', *Front. Plant Sci.*, 7, 1093.

Kumar, J., Kant. R, Kumar. S, Basu. P. S., Sarker, A. and Singh, N. P. (2016a), 'Heat tolerance in lentil under field conditions', *Legume Genom. Genet.*, 7, 1–11.

Kumar, J., Gupta, D. S., Kumar, S., Gupta, S. and Singh, N. P. (2016b), 'Current knowledge on genetic biofortification in lentil', *J. Agric. Food Chem.*, 64, 6383–96.

Kumar, J. (2015), 'Identification of early flowering genotype from wild taxa of lentil', *J. Food Leg.*, 28, 349–50.

Kumar, J., Basu, P. S., Srivastava, E., Chaturvedi, S. K., Nadarajan, N. and Kumar, S. (2012), 'Phenotyping of traits imparting drought tolerance in lentil', *Crop Pasture Sci.*, 63, 547–54.

Kumar, S., Gupta, S., Chandra, S. and Singh, B. B. (2004), 'How wide is the genetic base of pulse crops', pp. 211–21. In: M. Ali, B. B. Singh, S. Kumar and V. Dhar (Eds), *Pulses in New Perspective*. Kanpur, India: Indian Society of Pulses Research and Development.

Kumar, S., Imtiaz, M., Gupta, S. and Pratap, A. (2011), 'Distant hybridization and alien gene introgression', pp. 81–110. In A. Pratap and J. Kumar (Eds) *Biology and Breeding of Food Legumes*. Wallingford, UK: Comm. Agric. Bureaux.

Kumar, S., Rajendran, K., Kumar, J., Hamwieh, A. and Baum, M. (2015), 'Current knowledge in lentil genomics and its application for crop improvement', *Front. Plant Sci.*, 6, 1–13.

Kumar S., Sarker A., Erskine W., Geletu, B., Ahmed S. and Bayaa B. (2010), 'Valuable donors for Fusarium wilt resistance in lentil', p. 37. In *Book of Abstracts of International Food Legumes Research Conference (V) and European Conference on Grain Legumes (AEP VII)*, Antalya, Turkey.

Kumar, J., Srivastva, E., Singh, M., Kumar, S., Nadarajan, N. and Sarker, A. (2014a), 'Diversification of indigenous gene-pool by using exotic germplasm in lentil (*Lens culinaris* Medikus subsp. *culinaris*)', *Physiol. Mol. Biol. Plants*, 20, 125–32.

Kumar, S., Hamweih, A., Manickavelu, A., Kumar, J., Sharma, T. R. and Baum, M. (2014b), 'Advances in lentil genomics', pp. 111–30. In S. Gupta, N. Nadarajan and D. S. Gupta (Eds), *Legumes in Omics Era*. New York, NY: Springer.

Kumari, S., Sehgal, A., Kumar, J., Kumar, S., Singh, S., Siddique, K. H. M. and Nayyar, H. (2017), 'Identification of high temperature tolerant lentil *(Lens culinaris* Medik.) genotypes through leaf and pollen traits', *Front. Plant Sci.*, 8:744.

Materne, M., Leonforte, L., Hobson, K., Paull, J. and Gnanasambandan, A. (2011), 'Breeding for improvement of cool-season food legumes', pp. 49–62. In: A. Pratap and J. Kumar (Eds), *Biology and Breeding for Food Legumes*. Wallingford, UK: Comm. Agric. Bureaux.

McKenzie, B. A., Sherrell, C., Gallagher, J. N. and Hill, G. D. (1985), 'Response of lentils to irrigation and sowing date', *J. Agric. Sci.*, 8, 47–50.

Meier, K., McMurray, L., Willenborg, C. and Vandenberg, A. (2014), 'Investigating the genetics of improved metribuzin herbicide tolerance in lentil'. In: *Abstract in 6th International Food Legume*

Research Conference (IFLRCVI) and 7th International Conference on Legume Genetics and Genomics (ICLGGVII), 7–11 July 2014, Saskatoon, Canada.

Mekonnen, F., Mekbib, F., Kumar, S., Ahmed, S., Chahota, R. K. and Sharma, T. R. (2014), 'Identification of molecular markers associated with rust resistance genes in lentil (*Lens culinaris* sub sp. *culinaris*)', *Can. J. Plant Prot.*, 2, 27–36.

Muehlbauer, F. J., Mihov, M., Vandenberg, A., Tullu, A. and Materne, M. (2009), 'Improvement in developed countries', pp. 137–54. In: W. Erskine, F. J. Muehlbauer, A. Sarker and B. Sharma (Eds), *The Lentil—Botany, Production and Uses*. Wallingford, UK: Comm. Agric. Bureaux.

Negussie, T. and Pretorius, Z. A. (2012), 'Lentil rust: Present status and future prospects', *Crop Prot.*, 32, 119–28.

Podder, R., Banniza, S. and Vandenberg, A. (2013), 'Screening of wild and cultivated lentil germplasm for resistance to stemphylium blight', *Plant Genet. Resour.*, 11, 26–35.

Pouralibaba, H. R., Rubiales, D. and Fondevilla, S. (2016), 'Identification of pathotypes in *Fusarium oxysporum* f.sp. *lentis*', *Eur. J. Plant Path.*, 144, 539–49.

Quinn, M. A. (2009), 'Biological nitrogen fixation and soil health improvement', pp. 229–47. In: W. Erskine, F. J. Muehlbauer, A. Sarker and B. Sharma (Eds), *The Lentil—Botany, Production and Uses*. Wallingford, UK: Comm. Agric. Bureaux.

Rubeena, A., Ford, R. and Taylor, P. W. J. (2003), 'Construction of an intraspecific linkage map of lentil (*Lens culinaris* ssp. *culinaris*)', *Theor. Appl. Genet.*, 107, 910–16.

Rubeena, A., Taylor, P. W. J., Ades, P. K. and Ford, R. (2006), 'QTL mapping of resistance in lentil (*Lens culinaris* ssp. *culinaris*) to ascochyta blight (*Ascochyta lentis*)', *Plant Breed.*, 125, 506–12.

Rubiales, D., Pérez-De-Luque, A., Fernández-Aparicio, M., Sillero, J. C., Román, B., Kharrat, M., Khalil, S., Joel, D. M. and Riches, C. (2006), 'Screening techniques and sources of resistance against parasitic weeds in grain legumes', *Euphytica*, 147, 187–99.

Saha, G. C., Sarker, A., Chen, W., Vandemark, G. J. and Muehlbauer, F. J. (2010a), 'Inheritance and linkage map positions of genes conferring resistance to stemphylium blight in lentil', *Crop Sci.*, 50, 1831–9.

Saha, G. C., Sarker A., Chen, W. D., Vandemark, G. J. and Muehlbauer, F. J. (2010b), 'Identification of markers associated with genes for rust resistance in *Lens culinaris* Medik', *Euphytica*, 175, 261–5.

Saha, G. C., Sarker, A., Chen, W., Vandemark, G. J. and Muehlbauer, F. J. (2013), 'Inheritance and linkage map positions of genes conferring agromorphological traits in *Lens culinaris* Medik', *Int. J. Agron.*, 9, 618926.

Sarker, A., Erskine, W. and Singh, M. (2005), 'Variation in shoot and root characteristics and their association with drought tolerance in lentil landraces', *Genetic. Res. Crop Evol.*, 52, 89–97.

Sarker, A. and Singh, M. (2015), 'Improving breeding efficiency through application of appropriate experimental designs and analysis models: A case study of lentil (*Lens culinaris* Medik. subsp. *culinaris*) yield trials', *Field Crops Res.*, 179, 26–34.

Sarker, A., Aydogan, A., Chandra, S., Kharrat, M. and Sabaghpour, S. (2009), 'Genetic enhancement for yield and yield stability', pp. 102–20. In W. Erskine, F. J. Muehlbauer, A. Sarker and B. Sharma (Eds), *The Lentil–Botany, Production and Uses*. Wallingford, UK: Comm. Agric. Bureaux.

Savage, G. P. (1988), 'The composition and nutritive value of lentils (*Lens culinaris*)', *Nutr. Abs. Rev. (Series A)*, 58, 320–43.

Shafiqul Islam, Q. M., Monayem Miah, M. A. and Matiur Rahman, M. M. (2015), 'Assessment of Socioeconomic Impacts of Introducing Short Duration Lentil and Mungbean into Rice-based Cropping Systems in Western Bangladesh', Unpublished survey for ACIAR project – CIM/2009/038.

Sharpe, A. G., Ramsay, L., Sanderson, L. A., Fedoruk, M. J., Clarke, W. E. and Rong, L. (2013), 'Ancient orphan crop joins modern era, gene-based SNP discovery and mapping in lentil', *BMC Genomics*, 14, 1–13.

Singh, D., Singh, C. K., Taunk, J. and Tomar, R. S. (2016a), 'Genetic analysis and molecular mapping of seedling survival drought tolerance gene in lentil (*Lens culinaris* Medik.)', *Mol. Breed.*, 5, 1–2.

Singh, D., Singh, C. K., Tomar, R. S., Taunk, J., Singh, R., Maurya, S., Chaturvedi, A. K., Pal, M., Singh, R. and Dubey S. K. (2016b), 'Molecular assortment of *Lens* species with different adaptations to drought conditions using SSR markers', *PLoS ONE*, 1, e0147213.

Singh, D., Singh, C. K., Tomar, R. S. S., Chaturvedi, A. K., Shah, D., Kumar. A. and Pal, M. (2016c), 'Exploring genetic diversity for heat tolerance among lentil (*Lens culinaris* Medik.) genotypes of variant habitats by simple sequence repeat markers', *Plant Breed.*, 135, 215–23.

Singh, K. and Sandhu, T. S. (1988), 'Screening of cultivars of lentil for resistance to rust', *LENS Newslett.*, 15, 28–9.

Singh, M., Bisht, I. S., Kumar, S., Dutta, M., Bansal, K. C., Karale, M., Sarker, A., Amri, A., Kumar, S. and Datta, S. K. (2014) 'Global wild annual lens collection. A potential resource for lentil genetic base broadening and yield enhancement', *PLoS ONE*, 9(9), e107781.

Singh, M., Rana, M. K., Kumar, K., Bisht, I. S., Dutta, N., Gautam, N. K., Sarker, A. and Bansal, C. (2013), 'Broadening the genetic base of lentil cultivars through inter-sub-specific and interspecific crosses of Lens taxa', *Plant Breed.*, 132, 667–75.

Singh, S. J. and Sokhi, S. S. (1980), 'Pathogenic variability in *Uromyces viciae-fabae*', *Plant Dis.*, 64, 671–2.

Slinkard, A. E., Vandenberg, A. and Holm, F. A. (2007), 'Lentil plants having increased resistance to imidazolinone herbicides', US patent 7232942. http://www.uspto.gov/patft/index.html.

Sudheesh, S., Rodda, M. S., Davidson, J., Javid, M., Stephens, A., Slater, A. T., Cogan N. O. I., Forster, J. W. and Kaur, S., (2016) 'SNP-based linkage mapping for validation of QTLs for resistance to Ascochyta blight in lentil', *Front. Plant Sci.*, 7, 1604.

Tahir, M. and Muehlbauer, F. J. (1994), 'Gene mapping in lentil with recombinant inbred lines', *J. Hered.*, 85, 306–10.

Tar'an, B., Buchwaldt, L., Tullu, A., Banniza, S., Warkentin, T. D. and Vandenberg, A. (2003), 'Using molecular markers to pyramid genes for resistance to ascochyta blight and anthracnose in lentil (*Lens culinaris* Medik.)', *Euphytica*, 13, 223–30.

Taylor, P. W., Ades, P. K. and Ford, R. (2006), 'QTL mapping of resistance in lentil (*Lens culinaris* ssp. *culinaris*) to ascochyta blight (*Ascochyta lentis*)', *Plant Breed.*, 125, 506–12.

Tepe, I., Erman, M., Yazlık, A., Levent, R. and Ipek, K. (2005), 'Comparison of some winter-lentil cultivars in weed-crop competition', *Crop Prot.*, 24, 585–9.

Thapa Magar, D. B., Darai, R., Gauchan, D. and Sarker, A. (2014), 'Varietal Adoption and Marketing of Lentil in the Mid- and Far Western Terai region of Nepal', *Adv. Plants Agric. Res.* 1(5), 00026.

Thavarajah, D., Thavarajah, P., Casey, J. and Kumar, S. (2016), 'Lentil (*Lens culinaris* Medikus): A whole food rich in prebiotic carbohydrates to combat global obesity', pp. 193–208. In: A. K. Goyal (Ed.), *Grain Legumes*. InTech. http://dx.doi.org/10.5772/62567.

Tullu, A., Banniza, S., Bett, K. and Vandenberg, A. (2011), 'A walk on the wild side, exploiting wild species for improving cultivated lentil', *Grain Leg.*, 56, 13–14.

Tullu, A., Buchwaldt, L., Warkentin, T., Tar'an, B. and Vandenberg, A. (2003), 'Genetics of resistance to anthracnose and identification of AFLP and RAPD markers linked to the resistance gene in PI 320937 germplasm of lentil (*Lens culinaris* Medikus)', *Theor. Appl. Genet.*, 106, 428–34.

Tullu, A., Tar'an, B., Warkentin, T. and Vandenberg, A. (2008), 'Construction of an Intraspecific linkage map and QTL analysis for earliness and plant height in lentil', *Crop Sci.*, 48, 2254–64.

Tullu, A., Bett, K., Banniza, S., Vail, S. and Vandenberg, A. (2013), 'Widening the genetic base of cultivated lentil through hybridization of *Lens culinaris* 'Eston' and *L. ervoides* accession IG 72815', *Can. J. Plant Sci.*, 93, 1037–47.

Vail, S., Strelioff, J. V., Tullu, A. and Vandenberg, A. (2012), 'Field evaluation of resistance to *Colletotrichum truncatum* in *Lens culinaris, Lens ervoides*, and *Lens ervoides* × *Lens culinaris* derivatives', *Field Crops Res.*, 126, 145–51.

Verma, P., Goyal, R., Chahota, R. K., Sharma, T. R., Abdin, M. Z. and Bhatia, S. (2015) 'Construction of a genetic linkage map and identification of qtls for seed weight and seed size traits in lentil (*Lens culinaris* Medik.)', *PLoS ONE*, 10(10), e0139666.

Weeden, N. F., Muehlbauer, F. J. and Ladizinsky, G. (1992), 'Extensive conservation of linkage relationships between pea and lentil genetic maps', *J. Hered.*, 83, 123–9.

Wery, J., Silim, S. N., Knights, E. J., Malhotra, R. S. and Cousin, R. (1994), 'Screening techniques and sources of tolerance to extremes of moisture and air temperature in cool season food legumes', *Euphytica*, 73, 73–83.

Yenish, J. P., Brand, J., Pala, M. and Haddad, A. (2009), 'Weed Management', pp. 326–42. In: W. Erskine, F. J. Muehlbauer, A. Sarker, and B. Sharma (Eds), *The Lentil—Botany, Production and Uses*. Wallingford, UK: Comm. Agric. Bureaux.

Chapter 4

Improving cultivation of lentil

Fred J. Muehlbauer, Washington State University, USA; and Ashutosh Sarker, International Center for Agricultural Research in the Dry Areas (ICARDA), India

1 Introduction
2 Lentil description
3 Climate requirements for lentil cultivation
4 Land requirements for lentil cultivation
5 Nutrient requirements for lentil cultivation
6 Seed quality and treatments
7 Land preparation and planting
8 Innovations in lentil cropping systems
9 Conclusion and future trends
10 Where to look for further information
11 References

1 Introduction

Lentil is an important food legume in the semi-arid regions of the world where it can be grown successfully on limited soil moisture and in relatively poor soils. Despite successful production in marginal areas, food legumes are noted as being unresponsive to efforts to substantially increase yields when compared to the cereals. The inability of lentil and pulses in general to keep pace with cereals in terms of yield is often blamed on the fact that pulses and particularly lentils are most often produced in marginal areas characterized by poor soils and receive minimal inputs of land preparation, fertilizers, weed control, irrigation, mechanical planting and harvesting needed for optimum production. In addition, research efforts devoted to legume crops and particularly lentil are small when compared to more generously supported efforts in cereal crops. In spite of these constraints, steady progress has been made in yield per unit area; while production in extensive agricultural systems of North America and Australia has expanded world production of lentil. As Fig. 1 shows, lentil production held steady during the 1990s but increased substantially at the turn of the century with expanded area sown, with most of the increase coming from expanded production in Canada, Australia and the United States where yields are generally higher than the world average. Of these three countries, Canada now accounts for nearly 40%

http://dx.doi.org/10.19103/AS.2017.0023.23

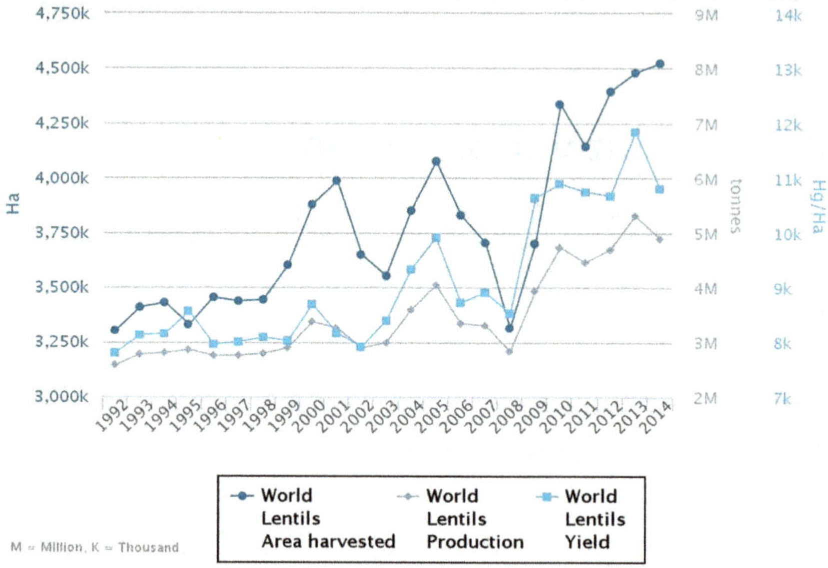

Figure 1 World Lentil area harvested, total production and yields from 1992 to 2014 (Source: FAOSTAT, 2016).

of world lentil production (Fig. 2). Production in these developed countries benefits from mechanized land preparation, sowing, effective weed control, timely harvesting and readily available storage facilities.

Lentils are nodulated following infection by *R. leguminosarum*, the same species that nodulates pea (*Pisum sativum*) and faba (*Vicia faba*) bean. Yield responses to artificial inoculation may or may not be favourable depending on soil type, availability of inorganic nitrogen or the previous crop. Like other legume crops, the ability to fix atmospheric nitrogen through symbiosis with *Rhizobium* species provides for development of protein in the seeds and also helps maintain and improve soil nutritional status.

Lentil provides a protein-rich food for populations in less developed regions of the world where along with other pulses it provides the needed proteins in diets.

This chapter will discuss agronomic practices used in lentil-producing countries and where significant improvements can be made. Most prominent are the traditional methods in developing countries with a heavy reliance on hand labour supplemented by animal power for everything from land preparation, sowing, harvesting, threshing and cleaning. Procedures used in developed countries where the crop is entirely mechanized are also included. Overall, there is substantial scope for improving lentil production agronomy; however, widespread adoption of more efficient methods are hindered in developing countries by relatively small farming operations and where mechanization is not easily applied. Production is often limited by poor seed stocks and inadequate seed supply systems. Nevertheless, practices for enhancing production potential are available and are becoming more widely used and include improved seed supply systems, better seed varieties, improved planting methods that ensure seed to soil contact for uniform emergence, better weed control and adoption of improved harvesting methods.

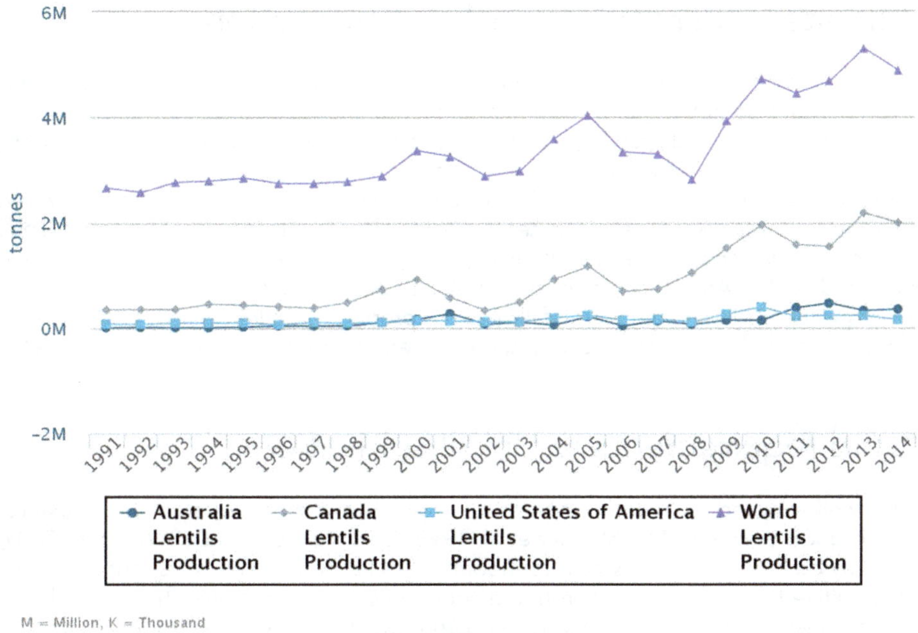

Figure 2 Effect of increased lentil production in Canada, Australia and the United States on total world production (1991–2014). (Source: FAOSTAT, 2016).

2 Lentil description

Lentil plants are slender, semi-erect annuals, usually between 30 and 45 cm tall (Muehlbauer et al. 1998). Individual plants and varieties vary from single stemmed to multi-branched and bushy forms. The lens-shaped seeds vary among varieties and range from 2 to 8 g/100 with seed colours varying from pale tan to completely black. Cotyledons can be red, yellow or green; however, red and yellow dominate among land races and varieties used for production. Pods generally contain one or two seeds. Protein concentrations average 26% and range from 22 to 28% depending on variety and region of production. In general, as with other pulses, lentils have small concentrations of sulphur-containing amino acids. Cereals, on the other hand, are relatively rich in sulphur-containing amino acids and also in tryptophan. The protein of lentil contains significant concentrations of lycine, an amino acid found in small quantities in cereal grains. Consequently, when cereals and lentils are consumed together they provide adequate amounts of amino acids for a healthy human diet (Muehlbauer et al. 1998). Aspects of nutrient concentrations and nutritional value of lentil are described in greater detail in another chapter.

Lentil plants have weakly upright stems that predispose the crop to pre-harvest lodging, making application of mechanized harvest difficult. The lodging problems in lentil crops are challenging to plant breeders who are focused on developing varieties with stiffer stems and remain upright at harvest maturity. Overcoming this plant structural problem and improving harvesting ease with tall upright varieties would be a major research advance.

3 Climate requirements for lentil cultivation

Lentil is well adapted to cool and dry temperate climates with low relative humidity where they can produce something of value where other legume crops often fail.

Lentil is adapted to cool growing conditions, and the young plants are tolerant of spring frosts. This allows for early spring planting dates.

Lentils have been grown extensively in the semi-arid parts of the world, where they have slightly lower yields, but have good seed quality. High humidity and excessive rainfall during the season encourages vegetative growth, which prevents good yield and can reduce seed quality. Annual rainfall of at least 250 to 300 mm will produce adequate yields of good quality seed while higher rainfall amounts produce excellent yields. Also excessive drought and/or high temperatures during the flowering and pod-fill period reduce yields.

4 Land requirements for lentil cultivation

Lentil is well adapted to a wide range of soil types but is more suited to light textured soils with good drainage (Muehlbauer et al. 1998; Cokkiygin et al. 2012; Ali et al. 2009). Lentils may be grown in slightly acid to moderately alkaline or saline soils, but are known to be sensitive to highly saline, boron and sodic soils where root growth is limited, thus reducing moisture and nutrient extraction. Lentil plants are sensitive to water-logging, which causes reduced oxygen to the roots and yellowing of the plants and often death in cases of prolonged exposure. While lentil can be gro wn over a wide range of pH, growth is severely limited at a pH of less than 5 or greater than 9. Therefore, applications of lime should be considered when soils are below pH 5.4 (Mahler 2015).

5 Nutrient requirements for lentil cultivation

Nitrogen provided to lentil crops through symbiosis with *Rhizobium leguminosarum* is generally sufficient for good yields. The presence of nitrogen-fixing bacteria is generally adequately present in most soils that have a history of long-term lentil production. In cases where soils are new to lentil production or where the crop has been absent for a number of years (usually 20 or more), inoculation with *R. leguminosarum* can be beneficial (Mahler et al. 1988; Mahler 2015). Since lentil receives most of its nitrogen requirements through symbiosis, applications of nitrogen to lentil crops is generally not beneficial and may, in fact, cause excessive vegetative growth at the expense of seed yield (Yadav et al. 2009).

Lentil has a high requirement for phosphorus and adequate amounts are needed to promote vigorous root systems, plant growth, and uniform crop maturity. Needs for phosphorus should be determined through periodic soil tests. Applications of phosphorus should be made by avoiding contact with the seed and should be made during tillage operations (Mahler 2015). Phosphorus applied in-row at the time of planting should be limited to the maximum safe rate of 22 kg/ha under good to excellent moisture conditions (Saskatchewan Pulse Growers 2016). Rates of seed-placed phosphate should be reduced if the soil has less than ideal moisture.

Lentil also has a high demand for potassium and actual needs should be determined through soil tests. When potassium is shown to be 50 ppm or less, applications to correct the deficiency can be made by soil incorporation during seedbed preparation (Mahler 2015) or through in-row applications at the rate of 17 kg/ha. Most of the potassium taken up by lentil crops remains in residues and is not removed with the grain (Saskatchewan Pulse Growers 2016).

Sulphur is needed for adequate symbiotic nitrogen fixation and should be applied when soil tests show less than 10 ppm. Applications of 10–17 kg/ha can overcome the deficiency (Mahler 2015; Saskatchewan Pulse Growers 2016).

Limited micronutrient fertility studies have been conducted on lentil, but it is known to respond to small amounts of molybdenum when applied as a seed dressing for planting in molybdenum-deficient soils (Mahler 2015).

6 Seed quality and treatments

High-quality seed for planting is essential for good plant stands and yields. Seed lots with 90% or higher germination are generally chosen for planting lentil fields. However, if germination is less, seeding rates need to be adjusted to ensure adequate plant stands.

In general, farmers must use good quality seed with minimum standards of germination (greater than 90%), free of weed seeds and possibly treated with fungicides to prevent pre-emergence damping-off. In developed countries, seed registration and certification standards and procedures may be in place to ensure good quality seed is available to the farmers. However, in developing countries that account for a major share of world lentil production, seed certification and standards may be limited or non-existent. In such cases, village seed cooperatives which have been promoted are able to provide good quality seed at low cost to the farmers (Bushaw and van Gastel 2008; Witcombe et al. 2010).

Seed treatments with appropriate fungicides can prevent damping-off and ensure that good stands of healthy plants. Treatment with insecticides can also be beneficial for control of soil-borne insects such as wireworms and seed corn maggots. Molybdenum is often applied to seeds along with the other seed treatments. Also, Rhizobium may be applied to the seeds to ensure adequate inoculation.

7 Land preparation and planting

Land preparation and planting differs widely among the lentil-producing countries. It seems appropriate to discuss methods of production starting with traditional methods mostly used in developing countries and finally the methods used in developed countries.

Fields intended for lentils are usually chisel or mould board ploughed to aid water infiltration, control erosion and maximize the retention of crop residues.

Tillage along the contours of hills improves moisture infiltration and prevents excessive runoff and soil erosion. The crop is fall planted in most lentil growing countries and harvested in spring. In developed countries such as Canada and in the United States the crop is planted in spring when soils are sufficiently dry to allow tillage and planting. The general practice is to cultivate fields followed by spring-tooth harrowing to produce a firm and smooth surface for planting. Deep tillage is avoided to prevent loss of moisture. A well-prepared seedbed will have some crop residues on the surface, which improves water infiltration and reduces the erosive effects of rainfall. Soil temperatures increase quickly when fields are well prepared, resulting in rapid germination, good emergence and improved seedling growth. Immediately after planting, the fields are rolled to establish a smooth soil to facilitate mechanical cutting at harvest maturity.

More recently, reduced or no-till farming and direct seeding has been applied to lentil production to conserve soil moisture and reduce energy costs for land preparation and seeding (Kumar et al. 2011). This emerging practice relies on special equipment to open the soil using a disc or shoe type opener and placing the seed directly in contact with the soil at a predetermined depth. The practice has shown significant benefits by minimizing disturbance of the soil and residues. Overall these developing practices are becoming more widely used and have maintained or improved yields and returns to the farmers. There remain some significant problems with direct seeding of lentil relating to optimum placement of seed with existing seeding equipment; however, equipment capable of planting into previous crop residues and untilled soils are being developed and becoming more widely available. Of greater concern are weeds that are more difficult to control in untilled fields although weeds may be less of a problem as the practice is continued over a number of production cycles as weed seeds are not elevated from lower positions in the soil profile as is often the case with extensive tillage.

A major advantage of no-till and reduced till systems is the greater flexibility in planting times as the requirement of time-consuming land preparation is mostly eliminated. Adaptation of existing varieties to direct seeding into untilled fields is currently unknown and may prompt the rather obvious objective of breeding varieties specifically for reduced tillage systems.

7.1 Planting methods

In traditional farming systems used in many parts of West Asia, North Africa and South Asia, farmers mostly hand-spread the seeds on the soil surface followed by a tillage operation to incorporate and cover the seeds to ensure germination and emergence. This practice requires a high seeding rate as many of the seeds remain uncovered and do not germinate. Also, some seeds remain buried deep in the soil and are not able to emerge. These methods of planting leave an uneven soil surface which is not a problem in traditional systems because the crop is generally harvested by hand pulling. In many countries, the hand labour required for planting, harvesting, threshing and cleaning has become an expensive factor in lentil production and has led to reduced area devoted to lentil as farmers move to more enumerative crops such as cereal grains that are more easily harvested.

Studies indicate that a seeding depth of 4–5 cm is optimal for germination and growth, even though deeper plantings may have better access to soil moisture and improved protection from frost. Despite some success with deeper plantings, plant emergence may be poor due to soil compaction from farm machinery or heavy rains. Lentil seeds can germinate in the light or the dark and in constant or diurnally fluctuating temperature regimes. However, rates of germination, emergence and seedling growth are markedly affected by temperature. Optimum values for germination and growth vary with cultivar, age and size of seeds. Smaller seeded cultivars germinate more rapidly than larger ones at temperatures between 15°C and 25°C (Saint-Clair. 1972; Muehlbauer et al. 1998).

Great emphasis has been placed on adapting methods of field preparation, planting and mechanical harvest used in developed countries, such as Canada, the United States and Australia, to developing countries of the Near East, North Africa and South Asia.

Field preparation and planting with seed drills followed by a rolling operation is the key to mechanical harvest and is accomplished by rolling the fields after planting. While rolling ensures good soil-to-seed contact and more even germination and emergence, the operation provides a smooth and level surface for harvesting with cutter bars and direct combining. Rolling should be done as soon as possible after planting and when fields are dry. Rolling when the crop begins to emerge causes considerable damage to emerging seedlings and tends to spread disease. Where stones are present, rolling can effectively push stones into the soil where they are not a hindrance to the cutting operation.

7.2 Plant populations and spacing

Lentil yields are remarkably stable over a wide range of plant densities. Lentil plants are capable of filling available space by initiating lateral branches (Muehlbauer et al. 1998). Recommended seeding rates for lentil vary widely throughout the world and range from 30–40 kg/ha in South Asia to over 120 kg/ha in West Asia. Rates of 172–220 seeds/m^2 are recommended in North America. In Bangladesh, seeding rates of about 30–35 kg/ha are recommended for optimum yields (Rahman and Miah 1989).

Optimum row spacing and overall plant densities vary with growing area and conditions. In general, lentil yields were greater at the narrow (15 cm) between-row spacing and at the 1.5 cm spacing between plants within row (Wilson and Teare 1972). Overall, a plant population density of 200–220 plants/m^2 appears to be optimum for lentil yields.

It is clear that using seed drills saves on seed supplies as a reduced planting rate is required compared to broadcast seeding and it has the advantage of placing the seeds at optimum depth to promote germination and emergence.

7.3 Fallow replacement with annual cropping

Replacing fallow with annual cropping had a major effect on lentil production in Turkey (Acikgoz et al. 1994), where over three million hectares were converted from summer fallowing to production of pulse crops, particularly lentil and chickpea. Similarly, millions of hectares formerly devoted to summer fallowing and spring wheat production were converted to annual cropping in provinces of Central Canada, with most of the converted area being devoted to pulse crop production and particularly lentil (Muehlbauer and Slinkard 2000). The use of summer fallow land for lentil enabled greatly expanded production in that country as well as in Turkey. Benefits of replacing a fallow year with lentil produced the highest wheat production and profit when compared to the commonly used fallow-wheat sequence in Kazakhstan (Suleimenov et al. 2015). Replacing summer fallowing with annual cropping and the inclusion of lentil in rotation with wheat has added significantly to the economies of the regions where this has taken place. Also, the inclusion of an annual legume crop provides improved nutritional status of soils through the process of biological nitrogen fixation and the diversification of crops aids in disease and weed control.

8 Innovations in lentil cropping systems

Several innovations include planting lentil in various monocropping, double cropping, mixed cropping, intercropping and relay cropping systems to expand production. These systems are currently in widespread use in South Asia and were recently reviewed (Ali et al. 2009). These evolving systems have been responsible for expanded lentil production in previously unexploited situations. The exploitation of winter hardy lentil germplasm in developing varieties for cold highland areas of Turkey and the United States enables increased yields and more efficient use of soil and water resources.

Monocropping of lentil is the common practice where the crop is grown in rotation with cereal grains, mostly wheat or barley. Annual precipitation for successful lentil production ranges from 300 to 450 mm with barley grown at the lower rainfall average and wheat where there is sufficient rainfall (Ali et al. 2009). Temperatures are generally moderate in winter when most of the annual rainfall is received and hot in summers with diminishing rainfall as the crop develops toward dry seed maturity. The usual dry harvest period is an advantage for developing high-quality seeds and facilitates ease of harvest and threshing.

Double cropping in South Asia is becoming a common practice after harvest of the so-called *kharif* crops that are grown during the rainy season and include rice, maize, cotton, millet, sorghum, groundnut and possibly others. After the *kharif* season rice crop, farmers often leave the land fallow. However, the expanding practice of broadcasting lentil into harvested rice fields takes full advantage of available moisture and nutrients not used by the previous crop to produce an added crop of lentil in winter in India's eastern and northeastern neighbouring countries, Nepal and Bangladesh (Ali et al. 2009). Since pulses also have biological nitrogen fixation ability, the practice has the added benefit of improving soil nutrient status. Therefore, the practice is an enhancement of sustainable rice-based production systems prevalent in the region.

Mixed cropping is considered as an insurance against adverse biotic or abiotic stresses (Ali et al. 2009). In South Asia lentil seed is mixed with seeds of other crops such as wheat, mustard, barley or chickpea and broadcast sown. Yields of lentil were most adversely affected when mixed with wheat. In Turkey, mixtures of 70–80% lentil + 20–30% wheat gave the highest land equivalent ratio (LER) for straw and grain yields (Ciftci and Ulker 2005). A study of mixed cropping of lentil with five spring-sown companion crops including naked barley, wheat, oats, linseed and buckwheat was compared to monocropping of lentil in Germany (Wang et al. 2012). The LER was generally higher in mixed cropping than monocropping. Lentil–wheat and lentil–barley mixed cropping with a ratio of 3:1 resulted in the highest LER (ca. 1.50); lentil–linseed had the lowest LER for all ratios (Wang et al. 2012).

Intercropping systems use a system of distinct rows of lentil interspaced with rows of other crops in a predefined pattern. Common intercrops used in South Asia include wheat, barley, mustard and linseed (Ali et al. 2009). Studies have shown certain configurations to have greater productivity when compared to lentil alone. Intercropping with wheat and barley has not been particularly productive but intercropping with mustard and linseed has had positive returns (Ali et al. 2009).

Relay cropping is a common practice in South Asia, where lentil is sown into standing rice crops 10–15 days before rice harvest (Ali et al. 2009). The lentil crop develops using residual moisture and nutrients not taken up by the rice crop, thus providing an additional crop and utilizing unused inputs. Intercropping and relay cropping has shown potential for increasing monetary returns to farmers (Akter et al. 2004).

Winter lentil has shown great yield potential in cold highland areas. The development is based on germplasm originating in Turkey (Hamdi 1996; Spaeth and Muehlbauer 1991; Erskine et al. 1981) and has been used to develop winter hardy varieties for production in areas generally considered too cold and harsh for survival over winter. For such areas, breeders have developed varieties showing significant yield increases over traditionally grown spring types (Aydogan 2004). In addition, Turkish germplasm was used to develop a winter hardy variety for the U.S. Pacific Northwest, where significant yield advantages over spring type varieties were shown (Muehlbauer and McPhee 2007).

8.1 Weed control

Lentil plants are somewhat weak and minimally competitive with many grassy and broadleaf type weeds that commonly infest farmers' fields. Weed control in traditional farming systems is accomplished by hand weeding and sometimes by using the weeds as feed for livestock. In some dry areas, particularly Morocco, lentil is sometimes planted in widely spaced rows to facilitate mechanical cultivation between the rows (personal observation).

Control of weeds with herbicides is becoming a more widespread practice and includes the use of pre-plant soil incorporating herbicides that act to suppress germination and emergence of certain broadleaf and grass type weeds. Post-plant surface-applied pre-emergence herbicides are also used in production systems in developed countries. Herbicides to use in these applications are constantly changing and rates and methods of application are also evolving. Therefore, actual herbicides, rates of application and timing is not mentioned; however, a potential user will need to consult current recommendations and actual clearances prior to choosing a herbicide regime for control of weeds in lentil. For additional information see Yenish et al. (2009).

8.2 Harvesting

Harvesting lentil is a time-consuming and labour-intensive operation in many of the lentil-producing countries where hand harvesting continues to be the practice in many countries of the Middle East, North Africa and South Asia. The hand-pulled lentil plants dry in the fields and are collected and taken to an area where the lentils are threshed using animal-drawn implements or are fed into stationary threshers (Diekmann and Al-Saleh 2009).

Lentil harvest in North America and Australia is fully mechanized and generally consists of a swathing operation when the crop has reached physiological maturity (Diekmann 2009). The swathing operation begins when the crop is considered to be in prime quality condition and places the crop in 'wind-rows' to facilitate drying. After drying for 7–10 days depending on weather conditions the swathed crop is picked up from the 'wind-rows' by a combine and threshed to complete the harvest operation. However, direct combining of lentil is becoming more widely used in production areas of North America.

8.3 Drying and storage

Moisture content of lentil seed should be adjusted to 9–10% for safe storage (Ali et al. 2009). In most cases of lentil production, the moisture content will be at that level due to production in ordinarily dry areas. However, there are regions where this may not be the case requiring some form of artificial drying to ensure safe storage of the seed. When applying heated dryers, the temperature should be maintained at or below 43°C and the air movement should reach all the seeds uniformly. Drying of the seed is critical for maintaining viability for sowing and crop emergence.

9 Conclusion and future trends

Improving lentil production will require several complementary approaches including 1) choosing varieties for recommendation to farmers that are adapted to the region where they are intended to be grown; 2) improved availability of adequate supplies of high-quality seed possibly through development of local 'village'-based seed systems with the goal to provide affordable seed to farmers; 3) careful attention to best agronomic practices of land preparation, seeding rates and improved availability of planting equipment to ensure timely planting and good plant stands; 4.) improved weed control practices that reduce reliance on hand weeding and avoid plant injury; and 5) improving availability of harvesting equipment with the aim of reducing labour costs and providing a means of timely harvest of high-quality lentil crops.

10 Where to look for further information

The ICARDA website can provide additional references and discussions of research related to lentil agronomy at http://www.icarda.org/search/node/lentil%20Agronomy. The US industry website offers links to organizations that provide valuable information on lentil production and practices in the US. The website can be accessed at https://www.usapulses.org/resources. Similarly, the website of the Canadian pulse crop industry can be accessed for a wealth of information in lentil agronomy at http://www.pulsecanada.com. Also, the Australia pulse crop industries may be accessed at http://www.pulseaus.com.au. Extension Services can also be contacted for additional and location specific information on lentil agronomy and sources of seed adapted to particular areas.

11 References

Acikgoz, N., Karaca, M., Er, C. and Meyveci, K. 1994. Chickpea and lentil production in Turkey, pp. 388–98. In Muehlbauer, F. J. and Kaiser, W. J. (Eds), *Expanding the Production and Use of Cool Season Food Legumes*. Kluwer Academic Publishers: Dordrecht, The Netherlands, p. 989.

Akter, N., Alim, Md. A., Mahbulbul Islam, M., Naher, Z., Rahman, M. and Iqbal Hossain, A. S. M. 2004. Evaluation of mixed and intercropping of lentil and wheat. *J. Agron.* 3:48–51.

Ali, M., Singh, K. K., Pramanik, S. C., and Ali, M. O. 2009. Chapter 14. Cropping systems and production agronomy, pp. 213–28. In Erskine, W., Muehlbauer, F. J., Sarker, A. and Sharma, B. (Eds), *The Lentil: Botany, Production and Uses*. CABI: Wallingford, UK, p. 457.

Aydogan, A., Aydin, N., Kusmenoglu, I. and Karagoz, A. 2004. Improvement of winter hardy lentil varieties at high elevations of Turkey. *Tarla Bitkileri Merkez Araştırma Enstitüsü Dergisi*. 13: 1–3.

Bushaw, Z. and van Gastel, A. J. G. 2008. ICARDA's Seed-Delivery Approach in Less Favorable Areas Through Village-Based Seed Enterprises: Conceptual and Organizational Issues. ICARDA's Seed-Delivery Approach in Less Favorable Areas Through Village-Based Seed Enterprises: Conceptual and Organizational Issues. Published online: 11 Oct 2008.

Ciftci, V. and Ulker, M. 2005. Effect of mixed cropping lentil with wheat and barley at different seeding ratios. *J. Agron.* 1:1–4.

Cokkiygin, A. and Shtaya, M. J. Y. 2013. Lentil: Origin, cultivation techniques, utilization and advances in transformation. *Agric. Sci.* 1:55–62.

Dahmardeh, M. 2013. Intercropping barley (*Hordeum vulgare* L.) and lentil (*Lens culinaris* L.): yield and intercropping advantages. *J. Agric. Sci.* 5:209–13.

Diekmann, J. and Al-Saleh, Y. 2009. Chapter 16. Mechanization, pp. 248–61. In Erskine, W., Muehlbauer, F. J., Sarker, A. and Sharma, B. (Eds), *The Lentil: Botany, Production and Uses*. CABI: Wallingford, UK, p. 457.

Erskine, W., Meyveci, K. and Izgin, N. 1981. Screening a world lentil collection for cold tolerance. *LENS Newsletter* 8:5–8.

FAOSTAT. 2016. Food and Agriculture Organization of the United Nations Statistics Division. Rome.

Hamdi, A., Küsmenoĝlu, I. and Erskine, W. 1996. Sources of winter hardiness in wild lentil. *Genet. Resour. Crop Evol.* 43:63–7.

Kumar, S., Singh, R. G., Piggin, C., Haddad, A., Ahmed, S. and Kumar, R. 2011. *No-till Lentil: An Option for Profitable Harvest in Dry Areas*. ICARDA Publication: Grain Legumes No. 57: Aleppo, Syria.

Mahler, R. L. 2015. *Northern Idaho Fertilizer Guide, Lentils*. University of Idaho, Current Information Series 1083, p. 4.

Mahler, R. L., Saxena, M. C. and Aeschlimann, J. 1988. Soil fertility requirements of pea, lentil, chickpea and faba bean, pp. 279–89. In Summerfield, R. J. (Ed.), *World Crops: Cool Season Food Legumes*. Kluwer Academic Publishers: Dordrecht, The Netherlands, p. 1179.

Muehlbauer, F. J. and McPhee, K. E. 2007. Registration of 'Morton' winter hardy lentil. *Crop Sci.* 47:438–9.

Muehlbauer, F. J., Kaiser, W. J., Clement, S. L. and Summerfield, R. J. 1995. Production and breeding of lentil. *Adv. Agron.* 54:283–332.

Muehlbauer, F. J. and Slinkard, A. E. 2000. Regional reviews: Region 1: North America, pp. 67–70. In Knight, R. (Ed.), *Linking Research and Marketing Opportunities for Pulses in the 21st Century*. Kluwer Academic Publishers: Dordrecht, The Netherlands, p. 701.

Muehlbauer, F. J., Summerfield, R. J., Kaiser, W. J., Clement, S. L., Boerboom, C. M., Welsh-Maddux, M. M. and Short, R. W. 1998. *Principles and Practice of Lentil Production*. ARS-141. p. 20.

Rahman M. M. and Miah, A. A. 1989. Study on seedling mortality and plant competition of lentil at a wide range of plant population densities. In *Research Report on Pulse Agronomy*. Pulse Research Center, Bangladesh Agricultural Research Institute: Joydebpur, Gazipur, Bangladesh, pp. 27–33.

Saint-Clair. 1972. *Responses of Lens esculenta Moench to Controlled Environmental Factors*. Mededelingen Landbouwhogeschool: Wageningen, Nederland, Publication No. 72–12, pp. 1–84.

Sarker, A. and Kumar, S. 2011. *Lentils in Production and Food Systems in West Asia* and *Africa*. ICARDA Publication: Grain Legumes No. 56: Aleppo, Syria.

Saskatchewan Pulse Growers. 2016. Inoculation and fertility. http://saskpulse.com/growing/lentils/inoculation-and-fertility/

Spaeth, S. C. and Muehlbauer, F. J. 1991. Registration of three germplasm of winterhardy lentil. *Crop Sci.* 31:1395.

Suleimenov, M., Kiyas, A. and Kaskarbayev, Z. 2015. Pulses as replacement of summer fallow in semiarid steppes of Northern Kazakhstan. *International Journal of Agricultural Policy and Research* 3:351–8.

Urbano, G., Porres, J. M., Frias, J. and Vidal-Valverde, C. 2007. Chapter 5. Nutritional value. In Yadav, S. S., McNeil, D. and Stevenson, P. C. (Eds), *Lentil: An Ancient Crop for Modern Times*. Springer: Dordrecht, The Netherlands.

Wang, L., Gruber, S. and Claupein, W. 2012. Optimizing lentil-based mixed cropping with different companion crops and plant densities in terms of crop yield and weed control. *Org. Agric.* 2:79–87. doi:10.1007/s13165-012-0028-5.

Wilson, V. E. and Teare, I. D. 1972. Effects of Between- and Within-row Spacing on Components of Lentil Yield. *Crop Sci.* 12:507–10.

Witcombe, J. R., Devkota, K. P. and Joshi, K. D. 2010. Linking community-based seed producers to markets for a sustainable seed supply system. *Exp. Agric.* 46:425–37.

Yadav, S. S., McNeil, D. L., Andrews, M., Chen, C., Brand, J., Singh, G., Shivakumar, B. G. and Gangaiah, B. 2009. Chapter 13. Soil nutrient management, pp. 194–212. In Erskine, W., Muehlbauer, F. J., Sarker, A. and Sharma, B. (Eds), *The Lentil: Botany, Production and Uses*. CABI: Wallingford, UK, p. 457.

Yenish, J. P., Brand, J., Pala, M. and Haddad, A. 2009. Chapter 20. Weed Management, pp. 326–42. In Erskine, W., Muehlbauer, F. J., Sarker, A. and Sharma, B. (Eds), *The Lentil: Botany, Production and Uses*. CABI: Wallingford, UK, p. 457.

Chapter 5

The use of marker-assisted selection in developing improved varieties of soybean

Y.-C. Lee, R. Lemes Hamawaki, V. Colantonio, M. J. Iqbal and D. A. Lightfoot, Southern Illinois University, USA

1 Introduction
2 Genomes as intellectual property
3 Methods for MAS
4 Marker development
5 Choice of markers
6 Identification of polymorphism
7 Genetic and association map development
8 Marker-assisted recovery of recurrent parent genome
9 MAS in recurrent cross populations
10 Scoring of phenotypes
11 Use of molecular markers to select desirable traits in soybean
12 Future trends and conclusion
13 Where to look for further information
14 References

1 Introduction

Soybean geneticists and breeders aim to improve harvestable yield, reduce crop losses and reduce crop inputs by re-assortment among favourable haplotypes (Iqbal and Lightfoot 2004; Stefaniak et al. 2005; Lightfoot 2008, 2015) and the incorporation of new genes for new traits (Anand 1983; Arelli et al. 1994; Prabhu et al. 1999; Luckew et al. 2013). Rapid genetic gains can be made by incorporating genes for resistance to biotic and abiotic stresses. The positive effect of stress resistance is so reliable that journals rarely accept yield data for publication (Hnetkovsky 1994; Hnetkovsky et al. 1996). However, each stressor has an economic threshold, a point at which significant yield loss occurs (Gibson 1994), which is often contentious. For each stress it is important to determine the threshold and use resistance loci only when the threshold is exceeded. The reason for this

http://dx.doi.org/10.19103/AS.2017.0023.24

is the yield reduction caused by the incorporation of resistance genes in cases of absence of disease or other stressors.

DNA markers are used to select for alleles (forms) of genes that provide beneficial tendencies. Marker-assisted selection (MAS) with DNA markers allows for more rapid focus on the best genotypes (Lightfoot 2015). Molecular mapping and MAS with DNA markers for soybean have been costly and labour-intensive endeavours until recently (Prabhu et al. 1999; Luckew et al. 2013). Therefore, stress resistance traits with clear, easily scored phenotypes are often selected without markers or replications across years or multiple environments. However, traits with a bigenic or oligogenic inheritance pattern can be efficiently selected with DNA markers (Njiti et al. 2001; Kassem et al. 2006). Traits with a low-to-moderate heritability due to significant interactions with the environment are most efficiently both selected and prioritized by rank with DNA markers (Kassem et al. 2006; Luckew et al. 2013).

Many marker systems are available in crops (Shultz et al. 2007). Microsatellites have been used extensively over the past decades as the marker of choice for most breeders. They are co-dominant, and so can be used to identify (but not patent after the Myriad and Prometheus Supreme Court decisions) new loci, study dominance, identify hybrids, track contamination among lines, identify cultivars and select alleles. They are very abundant in all genomes, and if both short (type 1) and long (type II) microsatellites are considered, they approach one potential marker per 5–10 kbp. Microsatellites tend to show a high frequency of polymorphism, even among pairs of closely related cultivars. They have more than two alleles which makes them useful in many populations and for the selection of specific alleles. Since microsatellites use standard polymerase chain reaction (PCR) conditions and gel or capillary separation techniques that are scalable from dozens to thousands, they are both economical and technically easy to use. Microsatellite and other DNA markers will increasingly be used for selective genotyping approaches, including genotype ranking (indexing) based on partial phenotype data combined with favourable allele data, or bulked segregation event (segregant) analysis (BSA) (Darvasi and Solller 1992, 1994; Michelmore et al. 1991).

Single-nucleotide polymorphisms (SNPs) markers for major disease resistance loci have become increasingly important in soybean improvement (Meksem et al. 2001a; Afzal et al. 2013). The assay methods have become cheaper and more widely available. The Fluidigm approach has been widely adopted for TaqMan like assays. More recently, genotype by sequencing has become very common (Han et al 2015). However, SSRs (simple sequence repeats) still have clear uses.

Examples of the methods for developing and using DNA markers derived from genomic sequences are presented for monogenic, oligogenic and polygenic traits. Examples of successful mapping, fine mapping and gene isolation are given. When combined with high-throughput methods for genotyping and genome sequence, the use of association mapping with microsatellite markers will provide critical advances in the analysis of crop traits.

2 Genomes as intellectual property

A major use of DNA markers in soybean is to define regions encompassing resistance genes that can be used to claim and protect intellectual property (IP) (Webb et al. 1995, 1996; Hague et al. 2007; Lightfoot and Meksem 1997, 1999, 2010). Early patents in this area were quite broad, claiming large tracts of genome because they contained a single resistance gene. Recent actions by the patent office try to restrict the claims to single markers for single genes, making this type of patent more difficult and expensive to obtain. Some companies holding broad early patents have sought to broaden the claims still further to encompass selection for any locus in the region where a stress resistance gene is located (Webb et al. 1995, 1996, 2000, 2003). Here they seek to limit 'unintentional' selection of linked traits. However, the Supreme Court has ruled against such patents. Myriad (BARC1 and BARC2 rights owner) has rights over cDNA but not genes at those loci now. The implication for crops such as soybean is natural, genetic variations in genes cannot be patented. Therefore, each new project should start with due diligence, an examination of IP related to the trait and regions to be selected.

3 Methods for MAS

Screening for markers linked to a trait or recovery of recurrent parent genotype (RPG) can be carried out with leaf or seed samples. A high-throughput method of MAS is described (Fig. 2; Triwitayakorn et al. 2005; Ruben et al. 2006). Briefly, DNA can be isolated from plants or seeds. For sample collection from plants, a 1-cm^2 leaf disc from a young leaf was removed and stored in small notepads and kept on ice or dry ice. Alternately, seeds are placed on small filter paper discs soaked with sterile distilled water for 48 h for germination. About 1–2 mm of the tip of the radicals are carefully removed with sharp scalpel blades and the seeds are stored at 4°C until selections are ready. For dry seeds a small drill is used to obtain cotyledon tissue from a 1–2 mm hole in the seed. Leaf discs, radical tips or cotyledon dusts are placed in 96- or 384-well plates. Samples can be ground either by a matrix mill in the presence of 20–100 μl of 0.5 N sodium hydroxide or by 96- or 384-pin crusher (Fig. 2). After crushing, DNA is neutralized by an equal amount of acidic Tris. Other methods are popular including CTAB and propanol precipitations (see Xin et al. 2003). Alternately, a small sliver of seed leaf or radical tissue (0.1–0.2 mg) can be placed directly into the PCR if appropriate direct PCR (Phire) polymerases are used.

Once the DNA is isolated, a PCR with the selected primers is carried out in 96- or 384-well plates. Products can be electrophoresed in agarose gels when alleles differ by >3 bp (Fig. 3) or acrylamide gel or capillaries if the difference is 1–3 bp. For a higher throughput, SSR or SCAR primers can be labelled with fluorescence tags and multiplexed in the PCR or at the electrophoresis stage. The samples can then be electrophoresed using capillary- or gel-based DNA sequencers and analysed by fragment analysis software as provided by the fragment analysis equipment used. The shortcomings of gel electrophoresis can be avoided by developing TaqMan™ allelic discrimination (Fig. 1), or high-resolution melt (HRM) methods (Yuan et al. 2008) to polymorphisms within the microsatellite amplicons. Also considered for use are MALDI-TOF, PCR-OLA, molecular beacons, padlock probes and well fluorescence assays.

4 Marker development

For most crops there is a continued distributed community effort to add new markers to the existing linkage maps. The markers of choice include microsatellites and SNPs (Fig. 1–3; Song et al. 2004; Shultz et al. 2006a; Wu et al. 2010; Luckew et al. 2013). Many efforts have

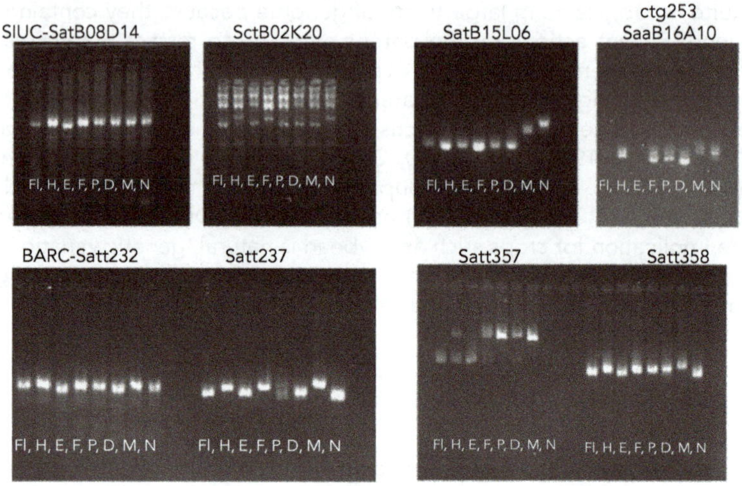

Figure 1 Screens for polymorphisms among soybean germplasm with BES-SSR markers. Eight Population Parents Flyer (Fl), Hartwig (H), Essex (E), Forrest (F), Pyramid (P), Douglas (D), Minsoy (M) and Noir1 (N) classified by their genotypes comparing SIUC- BES-derived and BARC-SSR microsatellite markers.

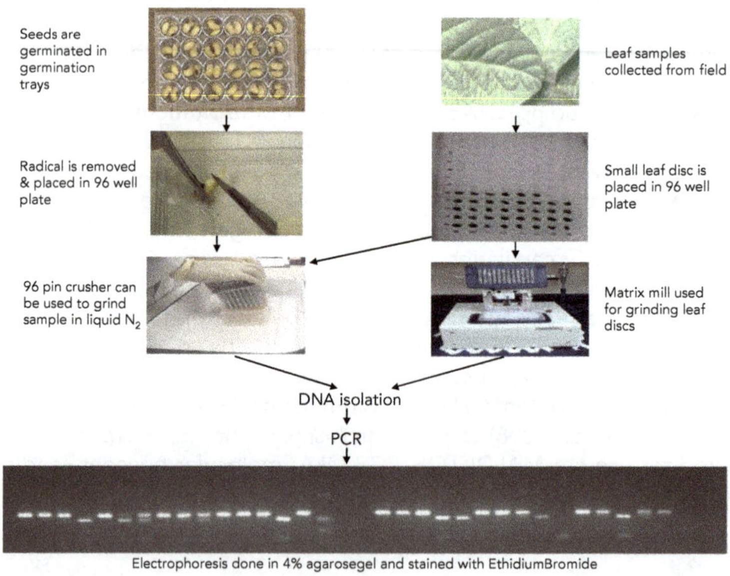

Figure 2 Steps involved in MAS. Gel electrophoresis can be replaced by other capillary-based electrophoresis techniques, a fluorescent-based assay such as TaqMan or invader or MALDI-TOF can be used for scoring polymorphism at the target allele.

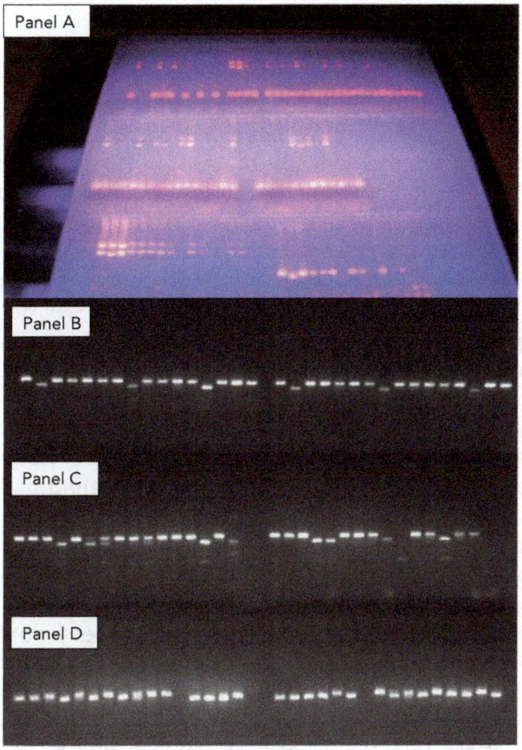

Figure 3 Segregation of SSR locus, Satt in a population. Individuals can be clearly defined as carrying the alleles of either parent 1 (*P1*), parent 2 (*P2*) or heterozygous. The 10-μl PCR was electrophoresed on a 4% metaphore agarose gel and stained with ethidium bromide. DNA was isolated from radical tissue in a 96-well plate. PCR was carried out in a 96-well plate and samples were loaded on gel with an 8-channel pipette. Panel A: A typical gel. Panel B–D. SattB54L22 scored among 92 individuals segregating for resistance to biotic stress. First two bands in A and last 2 in C were the parental DNAs.

been focused on regions that do not have many markers. Current work relies heavily on genome sequences, partially or fully re-sequenced genomes and physical maps (Wu et al. 2004; Shultz et al. 2006b; Schmutz et al. 2010; Lam et al. 2010; Wu et al. 2010; Han et al. 2015). In the future, development of new markers will rely on the genome sequences of many cultivars and the haplotype maps that will be developed from them (Zhu et al. 2003; Wu et al. 2010; Lam et al. 2010).

For map development renewable genetic resources will be necessary. There are many renewable collections of recombinant inbred lines (RILs). However, most lie in private hands and are not widely available. There is a community-wide effort to release certain key populations. For example, two soybean populations of 100 lines derived from the crosses of Essex and Forrest (Lightfoot et al. 2005) and Flyer by Hartwig (Kazi et al. 2007) were released. In addition, there are immortal collections of near isogenic line pairs available, with each pair capturing the small heterogeneous regions derived from a single RIL (Njiti et al. 1998, 2003; Lee et al. 2016). Similar resources will be released in many crops over the next decades as the industry pivots from QTL selection to gene identification and selection.

5 Choice of markers

The choice of marker depends upon the objective of the soybean selection programme, tools available to the individual/group researcher, markers available for a particular region or trait, and the information content associated with a particular marker system. Linkage disequilibrium must be considered. Linkage blocks can vary by over 100-fold (Zhu et al. 2003; Afzal et al. 2012). Hence comparatively distant markers work well in some regions but not others.

The first marker system to be used for soybean was the restriction fragment length polymorphism (RFLP), based on hypo-methylated genomic regions. RFLPs can be adapted to high-throughput markers by determination of the SNP detected (Schmidt et al. 2003). However, the paleopolyploid origin of soybean genome almost always leads to the detection of multiple DNA fragments with all RFLP probes. Single RFLP probes have been used to detect up to 19 independent loci (Mansur et al. 1996). The multiplicity of RFLP loci can make the locus identity ambiguous. Other factors that prevent the use of RFLP in mapping and marker-assisted breeding are the low levels of polymorphisms observed (Shoemaker and Specht 1995) and the high cost of automating the procedure for high-throughput screening. However, the IP is now not owned due to the recent Supreme Court decision against Myriad LLC and Promestheus LLC that natural genetic variation in genomic DNA cannot be protected.

Microsatellite markers use DNA sequence data and PCR to show polymorphisms within 5–9 (class I) or 10–20 (class II) simple tri- or di-nucleotide repeat motifs (motifs up to octo-exist but get rarer with size). These class II microsatellite markers have been shown to be highly polymorphic (Fig. 2; Akkaya et al. 1992; Morgante and Olivieri 1993; Shultz et al. 2007). Some highly polymorphic microsatellite loci can have more than 20 alleles (Maughan et al. 1995; Rongwen et al. 1995; Powell et al. 1996; Diwan and Cregan 1997). Such high levels of allelic diversity increase the possibility of detecting polymorphisms between parents of populations derived from the hybridization of adopted soybean genotypes.

Unfortunately, class II SSRs are rare (one per 20–50 Kbp) (Marek et al. 2002). Further, evidence of clustering has emerged from denser maps (Akkaya et al. 1995; Kassem et al. 2006; Schmutz et al. 2010; Wu et al. 2010). Analysis of genomic DNA sequences have shown the shorter class I microsatellites that contain less than ten simple tri- or di-nucleotide repeats (small simple sequence repeats or SSSRs) are more common than SSRs and are frequently polymorphic in soybean and other crops (Fig. 3; McCouch et al. 2002; Rota et al. 2005; Shultz et al. 2007). There are about ten to twenty such DNA markers (as defined) per 100 Kbp of soybean sequence. Class I markers can overcome the problem of marker clustering because of their abundance in the genome.

In addition, complex microsatellites (SSSRs and selectively amplified microsatellite polymorphic loci (SAMPL)) that contain two different tri- or di-nucleotide repeats are very common and are frequently polymorphic in the soybean genome (Witsenboer et al. 1998; Shultz et al. 2007). Selection for class I and II SSRs, SSSRs and SAMPLs in soybean can be made with standard gel or capillary electrophoresis (Cregan et al. 1999a; Prabhu et al. 1999), or electrophoresis-free assays. Therefore, microsatellite markers are currently the method of choice for MAS in most public university laboratories. This is based on their distribution in the genome, the polymorphism frequency and information content attached to each identified locus. The relative ease of automation for high-throughput screening programmes is another bonus.

SNP are single base pair changes (or 1–3 bp indels) between individuals or genotypes. They occur about twice as often in non-coding DNA compared to coding DNA. In coding

DNA, less than one-quarter of SNPs alter amino acid sequence, but this depends on the genes examined (Meksem et al. 2001b; Zhu et al. 2003; Ruben et al. 2006; Lam 2012). The observed SNP frequency in soybean is about 0.05–0.3% (Zhu et al. 2003). However, genes under intense selection, such as the major resistance genes *Rhg1*, *Rhg4* (resistance to *Heterodera glycines*), *Rfs1* and *Rfs2* (resistance to *Fusarium solani* f. sp. *glycines*) are encoded by 6–8 alleles as defined by SNPs (Meksem et al. 1999; Hauge et al. 2001; Triwitayakorn et al. 2005; Ruben et al. 2006; Liu et al. 2012) but only 3–4 alleles as defined by microsatellites (Afzal et al. 2012) or copy number variations (Cook et al. 2012), reflecting the extra SNP diversity imported by gene introgression of large genomic regions from non- and semi-domesticated soybeans (Han et al. 2015). Selection for SNPs in soybean can be made with standard gel or capillary electrophoresis, or electrophoresis-free assays such as TaqMan (Fig. 1; Meksem et al. 2001a,b), HRM, microarrays and MALDI-TOF. It has been observed that the SNP frequency in the amplicons of microsatellites and AFLP bands is tenfold higher than other regions of the genome (Meksem et al. 2001). Therefore, SNPs, AFLPs and microsatellites can be inter-converted if needed.

Deletions and insertions (indels) are multi-base pair changes between soybean genotypes (Han et al. 2015). They occur in non-coding DNA more often than incoding DNA. They are the basis of microsatellite and satellite polymorphisms. In coding DNA they alter amino acid sequence (Meksem et al. 2001b). The observed frequency in crops is about fivefold lower than for SNPs (Meksem et al. 2001a,b; Han et al. 2015). Some genes under intense selection like the major resistance gene *Rhg1* bear indels in introns for one of nine alleles (Meksem et al. 1999; Hauge et al. 2001; Ruben et al. 2006) in the 5' non-translated region and in both the 5' and 3' enhancer regions. Selection for indels can be made by non-PCR methods (Landegren et al. 2001) and isothermic assays like Invader (Mein et al. 2000). EnviroLogix assays have made significant advances in presence- and absence-type assays for GMO and major indels. The standard PCR-based gel or capillary electrophoresis, or electrophoresis-free assays such as TaqMan (Meksem et al. 2001a,b) or MALDI-TOF MS (Chen et al. 2003) can also be used (Fig. 1).

Among the electrophoresis-free assays, the TaqMan is one that has been efficiently used due to both the elimination of human errors and the fast and easy interpretation of output files (Meksem et al. 2001). This analysis is based on the use of two probes that have a quencher attached to the 3'-end and a fluorophore to the 5'-end. The quencher has the function of preventing the fluorescence of the fluorophore attached to the other end. In the process of DNA amplification, a TaqMan probe will hybridize to the targeted DNA region and when the *Taq* polymerase extends from the primer through the probe, the quencher and the fluorophore will be released from the probe. Thus, since the fluorophore is no longer close to the quencher, it emits a fluorescence and the signal is detected by a quantitative PCR thermal cycler. Currently, the technique is widely used to rapidly identify polymorphisms in SNP genotyping. TaqMan markers are often used in a very low-cost, miniaturized format using Fluidigm technology. KASPR is another format for single SNP selection with growing popularity (Kadam et al. 2016).

6 Identification of polymorphism

The first step in applying MAS to a soybean breeding programme is the identification of polymorphic loci between the parents that can be traced in the segregating population. The frequency of polymorphism is related to genetic distance between the parents and

Table 1 BES-SSR-derived polymorphic markers in ExF with their Williams 82 genome sequence contig series and numbers of different amplicons found among Essex Forrest Williams 82, Pyramid, Douglas, Minsoy and Noir1

Marker	Contig no. (V4)	Amplicon no.	Position on the chromosomes (bp)
SIUC_B10A18 SSR	9204	2	Gm02(47,894,592-47,894,411)bp
SIUC_B02B24 SSR	9102	2	Gm03(1,666,783-1,667,133)bp
SIUC_B08G14 SSR	9376	2	Gm01(54,075,573-54,074,935)bp
SIUC_B10P12 SSR	9036	4	Gm05(29,672,571-29,673,086)bp
SIUC_B08D14 SSR	9376	4	Gm12(4,290,147-4,290,380)bp
SIUC_B02K20 SSR	9102	2	Gm14(28,277,772-28,277,620)bp
SIUC_B13L17 SSR	9092	4	Gm02(20,925,818-20,924,917)bp
SIUC_B20C11 SSR	9320	6	Gm19(40,493,550-40,493,124)bp
SIUC_B20B01 SSR	9345	3	Scaffolds(4648-4240)bp
SIUC_B20K08 SSR	9320	1	Gm03(32,242,899-32,243,020)bp
SIUC_B27B08 SSR	9319	4	Gm11(36,595,894-36,596,083)bp
SIUC_B31L10 SSR	9201	5	Gm16(19,478,522-19,478,812)bp
SIUC_B35F13 SSR	9212	1	Gm08(5,918,300-5,917,909)bp
SIUC_H32G20 SSR	9284	1	Gm05(34,149,378-34,150,013)bp
SIUC_B35H07 SSR	8227	1	Gm01(34,958,752-34,958,988)bp
SIUC_B15I12 (or 21)	8131	2	Gm13(32,724,487-32,724,006)bp
SIUC_H100B10a SSR	8073	2	Gm20(17,813,803-17,814,735)bp
SIUC_B24H14 SSR	8128	2	Gm10(39,117,270-39,116,942)bp
SIUC_B32L16 SSR	8047	2	Gm10(31,284,953-31,284,711)bp
SIUC_B15J11 SSR	8131	6	Gm08(2,439,456-2,440,0640)bp
SIUC_B32K10 SSR	3195	4	Gm07(41,730,740-41,731,134)bp
SIUC_B13G15 SSR	2562	2	Gm13(23,414,927-23,415,175)bp
SIUC_B12C13 SSR	2475	1	Gm03(204,437-203,862)bp
SIUC_B29K15 SSR	2894	5	Gm07(14,767,796-14,767,619)bp
SIUC_B15B14 SSR	1632	4	Gm16(36,645,370-36,645,044)bp
SIUC_B31I19 SSR	1668	1	Gm02(13,639,758-13,639,455)bp
SIUC_B30A11 SSR	1536	6	Gm20(9,356,698-9,356,793)bp
SIUC_B14L17 SSR	1246 (V 3)	1	Gm18(306,561-306,380)bp
SIUC_B31B02 SSR	899	1	Gm16(11,256,058-11,255,731)bp
SIUC_H14J18 SSR	859 (v3)	1	Gm15(6,503,026-6,502,439)bp
SIUC_B21C09 SSR	810	4	Gm11(27,753,637-27,754,434)bp

Marker	Contig no. (V4)	Amplicon no.	Position on the chromosomes (bp)
SIUC_B15L05 SSR	628	4	Gm12(12,532,433-12,532,483)bp
SIUC_B15L06 SSR	628	2	Gm10(44,977,688-44,977,319)bp
SIUC_B15P23 SSR	244	3	Gm14(18,446,410-18,446,687)bp
SIUC_H32P20 SSR	187	1	Gm01(20,131,118-20,131,860)bp
SIUC_B30I21 SSR	175	1	Gm17(19,290,762-19,290,895)bp
SIUC_B30J24 SSR	175	6	Gm10(49,955,950-49,955,595)bp
SIUC_B34L17 SSR	167	3	Gm10(45,805,898-45,806,367)bp
SIUC_B34N08 SSR	167	3	Gm07(5,992,312-5,991,533)bp
SIUC_B29H12 SSR	142	1	Gm18(28,360,161-28,359,787)bp
SIUC_B17E19 SSR	158	6	Gm20(10,815,914-10,816,435)bp
SIUC_B31A06 SSR	137	4	Gm14(20,444,725-20,444,662)bp
SIUC-SAT138 SSR	Rhg1 ctg *	1	Gm08(41,986,016-41,986,075)bp
SIUC-SAT1 SSR	Rhg1 ctg *	1	Gm17(36,745,649-36,745,724)bp

* Ruben et al. 2006.

the type of marker used (Zhu et al. 2003). When low diversity or a paucity of markers is a problem, genome sequence-based marker identification is an efficient approach (Meksem et al. 1998; Cregan et al. 1999b; McCouch et al. 2000; Rota et al. 2005; Shultz et al. 2006, 2007; Lam et al. 2010; Schmutz et al. 2010; Wu et al. 2010). Screening of markers identified in genome sequences can be made bioinformatically. For instance, if two cultivars have been sequenced from different germplasm bases (e.g. Williams of northern US and Forrest of southern US), it is likely these markers will be polymorphic in other populations. Direct screening of markers without prior knowledge of polymorphism can be efficiently performed with 8–10 genotypes selected to encompass 75–90% of the crop variability (Fig. 1; Table 1). Monomorphic loci are discarded, with selection made for loci which are polymorphic in many genotypes (Shultz et al. 2007). BES-derived markers are named after the discovering institution (e.g. SIUC) and the BES from which they derive (e.g. B100B10). In a complete sequence they may be assigned a chromosome number (e.g. Gm18) and a base pair location (1 123 657) as identifiers.

7 Genetic and association map development

There is a wide variety of map-building and QTL analysis programmes available and many work well (Lander et al. 1996; Basten et al. 2001). At a minimum, the map should be built at a high LOD (logarithm of the odds) (4–5) then expanded at lower LODs to reduce transitive map errors. QTL should be detected by ANOVA, an interval map programme and a composite interval map programme. Map errors must be removed before IM and

CIM. An efficient error removal method is to look closely at the scores inferring double recombination events over short genetic distances.

Modern methods for early soybean selections based on an index weighted marker identified QTL for beneficial alleles tend to favour QTL at 2–3 fold over actual seed yields measured in diverse poorly replicated locations (Yuan et al. 2002; Lightfoot 2013). Phenotypes are measured with low replications in incomplete sets of environments during the normal breeding process. All that is required is a yield monitor and time to analyse data. Marker selections are efficiently carried out with 2–3 hundred microsatellite markers. Yield indexes so derived are highly predictive of seed yield in subsequent generations.

Across large germplasm collections, programmes such as TASSEL 3.0 work efficiently to identify QTL with about 5–30 thousand markers, a number easily achieved with genome sequence-based markers (Han et al. 2015). False associations may be identified within small groups of related germplasms. Many QTL of small effect are also discovered. The ability to detect more than two alleles can also present problems from over-partitioning. Associated loci with several alleles should be subject to a phylogeny analysis to determine relatedness among alleles that might allow reduction to two clades.

8 Marker-assisted recovery of recurrent parent genome

In plant breeding for resistance to stress, the accelerated backcross procedure was often used to transfer favourable alleles from donor genotypes into a recipient elite genotype (Hartwig and Epps 1973; Anand 1986). The donors often have many poor agronomic properties requiring several backcrosses to only partially eliminate them. The goal of this 'introgression' is to recover a transgressive segregation event (segregants) that yield more than the elite parent. To be worth the backcrossing effort yields must be increased by 1–2% per year for each year necessary to develop the introgression line. Only genes with a large effect on stress resistance can be selected in this way. In backcross gene introgression programmes, MAS has a great impact on increasing and accelerating the genome recovery of RPG (Tanksley et al. 1989). Individuals that are homozygous for the alleles of the recurrent parent at a large number of marker loci covering the entire genome are selected. Population size and marker density required in a background selection programme can be determined (Openshaw et al. 1994). The minimum recommended is the use of four markers per chromosome (80 markers spaced at ~200 cM) and a selection strategy for proximal recombinants of the target allele. The number of markers used for RPG recovery varies. About a 20-cM marker density represented an optimal trade-off between percentage recurrent parent recovery and management of data point throughput requirements (Frisch et al. 1999). Uniformly distributed markers are more effective than equal numbers per chromosome (Hospital et al. 1992), because smaller population sizes and fewer data points are required with equally spaced markers (Frisch et al. 1999). Therefore, selection of markers from a genome sequence or physical map rather than a less precise genetic map can improve the efficiency of RPG recovery.

9 MAS in recurrent cross populations

The majority of soybean breeding programmes make advances by re-assorting the entire genomes of two elite lines with favourable and complementary characteristics (Iqbal et al. 2001; Yuan et al. 2002; Kazi et al. 2008, 2009). The goal of shuffling the alleles was to recover transgressive segregants and epistatic interactions that yielded more than either parent. Again, the yield increase target has to be 1–2% per year for each year necessary to develop the line. Traditional recurrent selection cycles last from 5 to 7 years from crossing to release. The average cultivar is sold for 2–3 years before becoming obsolete. Therefore, marker selections that reduce development time, by stacking selections into a shorter time frame, will be very valuable. Further, since released cultivars appear to contain more recombination events than the population means, markers can be used to find well-shuffled genomes for phenotypic analysis (Stefaniak et al. 2005; Karangula et al. 2008).

MAS for stress resistance often competes with field scores, greenhouse assays and even low-cost laboratory assays of phenotypes, for price and effectiveness (Njiti et al. 1996, 2001, 2002). This competition is not always market driven, as investments in infrastructure made in the past determine economics in the present. Over time though, MAS has become widely adopted by most effective soybean breeders. MAS, as employed today, usually focuses on two to four genes per population with 25 to 6.25% of lines 'passing the test'. Samples are selected after preliminary visual selection for yield potential F3–F5. Rarely are F2 selections made, although early generation testing is being employed recently aiming to shorten the screening period while maintaining genetic gains (St. Martin and Geraldi, 2002). Hence, about 10% of a programme's recombinants will be tested in the MAS lab. Since a single lab will serve dozens of breeders each with a hundred thousand lines, the task can quickly reach millions of selections per year. Automation is very important in this situation (Prabhu et al. 1999; Lightfoot 2008, 2015).

10 Scoring of phenotypes

In order to avoid garbage-in, garbage-out (GIGO) QTL maps, stress resistance traits must be measured under controlled and/or highly replicated assays with clear and easily scored phenotypes. Traits with a bigenic or oligogenic inheritance pattern are most efficiently scored on a percentage of infection or incidence scale (Njiti et al. 2001). Traits with a low-to-moderate heritability due to significant interactions with the environment are most efficiently selected in greenhouse assays, but only if the greenhouse assay can be shown to correspond to the field results (Kassem et al. 2006). In pathogen assays, the use of low doses of standardized inocula were found to be particularly important (Arelli 1994; Njiti et al. 2001; Luckew et al. 2013). Trait distributions should be analysed before mapping to determine skewness and kurtosis, both are indicators of major gene effects and epistasis. Major morphological effects such as maturity dates and determinacy must be measured and correlated with the disease resistance traits to attempt to avoid selection of loci underlying such confounding traits (Prabhu et al. 1999; Njiti and Lightfoot 2006).

11 Use of molecular markers to select desirable traits in soybean

Thousands of QTL for hundreds of agronomic traits were mapped in crops such as soybean. Databases contain gene class, locus, alleles, phenotypes and two-point data. The availability of all this information and commercial drive to release new and improved varieties assist the use of MAS for new and improved variety development.

11.1 Soybean cyst nematode

One example of MAS in soybean is the selection for resistance to soybean cyst nematode (SCN). SCN, *H. glycines*, is a small plant-parasitic roundworm that attacks the roots of soybeans and causes significant crop losses in the infected fields. SCN resistance was introgressed from non-adapted, weedy and semi-domesticated types by backcrossing. Resistance was associated with both a linkage drag on yield and a genome load affecting yield (Fig. 4; Karangula et al. 2008). Two QTLs significantly contributing to soybean resistance of *H. glycines*, *rhg1* and *Rhg4*, were mapped on linkage groups G and A2 (chromosome 18 and 8) and the candidate genes isolated (Ruben et al. 2006; Liu et al. 2011; Srour et al. 2012; Cook et al. 2012). Cultivars selected for *rhg1* provide good resistance to *H. glycines* Hg type 0 and 7 (race 3). SSR marker Satt309 was mapped at a 0.5–2 cM distance from *rhg*1 (Cregan et al. 1999). It was shown to be 42–150 Kbp in physical distance from the *rhg1* region that was multigenic (Ruben et al. 2006; Srour et al. 2012; Cook et al. 2012). Satt309 was widely used for MAS for resistance to SCN race 3 (Mudge et al. 2002), but until 2015, it was used only for research purposes because of patent positions held privately.

With the sequencing of a large region of linkage group G of the soybean genome by Monsanto (St. Louis, MO), and the SIUC patent for perfect markers for SCN (Meksem et al. 1999; Hague et al. 2007; Ruben et al. 2006; Lightfoot and Meksem 2010; Afzal et al. 2011; Srour et al. 2012; Cook et al. 2012, 2014), it is now possible to select the four alleles for

Figure 4 The effects of gene stacking on resistance to SDS. On the left side is a cultivar with a stack of six resistance genes. To the right is a cultivar with few or no resistance genes showing the chlorosis and defoliation symptomatic of SDS.

resistance from *Rhg*1 alleles with 100% certainty. There are microsatellite markers for the promoter region, an indel in the intron and four SNPs adapted to TaqMan in the leucine-rich repeat region of the receptor-like kinase at *rhg1*. In total twenty-six SNP markers are directly located on the gene associated with resistance to SCN race 3, six alter protein allotype but do not alter *cis*-regulatory elements. There are indels in the 5′ and 3′ enhancer regions. Selection for recombination events within the gene have been made possible by saturation mapping based on three genome sequences of this region (2 susceptible and 1 resistant cultivar). Recombinants tend to be susceptible (Ruben et al. 2006; Hemmati et al. 2017; Cook et al., 2012; 2014). Recently, Lakssassi et al (2017) showed the SNAP gene at Rhg1 was important for resistance. There are no stable transgenics yet.

11.2 Sudden death syndrome

A different example of MAS in soybean is the selection for resistance to sudden death syndrome (SDS). SDS in the United States is caused by a member of the *Fusarium* sp. disease complex, *F. virguliforme* (e.g. *solani* f. sp. *glycines*) that attacks the roots of soybeans and causes significant crop losses in the infected fields (Lightfoot 2015). SDS resistance was found among elite adapted types and was exploited by gene stacking during inter-crossing (Luckew et al. 2013). Resistance to SDS was associated with no linkage drag on yield (Fig. 5), in fact it is a disease of high-yield genotypes. Among several resistance genes associated with resistance to SDS of soybean (Table 1), five were clustered with SCN resistance *Rhg1* (Meksem et al. 1999; Iqbal et al. 2001). The first *Rfs2* was formally proven to be pleiotropic with part of *Rhg1* (Afzal et al. 2011; Srour et al. 2012). The second region had 3 clustered genes, *Rfs1*, *Rfs* and *Rfs19* were responsible for part of a bigenic root resistance to *F. virguliforme* and can be selected for by a set of BAC-derived SSRs SIUC-Satt122 (Meksem et al. 1998; Shultz et al. 2007), a RAPD-derived SCAR (SIUC-OI03) or a candidate gene sequence (Lightfoot and Meksem 2010; Yuan 2013). The third region was near the telomere linked to BARC-Satt163. Mutagenesis suggests this gene is MIPs1a (Lightfoot et al. unpublished).

The second SCN and SDS root resistance gene can be selected for by SSR Satt594 (Kazi 2006; Kazi et al. 2008, 2009) and was linked to, but not pleiotropic with, *rhg5* needed for race 14 of Hg Type 1.2.5- resistance. Since linkage was in repulsion, the microsatellite markers have been used to identify recombinants of value to breeding programmes.

Each of the 18–30 resistance loci discovered for SDS resistance has associated a SSR marker (Table 2; Luckew et al. 2013). Each locus has been detected in multiple populations. Therefore, locus pyramids have been developed incorporating up to 6 loci, and locus pyramids of 10–12 genes were produced (Luckew et al. 2012). Locus pyramids have been shown as an effective technique in eliminating SDS symptoms from fields. In an attempt to understand resistance mechanisms, expression fingerprint for resistance to SDS and SCN has been extended into the proteome (Iqbal et al. 2003, 2005, 2011, 2016; Afzal et al. 2009). The expression profile and associated protein changes will be used to develop very-high-throughput assays for resistance to SDS.

However, the SDS resistance success story in Illinois in the 1990s lead to complacency among breeders, to the north, that was misplaced (Sherm and Yang 1995; Lightfoot et al. 2008, 2015). The fungus was still present in the soil and was under strong selective pressure to adapt to the partially resistant germplasm. Consequently, the disease has become severe in Iowa, Indiana, Minnesota and the Dakotas, and is resurgent in Illinois by 2016. As resistance increased, the danger of selecting for pathogen races also increased. Further, a second pathogen that causes SDS in the continent of south America, *Fusarium tucumanes* may yet infest the US fields (Aoki et al. 2005) and cannot be controlled by the same resistance gene set (Bashir 2008).

Table 2 SDS QTL found in adapted germplasm and different RIL populations

Linkage group	Soybase QTL name (linked markers)[a]	Populations	Probable locus names[b]	References
A2	SDS11-1 (Satt 187)	Ripley x Spencer	*qRfs7*	Hashmi 2004
C2	SDS 1-1 (0005-250)	Essex x Forrest	*cqRfs4*	Hnetkovsky et al. 1996
	SDS 4-2 (K455_1)	Essex x Forrest	*cqRfs4*	Njiti et al. 1998
	SDS 7-5 (Satt 371)	Essex x Forrest	*cqRfs4*	Iqbal et al. 2001
	SDS 8-2 (Satt 307)	Pyramid x Douglas	*cqRfs4*	Njiti et al. 2002
	SDS 8-3 (Satt 316)	Pyramid x Douglas	*qRfs9*	Njiti et al. 2002
	SDS 9-2 (Satt277)	Flyer x Hartwig	*cqRfs4*	Kazi 2005 (this work)
D2	SDS11-2 (Satt 528)	Ripley x Spencer	*cqRfs11*[c]	Hashmi 2004
	SDS 9-1 (Satt574)	Flyer x Hartwig	*cqRfs11*[c]	Kazi 2005 (this work)
F	SDS10-1 (Satt 160)	Essex x Forrest	*qRfs10*	Kassem et al. 2006
G	SDS 6-1 (Bng 122_1)	Essex x Forrest	*cqRfs1*[c]	Meksem et al. 1999
	SDS 3-1 (0G13_490)	Essex x Forrest	*cqRfs3*	Chang et al. 1996
	SDS 6-2 (Satt 309)	Essex x Forrest	*cqRfs2*	Meksem et al. 1999
	SDS 8-4 (0G01)	Pyramid x Douglas	*cqRfs3*	Njiti et al. 2002
	SDS 7-1 (Satt 214)	Essex x Forrest	*cqRfs2*	Iqbal et al. 2001
	SDS 7-2 (Satt 309)	Essex x Forrest	*cqRfs2*	Iqbal et al. 2001
	SDS 7-3 (Satt 570)	Essex x Forrest	*cqRfs1*[c]	Iqbal et al. 2001
	SDS 7-4 (OEO2_1000)	Essex x Forrest	*cqRfs3*	Iqbal et al. 2001
	SDS 4-1 (0I03-512)	Essex x Forrest	*cqRfs1*[c]	Njiti et al. 1998
	SDS 4-3 (Bng122_1)	Essex x Forrest	*cqRfs1*[c]	Njiti et al. 1998
	SDS 5-1 (Satt 038)	Flyer x Hartwig	*cqRfs1*[c]	Prabhu et al. 1999
	SDS 8-1 (Satt 163)	Pyramid x Douglas	*cqRfs2*	Njiti et al. 2002
	SDS9-3 (Satt427)	Flyer x Hartwig	*cqRfs3*	Kazi 2005 (this work)
I	SDS 7-6 (Satt 354)	Essex x Forrest	*qRfs5*	Iqbal et al. 2001
J	SDS10-2 (Satt 285)	Essex x Forrest	*cqRfs8*	Kassem et al. 2006
	SDS(Satt183)	GC87018-12-2B-1 x GC89045-13-1	*cqRfs8*	Sanitchon et al. 2004
N	SDS 2-7 (0C01-650)	Essex x Forrest	*cqRfs6*	Chang et al. 1996
	SDS 8-3 (Satt 080)	Pyramid x Douglas	*cqRfs6*	Njiti et al. 2002
	SDS (Satt 631)	Ripley x Spencer	*cqRfs6*[d]	Hashmi 2004

[a]QTL numbers are from Soybase where assigned.
[b]From 27 QTL detections there were 15 QTL counting each detection in a separate population once. Assuming QTL detected in common intervals in separate populations represents the alleles there are 11–12 loci.
[c]QTL associated with resistance to root infection.
[d]QTL detected in common intervals in separate populations or derived NILs are considered confirmed and suffixed with c under Soybean Genetics Committee recommendations (http://soybase.agron.iastate.edu/nomenclature/QTL.html).

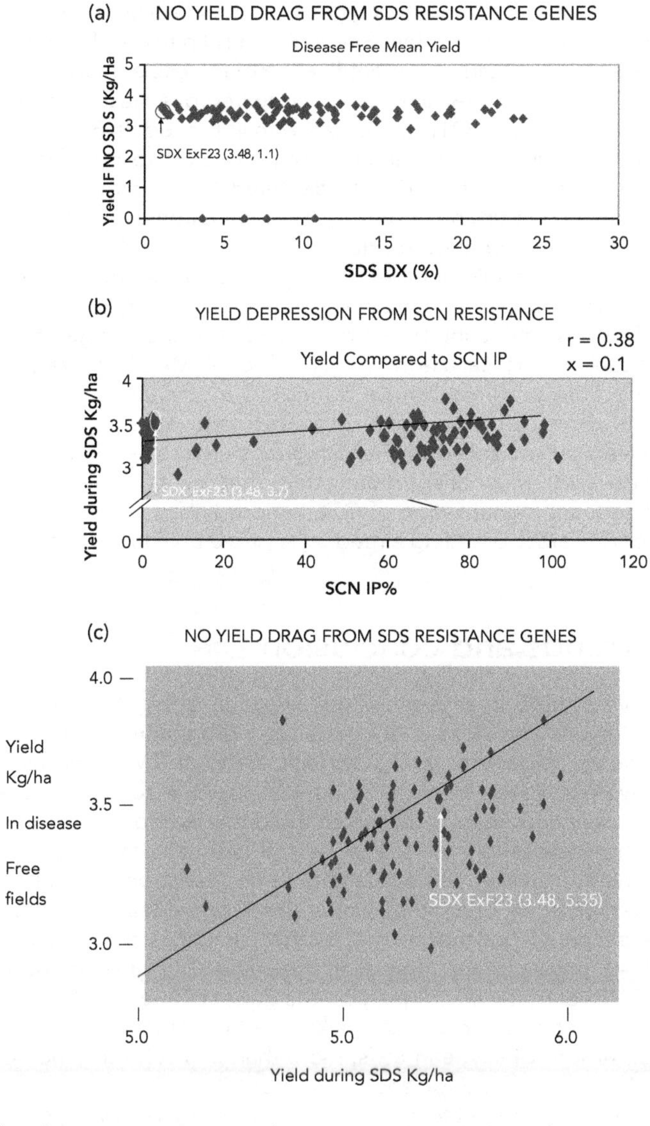

Figure 5 Yield drag from resistance to SCN but not SDS. (a) Correlation between resistance to SDS and yield. (b) Correlation between resistance to SCN and yield. (c) Correlation between yield during SDS infestation and yield in disease-free fields .

11.3 QTL for biological nitrogen fixation in soybean

Soybean is one of the most nitrogen (N)-requiring crops (Sinclair and Wit 1975). Removal of N in harvested soybean seeds is estimated to be around 280 lb per 80 bushels of soybean seeds (Salvagiotti et al. 2008). It is through the symbiotic association of this crop with the rhizobial bacteria that the plant obtains the major part of this N supply; on average the symbiosis contributes for 50–60% of the plant total N requirement (Salvagiotti et al. 2008).

Given the importance of this process to soybean yield and sustainability, breeding efforts have been pursued to improve this trait. Since biological nitrogen fixation (BNF) is highly affected by environmental factors, the identification of molecular markers for traits related to this characteristic might be a reliable approach to improve nitrogen fixation.

Presently, the identification of QTL associated with improved BNF have been addressed for the most part through the evaluation of traits related to nodulation. Using a RIL population (*n* = 157), Santos et al. (2013) identified QTL regions for nodule number, nodule weight/nodule number and shoot dry mass on chromosomes 15, 19, 11 and 20. Hwang et al. (2014) evaluated another RIL population (*n* = 97) in field conditions; they mapped several QTL for nodule number, average nodule weight, nodule size and total nodule weight. These are located in chromosomes 1, 4, 5, 6, 7, 8, 9, 15, 17, 18, 19 and 20. Although these findings have contributed to advance the knowledge of the host control over BNF in soybean, its practical use in breeding through MAS is still very limited. Among other reasons, this little progress can be attributed to the lack of information in what degree these traits affect yield in soybean. Therefore, innovative approaches that allow us to concurrently select for enhanced BNF and higher yield are needed. Ultimately, this will possibly include the assessment of agronomic traits that can indirectly affect or be affected by BNF. The subsequent identification of QTL regions linked to these traits will certainly serve as a tool to accelerate breeding aimed to improve BNF in soybean.

12 Future trends and conclusion

The main objective of MAS in soybean is the selection of highly desirable lines among thousands of genotypes and dozens of crosses per programme. Selections have to be made for monogenic, oligogenic and polygenic traits of low-to-moderate heritability. Selection can only be achieved by high-throughput, rapid, automated procedures for the detection of DNA polymorphisms attached to the desirable traits. With the availability of capillary electrophoresis-based genetic analysers, thousands of samples can be screened per day per machine. However, the availability of non-electrophoresis-based tools such as TaqMan (Landegren et al. 1998), the Invader-like assays (Mein et al. 2000) and MALDI-TOF for the detection of polymorphisms, further simplifies the screening procedures for large-scale MAS programmes. Moreover, the availability of gene markers or perfect markers such as SNPs or mutations directly on the genes of interest, derived from genome sequencing, transformation and mutation, makes the MAS procedures more efficient. With the availability of millions of soybean ESTs, more than 800 cultivar genome sequences, a physical map and many mutant populations (Lightfoot 2008, 2015), crops such as soybean have sufficient genomic resources for the rapid development of new markers for resistance to new biotic stresses. The application in new and improved varieties of marker-assisted breeding has become a widely accepted breeding tool. The next set of goals will be to maximize the benefits from these tools across the world.

13 Where to look for further information

Please visit the website http://ga4082@siu.edu for a complete list of BAC end sequence-derived microsatellite markers linked to a physical map and a reference genome sequence.

Be aware all genomes are not linear due to presence and absence, copy number and insertion variations.

14 References

Afzal, A. J., Srour, A., Hemmati, N., Saini, N., El Shemy, H. A. and Lightfoot, D. A. (2012). Recombination suppression at the dominant Rhg1/Rfs2 locus underlying soybean resistance to the cyst nematode. *Theor. Appl. Genet.*, 124: 1027–1039.

Akkaya, M. S., Bhagwat, A. A. and Cregan, P. B. (1992). Length polymorphism of simple sequence repeat DNA in soybean. *Genetics*, 132: 1131–1139.

Akkaya, M. S., Shoemaker, R. C., Specht, J. E., Bhagwat, A. A. and Cregan, P. B. (1995). Integration of simple sequence repeat DNA markers into a soybean linkage map. *Crop Sci.*, 35: 1439–45.

Anand, S. C. (1992). Registration of 'Hartwig' soybean. *Crop Sci.*, 32: 1060–70.

Aoki, T., O'Donnell, K., Homma, Y. and Lattanzi, A. R. (2003). Sudden-death syndrome of soybean is caused by two morphologically and phylogenetically distinct species within the *Fusarium solani* species complex–*F. virguliforme* in North America and *F. tucumaniae* in South America. *Mycologia*, 95: 660–84.

Arelli, A. P. (1994). Inheritance of resistance to *Heterodera glycines* race 3 in soybean accessions. *Plant Dis.*, 78: 898–900.

Ashfield, T., Bocian, A., Held, D., Henk, A., Marek, L., Danesh, D., Penuela, S., Meksem, K., Lightfoot, D. A., Young, N., Shoemaker, R. and Innes, R. (2003). Genetic and physical mapping of the soybean *Rpg1-b* disease resistance gene reveals a complex locus containing several tightly linked families of NBS/LRR genes. *Mol. Plant Microbe Interact.*, 16: 817–26.

Basten, C. J., Weir, B. S. and Zeng, Z. (2001). *QTL Cartographer Version 2.0*. Raleigh, NC: Department of Statistics, North Carolina State University, USA.

Chen, C. H., Potter, N. T. and Taranenko, N. T. (2003). Detection of trinucleotide repeat containing genes by matrix-assisted laser desorption/ionization (MALDI) mass spectrometry. *Methods Mol. Biol.*, 217: 91–100.

Cook, D. E., Lee, T. G., Guo, X., Melito, S., Wang, K., Bayless, A. M., Wang, J., Hughes, T. J., Willis, D. K., Clemente, T. E., Diers, B. W., Jiang, J., Hudson, M. E. and Bent, A. F. (2012). Copy number variation of multiple genes at rhg1 mediates nematode resistance in soybean. *Science*, 338: 1206–9.

Cook, D. E., Bayless, A. M., Wang, K., Guo, X., Song, Q., Jiang, J. and Bent, A. F. (2014). Distinct copy number, coding sequence, and locus methylation patterns underlie rhg1 mediated soybean resistance to soybean cyst nematode. *Plant Physiol.*, 165(2): 630–47.

Cregan, P. B., Mudge, J., Fickus, E. W., Danesh, D., Denny, R. and Young, N. D. (1999). Two simple sequence repeat markers to select for soybean cyst nematode resistance conditioned by the *rhg1* locus. *Theor. Appl. Genet.*, 99: 811–18.

Darvasi, A. and Soller, M. (1992). Selective genotyping for determination of linkage between a marker locus and a quantitative trait locus. *Theor. Appl. Genet.*, 85: 353–9.

Darvasi, A. and Soller, M. (1994). Selective DNA pooling for determination of linkage between a molecular marker and a quantitative trait locus. *Genetics*, 138: 1365–73.

Frisch, M., Bohn, M. and Melchinger, A. E. (1999). Comparison of selection strategies for marker-assisted backcrossing of a gene. *Crop Sci.*, 39: 1295–301.

Han, Y., Zhao, X., Liu, D., Li, Y., Lightfoot, D. A., Yang, Z., Zhao, L., Zhou, G., Wang, Z., Huang, L., Zhang, Z., Qiu, L., Zheng, H. and Li, W. (2015). Domestication footprints anchor genomic regions of agronomic importance in soybeans. *New Phytol.*, 209: 871–84. doi:10.1111/nph.13626.

Hauge, B. M., Wang, M. L., Parsons, J. D. and Parnell, L. D. (2001). Nucleic acid molecules and other molecules associated with soybean cyst nematode resistance. US patent: WO 0151627-A 19-JUL-2001

Hnetkovsky, N. (1994). Sudden death syndrome resistance genes mapped by molecular markers. MS Thesis, PSGA, SIUC, Carbondale IL, p.144.

Hnetkovsky, N., Chang, S. J. C., Doubler, T. W., Gibson, P. T. and Lightfoot, D. A. (1996). Genetic mapping of loci underlying field resistance to soybean sudden death syndrome (SDS). *Crop Sci.*, 36: 393–400.

Hwang, S., Ray, J. D., Cregan, P. B., King, C. A., Davies, M. K. and Purcell, L. C. (2014). Genetics and mapping of quantitative traits for nodule number, weight, and size in soybean [Glycine max L. (Merr.)]. *Euphytica*, 195: 419–34.

Iqbal, M. J., Meksem, K., Njiti, V. N., Kassem, My. A. and Lightfoot, D. A. (2001). Microsatellite markers identify three additional quantitative trait loci for resistance to soybean sudden death syndrome (SDS) in Essex × Forrest RILs. *Theor. Appl. Genet.*, 102: 187–92.

Iqbal, M. J., Yaegashi, S., Njiti, V. N., Ahsan, R., Cryder, K. L. and Lightfoot, D. A. (2002). Resistance locus pyramids alter transcript abundance in soybean roots inoculated with *Fusarium solani* f.sp. *glycines. Mol. Genet. Genomics.*, 268: 407–17.

Iqbal, M. J., Afzal, J., Yaegashi, S., Ruben, E., Triwitayakorn, K., Njiti, V. N., Ahsan, R., Wood, A. J. and Lightfoot, D. A. (2002). A pyramid of loci for partial resistance to *Fusarium solani* f.sp. *glycines* maintains myo-inositol-1-phoshate synthase expression in soybean roots. *Theor. Appl. Genet.*, 105: 1115–23.

Iqbal, M. J. and Lightfoot, D. A. (2004). Application of DNA markers: soybean improvement. In Lörz, H. and Wenzel, G. (Eds), *Molecular Marker Systems in Plant Breeding and Crop Improvement.* New York, USA: Springer, 475.

Iqbal, M. J, Yaegashi, S, Ahsan, R, Shopinski K. L. and Lightfoot, D. A. (2005). Root response to *Fusarium solani* f. sp. *glycines*: temporal accumulation of transcripts in partially resistant and susceptible soybean. *Theor. Appl. Genet.*, 110: 1429–38.

Iqbal, M. J., Majeed, M., Humayun, M., Lightfoot, D. A. and Afzal, A. J. (2016). Proteomic profiling and the predicted interactome of host proteins in compatible and incompatible interactions between soybean and *Fusarium virguliforme. Appl. Biochem. Biotechnol.*, 180: 1657–74.

Kadam, S., Vuong, T. D., Qiu, D., Meinhardt, C. G., Song, L., Deshmukh, R. and Nguyen, H. T. (2016). Genomic-assisted phylogenetic analysis and marker development for next generation soybean cyst nematode resistance breeding. Plant Science, 242, 342–350.

Karangula, U. B., Kassem, M. A., Gupta, L., El-Shemy, H. A. and Lightfoot, D. A. 2009. Locus interactions underlie seed yield in soybeans resistant to *Heterodera glycines. Curr. Issues Mol. Biol.*, 11. (Suppl 1): i73–84. https://doi.org/10.21775/cimb.011.i73

Kassem, M. A., Shultz, J., Meksem, K., Cho, Y., Wood, A. J., Iqbal, M. J. and Lightfoot, D. A. (2006). An Updated 'Essex' by 'Forrest' Linkage Map and First Composite Interval Map of QTL Underlying Six Soybean Traits. *Theor. Appl. Genet.*, 113: 1015–26.

Kazi, S., Njiti, V. N., Doubler, T. W., Yuan, J., Iqbal, M. J., Cianzio, S. and Lightfoot, D. A. (2007). Registration of the Flyer by Hartwig recombinant inbred line mapping population. *J Plant Reg.*, 1: 175–8.

Kazi, S., Shultz, J., Bashir, R., Afzal, J., Njiti, V. and Lightfoot, D. A. (2008). Separate loci underlie resistance to soybean sudden death syndrome in 'Hartwig' by 'Flyer'. *Theor. Appl. Genet.*, 116: 967–77.

Kazi, S., Shultz, J., Afzal, J., Hashmi, R., Jasim, M., Bond, J., Arelli, P. R. and Lightfoot, D. A. (2010). Iso-lines and inbred-lines confirmed loci that underlie resistance from cultivar 'Hartwig' to three soybean cyst nematode populations. *Theor. Appl. Genet.*, 120: 633–40.

Lakhssassi, N., Liu, S., Bekal, S., Zhou, Z., Colantonio, V., Lambert, K., Barakat, A., and Meksem, K. (2017). Characterization of the Soluble NSF Attachment Protein gene family identifies two members involved in additive resistance to a plant pathogen. *Scientific Reports*, 7: 45226.

Lam, H. M., Xu, X., Liu, X. and et al. (2010). Resequencing of 31 wild and cultivated soybean genomes identifies patterns of genetic diversity and selection. *Nature Genet.*, 42: 1053–62.

Landegren, U., Nilsson, M. and Kwok, P.W. (1998). Reading bits of genetic information: methods for single nucleotide polymorphism analysis. *Genome Res.*, 8: 769–76.

Lander, E., Green, P., Abrahamson, J., Barlow, A., Daley, M., Lincoln, S. and Newburg, L. (1987). MAPMAKER: An interactive computer package for constructing primary genetic linkage maps of experimental and natural populations. *Genomics*, 1: 174–81.

Lee, Y. C., Iqbal, M. J., Njiti, V. N., Kantartzi, S. K. and Lightfoot, D. A. (2016). Fine mapping quantitative trait loci that underlie resistance to soybean sudden death syndrome using NILs and SNPs. *Crop Sci.*

Lightfoot, D. A. (2001). Soybean sudden death syndrome resistant soybeans, soybean cyst nematode resistant soybeans and methods of breeding and identifying resistant plants. United States Patent 6,300,541.

Lightfoot, D. A. and Meksem, K. (2011). Isolated soybean cyst nematode and sudden death syndrome polypeptides. US Patent 7,902,337.

Lightfoot, D. A., Njiti, V. N., Gibson, P. T., Kassem, M. A., Iqbal, J. M. and Meksem, K. (2005). Registration of Essex x Forrest recombinant inbred line (RIL) mapping population. *Crop Sci.*, 45(4): 1678–81.

Lightfoot, D. A. (2008). Soybean genomics: developments through the use of cultivar 'Forrest'. *Int. J. Plant Genomics*, 2008: 1–22. doi:10.1155/2008/793158.

Lightfoot, D. A. (2015). Two decades of molecular marker-assisted breeding for resistance to soybean sudden death syndrome. *Crop Sci.*, 55: 1460–84.

Liu, X., Liu, S., Jamai, A., Bendahmane, A., Lightfoot, D. A., Mitchum, M. G. and Meksem, K. (2011). Soybean cyst nematode resistance in soybean is independent of the Rhg4 locus LRR RLK gene. *Funct. Integr. Genomics.*, 11: 539–49. doi:10.1007/s10142-011-0225-4.

Luckew, A., Leandro, L., Bhattacharyya, M., Nordman, D., Lightfoot, D. A. and Cianzio, S. (2013). Usefulness of ten genomic regions in soybean associated with sudden death syndrome resistance. *Theor. Appl. Genet.*, 126: 2391–403.

Mansur, L. M., Orf, J. H., Chase, K., Jarvick, T., Cregan, P. B. and Lark, K. G. (1996). Genetic mapping of agronomic traits using recombinant inbred lines of soybean. *Crop Sci.*, 36: 1327–36.

Marek, L. F., Mudge, J., Darnielle, L., Grant, D., Hanson, N., Paz, M., Huihuang, Y., Denny, R., Larson, K., Foster-Hartnett, D., Cooper, A., Danesh, D., Larsen, D., Schmidt, T., Staggs, R., Crow, J. A., Retzel, E., Young, N. D. and Shoemaker, R. C. (2001). Soybean genomic survey: BAC-end sequences near RFLP and SSR markers. *Genome*, 44: 572–81.

Martin, S. K., and Geraldi, I. O. (2002). Comparison of three procedures for early generation testing of soybean. *Crop Sci*., 42:705–9.

McCouch, S. R., Teytelman, L., Xu, Y., Lobos, K. B., Clare, K., Walton, M., Fu, B., Maghirang, R., Li, Z., Xing, Y., Zhang, Q., Kono, I., Yano, M., Fjellstrom, R. J., DeClerck, G., Schneider, D., Cartinhour, S., Ware, D. and Stein, L. (2002). Development and mapping of 2240 new SSR markers for rice (*Oryza sativa* L.). *DNA Res.*, 9: 199–207.

Mein, C. A., Barratt, B. J., Dunn, M. G., Siegmund, T., Smith, A. N., Esposito, L., Nutland, S., Stevens, H. E., Wilson, A. J., Philips, M. S., Jarvis, N., Law, S., de Arruda, M. and Todd, J. A. (2000). Evaluation of single nucleotide polymorphism typing with invader on PCR amplification and its automation. *Genome Res.*, 3: 330–43.

Meksem, K., Doubler, T. W., Chang, S. J. C., Chancharoenchai, K., Suttner, R., Cregan, P., Rao Arelli, P., Gibson, P. T. and Lightfoot, D. A. (1999). Clustering among genes underlying QTL for field resistance to Sudden Death Syndrome and cyst nematode race 3. *Theor. Appl. Genet.*, 99: 1131–42.

Meksem, K., Hyten, D., Ruben, E. and Lightfoot, D. A. (2001a). High-throughput genotyping for a polymorphism linked to soybean cyst nematode resistance gene *Rhg*4 by using Taqman probes. *Mol Breed*, 77: 63–71.

Meksem, K., Ruben, E., Hyten, D., Triwitayakorn, K. and Lightfoot, D. A. (2001b). Conversion of AFLP bands into high-throughput DNA markers. *Mol. Genet. Genom.*, 265: 207–14.

Michelmore, R. W., Paran, I. and Kesseli, R. V. (1991). Identification of markers linked to disease-resistance genes by bulked segregant analysis: a rapid method to detect markers in specific genomic regions by using segregating populations. *Proc. Natl. Acad. Sci. U.S.A.*, 88: 9828–32.

Morgante, M. and Olivieri, A. M. (1993). PCR-amplified microsatellites as markers in plant genetics. *Plant J.* 3: 175–82.

Njiti, V., Gray, L. and Lightfoot, D. A. (1997). Rate-reducing resistance to *Fusarium solani* f.sp. *phaseoli [nee: glycines]* underlies field resistance to soybean sudden-death syndrome (SDS). *Crop Sci.*, 37: 1–12.

Njiti, V. N., Doubler, T. W., Suttner, R. J., Gray, L. E., Gibson, P. T. and Lightfoot, D. A. (1998). Resistance to soybean sudden death syndrome and root colonization by *Fusarium solani* f. sp. *glycines* in near-isogeneic lines. *Crop Sci.*, 38: 472–7.

Njiti, V., Johnson, J. E., Torto, T. A., Gray, L. E. and Lightfoot, D. A. (2001). Inoculum rate influences selection for field resistance to soybean sudden death syndrome in the greenhouse. *Crop Sci.*, 41: 1726–31.

Njiti, V. N., Meksem, K., Iqbal, M. J., Johnson, J. E., Kassem, M. A., Zobrist, K. F., Kilo, V. Y. and Lightfoot, D. A. (2002). Common loci underlie field resistance to soybean sudden death syndrome in Forrest, Pyramid, Essex, and Douglas. *Theor. Appl. Genet.*, 104: 294–300.

Njiti, V., Myers O., Schroeder D. and Lightfoot, D. A. (2003). Glyphosate on Roundup Ready soybean: Effects on root infection by *Fusarium solani* f. sp. *glycines* and Sudden Death Syndrome. *Agron. J.*, 95: 1140–5.

Njiti, V. N. and Lightfoot, D. A. (2006). Genetic Analysis Infers Dt loci Underlie Resistance to SDS caused by *Fusarium virguliforme* in Indeterminate Soybeans. *Can. J. Plant Sci.*, 41: 83–9.

Powell, W., Morgante, M., Andre, C., Hanafey, M., Vogel, J., Tingey, S. and Rafalski, A. (1996). The comparison of RFLP, RAPD, AFLP, and SSR (microsatellite) markers for germplasm analysis. *Mol Breed*, 2: 225–38.

Prabhu, R. R., Njiti, V. N., Bell-Johnson, B., Johnson, J. E., Schmidt, M. E., Klein, J. H. and Lightfoot, D. A. (1999). Selecting soybean cultivars for dual resistance to soybean cyst nematode and sudden death syndrome using two DNA markers. *Crop Sci.*, 39: 982–7.

Rongwen, J., Akkaya, M. S., Bhagwat, A. A., Lavi, U. and Cregan, P. B. (1995). The use of microsatellite DNA markers for soybean genotype identification. *Theor. Appl. Genet.*, 90: 43–8.

Rota, M. L., Kantety, R. V., Yu, J. K. and Sorrells, M. E. (2005). Nonrandom distribution and frequencies of genomic and EST-derived microsatellite markers in rice, wheat, and barley. *BMC Genomics*, 6: 23–32.

Ruben, E., Aziz, J., Afzal, J., Njiti, V. N., Triwitayakorn, K., Iqbal, M. J., Yaegashi, S., Arelli, P., Town, C., Meksem, K. and Lightfoot, D. A. (2006). Genomic analysis of the 'Peking' *rhg1* locus: candidate genes that underlie soybean resistance to the cyst nematode. *Mol. Genet. Genomics*. 276: 320–30.

Salvagiotti, F., Cassman, K. G., Specht, J. E., Walters, D. T., Weiss, A. and Dobermann, A. (2008). Nitrogen uptake, fixation and response to fertilizer N in soybeans: a review. *Field Crop. Res.*, 108: 1–13.

Santos, M. A., Geraldi, I. O., Garcia, A. A., Bortolatto, N., Schiavon, A. and Hungria, M. (2013). Mapping of QTLs associated with biological nitrogen fixation traits in soybean. *Hereditas*, 150: 17–25.

Scherm, H. and Yang, X. B. (1996). Development of sudden death syndrome of soybean in relation to soil temperature and soil water potential. *Phytopathology*, 86, 642–9. doi:10.1094/Phyto-86-642.

Schmutz, J., Cannon, S. B., Schlueter, J., Ma, J., Mitros, T., Nelson, W., Hyten, D. L., Song, Q., Thelen, J. J., Cheng, J., Xu, D., Hellsten, U., May, G. D., Yu, Y., Sakurai, T., Umezawa, T., Bhattacharyya, M. K., Sandhu, D., Valliyodan, B., Lindquist, E., Peto, M., Grant, D., Shu, S., Goodstein, D., Barry, K., Futrell-Griggs, M., Abernathy, B., Du, J., Tian, Z., Zhu, L., Gill, N., Joshi, T., Libault, M., Sethuraman, A., Zhang, X., Shinozaki, K., Nguyen, H. T., Wing, R. A., Cregan, P., Specht, J., Grimwood, J., Rokhsar, D., Stacey, G., Shoemaker, R. C. and Jackson, S. A. (2010). Genome sequence of the palaeopolyploid soybean. *Nature*, 463: 178–83.

Shoemaker, R. C. and Specht, J. E. (1995). Integration of the soybean molecular and classical genetic linkage groups. *Crop Sci.*, 35: 436–46.

Shultz, J. L., Jayaraman, D., Shopinski, K. L., Iqbal, M. J., Kazi, S., Zobrist, K., Bashir, R., Yaegashi, S., Lavu, N., Afzal, A. J., Yesudas, C. R., Kassem, M. A., Wu, C., Zhang, H. B., Town, C. D., Meksem, K. and Lightfoot, D. A. (2006). The soybean genome database (SoyGD): a browser for display of duplicated, polyploid, regions and sequence tagged sites on the integrated physical and genetic maps of *Glycine max*. *Nucleic Acids Res.*, 34: D758–65.

Shultz, J. L., Kazi, S., Afzal, J. A., Bashir, R. and Lightfoot, D. A. (2007). The development of BAC-end sequence-based microsatellite markers and placement in the physical and genetic maps of soybean. *Theor. Appl. Genet.*, 114: 1081–90.

Sinclair, T. R. and Wit, C. T. de. (1975). Photosynthate and nitrogen requirements for seed production by various crops. *Science*, 189: 565–7.

Song, Q. J., Marek, L. F., Shoemaker, R. C., Lark, K. G., Concibido, V. C., Delannay, X., Specht, J. E. and Cregan, P. B. (2004). A new integrated genetic linkage map of the soybean. *Theor. Appl. Genet.*, 109: 122–8.

Srour, A., Afzal, A. J., Saini, N., Blahut-Beatty, L., Hemmati, N., Simmonds, D. H., El Shemy, H., Town, C. D., Sharma, H., Liu, X., Li, W. and Lightfoot, D. A. (2012). The receptor like kinase transgene from the *Rhg1/Rfs2* locus caused pleiotropic resistances to soybean cyst nematode and sudden death syndrome. *BMC Genomics*, 13: 368.

Stefaniak, T. R., Hyten, D. L., Pantalone, V. R., Klarer, A. and Pfeiffer, T. W. (2006). Soybean cultivars resulted from more recombination events than unselected lines in the same population. *Crop Sci.*, 46: 43–51.

Tanksley, S. D., Young, N. D., Paterson, A. H. and Bonierbale, M. W. (1989). RFLP mapping in plant breeding: new tools for an old science. *Bio/technology*, 7: 257–64.

Triwitayakorn, K. (2002). Positional cloning of the *Rfs* loci. PhD Thesis, SIUC Carbondale, 232pp.

Triwitayakorn, K., Njiti, V. N., Iqbal, M. J., Yaegashi, S., Town, C. and Lightfoot, D. A. (2005). Genomic analysis of a region encompassing *QRfs1* and *QRfs2*: genes that underlie soybean resistance to sudden death syndrome. *Genome/Génome*, 48: 125–38.

Webb, D. M., Baltazar, B. M., Rao-Arelli, A. P., Schupp, J., Keim, P., Clayton, K., Ferreira, A. R., Owens, T. and Beavis, W. D. (1995). QTLs affecting soybean cyst-nematode resistance. *Theor. Appl. Genet.*, 91: 574–81.

Webb, D. M. (1996). Soybean cyst nematode resistant soybeans and methods of breeding and identifying resistant plants. United States Patent 5,491,081.

Webb, D. M. (2000). Positional cloning of soybean cyst nematode resistance genes. United States Patent 6,162,967.

Webb, D. M. (2003). Quantitative trait loci associated with soybean cyst nematode resistance and uses thereof. United States Patent 6,538,175.

Witsenboer, H., Vogel, J. and Michelmore, R. W. (1998). Identification, genetic localization and allelic diversity of selectively amplified microsatellite polymorphic loci (SAMPL) in lettuce and wild relatives (Lactuca spp.). *Genome*, 40: 923–36.

Wu, X., Ren, C., Joshi, T., Vuong, T., Xu, D. and Nguyen, H. T. (2010). SNP discovery by high-throughput sequencing in soybean. *BMC Genomics*, 11: 469.

Xin, Z., Velten, J. P., Oliver, M. J. and Burke, J. J. (2003). High-throughput DNA extraction method suitable for PCR. *BioTechniques*, 34: 820–6.

Yuan, J., Haroon, M., Lightfoot, D. A., Pelletier, Y., Liu, Q., Bizimungu, B. and Li, X.-Q. (2009). High-resolution DNA melting analysis of allelic expression. *Curr. Issues Mol. Biol.*, 11(S1): 1–9.

Yuan, Z., Njiti, V., Meksem, K., Iqbal, M. J., Triwitayakorn, K., Kassem, M. A., Davis, G. T., Schmidt, M. E. and Lightfoot, D. A. (2002). Identification of yield loci in soybean populations that segregate for disease resistance. *Crop Sci.*, 42(1): 271–7.

Zhu, Y. L., Song, Q. J., Hyten, D. L., van Tassell, C. P., Matukumalli, L. K., Grimm, D. R., Hyatt, S. M., Fickus, E. W., Young, N. D. and Cregan, P. B. (2003). Single-nucleotide polymorphisms in soybean. *Genetics*, 163: 1123–34.

Chapter 6

Improving cultivation practices for soybeans in sub-Saharan Africa

Frederick P. Baijukya and Harun M. Murithi, International Institute of Tropical Agriculture (IITA), Tanzania; and Fred Kanampiu, International Institute of Tropical Agriculture (IITA), Kenya

1 Introduction
2 Trends in soybean production and consumption in SSA
3 Best practices to enhance soybean yields: inoculation with rhizobia and use of improved varieties
4 Best practices to enhance soybean yields: improved crop management
5 Other considerations to improve soybean productivity in SSA
6 Conclusion
7 Where to look for further information
8 References

1 Introduction

Soybean (*Glycine max* L Merr.) now constitutes an important component of the cropping systems in sub-Saharan Africa (SSA). Various governments together with their development partners are promoting soybean cultivation among small-scale farmers because of three fronts that the crop provides, which are improving health and nutrition for human and for livestock, increasing rural incomes, and sustainable intensification of farm production by improving the fertility of the soils (Crutzen et al., 2007; Giller, 2001). The rapidly growing feed industry for poultry and aquaculture (of which more than 20% of raw material is soybean) and the recognition of healthy eating among large population have increased the commercial value of the crop in SSA (FAO STAT, 2015).

Soybean is an attractive crop that is well suited in most parts of SSA (Fig. 1). However, SSA currently produces <1% of global soybean output and is a significant net importer of soybean products, importing more than 55% of its total demand. The current soybean yield in SSA estimated at 1.2 tons ha^{-1}, which is 30% lower than the global average yields of nearly 2.5 tons ha^{-1} (Abate et al., 2011). Several studies have attributed low soybean yields in SSA to a number of factors including use of low-yielding varieties, inadequate use of fertilizers, limited use of rhizobia inoculants in soils with no history of inoculation or soybean production (Sanginga and Woomer,

http://dx.doi.org/10.19103/AS.2017.0023.25

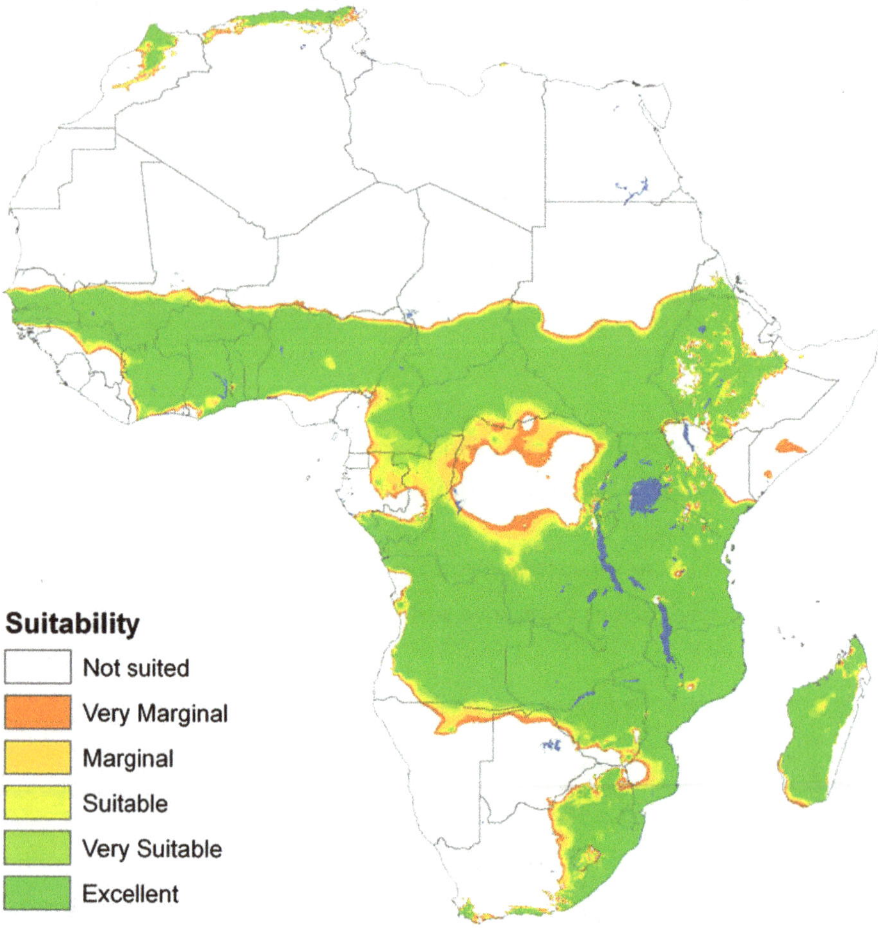

Figure 1 Soybean suitability map for Africa (IITA, 2008).

2009), and poor crop management (Tittonell and Giller, 2013). Other factors include lack of extension services, poor access to input and output markets and frequent drought due to changing climate (Abate et al., 2011).

Farmers in SSA need to improve soybean cultivation practices and achieve close to 2.0 tons ha^{-1}, if they have to make a profit from its production (Technoserve, 2011). SSA has favourable climatic conditions and 80% of its favourable agricultural land remains unutilized, presenting great opportunity to increase both the yield and the global market share of soybean (Hartman et al., 2011). In this chapter, we review the trend of soybean production and consumption in SSA and take stock of available improved practices, which if adopted, can result in increased soybean yields, giving examples of how the practices have improved soybean yield and production in countries such as Democratic Republic of the Congo (DRC), Kenya, Nigeria, Tanzania, Uganda and Zimbabwe. We conclude by proposing an integrated approach for sustainable soybean cultivation that recognizes and embraces the wide diversity of farms and farming systems found in SSA to tailor soybean technologies to different farming households.

2 Trends in soybean production and consumption in SSA

The past three decades have witnessed an increase in soybean yield and overall production across SSA. Soybean yield in SSA has increased by 3%, from less than 1 ton ha^{-1} before 2000 to about 1.2 tons ha^{-1} in 2011–13 (FAO STAT, 2015). During the same period, soybean production more than tripled from about 0.5 million tons to 2 million tons, with a noticeable increase in the total area under soybean by 3.5%. Both yield increase and area expansion have a roughly comparable contribution to the observed growth in soybean production (Fig. 2).

Nevertheless, this progress is still not sufficient to meet the growing demand over the past decades. Analysis across SSA reveal an increase in soybean consumption of 7.4% between 1990 and 2011, that outpaced production, with more than 50% of the production gap filled by importation (Akibode and Maredia, 2011). Going by individual countries, it is only Uganda that has satisfied its local demand and has managed to trade the surplus (Fig. 3). Strong public sector research programme, which is involved in the supply of improved soybean varieties to the seed companies, is a key factor for successful soybean development in Uganda.

As of 2015, the soybean demand gap for SSA stood at more than 4 million metric tons valued at US$2 billion (FAO STAT, 2015). The demand gap is expected to increase

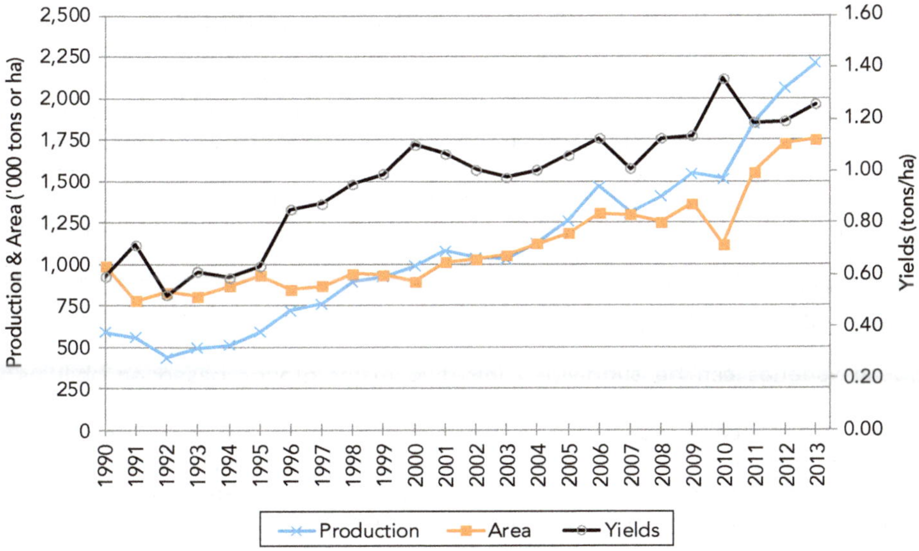

Figure 2 Trends in soybean production, area and yields in sub-Saharan Africa, 1990–2013 Source: FAOSTAT (2015, www.faostat.org).

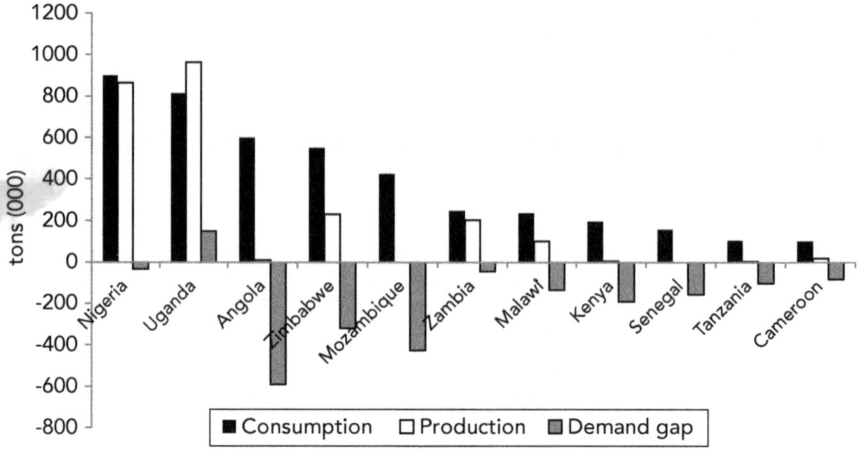

Figure 3 Estimated demand production and demand gap for some SSA countries (FAO STAT, 2015).

to 30 million metric tons or US$13.5 billion by 2030, given the current population and income growth rates in SSA. This will cause a substantial outlay of scarce foreign exchange. Akibode and Maredia (2011) propose an increase of 0.37% $yr^{-1}$of land under soybean and grain yield of 50%, if SSA is to meet the projected soybean demand in 2030.

3 Best practices to enhance soybean yields: inoculation with rhizobia and use of improved varieties

Higher soybean yields can be achieved by use of improved varieties with enhanced potential for biological nitrogen fixation (BNF), inoculation with high-quality rhizobial inoculants and targeting these soybean–inoculant combinations within different agro-ecologies. These practices are described in this section. Section 4 then reviews improved crop management practices for soybean cultivation. The translated combination of these practices has been proven to increase the total nitrogen (N) fixed by the soybean and ultimately the yield (Giller and Cadisch, 1995).

3.1 Inoculation with rhizobia

Soybean varieties can be subdivided into two major groups based on abilities in which they host *Rhizobium* bacteria. One group consists of soybean varieties that nodulate with a specific type of rhizobial strains and that are referred to as specific in the rhizobia requirements. Majority of these genotypes were bred in North America where indigenous rhizobia compatible with soybean were absent (Giller, 2001). The other group is for soybean varieties that nodulate with a wide range of rhizobia strains and is referred to as non-specific (or promiscuous). Soybean varieties that have specific requirements for rhizobia bacteria need inoculation using commercially available inoculants (Giller, 2001).

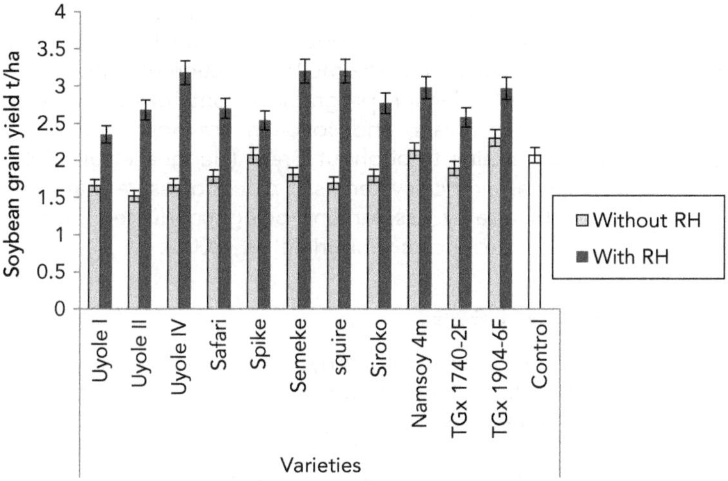

Figure 4 Response of soybean varieties to *Rhizobium* inoculation in southern highlands of Tanzania. Data are averages of four sites over three seasons. Semeki, Squire and Siroko are specific varieties, the rest of them are promiscuous. A control was Uyole II variety grown with nitrogen fertilizers applied at 40 kg/ha (Baijukya et al., Unpublished).

BNF in soybean is a proven success, with strong synergies between the N fixed and yield of soybean. Studies conducted in Nigeria and Zimbabwe revealed that soybean fixed more than 150 kg and 120 kg N ha^{-1}, respectively, when grown in association with effective bradyrhizobia (Sanginga et al., 2001; Mapfumo et al., 2005). In Tanzania, both specific and promiscuous varieties are reported to respond positively with rhizobia inoculation (Fig. 4).

Soybean varieties with specific requirements for *Rhizobium* respond more to inoculation by yielding more grains when compared to promiscuous varieties. The promiscuous varieties also responded positively to inoculation though they have the ability to be grown in certain locations without inoculation. Similar results have been reported in Zimbabwe (Mpepereki et al., 2002) and Mozambique (Gyogluu, 2011). Field evidence also shows that rhizobia inoculation increase N_2 fixation in the range of 35–95 kg ha^{-1}, resulting in yield benefits worth US$50–75 for an investment of only US$3–5 but with small amounts of phosphorus (P) fertilizer to ensure acceptable legume yields (Woomer et al., 2014). However, inoculant use in smallholder farming of SSA is very limited because of poor awareness, availability and undeveloped distribution system.

How inoculants have been used to improve soybean yield in Brazil

Success stories on the wide-scale use of rhizobium inoculants on soybean are reported from Brazil. Soybean was first introduced in Brazil near the end of the nineteenth century but widespread adoption did not begin until the 1960s in Brazil with rapid expansion between 1980's and 2010 such that today Brazil is responsible for about 25% of the world's soybean production (Hungria et al., 2005). Presently, soybean is grown in over 22 million ha in Brazil and the economic returns of using BNF instead of minerals N fertilizer is US$6 billion per year (Hungria et al., 2006). Compatible rhizobia for soybean are rarely present in soils not planted with soybean before, thus inoculation with rhizobia is recommended (Döbereiner and Duque, 1982). Initial problem with imported inoculants in the acid cerrado soils was overcome by selection of strains adapted to the local conditions, and the strains' selection programme continues to achieve increasing demands of more productive cultivars, and soybean are now routinely grown with inoculation carrying selected strains throughout Brazil (Hungria et al., 2005, 2006). A strong regulatory and quality assurance system is in place to assure that the best strains are offered to farmers, with quality assessment being conducted independently of country's commercial inoculant producers (Hungria et al., 2006).

3.2 Use of improved varieties

In SSA, soybean is grown across a range of biophysical and socio-economic environment, hence the need to target varieties to the environment where they can best perform. The key environmental factors that restrict on-farm soybean yield are low soil fertility, pests and diseases as well as low soil moisture as a result of mid- or end-season drought. Therefore, variety choice for particular agro-ecology should be made in consideration of such problems. There is a large number of soybean varieties with growth duration between 60 days (early maturity) and more than 150 days (late maturity) that are ready to deploy in different agro-ecologies in SSA (Table 1). The early maturing varieties are better suited to low altitude, warmer and drier climate, or in situations where seasons are predicted to be short while the long maturing varieties are better suited in medium to higher altitude areas with more reliable rains.

IITA has successfully developed improved varieties of soybean that are promiscuous, fix more N, are high yielding, are resistant to pod-shattering and have long shelf-life. Most of these cultivars have some degree of enhanced disease resistance, especially against soybean rust (*Phakopsora pachyrhizi*), frogeye leaf spot and bacterial pustule, which are now endemic in soybean growing areas of west, east and southern Africa (Hailu, 2011; Kumudini et al., 2008). Proactive seed companies, for example, seed companies like Seed Co and National Breeding Programmes (e.g. Makerere University) have, over time, developed and released soybean varieties high in oil contents and with tolerance to soybean rust disease. These varieties have been widely tested and found suitable in many agro-ecologies that they were developed for (Woomer et al., 2014).

Wide availability of soybean seeds, however, remains a limitation to exploiting soybean in the smallholder farming sector in SSA. Unlike cross-pollinated crops such as maize, variety improvement in self-pollinated crops such as soybean is hindered by the intellectual property (IP) position (access and ability to protect elite germplasm and technology by the private sector). Over the past decade, IP in soybean has been a contentious issue leading to termination of conventional soybean breeding programmes by the major multinational seed companies in Africa to concentrate into GM soybean specifically for the South African market (Alene et al., 2015).

How improved varieties have contributed to soybean yield in Nigeria

Identification and release of high-yielding promiscuous varieties from the breeding programme at IITA had led to their widespread use by farmers in Nigeria. Uptake of

Table 1 The common soybean varieties available in sub-Saharan Africa

Region	Variety name	Common name	Country	Year of release	Source	Maturity period	Reference
East and Central Africa	TGx-1740-2F	SB19	Kenya	2010	IITA	Early	Rusike et al., 2013
		SB19	DRC	2005	IITA		Rusike et al., 2013
		SB19	Rwanda	2007	IITA		RAB Report,, 2013
	Maksoy 1N	SB24	DRC	2005	NARO/ MAK	Early	Rusike et al., 2013
		SB24	Rwanda	2013	IITA		RAB Report, 2013
			Uganda	2004	IITA		
	Namsoy 4M	–	Uganda	2004	NARO/ MAK	Early	
		–	Rwanda	2009	Uganda		Rusike et al., 2013
	PEKA 6	–	DRC, Rwanda	2005	India	Medium	Rusike et al., 2013
	Yezumutima	–	Burundi	2007	Unknown		
		–	Rwanda	2000	Unknown		
	Soprosoy	–	Burundi, Rwanda	2007	India	Early	
	SC Squire and SC Sequel	–	Rwanda	2013	Seed Co	Medium	RAB Report, 2013
	SC Saga	–	Kenya	2007	Seed Co	Medium	RAB Report
	TGx 1485-1D	–	Burundi	Informal	IITA	Early	
	TGx 1835-10E		Uganda	2004	IITA	Early	Hailu, 2011
Southern Africa	TGx 1740-2F	Tikolore, Wamini	Malawi, Mozambique	2011	IITA	Early	Rusike et al., 2013
		Kafue	Zambia	2015	IITA		IITA Bulletin, 2015
	TGx 1937-1F	Olima	Mozambique	2011	IITA	Early	Rusike et al., 2013
		Mwembeshi	Zambia	2015	IITA		IITA Bulletin, 2015
	SC Safari		Zambia	2004	Seed Co	Early	Rusike et al., 2013
			Zimbabwe	2001	Seed Co		Rusike et al., 2013
	SC Storm		Zambia	2001	Seed Co	Medium	Rusike et al., 2013
			Zimbabwe	2000	Seed Co		Rusike et al., 2013
	SC Squire, SC Saga		Zimbabwe	2008	Seed Co	Medium	Rusike et al., 2013
	SC Sequel			2009	Seed Co	Medium	Rusike et al., 2013
Western Africa	TGX 1740-2F		Nigeria	1992	IITA/ NACRI	Early	Hailu, 2011
			Benin and Togo	1998	IITA		
	TGx-1448-2E		Cameroon	1994	IITA	Late	Hailu, 2011

(Continued)

Table 1 *(Continued)*

Region	Variety name	Common name	Country	Year of release	Source	Maturity period	Reference
			Ghana, Nigeria	1992	IITA		
			Togo	1998	IITA		
	TGx 1485-1D		Nigeria	1992	IITA	Early	Hailu, 2011
			Benin and Togo	1998	IITA		Hailu, 2011
	TGx 1440-1E		Nigeria	1992	IITA	Late	Hailu, 2011
			Benin and Togo	1998	IITA		
	TGx 1835-10E		Nigeria	2004	IITA	Early	Hailu, 2011

the new promiscuous gained rapid momentum as they become more widely known to farmers so that the new varieties were being grown by 75% of male farmers and 62% of women farmers by 1996 (Sanginga et al.,1999). Varieties developed more recently are even more promiscuous and are likely to be widely used by smallholders in future (Sanginga et al., 2001). The success of soybean in Nigeria was linked to training in household utilization and the presence of small industries for soybean processing providing a ready market for soybean products, and stimulated production and consumption of soybean products (Sanginga et al., 1999). Partnerships were formed with government, social, voluntary agencies, and NGOs to incorporate soybean utilization into their activities. Hospitals were also involved and several child-weaning foods were made with the incorporation of soybean. This explains how strategic alliances of all important stakeholders, training and capacity building as well as awareness creation catalysed the whole process.

Earlier field studies performed in Nigeria have shown that soybean derived about 60% of its N from fixation with a contribution to the N balance in the cropping system ranging from -8 to 43 kg N ha^{-1} depending on the soybean cultivars. In a rotation maize and soybean system where legume residues are retained and mineral fertilizer is applied (15 kg Pha^{-1} applied to soybean and 45 kg N ha^{-1}applied to maize), soybean yields were 2.5 tons of grain ha^{-1}with N fixation contributing about 50 kg N ha^{-1}. The maize following soybean had 75% greater yield than maize following maize (Sanginga et al., 2001). The big success that has been reported for the research and development activities related to soybean production, processing, marketing and utilization in Nigeria were possible because of a group of highly motivated people from many organizations in the country.

Small to medium size seed companies usually do not have breeding programmes and they rely heavily on public organizations, such as IITA and NARS such as Makerere University, for breeders and foundation seed. A strong public sector variety improvement programme for self-pollinated crops will thus enhance public-private partnerships in which the seed companies will source the breeder's and foundation seed for certified seed production, thus ensuring that the farmers have access to improved seed.

4 Best practices to enhance soybean yields: improved crop management

4.1 Nutrient management

Soybeans do not require supplemental N fertilizer because they can fix N through a symbiotic relationship with *Bradyrhizobium* bacteria. If used, band-placed fertilizer is recommended at least 5 cm to the side and 5 cm below the seed. Placing any fertilizer in contact with the soybean seed may cause damage. P is especially the nutrient limiting the efficiency of soybean to fix N, hence yield in most soils (Giller, 2001; Kamara et al., 2008; Ronner, et al., 2015). In a study of N x P interactions on soybean productivity in northeast Nigeria, N fertilizer did not affect grain yield while P fertilizer increased grain yield by 46–95% at 20 kg P ha^{-1} and by 51–316% at 40 kg ha^{-1} (Kamara et al., 2008). A similar observation has been made in Ghana, Rwanda, DRC, Kenya, Malawi, Mozambique and Zimbabwe (Woomer et al., 2014).

Adaptation studies conducted by N2-Africa project (Giller et al., 2013) and spanned across SSA revealed that application of inoculants in conjunction with application of P at lower rate (20–30 kg ha^{-1}) greatly increased soybean nodulation (Fig. 5) and yields of between 5 and 50% compared to no-inoculated control. However, same studies confirmed the existence of huge variability in response to inoculation and P application across and within farms in the same agro-ecologies (Fig. 6). In control plots sown at the

Figure 5 Response to P and inoculation with soybean in DRC, observed from one N_2 Africa adaptation plots in Mushomo village, Sud Kivu, DRC, during 2010 short rains season.

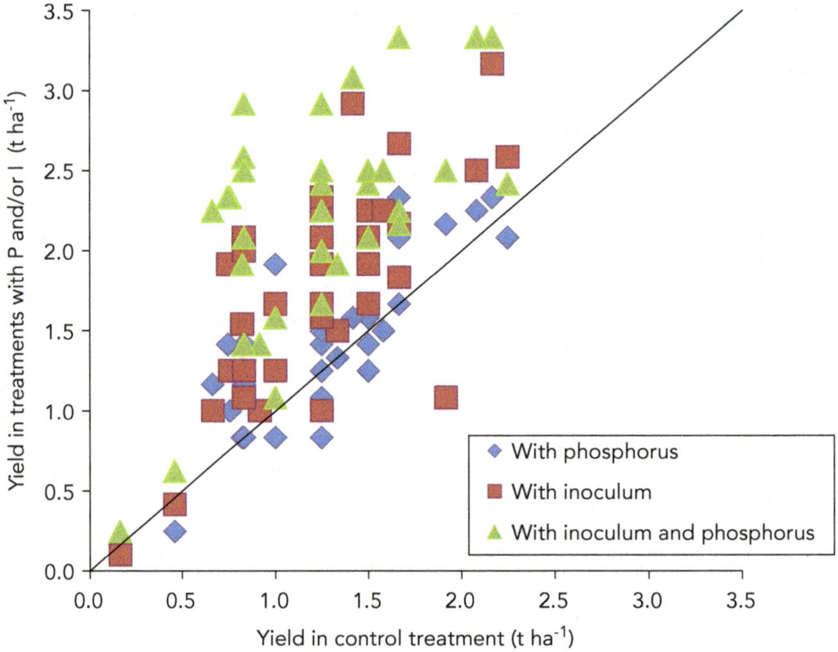

Figure 6 Soybean yield with fertilizer-P and rhizobia inoculant against control in 34 demonstration sites in Southern Kivu, DRC (Giller et al., 2014).

same time, yield varied between 0.2 and 2.5 tons ha^{-1}. Addition of P alone gave variable yield whereas the combinations of two gave the highest yields with many plots yielding above 3.0 tons ha^{-1}. According to Giller (2001), strong positive interaction results from the increased demand for P when the plants grow much more strongly. Lack of response to P and inoculant application in some farms presented in Fig. 6 indicate that other nutrients than N and P are limiting. For example, on depleted sandy soils of Zimbabwe, crop response to N, P and inoculant application was limited by deficiencies of Zn, Ca, Mg and K (Zingore et al., 2008). Deficiency symptoms of secondary nutrients, for example, Mg, S and micronutrients are increasingly becoming visible, especially in highly populated areas where land is intensively cultivated with little nutrients (Vanlauwe et al., 2015).

4.2 Optimizing plant population

Seeding rate, plant population and row spacing are tied together and determine the yield a farmer will get. If the population is too high, plants compete with each other and often lodge. If the population is too low, a farmer is wasting growing space and lowering yield. The latter is true in SSA, where majority of farmers grow soybean at wider spacing. Overall soybean planting spacing of 45 cm between rows and 5 cm within rows has been found suitable across SSA and to the majority of soybean varieties (Woomer et al., 2014). In Kenya, for example, biomass accumulation by soybean ranged between 2.5 and 4.0 tons ha^{-1} at a narrow spacing of 45x5 cm against 2.1 and 3.1 tons ha^{-1} at a wide spacing of 75x5 cm, depending on the variety (Table 2). Consequently, the mean N fixed

Table 2 Biomass accumulation, N fixation and grain yield of eleven soybean varieties grown at two spacings (45x5 m and 45x5 cm). Data are observation over two seasons (2012 long rains and short rains of 2012/2013) and from three sites: Kambare, Nyabeda and Ekitale in western Kenya. Source Wekesa, 2015

Soybean variety	Biomass accumulation (tons ha−1)		Nitrogen fixed (kg ha−1)		Grain yield (tons ha−1)	
	45x5	75x5	45x5	75x5	45x 5	75 x 5
835/5/30	2.9	2.2	65	47	2.2	1.8
EAI3600	3.1	1.9	77	43	2.1	1.5
SBH3/7/4	2.5	2.2	69	50	2.2	1.3
SC 23–6-16	3.0	2.5	72	51	1.9	1.9
SC Sequel	2.9	2.2	74	49	2.3	2.0
SC Squire	2.9	2.1	66	46	3.0	2.2
TGx1740-2F	3.8	2.5	90	56	2.5	1.9
TGx1904-6F	3.5	2.4	82	46	2.6	2.0
TGx1987-10F	4.6	3.1	120	81	2.4	2.5
TGx1987-18F	3.9	3.1	96	69	2.2	2.0
TGx1987-62F	4.0	2.9	100	69	2.4	1.5
Mean	3.4	2.5	83	55	2.3	1.9
LSD_5%	0.4		34		0.6	

was highest (120 kg ha^{-1}) at a spacing of 45 cm compared to 81 kg ha^{-1} at spacing of 75 cm. A similar trend was observed with the grain yield (1.9 to 3.0 tons ha^{-1}) at spacing of 45x5 cm against 1.3–2.2 tons ha^{-1} at spacing of 75x5 cm, respectively. Planting of soybean is labour intensive and placement of seeds at proper depth is key to achieving well-established crop at optimal plant population. However, farm power appears to be a 'forgotten resource'. A consequence of low farm mechanization is high labour drudgery throughout the production cycle. Sustainable intensification in soybean production in SSA will require an increase in power supply via improved access to mechanization and/or a decreased power demand via energy saving technologies such as conservation agriculture (CA).

4.3 Managing diseases and insect pests

Apart from the abiotic constrains, biotic factors such as diseases, insect pests and weeds contribute to significant yield losses in soybean. The major soybean diseases in Africa include bacterial blight, bacterial pustule, frogeye leaf spot, red leaf blotch, downey mildew, soybean mosaic virus, root rots and soybean rust (Levy, 2005; Kawuki et al., 2003; Murithi et al., 2015). Soybean rust in particular is a major threat to soybean production worldwide due to the ease of spread through wind currents. The disease caused by a biotrophic pathogen *P. pachyrhizi* results in yield losses between 10 and 90% (Levy et al., 2005; Pretorius et al., 2001). Breeding for resistance is considered as the best cost-effective and sustainable option for controlling soybean rust; however, there are no commercially available resistant lines due to the variability of this pathogen across the globe (Hartman et al., 2005).

Fungicides are primarily used for controlling soybean rust. In South Africa and Zimbabwe, for example, application of fungicides reduced the disease severity and significantly increased soybean grain yield (Levy, 2005; Miles, 2007). Yield stability approach has been used for developing tolerant varieties based on the selection of the best yielding lines in severe soybean infestation with and without fungicide application. Seed Co, Zimbabwe, used this method to develop tolerant varieties such as SC Squire and SC Saga (Tichagwa, 2004) that have been released in Rwanda, Zambia and Zimbabwe. A soybean rust-resistant IITA breeding line was released as Maksoy 1N in Uganda in 2004 and as TGx 1835–10E in Nigeria in 2008. The line was also resistant to soybean rust in Mozambique and Malawi (Hartman et al., 2011). In addition, Makosy 2N was developed and released as a tolerant variety in Uganda in 2008 (Tukamuhabwa and Maphosa, 2012).

UG5 cultivar from Uganda is resistant to soybean rust isolates from different countries; the cultivar is currently used for developing resistant varieties in Uganda and Nigeria (Tukamuhabwa and Maphosa, 2012). In 2010, cultivar TGx 1987-10F and TGx 1987-62F with a UG 5 background were released in Nigeria (Hartman et al., 2011). Additional resistant varieties namely NCRISOY-1 (TGx1988-5F) and NCRISOY-2 (TGx1989-19F) were released in Nigeria in 2014. Their potential yield ranges between two and three tons ha^{-1} (IITA Bulletin, 2015). The release of soybean rust-resistant varieties from breeding programmes in Africa could lead to greater adoption that would have a positive impact on production.

Common insect pests on soybean in Africa include African bollworm, leaf miners, pod borers, soybean leaf beetle, aphids and stinkbugs (Dugje et al., 2009). Currently, small-scale farmers do not use insecticide on soybean, as this may be impacting on yield despite the effort geared into soil improvement. Integrated pest management (IPM) is the single most efficient way of controlling pests in soybean. The IPM approaches such as rotation, early planting and biological control in addition to the regulated use of chemicals could significantly reduce pest infections on grain legumes including soybean (Belmain et al., 2013).

New pests emerge through invasion and control of primary pests leads to the emergence of secondary pests, as pests develop resistance to control strategies (Murithi et al., 2014). Changes in weather can make pest outbreaks more unpredictable and/or severe, and often the scaling up of production via higher yielding varieties, more intensive production or through the increase in land area planted leads to more persistent pest problems (Murithi et al., 2014). Therefore, more integrated pest control that improves quality and quantity of crop production for less cost is critical to attaining improved productivity. Realizing agro-ecological intensification of soybean production in Africa requires a re-assessment of diseases and insect pest management, looking at current and potential options and how future research, specifically focused on soybean production in Africa will be structured.

5 Other considerations to improve soybean productivity in SSA

5.1 Integrated approaches towards soybean yield

Integrated approaches to improve the interaction between soybean genotype (G_L) x rhizobia inoculant genotype (G_R) × environment (E) × management (M) is now a widely accepted approach to improving soybean production (Giller et al., 2013). However, crop response to these inputs is highly variable across small distances within farms and farming systems and across countries, emphasizing the need to understand variability, in order to tailor legume-based technologies to match the socio-ecological niches and the needs of a diverse group of farmers (Ojiem et al., 2007).

Future soybean production is expected to increase in SSA due to expanded production area and higher yields. An integrated approach to soybean cultivation is therefore required to ensure sustainable yields. Improved varieties that are highly adapted to multiple environments in combination with appropriate nutrient management and important cultural practices such as correct spacing are necessary for sustainable soybean cultivation. For example, areas in western Kenya that have higher incidences of soybean rust disease and soils depleted of nutrients net returns to investments strongly vary with management decisions. Farmers in west Kenya have adopted planting of soybean variety TGx1740-2F (SB 19) and use of rhizobia inoculants. Large economic returns are not realized with SB 19, which is extremely susceptible to rust, without investing in the fertilizer blended for legumes (Fig. 7). Switching over to variety SC Saga (which is tolerant to soybean rust disease) offers an immediate solution. Clearly, farmers must know their crops and land to manage them most effectively.

Diseases and insect pests that are a major impediment to soybean production require an integrated control approach. Utilization of resistant and/or tolerant varieties in

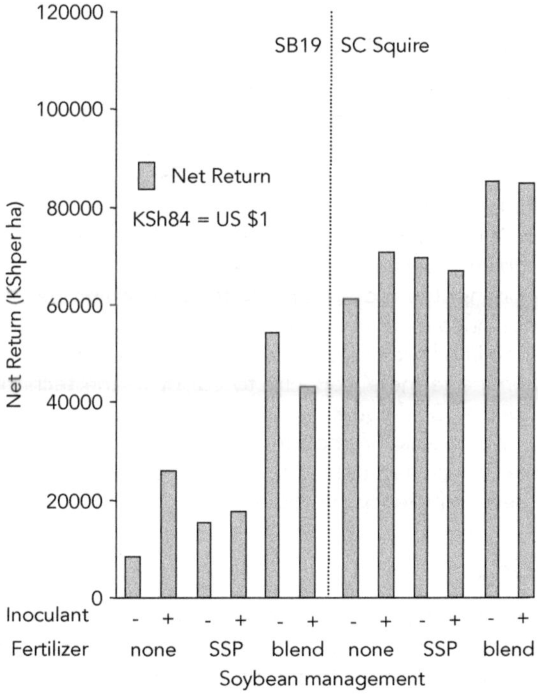

Figure 7 Management response testing of soybean in Butere, west Kenya, illustrates the economic benefits of varietal choice and more complete fertilization. KSh 85 = US$1 (Woomer et al., 2014).

combination with appropriate fungicide application regimes will improve productivity and may reduce costs of production. Monitoring of soybean pests and diseases through timely scouting will guide the application of appropriate control measures.

5.2 Climate change impacts on soybean production

Climate change will have a considerable impact on agriculture as new patterns of weather emerge, causing a shift in precipitation, temperature and climate extremes that may affect agro-ecological zones. Changes in average temperature and rainfall will have direct and indirect impact on plant productivity. Any changes in the future climate scenario will affect soybean production globally. Zheng et al. (2009) demonstrated that exposure of soybean to high temperature regime had substantial negative impact on yield in northeast China. Increases in soybean yields with increasing CO_2 levels have been reported (Sakurai et al., 2014). To counter these challenges, a climate-smart approach such as the use of heat-tolerant varieties, early planting, soil water management and use of early duration materials will be required in the farming systems to maintain and enhance productivity and resilience in agricultural ecosystems. Drought-tolerant and rust-resistant varieties including TGx 1485-1D, TGx 1740-2F, TGx 1937-1F, TGx 1904-6F, TGx 1988-22F, TGx 1989-60F,, TGx 1988-18F and TGx 1908-8F are available from IITA (IITA, 2012, 2016) that farmers can use to counter the effect of climate change. These varieties were released in several African countries (Tukamuhabwa and Maphosa, 2012, IITA, 2016).

5.3 Improved access to inputs

The most important inputs for soybean production are seeds, fertilizers, inoculants, crop protection products, equipment and tools. Limited access and inadequate knowledge on the utilization of these inputs is a key constraint to the development of soybean sector in SSA (Wilson, 2015). In the majority of SSA countries, such inputs are poorly available in rural areas and when available they tend to be expensive as most are imported, and distribution costs tend to be high. It is generally recognized that SSA displays a combination of high soil nutrient deficits and very low fertilizer use (7 kg ha^{-1} vs >150 kg ha^{-1} in Asia) that comes from a set of failures on input and output markets (Druilhe and Barreiro-Hurlé, 2012). On the demand side, poor price incentives, highly seasonal and variable production due to increasing rainfall variability, lack of liquidity, credit or insurance and lack of knowledge about fertilizers undermine farmers' capacity to adopt to the technology or to reap the benefits of technology use. Consequently, farmers opt to recycle the seed and apply no inputs, hence attain lower yields. Promotion of collective input purchase combined with smart fertilizer subsidies is reported to lead to timely access to inputs such as quality seeds and fertilizers and increase efficiency in production.

5.4 Input financing

One of the key reasons that farmers in SSA fail to invest in soybean production is that they lack access to farming inputs such as seeds, fertilizers, insecticides, fungicides and other agricultural resources. Access to affordable credits could enhance smallholder soybean production as this would enable them to acquire quality seeds and fertilizers. This has been achieved in some countries, for example Kenya, through various contracting schemes between smallholder farmers and soybean processors or traders

(Woomer, 2014). In this way the farmers can access loans to finance their production and working capital. The increase of soybean production by more than 166% in Nigeria, from 1988 to 1998 was attributed to subsidized mineral fertilizers and bank credit to farmers among other factors (Chianu et al., 2008). For sustainability, governments should provide policies to guide subsidies for agricultural production and play an active role in soybean price regulation.

5.5 Agricultural extension and advisory services

Agricultural extension and advisory services are a vital element of the array of market and non-market entities and agents that provide critical flows of information that can improve farmers' and other rural peoples' welfare. These services could play a significant role in stimulating economic growth, reduction of poverty and improving food security in SSA. Soybean is a relatively new crop in SSA, therefore extension services are required to train farmers in agronomy, market potential and value addition for household utilization. There is need to revitalize and to deploy innovative methods to facilitate access of farm families to relevant technologies and relevant knowledge and information. Decentralization of extension services to district level coupled with market-driven extension could enhance ease of access. Moreover, participation of the private sector in providing extension services is one of the ways of making them accessible.

5.6 Access to output market

One of the key strategies for a successful integration of grain legumes in smallholder farming systems in Africa is market linkage (Woomer et al., 2004). The extent to which soybean farmers participate in output markets determines the net income benefits. Success of the soybean industry achieved in soybean production, for example, in Nigeria, Uganda and Zimbabwe, was attributed to the incorporation of smallholder soybean producers in the soybean supply value chain (Adelodun, 2011). Value addition through household- and community-level processing played a role in expanding the markets (Chianu et al., 2008). Some of the determinants of farmer's participation in soybean market participation include gender of household head, distance to markets, access to formal sources of market information and cacess to agricultural extension services (Zamasiya et al., 2014). In addition, use of improved varieties and inoculants improved productivity and therefore marketable surplus. Market access can be enhanced through bulking of produce through farmer groups, value addition and access to market information.

6 Conclusion

Increasing soybean yields in SSA will require farmers to adopt improved practices and management that addressed problems of degraded soil fertility and increased disease and insect pressure. Matching soybean varieties to the specific requirements of individual fields is key to maximizing yields. Increased access to new varieties, fertilizers and inoculants will also have a significant impact on overall farm yields. Planting practices, including optimized planting densities and seeding rate, also contribute significantly to soybean yield. Common problems that impede intensification of soybean production relate to the consolidation of smallholder production into sufficiently large enough quantities, timely availability of inputs and human capacity. The greatest successes have been observed where well-developed institutions support robust legume technologies.

7 Where to look for further information

- N2Africa website: www.n2africa.org
- Soybean Innovation Lab (SIL): soybeaninnovationlab@illinois.edu.

8 References

Abate, T., Alene, A. D, Bergvinson, D., Silim, S., Orry, A. and Asfaw, S. (2011). Tropical Legumes in Africa and South Asia. Knowledge and opportunities. TL II Research Report No. 1, ICRISAT, Nairobi, 104p.

Adelodun, K. L. (2011). Soybean: Africa's Potential Cinderella Food Crop. In: Tzi-Bun, N. G. (ed.), *Soybean-Biochemistry, Chemistry and Physiology*, p. 1351. http://www.intechopen.com/books/soybean-biochemistry-and-physiology

Akibode, S. and Maredia, M. (2011). Global and regional trends in production, trade and consumption of food legume crops. Report submitted to CGIAR special panel on impact assessment 27 March 2011. 83p. http://impact.cgiar.org/sites/default/files/images/Legumetrendsv2.pdf

Alene, A. D., Abdoulaye, T., Rusike, J., Manyong, V. and Walker, T. S. (2015). The effectiveness of plant improvement programs from the perspective of varietal output and adoption: cassava, cowpea and soybean and yam in sub-Saharan Africa and maize in west and central Africa. In: Walker, T. S. and Alwang, J. (eds), *Crop improvement, adoption and impact of improved varieties in food crops in sub-Saharan Africa*. CABI. pp. 74–122.

Belmain, S. R., Haggar, J., Holt, J. and Stevenson, P. C. (2013). Managing legume pests in sub-Saharan Africa: Challenges and prospects for improving food security and nutrition through agro-ecological intensification. Chatham Maritime (United Kingdom): Natural Resources Institute, University of Greenwich. 34p.

Chianu, J. N., Ohiokpehai, O., Vanlauwe, B., Adesina, A., de Groote, H. and Sanginga, N. (2008). Promoting a versatile but yet minor crop: Soybean in the farming systems of Kenya. *Journal of Sustainable Development in Africa* 10(2), 722–47.

Crutzen P. J., Mosier A. R., Smith K. A. and Winiwarter W. (2007). N_2O release from agro-biofuel production negates global warming reduction by replacing fossil fuels. *Atmospheric Chemistry and Physics* 8, 389–95.

Döbereiner, J. and Duque, F. F. (1982). Contribuçäo da pescuisa en fixaçäo biologica de nitreogénio para disenvolimento de Brasil. *Rural nomica Rural Brasilia* 18, 447–60.

Druilhe, Z. and Barreiro-Hurlé, J. (2012). Fertilizer subsidies in sub-Saharan Africa. ESA Working paper No. 12–04. Rome, Italy, FAO.

Dugje, I. Y., Omoigui, L. O., Ekeleme, F., Bandyopadhyay, R., Lava Kumar, P. and Kamara, A. Y. (2009). *Farmers' Guide to Soybean Production in Northern Nigeria*. International Institute of Tropical Agriculture, Ibadan, Nigeria. 21 p.

FAOSTAT, (2015). Food and Agriculture Organization of United Nations (FAO) database. http://faostat3.fao.org/download/Q/QC/E. Accessed on 17 July 2016.

Giller, K. E. (2001). Nitrogen Fixation in Cropping Systems. 2nd Edition. CAB International, Wallingford, UK, 423p.

Giller, K. E. and Cadisch, G. (1995). Future benefit from biological nitrogen fixation: An ecological approach to agriculture. *Plant and Soil* 174, 255–77.

Giller, K. E., Franke, A. C., Abaidoo, R., Baijukya, F., Bala, A., Boahen, S., Dashiell, K., Kantengwa, S., Sanginga, J. M., Sanginga, N., Simmons, A., Turner, A. D., De Wolf, J., Woomer, P. L. and

Vanlauwe, B. (2013). N2Africa. Putting nitrogen fixation to work for smallholder farmers in Africa. In: Vanlauwe, B., Van Asten, P. and Blomme, G. (eds), 2013. *Agro-ecological intensification of agricultural systems in the African highlands*. Routledge, London and New York, pp. 156–75.

Gyogluu, C. (2011). Plant Growth and symbiotic functioning of promiscuous- nodulating soybean genotypes inoculated with *Bradyrhizobium japonicum* strain WB 74. MSc thesis. Tshwane University of Technology, South Africa.

Hailu, T. (2011). Breeding for Promiscuous Soybeans at IITA, In: Sudaric, A. (ed.), 2011. Soybean - Molecular Aspects of Breeding, Available from: http://www.intechopen.com/books/soybean-molecular-aspects-of-breeding/breeding-for-promiscuous- soybeans-at-iita

Hartman, G. L, Hill, C. B., Twizeyimana, M., Miles, M. R. and Bandyopadhyay, R. (2011). Interaction of soybean and *Phakopsora pachyrhizi*, the cause of soybean rust. CAB Reviews: Perspectives in Agriculture, Veterinary Science, *Nutrition and Natural Resources* 6(25).

Hartman, G. L., Miles, M. R. and Frederick, R. D. (2005). Breeding for resistance to soybean rust. *Plant Disease* 89, 664–6.

Hungria, M., Franchini, J. C, Campo, R. J. and Graham, P. H. (2005). The importance of nitrogen fixation to soybean cropping in South America. In: Warner, D. and Newton, W. F. (eds), 2005. *Nitrogen Fixation in Agriculture, Forestry, Ecology and Environment*. Springer, The Netherlands, pp. 22–42.

Hungria, M., Campo, R. J., Mendesl. C. and Graham, P. J. (2006). Contribution of nitrogen fixation to the N Nutrition of grain crops in the tropics: The success of Soybean (Glycine max L. Merry) in South America. In: Singh, R. P., Shankar, N. and Jaiwal, P. K. (eds), 2006. *Nitrogen, Nutrition in Plant Production*. Studium Press/LLC, Hauston, Texas, pp. 43–93.

IITA, (2016). Gearing up for impact. IITA 2015 Annual Report. Ibadan, Nigeria. International Institute of Tropical Agriculture (IITA). Available online at www.iita.org/annual-reports

IITA, (2012). Annual Report. Ibadan Nigeria. International Institute of Tropical Agriculture (IITA). Available online at www.iita.org/annual-reports

InternationalInstitute of Tropical Agriculture (IITA) Bulletin, (2015). No.2300. IITA Nigeria. http://www.iita.org/c/document_library/get_file?uuid=e2d82103-bba4-4785-9276-2fe568e12599&groupId=25357

International Institute of Tropical Agriculture (IITA), (2008). Thirty years R4D in soybean: what's next? R4D Review, 21 Sept. 2008. http://r4dreview.org/2008/09/30-years-r4d-in-soybean-what's-next/

Kamara, A. Y., Kwari, J. D, Ekeleme, F., Amoigu L. and Abaidoo R. (2008). Effect of phosphorus application and soybean cultivar on grain and dry matter yield of subsequent maize in the tropical savanna of north-eastern Nigeria. *African Journal of Biotechnology* 7, 2563–99.

Kawuki, R., Adipala, E., Lamo, J. and Tukamuhabwa, P. (2003). Responding to the soybean rust epidemic in sub-Saharan Africa. A Review. *Africa Crop Science Journal* 11, 301–18.

Kumudini, S., Kennedy, B., Godoy, C. V., Omielan, J., Prior, E., Hershman, D. and Tollenaar, M. (2008). Impact of *Phakopsora pachyrhizi* infection on soybean leaf photosynthesis and radiation absorption. *Crop Science* 48, 2343–2.

Levy, C. (2005). Epidemiology and Chemical Control of soybean rust in Southern Africa. *Plant Disease* 89, 669–74.

Mapfumo, P., Mtambanengwe, F., Giller, K. E. and Mpepereki, S. (2005). Tapping indigenous herbaceous legumes for soil fertility management by resource-poor farmers in Zimbabwe, *Agriculture, Ecosystems and Environment* 109, 221–33.

Miles, M. R., Levy, C., Morel, W., Mueller, T., Steinlage, T., van Rij, N., Frederick, R. D. and Hartman, G. L. (2007). International fungicide efficacy trials for the management of soybean rust. *Plant Disease* 91, 1450–8.

Mpepereki, S., Javaheri, F., Davis, J. P. and Giller, K. (2002). Soybean and sustainable agriculture. Promiscuous soybeans in southern Africa. *Field Crops Research* 65, 137–49.

Murithi, H. M., Beed, F., Tukamuhabwa, P., Thomma, B. P. H. J. and Joosten, M. H. A. J. (2015). Soybean production in eastern and southern Africa and threat of yield loss due to soybean rust caused by *Phakopsora pachyrhizi*. *Plant Pathology*, 65, 176–88.

Murithi, H., Fen, B. and Abubakari, K. (2014). *Integrated Pest Management (IPM) of Soybean*. IITA, Ibadan, Nigeria, 36p.

Ojiem, J. O., Vanlauwe, B., de Ridder, N. and Giller, K. E. (2007). Niche-based assessment of contribution of legumes to the nitrogen economy of western Kenya smallholder farms. *Plant and Soil* 292, 119–35.

Pretorius, Z. A., Kloppers, F. J. and Frederick, R. D. (2001). First report of soybean rust in South Africa. *Plant Disease* 85, 1288.

Ronner, E., Franke, A. C. Vanlauwe, B., Diandad, M., Edehe, E,. Ukeme, B., Balaf, A., van Heerwaardena, J. and Giller, K. E. (2015). Understanding variability in soybean yield and response to P-fertilizer and rhizobium inoculants on farmers' fields in northern Nigeria. *Field Crops Res.* http://dx.doi.org/10.1016/j.fcr.2015.10.023

Rusike J., van den Brand, G., Boahen S., Dashiell, K., Kantengwa, S., Ongoma, J., Mongane, D. M., Kasongo, G., Jamagani, Z. B., Aidoo, R. and Abaidoo R. (2013). Value chain analyses of grain legumes in N2Africa: Kenya, Rwanda, eastern DRC, Ghana, Kenya, Mozambique, Malawi, and Zimbabwe, www.N2Africa.org, 96p.

Rwanda Agriculture Board (RAB) report, (2013). http://www.minagri.gov.rw/index.php?id=469&tx_ttnews%5Btt_news%5D=173&cHash=1be6b60ef3c3f37012c41dd782aa7a5a.

Sakurai, G., Iizumi, T., Nishimori, M. and Yokozawa, M. (2014). How much has the increase in atmospheric CO2 directly affected past soybean production? *Scientific Reports*, 4, 4978. http://doi.org/10.1038/srep04978

Sanginga, N. and Woomer, P. L. (2009). Integrated Soil Fertility Management in Africa: Principles, practices and Development Process. Tropical Soil Biology and Fertility Institute of the International Centre for Tropical Agriculture. Nairobi, 263p.

Sanginga, P. C., Adesina, A. A., Manyong, V. M., Otie, O. and Dashiell, K. (1999). Social impact of Soybean in Nigeria's Southern Guinea Savanna. IITA, Ibadan, Nigeria. 32p.

Sanginga, N., Okogun, J. A, Vanlauwe, B., Dield, J. and Dashiell K. (2001). Contribution of nitrogen fixation to the maintenance of soil fertility with emphasis on promiscuous soybean maize-based cropping systems in the moist savanna of West Africa. In: Tian, G., Ishida, F. and Keatinge, J. D. H. (eds), 2001. *Sustaining Soil Fertility in West Africa.* ASA, Wisconsin, pp. 157–78.

Technoserve, (2011). Sothern Africa Regional Soybean Roadmap. Final report. www.technoserve.org/files/../technoserve-bmgf-regional-report.

Tichagwa, J. S. (2004). Breeding for resistance to soybean rust in Zimbabwe. In: Moscardi, F., Hoffmann-Campo, C. B., Saraiva, O. F., Galerani, P. R., Krzyzanowski, F. C. and Carra~o-Panizzi, M. C. (eds), *Proceedings of VII World Soy bean Research Conference, IV International Soybean Processing and Utilization Conference, III Congresso Mundial de Soja (Brazilian Soybean Conference).* Londrina, Brazil: Embrapa Soybean, pp. 349–53.

Tittonell, P. and Giller, K. E. (2013). When yield gaps are poverty traps: The paradigm of ecological intensification in African smallholder agriculture. *Field Crops Research* 143, 76–90.

Tukamuhabwa, P. and Maphosa, M. (2012). State of Knowledge in Breeding for Resistance to Soybean Rust in the Developing World. Rome, Italy: FAO. FAO Plant Production and Protection Paper 204.

Vanlauwe, B., Descheemaeker, K., Giller, K. E., Merckx, R., Nziguheba, G., Wendt, J. and Zingore, S. (2015). Integrated Soil Fertility Management in sub-Saharan Africa. Unraveling local adaptation. *Soil*, 1, 491–508, www.soil-journal.net/1/491/2005/doi:10.5194/soil-1-4-2015.

Waswa, W. (2015). Management Practices Affecting Soybean Production in Western Kenya. Msc Thesis. University of Eldoret, Kenya. 120p.

Woomer, P. L., Huising, J., Giller, K. E. et al. (2014). N2Africa Final Report of the First Phase 2009–2013, www.N2Africa.org, 138p.

Zamasiya, B., Mango, N., Nyikahadzoi, K. and Siziba S. (2014). Determinants of soybean market participation by smallholder farmers in Zimbabwe. *Journal of Development and Agricultural Economics* 6(2), 49–58.

Zheng, H. F, Chen, L. D. and Han, X. Z. (2009). The effects of global warming on soybean yields in a long-term fertilization experiment in Northeast China. *Journal of Agricultural Science* 147, 569–80.

Zingore, S., Delve, R. J., Nyamangara, J. and Giller, K. E. (2008). Multiple benefit of manure: The key to maintenance of soil fertility and restoration of depleted sandy soils of African smallholder farms. *Nutrient Cycling in Agroecosystems* 80, 267–82.

Chapter 7

Developing improved varieties of groundnut

C. Michael Deom, University of Georgia, USA; and David Kalule Okello, National Semi-Arid Resources Research Institute, Uganda

1 Introduction
2 Genetic resources for trait mining
3 Breeding targets
4 Groundnut oil quality and content
5 Case study: groundnut improvement in Uganda
6 Conclusion and future trends
7 Where to look for further information
8 Acknowledgements
9 References

1 Introduction

Increasingly, the health benefits of groundnuts (peanuts) have been illustrated with frequent consumption linked to a reduced risk of cardiovascular disease, type 2 diabetes mellitus, hypertension, and numerous other major chronic diseases (Ros, 2010; Bao et al., 2013). Groundnut is a nutrient-dense food rich in digestible protein, unsaturated fatty acids (e.g. oleic acid), minerals (e.g. copper and manganese), vitamins (e.g. biotin, niacin, folate, B1), fibre, and polyphenolic antioxidants (e.g. p-coumaric acid and resveratrol) (Ros, 2010; Settaluri et al., 2012; Craft et al., 2010). Indeed, Plumpy'nut, a groundnut-based paste, is one of the most widely used ready-to-use therapeutic foods to treat severely malnourished children in developing countries (Guiman and Guiman, 2012). New groundnut technologies with improved traits are routinely introduced into production and can be rapidly scaled up to meet demand.

Groundnuts are cultivated worldwide in tropical and subtropical regions, with production of ~42 million metric tons in 2014 (Fig. 1) from ~26 million hectares (Food and Agricultural Organization of the United Nations [FAOSTAT], 2016). In developing countries, groundnut is an important source of cash income that contributes significantly to food security and poverty alleviation (Smartt, 1994; Moyo et al., 2007; Kessie et al., 2011). In many developing countries where women predominantly grow and manage the crop (Fig. 2 and 3), groundnut

http://dx.doi.org/10.19103/AS.2017.0023.26

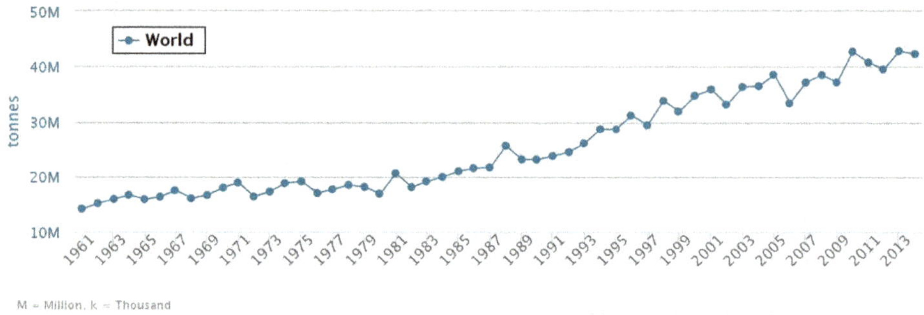

Figure 1 Groundnut (shell) production trends worldwide from 1961 to 2014 (Food and Agricultural Organization of the United Nations [FAOSTAT], 2016). Tonnes = metric tons.

Figure 2 Women harvesting groundnut in the Serere District of Eastern Uganda (top left) and shelling and sorting groundnut in the Pader District of Northern Uganda (top right). Farmers stripping harvested groundnut in the Serere District of Eastern Uganda (bottom left) and a mother/farmer digging groundnut during dry season harvest.

cultivation has a direct bearing on the overall nutritional status and financial well-being of women and children.

As a legume, groundnuts fix nitrogen (Peng et al., 2017), which is especially useful in increasing productivity in low fertility soils that occur naturally, as a result of deforestation, or from extensive and improper agricultural practices. Groundnut requires few inputs, making it appropriate for cultivation by farmers in low-input agriculture environments. Smallholding farmers typically have holdings of less than one hectare and grow different crops, practice non-mechanized farming, and have limited resources for agricultural operations. In developing countries, groundnuts are grown by smallholding farmers as a subsistence/cash crop under rain-fed conditions, in one or two growing seasons a

Figure 3 Drying harvested groundnut in Uganda (top). Woman farmer in the village of Kibong in the Pader District of Northern Uganda showing an improved variety of Erudurudu, a popular landrace in Uganda (bottom).

year, depending on rainfall patterns. In contrast, in developed countries, groundnuts are generally grown on large family or commercial farms under irrigation with sufficient resources to maximize yields.

Yield and factors that affect yield are the most heavily targeted traits in groundnut improvement programmes worldwide (Janila et al., 2013, 2016a); this is especially true in developing countries. Yield is determined by varietal genetics, biotic and abiotic factors, and management practices. Therefore, primary challenges in developing improved varieties of groundnut are in identifying high-yielding varieties for the specific agro-ecological zones being targeted and in addressing factors such as resistance to biotic stresses (plant pathogens) and tolerance to abiotic stresses (drought, including early maturity adaption to special cropping systems, and heat stress). Additional traits that do not affect yield *per se*, but are currently important in improving quality, are high and/or improved oil content and resistance to mycotoxin-producing fungi . Mitigating mycotoxin contamination is a priority because of the impact of the toxins on food security, including human and animal health, and on marketability of the crop. This chapter reviews advances in groundnut improvement and includes a case study to provide a real-life example of how a groundnut improvement programme is accomplishing this goal.

2 Genetic resources for trait mining

It is important for breeders to have access to a diverse array of germplasm in developing improved groundnut varieties. A number of *ex situ* germplasm collections are available as genetic resources for breeders (Barkley et al., 2016, Janila et al., 2013). The largest is at the International Crops Research Institute for the Semi-Arid Tropics (ICRISAT) in India. As a member of the Consultative Group for International Agricultural Research Consortium, ICRISAT has a mandate for research and development on groundnut improvement. ICRISAT has a collection of 15446 accessions, of which 14968 are cultivated groundnut and 478 are wild species (Barkley et al., 2016). Two smaller, but substantial, collections are also available as a genetic resource at the USDA Agricultural Research Service (ARS) Plant Genetic Resources Conservation Unit in the United States (~9321 cultivated accessions and 655 wild accessions) and the Oil Crops Research Institute (OCRI) in China (~8439 cultivated accessions) (Barkley et al., 2016). Additional smaller collections also exist in Argentina, Brazil and Indonesia. Standardized descriptors are available for most of the accessions in the collections to aid in selection.

Since the collections are too large to efficiently mine for traits, the core collection concept was developed (Frankel, 1984) and has been applied to the groundnut collections (Holbrook et al., 1993; Upadhyaya et al., 2003; Jiang et al., 2008; Barkley et al., 2016). A core collection is approximately 10% of the main collection, which is selected to represent the genetic diversity of the crop. Since core collections of the large repositories are still difficult to manage in breeding programmes, the mini-core concept was subsequently established (Upadhyaya and Ortiz, 2001; Holbrook and Dong, 2005; Jiang et al., 2010). Mini-cores comprise about 10% of the core collection (approximately 1% of the primary collection), are selected to represent the genetic diversity of the crop, and are especially useful for breeders to facilitate crop improvement programmes. The core/mini-core concepts have been productively utilized to identify a number of diverse agronomic traits, resulting in increased groundnut yield and quality (Upadhyaya, 2005; Upadhyaya et al., 2012, 2014; for a review, see Barkley et al., 2016).

3 Breeding targets

3.1 Yields

While breeding for improved varieties varies depending on the particular objectives of the breeding programme, the ultimate goal of any groundnut improvement programme is to produce high-yielding cultivars. Yield is a quantitative trait that is greatly influenced by genotype and environmental interactions (G x E interactions) (Zhang et al., 2011). Physiology traits, such as crop growth rate, length of reproductive growth and the amount of the growth rate allocated to pod yield (Janila et al., 2013) affect pod number, seed size and seed weight, which are important parameters contributing directly to yield and are targeted in breeding programmes (Rathnakymar et al., 2012; Janila et al., 2013; Lal et al., 2014; Chen et al., 2016; Janila et al., 2016a). However, numerous other genetic traits (e.g. maturity, seed dormancy and peg strength) have an indirect but significant effect on yields (Janila et al., 2013), even under optimal environmental conditions.

Groundnut is susceptible to a number of biotic and abiotic stresses (Dwivedi et al., 2003; Okello et al., 2013). These stresses prevent the crop from attaining its full genetic yield potential and are extensively involved in determining the difference between the potential yield and actual yield. On-station yields in developing countries, where growth conditions are generally optimized, are typically much higher than on smallholder plots, where few if any inputs are used to control environmental influences (biotic or abiotic stresses).

Therefore, while physiological traits that increase yields are targeted in breeding programmes, overcoming biotic and abiotic stresses can have huge effects on groundnut yield and quality, and these traits have drawn considerable attention in groundnut improvement programmes worldwide in both developing and developed countries. Indeed, much of the discussion in this review on yield will focus on closing yield gaps caused by biotic and abiotic stresses. This means identifying and moving resistance/tolerance traits into breeding lines. In some cases, such as disease resistance in groundnuts, there is a fitness cost with increased disease resistance coming at the expense of yield (Nigam et al., 1991; Knauft and Wynne, 1995). The fitness cost of resistance against pathogens and pests has been shown in other plant-pathogen systems (Mauricio, 1998; Tian et al., 2003; Cipollini and Heil, 2009; Todesco et al., 2010; Vos et al., 2013; Cipollini et al., 2014; Bally et al., 2015). While less is known on fitness costs of resistance to abiotic stress, it has been shown that abiotic stress can up-regulate R-genes and can therefore result in a fitness cost to the effected plant, even more so in the absence of pathogens (MacQueen and Bergelson, 2016). Therefore, when developing improved groundnut varieties with resistance/tolerance to biotic and abiotic stresses to optimize yield, cost-benefit trade-offs may result in a compromise between yield potential and the level of resistance/tolerance.

3.2 Biotic stress resistance

Recent estimates suggest that crop losses worldwide due to plant pathogens range from 10 to 16% (Strange and Scott, 2005; Oerke, 2006) and 13 to 18% for insects (Pinstrup-Andersen, 2001; Oerke, 2006). As expected, a number of plant pathogens and pests also place significant production constraints on groundnut. Important groundnut pathogens worldwide include viruses (e.g. groundnut rosette disease [GRD], tomato spotted wilt virus [TSWV] and peanut bud necrosis virus [PBNV]); fungi (e.g. early leaf spot, ELS [*Passalora arachidicola*], late leaf spot, LLS [*Passalora personata*], rust [*Puccinia arachidis*], stem rot [*Sclerotium rolfsii*], and aflatoxin contamination (*Aspergillus flavus* and *Aspergillus parasiticus*); and bacteria (bacterial wilt [*Ralstonia solanacearum*]). If not controlled, these pathogens can cause significant loss to yield and quality in groundnut. Insect pests include aphids and thrips (which, in addition to causing feeding damage, vector GRD and tospoviruses, respectively) and groundnut leaf miner. The most feasible, efficient, and safest approach to controlling pathogens and pests is to breed for resistance. This is especially true in developing countries where the cost of pesticides is prohibitive for resource-poor smallholder farmers.

While a number of pathogens are important worldwide, others have more regional impact. To keep the discussion focused, we target a few of the groundnut pathogens that have widespread impact or that have been extensively targeted in breeding programmes.

3.2.1 Viruses

GRD is the most destructive viral disease of groundnut in sub-Saharan Africa (SSA), and arguably the most important pathogen disease on groundnut, with yearly crop losses of approximately $156 million (Naidu et al., 1999). The disease is endemic to SSA and is a major constraint to groundnut production. GRD can cause up to 100% yield losses if infection occurs before flowering and during severe epidemics (Naidu et al., 1999; Alegbejo and Abo, 2002; Okello et al., 2010a).

GRD results from an intimate interaction between groundnut rosette assistor virus (GRAV, member of the genus *Enamovirus*), groundnut rosette virus (GRV, member of the genus *Umbravirus*), and the satellite RNA (satRNA) of GRV. Together the three components result in a complex aetiology that is crucial to the development and transmission of the disease. GRV acts as a helper virus for replication of satRNA (Taliansky et al., 1996; Murant et al., 1988). SatRNA is required for aphid transmission of GRV (Murant, 1990) and is primarily responsible for inducing the different symptoms of GRD (green rosette and chlorotic rosette, although variations between the two occur; Fig. 4) (Murant et al., 1988; Murant and Kumar, 1990; Taliansky et al., 1997). GRAV acts as a helper virus for aphid transmission in that GRV, which does not encode for a coat protein (CP), and satRNA are packaged in the CP of GRAV to form particles that are aphid transmissible (Murant, 1990; Robinson et al., 1999). The disease is transmitted by the aphid, *Aphis craccivora* Koch. For the

Figure 4 Stunted plants with groundnut rosette disease having chlorotic symptoms (top left) and green symptoms (top right). Groundnut rosette disease susceptible variety in the foreground and resistant variety in the background (bottom).

disease to develop and be transmissible, all three disease agents must be transmitted together by a single aphid. Therefore, disease resistance mechanisms would be expected to interfere with the life cycles of GRV, GRAV or both viruses simultaneously.

Because of the ubiquitous and virulent nature of GRD, the breeding programmes in SSA insure that most newly released varieties have some level of GRD resistance. Resistance to GRD was initially identified in landraces of late-maturing Virginia types from West Africa (Naidu et al., 1999), which were used in the early GRD resistance breeding programmes. GRD resistance identified in cultivated groundnut is directed against GRV and the satRNA, is effective against both green and chlorotic symptom types, and results from two independent recessive genes (Nigam and Bock, 1998). Numerous sources of resistance to GRD have been reported, including 101 late-maturing Virginia types and 15 early maturing Spanish types (Subrahmanyam et al., 1998). In another study, 65 accessions, including seven early maturing types, were reported to have field resistance (Olorunju et al., 2001). The source of resistance for a majority of the resistant accessions from both studies originated from the resistant lines originally identified in West Africa (Subrahmanyam et al., 1998; Olorunju et al., 2001). Presently, details are not available on how the resistance mechanism(s) functions. While breeding programmes in SSA have moved GRD resistance into all maturing types, they have especially focused on early maturing (90–110 days), high-yielding Spanish and Virginia types that are adapted to the variable climates and agro-ecological zones of SSA.

A number of GRD-resistant cultivars have been released in Africa (Singh and Nigam, 2016), including the SAMNUT series from Nigeria and the Serenut series from Uganda (Nagaraji, 2014; Okello et al., 2015). In addition, a number of National Breeding programmes continue to evaluate and directly release primary introductions from the ICRISAT germplasm collection with resistance to GRD, or include the lines in their breeding programmes to develop improved varieties with GRD resistance.

To date, GRAV resistance has not been detected in cultivated groundnut. However, resistance to GRAV has been found in wild *Arachis* species along with resistance only to GRV and satRNA as well as resistance to all three disease components (Subrahmanyam et al., 2001). Moving any of the resistance traits from wild *Arachis* into cultivated groundnut would increase the presently narrow genetic base of resistance. GRAV resistance in cultivated groundnut could be used to generate more durable and robust resistance to GRD by stacking the GRAV resistant mechanism(s) onto the GRV resistance mechanism(s). In addition, GRAV infection of groundnut in the absence of GRV and satRNA causes significant vigour and yield losses (Naidu and Kimmins, 2007) that would be mitigated with GRAV resistance.

Another approach that has been exploited to control GRD is the use of aphid resistant accessions. The landrace ICG 12991 has resistance to the aphid vector, but does not have resistance to GRV or GRAV. The mechanism was suggested to be an inducible hypersensitive response at the aphid feeding site that inhibits phloem feeding, resulting in limited aphid population growth and increased aphid mortality (Willekens, 2003). However, resistance in ICG 12991 can be broken under high aphid pressure in the field. The landrace has been used extensively to manage GRD in Malawi (released in 2001 as 'Baka'; Deom et al., 2006), Mozambique, Uganda (release in 2002 as 'Serenut 4T'; Deom et al., 2006), and Zambia. Resistance is controlled by a single recessive gene (Herselman et al., 2004).

Tomato spotted wilt disease, which is caused by TSWV, is a major constraint to production on groundnut in the United States. The disease is especially significant in the southeastern USA, which produces approximately 85% of the crop. Over a 10-year period ending in

2011, annual average losses in Georgia alone were estimated at $12.3 million (Riley et al., 2011). TSWV is vectored by two thrips species, *Frankliniella fusca* Hinds (tobacco thrips), considered the more important species for transmission, and *F. occidentalis* (western flower thrips) (Riley et al., 2011). To date, there is no known resistance in groundnut to the thrips vector. Field resistance to TSWV is not immunity and is generally characterized by less symptomatic plants and higher yields (Culbreath et al., 2003), suggesting the resistance is quantitative. The most important tool in managing the disease is TSWV-resistant cultivars; however, optimal yields require resistant cultivars in concert with a suite of cultural practices (i.e. planting date, planting density, etc.) (Kemerait et al., 2016).

Since the rise in TWSV incidence in the 1980s, considerable effort has gone into breeding programmes to develop high-yielding cultivars with TSWV resistance, with the majority of released cultivars having some level of TSWV resistance. The first generation of TSWV-resistant cultivars, which had moderate levels of field resistance to TSWV, was released from the mid-1980s to approximately 1998 and included Southern Runner, Georgia Green (grown extensively) and Georgia Browne (Sundaraj et al., 2014). Further breeding efforts resulted in a second generation of cultivars that typically had higher levels of TSWV resistance and included Georgia-02C, Georgia-06G (a widely-grown variety), Tifguard, Georgia Greener and Georganic (Sundaraj et al., 2014). More recently, a third generation of high-yielding cultivars has been released with a trend for higher levels of TSWV resistance than the second generation, although levels vary depending on various other traits being targeted in the breeding programmes. Cultivars in this group include Georgia-07W, -09B (high oleic), -10T, -12Y (the highest-rated level of resistance for this group), -13M (high oleic) and -14N, Florida-07 (high oleic), FloRun 107, TUFRunner 511 (high oleic), TUFRunner 727 (high oleic), and TUFRunner 297 (high oleic) (Branch and Brenneman, 2008; Branch, 2010; Branch and Culbreath, 2011; Branch, 2013; Branch, 2014; Branch and Brenneman, 2015; Gorbet and Tillman, 2009; Tillman and Gorbet, 2015; Hollis, 2015). The mechanism functioning in field resistance is unknown, but is likely similar, if not identical, for all the above-mentioned resistant cultivars. The source of TSWV resistance in all the above-cited cultivars comes from C-99R, a descendant of PI 203396. No TSWV strains have been identified that break TSWV resistance in cultivated groundnut. This might be explained by work on second-generation cultivars, suggesting that the levels of resistance might not exert significant levels of selective pressure on the TSWV population (Sundaraj et al., 2014). Additional sources of TSWV resistance that are not derived from C-99R have been identified and would be welcome, especially those with different resistance mechanisms to increase the genetic base (Lyerly et al., 2002; Culbreath et al., 2005).

Peanut bud necrosis disease is a major constraint to groundnut production in South and Southeast Asia. The disease is caused by PBNV and is transmitted by *Thrips palmi*. Yearly crop losses of approximately $89 million (Reddy et al., 1995) can occur with over 50% yield losses (Dwivedi et al., 1995). Large-scale screening of the ICRISAT germplasm collection has identified several accessions as well as a number of breeding lines/cultivars with PBNV resistance/tolerance (Nigam et al., 2012; Srinivasaraghavan et al., 2013; Singh and Nigam, 2016). Two breeding lines – ICGV 86031 and 86388 – were found to be resistant to both the virus and the *Thrips* vector, while a number of accessions have resistance only to the vector (Dwivedi et al., 1995; Singh and Nigam, 2016). It would be interesting to determine if resistance to *T. palmi* is functional against the two thrips species that transmit TSWV, *F. fusca* and *F. occidentalis*. In Thailand, two groundnut lines (ICGV 86388 and IC 10) were found to be resistant to PBNV (Pensuk et al., 2004). Eight PBNV-resistant accessions were identified in four wild *Arachis* species (Nigam et al., 2012), which could be a source of genetic material to widen the genetic base of resistance. At least ten resistant cultivars have been released in India and one in Thailand (Singh and Nigam, 2016).

3.2.2 Aflatoxin

Aflatoxins are a class of mycotoxins mainly produced by two genera of *Aspergillus*, *A. flavus* (predominant in Africa and Asia) and *A. parasiticus* (predominant in the USA), during drought stress and extreme temperatures. Breeding for resistance to aflatoxin contamination in groundnut has received considerable attention in recent years due to the adverse effects of the toxins on food safety and human health. The toxins cause liver damage and cancer in humans and in some animals, childhood stunting and arrested development, and immunosuppression of cell-mediated immunity (UNEP, 2016; Jolly et al., 2008). Aflatoxins affect approximately 4.5 billion people in developing countries. In developed countries, maximum allowable amounts of aflatoxin in food and feeds (internally derived and imported) have been established by law and are closely monitored, which is not the case in most developing countries. In addition to the toll on human and animal health, it is estimated that Africa loses approximately $450 million annually to crops that cannot be sold because of contamination (Nowakowski, 2015).

To date, some resistance to aflatoxin has been identified, but no immunity has been detected in groundnut. Three resistance mechanisms could lead to aflatoxin control and could be components of a breeding programme (Singh and Nigam, 2016). The first is pod shell structure, which is the primary barrier to *Aspergillus* infection. For economic reasons, this approach has not been pursued by ICRISAT, since the ease of shelling is an important trait in groundnut processing and since interfering with this trait could negatively affect the acceptance of a variety (Nigam et al., 2009). The second is the seed coat barrier to *Aspergillus*. Liang et al. (2003) found a correlation between thicker waxy seed coat content and resistance to *A. flavus*, suggesting the seed coat can act as an effective physical barrier. Asis et al. (2005) found that resistance to *Aspergillus* colonization and aflatoxin accumulation correlated with seed coat integrity in three *Aspergillus*-resistant accessions (PI 337394, PI 337409 and J11) and Jiang et al. (2006) showed that seed coat integrity was important for resistance to Aspergillus infection in two accessions (G845 and G8). The third resistant mechanism is immune responses to the pathogens that occur in the cotyledon, the site of aflatoxin production, such as PAMP-triggered immunity (PTI; Zipfel and Robatzek, 2010) and effector-triggered immunity (ETI; Cui et al., 2015). Since the three approaches would target distinct mechanisms that are independently inherited, they could be stacked onto each other to provide a more durable and higher level of resistance (Upadhyaya et al., 2002). However, there is limited information on the genetics of the resistance mechanisms, and reliable screening techniques are lacking since aflatoxin contamination is extremely variable. To obtain a better understanding of the molecular interaction between *A. flavus* and groundnut, Wang et al. (2016) analysed differences in the RNA-seq/transcriptome profiling data of *A. flavus* on infected seed from resistant (Zhonghua 6) and susceptible (Zhonghua 12) cultivars. Differential expressed genes associated with mycelial growth, conidial development and aflatoxin biosynthesis were up-regulated in *A. flavus* infecting the susceptible host compared with the resistant host, suggesting that *A. flavus* mycelia more easily penetrate cell walls of seed from the susceptible cultivars and produce more aflatoxin than in resistant plants (Wang et al., 2016). The lower levels of aflatoxin in resistant seed might result from immune resistance within the cotyledons, where aflatoxin is produced.

Because aflatoxin contamination is determined by multiple factors, limited resistance provided by the various resistance components, and high G x E interactions, controlling aflatoxin contamination is complex and progress in developing resistant/tolerant plants has been limited (Nigam et al., 2009; Bandyopadhyay et al., 2016). Progress in breeding for resistance to aflatoxin, as well as some promising accessions, has been reviewed by Nigam et al. (2009) and Singh and Nigam (2016) for India and Africa, Liang et al. (2009) for China, and Holbrook et al. (1993, 2009) for the United States. Because drought and heat stress predispose groundnut to aflatoxin contamination, resistance/tolerance to these two abiotic stresses may help in mitigating aflatoxin contamination (see drought and heat stress below). The roles that environmental stress, the accumulation of reactive oxygen species and aflatoxin play in *A. flavus* – host interactions – as well as the current status of molecular breeding and marker development for use in following *A. flavus* resistance in groundnut – have been reviewed (Fountain et al., 2015). Since numerous factors influence pre- and post-harvest aflatoxin contamination throughout the groundnut supply chain (Diao et al., 2015), it is likely that any aflatoxin mitigation breeding programme will also require optimizing pre- and postharvest management practices to minimize conditions that favour aflatoxin contamination independent of plant genetics (Okello et al., 2010b; Waliyar et al., 2008; Monyo et al., 2012; Waliyar et al., 2015).

An approach taken to begin identifying aflatoxin-resistant genes in maize that may be useful in groundnut was to compare proteomic data between susceptible and resistant cultivars (Chen et al., 2009). A number of resistance-associated genes were identified. When RNAi was used to silence two of the resistance-associated genes in transgenic resistance lines, some of the lines were significantly more susceptible to *A. flavus* infection and toxin production. Another approach used an expressed sequence tag (EST)/microarray strategy to identify 62 genes that were up-regulated in a resistant groundnut cultivar following *Aspergillus* infection (Guo et al., 2011). Twenty-two putative *Aspergillus*-resistant genes were constitutively expressed in the resistant cultivar compared to the susceptible cultivar. The homologues of some of the genes identified were previously shown to be involved in fungal resistance in groundnut, soybean and maize (Guo et al., 2011). Recent transgenic approaches to controlling *A. flavus*, as well as foliar pathogens, appear promising as a source of information towards understanding resistance mechanisms, which could provide insights and be useful in directing breeding programmes. Prasad et al. (2013) illustrated that expression of the rice chitinase gene, *Rchit*, in transgenic groundnut resulted in transformants with significantly greater disease resistance to *A. flavus*, LLS and rust compared to control plants. The premise was that chitinase-directed degradation of chitin in fungal hyphae, which delays fungal growth, induced the release of elicitors that stimulate the plant immune system. Sundaresha et al. (2010) showed that overexpression of pathogenesis-related protein 2 (β 1-3 glucanase) from tobacco in transgenic groundnut confers resistance to ELS and *A. flavus*. The resistance to *A. flavus* was characterized by inhibition of hyphal growth and aflatoxin accumulation. Similarly, co-overexpression of two antifungal genes, *non-expressor of Pathogenesis-Related genes-1* (*NPR-1*), which regulates the onset of systemic acquired resistance, and *defensin*, in transgenic groundnut resulted in resistance to *A. flavus* and ELS (Sundaresha et al., 2016).

3.2.3 Foliar diseases

Globally, ELS, LLS and rust are the most widespread foliar diseases in groundnut. ELS and LLS can occur simultaneously, but the prevalence of each depends on the cultivar, cropping system and environment. The three foliar pathogens negatively affect yields, as well as haulm and seed quality, primarily by decreasing photosynthetic capacity and increasing defoliation. Yield losses for ELS and LLS in West and central Africa ranged from 20 to 50% and greater than 50% in combination with groundnut rust (Waliyar et al., 1994). Components of resistance that are generally measured to identify lines that are resistant to the foliar diseases are lower infection frequencies, lower sporulation rates, small number of lesions and small lesion size, and longer incubation periods (Singh et al., 1997). Of the three, LLS is the most widespread and well studied (Fig. 5).

Scientists from ICRISAT and China identified 69 and 53 accessions, respectively, with resistance to LLS and 169 and 92 accessions, respectively, with resistance to rust (Subrahmanyam et al., 1989, Subrahmanyam et al., 1995; Singh et al., 1997, 42; Liao, 2014). From the ICRISAT screening program, 42 accessions were found to be resistant to LLS and rust (Singh et al., 1997). Accessions with resistance to LLS and rust occurring either independently or together were predominantly found in var. *fastigiata* and var. *peruviana*, or from wild species (Subrahmanyam et al., 1989, Subrahmanyam et al., 1995;

Figure 5 Late leaf spot disease symptoms on leaves (top). Severely affected and defoliated groundnut plants showing late leaf spots symptoms (bottom).

Singh et al., 1997; Chattopadhyay et al., 2016). For LLS and rust resistances, a linkage was detected between resistance and poor agronomic traits, including low yields, undesirable pod traits, late maturity, low shelling outturn and poor adaptability, resulting in many of the resistance lines not being useful in breeding programmes. Groundnut lines with resistance or moderate resistance to ELS, LLS, ELS and LLS, and LLS and rust that are currently used in various growing regions of the world, largely India and the United States, have been described (Singh and Nigam, 2016; Chattopadhyay et al., 2016). High levels of resistance to ELS, LLS and rust are found in wild species of *Arachis*. Several interspecific populations have resulted in lines with acceptable levels of resistance to LLS and rust with more favourable traits for use in breeding programmes (Singh et al., 1997; Liao, 2014). Interspecific hybridization is being used to introgress resistance from diploid species of *Arachis* that are cross-compatible into the narrow genetic base of cultivated groundnut. In addition, synthetic amphidiploids have recently been used to introgress resistance genes from genetically diverse wild species into cultivated groundnut (Kuman et al., 2014). As discussed above (see *Aflatoxin*), transgenic technology used to generate plants with resistant to *Aspergillus* has also provided information about the mechanisms of resistance to ELS, LLS and rust (Sundaresha et al., 2010; Prasad et al., 2013; Sundaresha et al., 2016). Another study generated transgenic groundnut plants that constitutively co-expressed two antifungal proteins from black nightshade and radish (*Solanum nigrum* osmotin-like protein [*SniOLP*] and *Raphanus sativus* antifungal protein-2 [*Rs-AFP2*] genes, respectively) (Vasavirama and Kirti, 2012). The transgenic plants were found to have significantly higher levels of resistance to LLS.

Inheritance studies suggest that resistance to ELS, LLS and rust are complex and quantitative traits controlled by multiple recessive genes (Singh et al., 1997; Khedikar et al., 2010). One major quantitative trait locus (QLT) has been recently identified for rust resistance (Khedikar et al., 2010) and work is in progress to identify additional molecular markers for LLS and rust resistance to use in future breeding programmes (Leal-Bertioli et al., 2009; Varshney et al., 2014; Sukruth et al., 2015; Kolekar et al., 2016).

Finally, because of limited foliar disease resistance in cultivated groundnut, control has required, and will continue to require, an integrated approach involving best cultural practices (i.e. rotation of groundnuts with other crops and management of groundnut residue following harvest to minimize primary inoculum in the field) and fungicide application programmes in concert with resistant varieties to optimize management of foliar diseases.

3.3 Abiotic stress resistance

3.3.1 Drought and heat stress

Drought and extreme temperatures are the most important abiotic stresses that affect groundnut and are serious constraints to yield and seed quality. The outcome of drought stress is dependent on when it occurs relative to various growth stages of the plant (Holbrook et al., 2016). Moderate drought stress during the early stage (pre-flowering, vegetative) has the least detrimental effect on groundnut yield and quality. In contrast, mid-season drought (flowering and pod development) and late-season drought (pod fill) result in yield losses, with late-season being the most severe (Holbrook et al., 2016). These stresses are expected to become even more of a constraint on production with rising temperatures and erratic weather patterns anticipated as a result of climate change (Lobell et al., 2007; Schlenker and Roberts, 2009; Ray et al., 2015).

Efficient use of soil moisture in drylands with unpredictable seasonal rainfall is an important component of drought tolerance and can be expressed as water-use efficiency (WUE) or as the genetic component of WUE, transpiration efficiency (TE) (Blum, 2005; Vadez et al., 2014). TE is difficult to measure, but a number of easy-to-measure proxy traits, especially specific leaf area (SLA) and soil plant analysis development (SPAD) chlorophyll meter reading (SCMR), have been correlated with TE in some studies and have been used to screen groundnut accessions for drought tolerance (Valez et al., 2014: Holbrook et al., 2016). However, there are limitations to using the surrogate traits (Valez et al., 2014) and there is research suggesting that SCMR and SLA may not be relevant predictors of drought tolerance in all cases, but may be genotype specific (Hamidou et al., 2012; Holbrook et al., 2016 and references within).

The ICRISAT mini-core collection was screened for drought tolerance and 18 accessions were identified with favourable SLA and SPAD values (Upadhyaya, 2005). In addition, four high-yielding accessions were selected under water-stress conditions from an ICRISAT reference collection of 247 accessions (Hamidou et al., 2012). In the later study, no correlation was observed with SLA and SCMR, but a correlation was observed between yield, harvest index and pod growth rate with drought. Genotype effects as well as genotype x irrigation interactions were significant, suggesting that it is important to screen for drought tolerance traits under drought stress conditions. In contrast, screening of the US mini-core has resulted in the selection of 18 accessions based on SLA and SCMR values relative to controls and these accessions are available for use in improvement programmes for drought tolerance (Selvaraj et al., 2010).

Drought tolerance in plants is a complex trait that has evolved to encompass multiple mechanisms at the morphological, cellular, molecular and biochemical levels (Fang and Xiong, 2015). Depending on the plant, different mechanisms function at different developmental stages. For an overview of drought resistance in crops, Hu and Xiong (2014) provides an extensive general review covering the genetics, functional genomics, genetic engineering and breeding of drought-tolerant crops. Some of the information developed in well-established crops (e.g. genes providing drought-resistant traits) will likely be applicable to groundnut.

In groundnut, a number of studies sequencing transcriptomes of drought-stressed versus non-stressed plants have identified genes involved in drought tolerance (Sun et al., 2013; Holbrook et al., 2016 and references within). Some of the drought-tolerance-associated genes, such as dehydration-responsive element binding (*DREB*) transcription factors and isopentenyltransferase (*IPT*) have been overexpressed in transgenic plants and have provided encouraging results about the mechanisms of drought tolerance (Pruthvi et al., 2014; Sarkar et al., 2014; Sarkar et al., 2016). *DREB* transcription factors regulate many abiotic-stress-inducible genes and have been shown to be critical in improving abiotic-stress tolerances (Lata and Prasad, 2011). *IPT* plays a critical role in cytokinin biosynthesis and can delay drought-induced senescence when expressed in transgenic plants (Smart et al., 1991; Gan et al., 1995). When expressed under the regulatory control of a water-deficit inducible promoter, *IPT* overexpression induced extreme drought tolerance in transgenic *Nicotiana tabaccum* cv. SR1 (Rivero et al., 2007) and significantly reduced drought-induced yield losses in groundnuts (Qin et al., 2011). As more drought-tolerance-associated genes are identified and studied, additional approaches to mitigating drought will be an option.

Screening for yield under drought stress will continue to be an approach many breeders will use for identifying tolerant lines as well as including accessions with drought-tolerant traits into breeding programmes. However, one of the most useful approaches is early maturity. This trait is extremely useful in rain-fed growing systems of the semi-arid tropics characterized by low and erratic rainfall and where growing seasons are short. Therefore, early-maturity lines represent a means of escaping drought. Prior to 2006, the early maturing variety that was primarily used in breeding programmes was Chico (*vulgaris*), although Gangapuri (*fastigiata*) and JL24 (*vulgaris*) were also used to a lesser extent (Upadhyaya et al., 2006). To overcome the narrow genetic base and to provide improved agronomic traits, the ICRISAT core collection was screened for the early-maturing trait (Upadhyaya et al., 2006), identifying 21 accessions and four landraces. Early flowering in groundnut is a simple inherited trait (Upadhyaya and Nigam, 1994), which could be easily selected in breeding programmes. In addition to early maturity, a number of accessions with drought-tolerance traits including high TE, high harvest index, high root mass and high SPAD have been studied to determine the biological bases for the tolerance and could be used to enhance groundnut under drought-stressed environments (Holbrook et al., 2016).

Heat stress, alone or in combination with drought, can have a significant effect on plant growth, development and reproduction. Effects of heat stress can be overcome by adaptive adjustments in several biochemical pathways and physiological functions, resulting in thermotolerance (Maestri et al., 2002). Plants have an inherent or basal tolerance to survive heat stress and an acquired thermotolerance (ATT, acclimation) to high/lethal temperatures, which is ubiquitous in different crop plants (Senthil-Kumar et al., 2007). The best characterized ATT response is the production of heat shock proteins (reviewed in Senthil-Kumar et al., 2007). Very little is known about ATT in groundnuts, likely due to the lack of a high-throughput screening procedure for following the multigenic trait in segregating populations (Senthil-Kumar et al., 2007). However, a recently developed leaf-disc bioassay which detects changes in temperature sensitivity of chlorophyll accumulation has shown promise as a surrogate of ATT and as a tool to identifying heat-tolerant groundnut lines (Selvaraj et al., 2011).

4 Groundnut oil quality and content

One of the most widely promoted quality traits in groundnut in recent years is high oleic chemistry (high oleic to linoleic ratio). Increasing the oleic fatty acid content from 45–55% to approximately 80% results in improved oil quality that increases the shelf-life of groundnuts, helps maintain desirable flavour attributes during storage, and improves health benefits by reducing the risk of cardiovascular disease.

The discovery of high oleic acid groundnut, F435 (Norden et al., 1987), and the availability of this germplasm to US breeders have immensely facilitated the development of groundnut cultivars with high oleic chemistry. Additional cultivars with the high oleic trait are derived from F435 by traditional breeding programmes or through chemical/radiation mutagenesis followed by inclusion of the mutants into breeding programmes (Chu et al., 2009). High oleic acid results from loss-of-function mutations in the fatty acid Δ^{12} desaturase enzymes in groundnut, the *ahFAD2A* gene (from the A subgenome) and/or *ahFAD2B* gene (from the B subgenome) (Chu et al., 2009). A number of high oleic cultivars have been released in Spanish, Virginia, Runner and Valencia types (Chu et al., 2009; Gorbet and Tillman, 2009; Branch, 2010; Branch, 2014; Holbrooke et al., 2016; Hollis, 2015; Janila et al., 2016b). Extensive knowledge of fatty acid synthesis in plants and the characterization of mutants that result in the high oleic trait have led to the development of molecular markers and protocols for marker-assisted selection (MAS) as early as the F2 stage, which optimizes breeding time and resources (Barkley et al., 2010; Chen et al., 2010; Janila et al., 2016b).

High oil content in groundnut is important for the oil processing industry, especially in countries where the use of groundnut oil for cooking is popular. Most groundnut accessions with an oil content greater than 55% are in the Spanish genetic background (Liao, 2014). Large-scale screening of groundnut accessions in China identified accessions with oil content up to 60.2% (Liao, 2014). Additional screening of 87 wild *Arachis* accessions identified 12 accessions with oil content higher than 58%. Unfortunately, high oil content cultivars appear to be less resistant to pathogen infection, such as LLS and *A. flavus* (Liao and Holbrooke, 2007; Liao, 2014). Recent research has shown that genetic manipulation of the fatty acid biosynthetic pathway can be used to facilitate improvements in oil production in plants. Ectopic expression of the groundnut lysophosphatidic acid acyltransferase gene (*AhLPAT2*) in transgenic Arabidopsis seed resulted in higher levels of total fatty acids and the proportion of unsaturated fatty acids in Arabidopsis seed (Chen et al., 2015). In addition, overexpression of the *AhLPAT2* gene resulted in increased levels of acyl carrier protein 1 and acetyl-CoA subunit, two enzymes involved in *de novo* fatty acid synthesis.

5 Case study: groundnut improvement in Uganda

In this review, the case study focuses on collaborative research that has resulted in the release and ongoing development of high-yielding improved groundnut varieties with pathogen/pest resistance, drought tolerance and high oleic content in Uganda. The outcome of the research has and will continue to result in the development and adoption of improved cultivars that will provide increased yields and increased quality.

New sources of resistances to a range of stresses are continuously being identified at the National Semi-Arid Resources Research Institute (NaSARRI) in Uganda in their groundnut improvement programme. Much of the groundnut germplasm used in Uganda and Southern Africa was initially screened for improved traits at ICRISAT-Malawi and were provided as primary introductions. Targeted lines were disseminated to SSA countries for screening in the specific agro-ecological zones of each country. Uganda is a known hotspot for GRD, LLS and leaf miner infestations. NaSARRI routinely screens regional and international groundnut germplasm for responses to those stresses under natural conditions in known hotspots. Desirable lines may then be eventually released as such

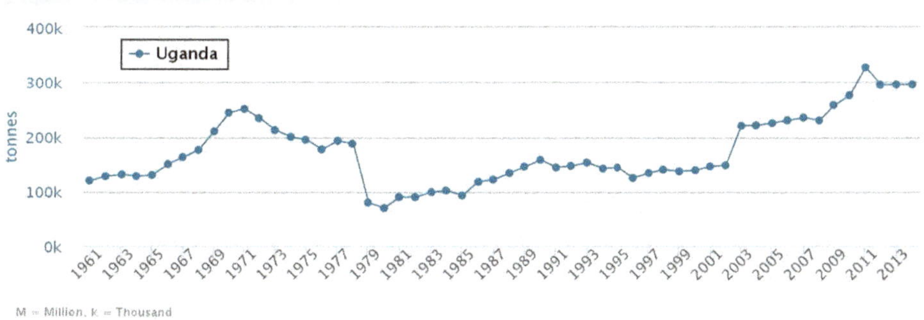

Figure 6 Groundnut (shell) production trends in Uganda from 1961 to 2014 (Food and Agricultural Organization of the United Nations [FAOSTAT], 2016). Tonnes = metric tons.

Table 1 Origin, botanicals, breeding methods and cultivars' attributes

Varieties	Origin (year released)	Botanical	Breeding[1] method	Shortcomings	Positive attributes	Special trait
Serenut 1R	ICRISAT Malawi (1998)	Virginia	Primary introduction	GRD susceptible Leaf miner susceptible *Aspergillus flavus* susceptible (aflatoxin contamination)	Large red seeds Resistance to leaf spot diseases Drought tolerant	Confectionery type Soft shell (easy to shell) Stay green
Serenut 2	ICRISAT Malawi (1998)	Virginia	Primary introduction	Leaf miner susceptible Difficult to shell	Large tan seeds Resistance to leaf spot diseases and GRD Good for condiments, flour	Highly drought tolerant Tolerant to aflatoxin contamination
Serenut 3R	ICRISAT Malawi (2002)	Spanish	Primary introduction	Long dormancy (45 days) Weak pegs at harvest (pods remain in soil) Bitter to taste when raw Susceptible to aflatoxin contamination	Resistance to leaf spot diseases Tolerant to drought	High oil Good for butter
Serenut 4T	ICRISAT Malawi (2002)	Spanish	Primary introduction	Susceptible to leaf miner, leaf spot diseases and GRD Small seeds Difficult to shell Susceptible to aflatoxin contamination	Good for confectionery and butter	Highest shelling return (73%)
Serenut 5R	ICRISAT Malawi (2010)	Virginia	Primary introduction[1]	Leaf miner susceptible Susceptible to aflatoxin contamination	Medium to large seeds Tolerant to drought GRD resistance	Short dormancy (about 30days) Confectionery type Red - attractive market colour
Serenut 6T	ICRISAT Malawi (2010)	Spanish	Primary introduction	Leaf miner susceptible Leaf spot susceptible Susceptible to aflatoxin contamination	GRD resistant Medium to big seeds Easy to shell Early maturing (90 days)	Very tall bunchy groundnut Very easy to harvest Confectionery type
Serenut 7T	NaSARRI (2011)	Virginia	Hybridization and selection	Leaf miner susceptible Susceptible to aflatoxin contamination	Drought tolerant Uniform mat-type growth Resistance to GRD and leaf spot diseases Large pods and seeds Stay green	Easy to shell (soft pods) Confectionery type Good for condiments, flour

Serenut 8R	NaSARRI (2011)	Virginia	Hybridization and selection	Leaf miner susceptible Susceptible to aflatoxin contamination	Drought tolerant Uniform mat-type growth Resistance to GRD and leaf spot diseases Stay green	Bunchy groundnut Very easy to harvest Deep red with faint dark stripes attractive market colour Good for butter
Serenut 9T	NaSARRI (2011)	Virginia	Hybridization and selection	Leaf miner susceptible	Very sweet Drought tolerant Uniform mat-type growth Resistance to GRD and leaf spot diseases Easy to shell Stay green	Tolerant to both *Aspergillus* colonization and aflatoxin contamination Tolerant to thrips feeding Best tan confectionery
Serenut 10R	NaSARRI (2011)	Virginia	Hybridization and selection	Colour intermediate tan and red Premium market prefers red seed coat Susceptible to aflatoxin contamination	Resistance to GRD and leaf spot diseases Drought tolerant Stay green	Leaf miner tolerant
Serenut 11T	NaSARRI (2011)	Virginia	Hybridization and selection	Leaf miner susceptible Susceptible to aflatoxin contamination	Drought tolerant Resistance to GRD and leaf spot diseases Easy to shell Stay green	Giant pod and seeds Confectionery type
Serenut 12R	NaSARRI (2011)	Virginia	Hybridization and selection	Leaf miner susceptible Medium seed size Susceptible to aflatoxin contamination	Drought tolerant Resistance to GRD and leaf spot diseases Stay green	Very tall bunchy groundnut Very easy to harvest Deep red attractive market colour Confectionery type
Serenut 13T	NaSARRI (2011)	Virginia	Hybridization and selection	Leaf miner susceptible Susceptible to aflatoxin contamination	Drought tolerant Resistance to GRD Easy to shell (soft pods) Good for butter and condiments Stay green	Uniform light tan attractive seeds
Serenut 14R	NaSARRI (2011)	Virginia	Hybridization and selection	Leaf miner susceptible Susceptible to aflatoxin contamination	Drought tolerant Resistance to GRD and leafspot diseases Stay green	Confectionery type

[1]Primary Introduction: The introduced groundnut lines are evaluated for adaptability and are released without any further improvement.
In Uganda, the Western, Central and Southern regions prefer red (R) seed coat groundnuts, whereas tan (T) seed coat groundnuts are preferred in the Eastern and Northern regions. Therefore, two seed coat phenotypes are normally released to meet the regional preferences.

or may enter a breeding programme for further improvement. More recently, germplasm from different collections within the United States through collaborations between breeders in the Peanut and Mycotoxin Innovation Laboratory (PMIL) of the United States Agency for International Development (USAID) is also being screened or being used to target specific traits by breeders within SSA. Additional sources of germplasm diversity are locally adapted germplasm/landraces, germplasm from USDA and New Mexico State University (USA), Centre de Coopération Internationale en Recherche Agronomique pour le Développement/Institut Senegalais de Recherche Agricole (CIRAD/ISRA), and from the National Agricultural Research Institutes of South Sudan and Zambia.

Prior to 2006, political instability and the resulting civil conflicts had posed serious constraints to improving and maintaining a groundnut improvement programme in Uganda (Fig. 6; Okello et al., 2010a). However, by the late 1990s to early 2000s, targeted primary introductions selected for adaption resulted in the release of the Serenut 1–4 series (Table 1) to boost groundnut production. Since 2006, the most important traits identified and targeted in the groundnut improvement programme have included biotic (GRD, LLS, aflatoxin contamination, groundnut leaf miner) and abiotic stresses (drought tolerance). The quality traits include oil levels and oil content, as well as seed size and seed colour. Developing early maturity varieties (90–110 days) was also a high priority. All these traits have been, or are being, introduced into the three common botanicals of Spanish, Virginia and Valencia.

5.1 Breeding, networking and awareness

Once the desired traits were defined, NaSARRI set up an active participatory trait based introgression programme targeting the key identified traits in a participatory manner. Key trait hotspots, mega and minor environments, regional sites for National Performance Trials (NPT) and on-farm trials were identified. Groundnut farmer groups and extension staffs were trained, and strong networks were built with the participating local initiatives (non-governmental organizations [NGOs], community-based organizations [CBOs], local leaders and policy makers). Media outlets are routinely present at all events.

5.2 Improved groundnut varieties

5.2.1 Improvements for resistance to GRD

Two approaches have been used in developing improved varieties with resistance to GRD in Uganda and East Africa. The first is to obtain GRD-resistant varieties from ICRISAT-Malawi (primary introductions) followed by on-station and on-farm evaluation for adaption, including yields and relative strength of GRD resistance, in various agro-ecological zones of Uganda. This approach led to the release of the Serenuts 1–6 series (Table 1).

The second approach has been to introgress GRD resistance into desired varieties. The progenies are tested in nurseries in hotspots (GRD appears as early as 2 weeks post-emergence) and the disease pressure is augmented by using known susceptible checks. Superior lines are advanced amidst selection until they attain uniformity at Filial generation 6 (F_6). Yield trials (preliminary, intermediate and advance) are performed on-station amidst high disease pressure. Superior lines are then evaluated at multiple NPT locations for wide adaptability over three growing seasons. This approach led to the development of the current series designated Serenuts 7–14 (Table 1). These commercial varieties belong to Spanish and Virginia botanicals. Valencia botanicals are susceptible to GRD and to date only one resistance source has been identified in South Sudan, the landrace designated Abutalata. Numerous Valencia crosses with GRD-resistant lines are being advanced.

5.2.2 Genesis of Serenut 1–14 groundnut cultivar series (1998–2016)

The Serenut 1–4 series, ICRISAT-Malawi primary introductions, were selected and released as a short-term strategy to restore the defunct groundnut seed sector that occurred between the early 1990s and early 2000s due to political instabilities. Serenut 1R and Serenut 2 were released in 1998 as GRD-resistant and high-yielding materials. However, Serenut 1R GRD resistance broke down within 1 year of post-release and farmers disliked the 110-day maturity of Serenut 1R and 2. Serenut 2 also has a hard shell, making shelling difficult. Researchers responded by releasing short-maturing Serenuts 3R and 4T (90 days) in 2002. However, Serenut 3R has a fresh seed dormancy period of about 45 days, making it unsuitable for the two rain-fed cropping cycles in most groundnut-growing areas of Uganda (the gap between the 1st and 2nd rains is usually 30 days). In addition, the pegs were also very weak at maturity with many pods remaining in the soil at harvest. While Serenut 4T was resistant to the aphid vector (*Aphis craccivora*) of GRD, the resistance breaks down under high aphid pressure and the tiny seeds are difficult to shell – two factors that make Serenut 4T less preferred by farmers. Researchers responded by releasing Serenut 5R and 6T in 2010 to address the weaknesses of Serenuts 3R and 4T, respectively (Table 2). Serenut 5R and 6T are ICRISAT-Malawi primary introductions. Serenut 5R has a short dormancy of about 30 days, making it fit the two-groundnut cropping cycle and the pegs are stronger than Serenut 3R, thereby reducing pod loss at harvest. Serenut 6T is resistant to GRD, has medium to big tan seeds, and is as early maturing (90 days) as Serenut 4T. Researchers also responded by improving Serenut 2 (ease of shelling) and Serenut 1R (GRD resistant) while maintaining their preferred traits. The ease of shelling trait from Serenut 1R was introgressed into Serenut 2 and GRD resistance from Serenut 2 into Serenut 1R. Through selections of segregating progenies both red and tan seed coat types phenotypically identical to their progenitors were identified as replacement options for Serenut 1R and 2. These lines were evaluated for yields, disease resistances and wide adaptability in major agro-ecologies and were released as Serenut 7–14 groundnut varieties (Table 2).

Table 2 Groundnut varieties and replacements

No	Old variety	Replaced by
1	Serenut 1R	Serenut 8R, Serenut 10R, Serenut 12R, Serenut 14R,
2	Serenut 2	Serenut 7T, Serenut 9T, Serenut 11T, Serenut 13T
3	Serenut 3R	Serenut 5R
4	Serenut 4T	Serenut 6T

T, tan seed coat. R, red seed coat.

5.2.3 Regional improvements for GRD resistance

Presently, seven South Sudanese landraces have been improved in Uganda for GRD and LLS resistances and yields. The improved lines are being used by the South Sudanese breeder to initiate a breeding programme. In addition, eleven preferred Zambian varieties are being improved for GRD, LLS and leaf miner resistance, as well as early maturity for reintroduction into the Zambian breeding programme. To date, germplasm from the Uganda breeding programme has been requested and provided to scientists in Ethiopia, Ghana, Haiti, Malawi, Mali, Mozambique, South Sudan and the United States. Additional germplasm requests have been received from Burkina Faso, Burundi, Democratic Republic of the Congo, Ivory Coast, Kenya, Malawi, Mali, Senegal and Zambia.

5.2.4 Improvements for leaf spot resistance

Leaf spot resistance is an important targeted trait in Uganda and has subsequently been introgressed into the Serenut 7–14 series (Table 1). Since these materials mature when they are still green (stay green), we have embarked on using days-to-maturity as the major maturity indicator. Recently, an improvement programme to introgress LLS resistance into landraces and susceptible released materials has been initiated.

5.2.5 Improvements for resistance to groundnut leaf miner

Groundnut leaf miner (LM, *Aproaerema modicella* Deventer) is an important pest of groundnuts in both the rainy and post-rainy seasons and is regarded as the most important pest threatening groundnut production in Uganda. It has also had a significant impact on yields in the Democratic Republic of Congo (DRC), Malawi, Mozambique and South Africa. LM is much more damaging during the short rainy cycle when long droughts precede rains. So far only one commercial variety, Serenut 10R, has moderate resistance (Okello et al., 2013). Four advanced lines (SGV 0023, SGV 002, SGV 0053 and SGV 0084) have shown high levels of resistance (greater than that in Serenut 10R). The advanced lines will provide much-needed sources of LM resistance for introgression into desirable varieties. In addition to host-plant resistance, other management options include pesticide sprays, crop rotation and use of trap crops to control leaf miner damage in groundnut.

5.2.6 Improvements for high oleic trait

This trait has gained the attention of our breeding programme because of its superior oil chemistry and associated health benefits. The challenge is that the high oleic trait was previously bred into varieties in an environment devoid of GRD. All of the known high oleic materials are susceptible to GRD. Therefore, introgression of the high oleic trait into GRD-resistant and adapted materials is a priority. Presently, populations with GRD-resistant backgrounds are being generated using marker-assisted backcrossing to increase the efficiency and speed of the breeding programme.

5.2.7 Improvements for confectionery groundnut

This market class is becoming popular in Uganda and East Africa. Typically, preferred attributes of confectionery groundnuts are high sugar, low oil and an attractive seed size (Janila et al., 2016a). Indeed, a low oil content with a high oleic/linoleic ratio will be targeted in the future. Advanced lines are routinely screened for 100 seed weight ≥ 55 g and low oil content. This is done in both tan and red skin types. The Serenut varieties 5R, 6T, 7T, 9T, 11T, 12R and 14R qualify as the confectionery type (Table 1).

5.2.8 Landrace improvements

An active programme in Uganda is ongoing to improve popular landraces such as Red Beauty, Acholi white, Egoromoit, Agrikaca, Gwerinut and Erudurudu. Advanced progenies are being evaluated for yield (pod and seed) superiority and disease resistances (GRD and LLD). Stabilized superior lines will be selected in a participatory manner and subsequently released. Improvements in landraces are important in rapid acceptance by farmers and consumers since historically they are preferred varieties.

6 Conclusion and future trends

Considerable progress has been made in increasing groundnut yields and quality in the last two decades. Much of this has focused on narrowing production gaps by minimizing losses resulting from biotic and abiotic stresses, increasing quality by primarily selecting for desirable oils and oil content, and controlling aflatoxin contamination. Advances have been made through screening for primary introductions, with further genetic enhancements occurring through breeding programmes. Screening of germplasm collections for traits to increase yield and quality has been very successful to date, with the most limiting factor being the narrow genetic base of cultivated groundnut.

One of the primary goals in groundnut biology is to identify the genes and gene combinations that will lead to significant crop improvement and groundnut productivity. This goal requires a multidisciplinary and integrated approach involving plant and molecular biologists, molecular and traditional breeders, knowledge of genomic organization and gene functions, robust statistical and biometrical approaches to estimate genetic effects, and the ability to manage and mine large diverse datasets (Moose and Mumm, 2008).

Genetic diversity in cultivated groundnut is very narrow, but ample diversity has been found in wild *Arachis* species, especially for resistance to groundnut pathogens and insect pests (Stalker et al., 2016). Therefore, wild groundnut species represent a wealth of genetic diversity that, if tapped, could be extensively used to enhance the genetic potential of cultivated groundnut. However, differences in ploidy levels (cultivated groundnuts are tetraploid and wild *Arachis* species are diploid) limit the use of wild relatives in expanding cultivated groundnut gene pools. Some success has been achieved using interspecific hybridization to move traits from wild groundnut into cultivated groundnut, but the unconventional breeding strategies remain difficult and unpredictable (Garcia et al., 2006; Seijo et al., 2007; Nautiyal et al., 2008; Fonceka et al., 2009; Bertioli et al., 2011; Leal-Bertioli et al., 2015 and Stalker et al., 2016). Developing more efficient methods for the introgression of genes from wild groundnut relatives into cultivated groundnut will be welcome and will provide a source of needed traits for improving cultivated groundnut.

Expanding the development of molecular markers for use in MAS and genomic selection (GS), a more robust form of MAS for following quantitative trait loci (QTL), will rapidly accelerate the selection of desirable traits in conventional as well as unconventional breeding programmes (Collard and Mackill, 2008; Jannink et al., 2010; Pandey et al., 2014; Holbrook et al., 2016). Developing an improved variety can take years by phenotype selection, while

MAS can be highly efficient and precise in comparison. In the last decade, the science of genomics in groundnut has advanced considerably with a large number of high-throughput DNA markers (SSRs and SNPs), high-density genetic maps, marker traits and genomic regions associated with agronomically beneficial traits identified (Guo et al., 2016). Directed efforts are underway to integrate these QTLs to enhance the genetic potential of groundnut. In the future, MAS and GS will become more effective tools of genetic improvement even in developing countries, which presently lag in the use of MAS in part due to inadequate infrastructure, high genotyping costs and inadequate human capacities. With the collective efforts of the broad community of scientists committed to groundnut biology and crop improvement, molecular breeding will further expand its contributions and impacts to meeting increasing needs for sustainable increases in groundnut productivity.

Finally, the use of transgenic plant technology in the future could greatly expand and accelerate groundnut breeding programmes. While this technology may not be socially practical at the present time, it has enormous current value in providing proof-of-concept for identifying mechanisms responsible for biotic and abiotic resistances and groundnut quality that could guide traditional breeders in groundnut improvement programmes. A number of reviews that discuss the practicality and potential impact of transgenic technology in groundnut improvement programmes have been published (Sun et al., 2013; Hu and Xiong, 2014; Krisha et al., 2015; Mallikarjuna et al., 2016). Examples cover improvement for biotic and abiotic stresses, improved quality, minimizing pesticide impact on the environment, and improving nutrition and food safety. In addition, transgenic technology could provide information for groundnut improvement to increase diversity as changes in global weather patterns put constraints on traditional cropping areas and marginal varieties.

The world population is predicted to increase from 7.2 billion people in 2012 to 9.6–12.3 billion in 2100, with much of the increase occurring in Africa (Gerland et al., 2014). Considering the unpredictable nature of climate change and the challenges it will place on agriculture, the demand on world food production will be enormous. Because of the health-enhancing, energy-rich, highly nutritional values of groundnut as a food source and the ability of the legume to improve soil quality, groundnut will be an important component for achieving food security in the future. For a discussion of the potential role of groundnuts in addressing global food security, see Valentine (2016).

7 Where to look for further information

Two books that readers will want to explore for more information on topics covered in this review are *Peanuts: Genetics, Processing and Utilization*, 2016, Waltham, MA: Academic Press and AOCS Press and *Genetics, Genomics and Breeding of Peanuts*, 2016, Boca Raton: CRC Press. Review articles include those in the reference section by Barkley et al., 2016a (Global resources of genetic diversity in peanut), Holbrooke et al., 2016 (Recent advances in peanut breeding and genetics), Janila et al., 2013 (Groundnut improvement: use of genetic and genomic tools), Janila et al., 2016 (Genomic tools in groundnut breeding program: status and perspective), Moose and Mumm, 2008 (Molecular plant breeding as the foundation for 21st century crop improvement), Mallikarjuna et al., 2016 (Genetic engineering for genetic improvement: current status and prospects), Stalker et al., 2016 (Biology, speciation, and utilization of peanut species), and Valentine, 2016 (The role of peanut in global food security).

8 Acknowledgements

The authors would like to thank Dr Sangam Dwivedi and Dr Naveen Puppala for valuable comments on the manuscript. This publication was made possible through support provided by the Office of Agriculture, Research and Policy, Bureau of Food Security, U.S. Agency for International Development, under the terms of Award No. AID-ECG-A-00-07-0001 to the University of Georgia as management entity for the U.S. Feed the Future Innovation Lab on Peanut Productivity and Mycotoxin Control. The opinions expressed herein are those of the author(s) and do not necessarily reflect the views of the U. S. Agency for International Development.

9 References

Alegbejo, M. D. and Abo, M. E. (2002). Etiology, ecology, epidemiology and control of groundnut rosette disease in Africa. *J. Sustain. Agr.* 20: 17–29. Doi: 10.1300/J064v20n02_04.

Asis, R., Barrionuevo, D. L., Giorda, L. M., Nores, M. L. and Aldao, M. A. (2005). Aflatoxin production in six peanut (*Arachis hypogaea* L.) genotypes infected with *Aspergillus flavus* and *Aspergillus parasiticus*, isolated from peanut production areas of Cordoba, Argentina. *J. Agric. Food Chem.* 53: 9274–80. Doi: 10.1021/jf051259+.

Bally, J., Nakasugi, K., Jia, F., Jung, H., Ho, S. Y. W., Wong, M., Paul, C. M., Naim, F., Wood, C. C., Crowhurst, R. N., Hellens, R. P., Dale, J. L. and Waterhouse, P. M. (2015). The extremophile *Nicotiana benthamiana* has traded viral defense for early vigor. *Nat. Plants* 1: 15165. Doi: 10.1038/NPLANTS.2015.165.

Bandyopadhyay, R., Ortega-Beltran, A., Akande, A., Mutegi, C., Atehnkeng, J., Kaptoge, L., Senghor, A. L., Adhikari, B. N. and Cotty, P. J. (2016). Biological control of aflatoxins in Africa: current status and potential challenges in the face of climate change. *World Mycotoxin J.* 9: 771–89. Doi: 10.3920/WMJ2016.2130

Bao, Y., Han, J., Hu, F. B., Giovannucci, E. L., Stampfer, M. J., Willett, W. C. and Fuchs, C. S. (2013). Association of nut consumption with total and cause-specific mortality. *New Engl. J. Med.* 369: 2001–11. Doi:10.1056/NEJMoal1307352.

Barkley, N. A., Chenault Chamberlin, K. D., Wang, M. L. and Pittman, R. N. (2010). Development of a real-time PCR genotyping assay to identify high oleic peanuts (*Arachis hypogaea* L.). *Mol. Breed.* 34: 638–43. Doi: 10.1007/s11032-009-9338-z.

Barkley, N. A., Upadhyaya, H. D., Liao, B. and Holbrooke, C. C. (2016). Global resources of genetic diversity in peanut. In Stalker, H. T and Wilson, R. F. (Eds), *Peanuts: Genetics, Processing and Utilization*. Waltham, MA: Academic Press and AOCS Press, pp. 67–109. Doi: 10.1016/B978-1-63067-038-2.00003-4.

Bertioli, D. J., Seijo, G., Freitas, F. O., Valls, J. F. M., Leal-Bertioli, S. C. M. and Moretzohn, C. 2011. An overview of peanut and its wild relatives. *Plant Genet. Resour.* 9: 134–149. Doi: 10.1017/S1479262110000444.

Blum, A. (2005). Drought resistance, water-use-efficiency, and yield potential – are they compatible, dissonant, or mutually exclusive? *Aust. J. Agric Res.* 56: 1159–68. Doi: 10.1071/AR05069.

Branch, W. D. and Brenneman, T. B. (2008). Registration of 'Georgia-07W' peanut. *J. Plant Reg.* 2: 88–91. Doi: 10.3198/jpr2007.12.0682crc.

Branch, W. D. (2010). Registration of 'Georgia-09B' peanut. *J. Plant Reg.* 4: 175–8. Doi: 10.3198/jpr2009.12.0693crc.

Branch, W. D. and Culbreath, A. K. (2011). Registration of 'Georgia-10T' peanut. *J. Plant Reg.* 5: 279–81. Doi: 10.3198/jpr2010.11.0635crc.

Branch, W. D. (2013). Registration of 'Georgia-12Y' peanut. *J. Plant Reg.* 7: 151–3. Doi: 10.3198/jpr2012.11.0048crc.

Branch, W. D. (2014). Registration of 'Georgia-13M' peanut. *J. Plant Reg.* 8: 253–6. Doi: 10.3198/jpr2013.11.0071crc.

Branch, W. D and Brenneman, T. B. (2015). Registration of 'Georgia-14N' peanut. *J. Plant Reg.* 9: 159–61. Doi: 10.3198/jpr2014.11.0082crc.

Chattopadhyay, C., Kolte, S. J. and Waliyar, F. (2016). *Diseases of Edible Oilseed Crops*. Boca Raton: CRC Press, pp. 13–100. ISBN: 13-978-1-4665-9566-8.

Chen, Z. Y., Brown, R. L., Guo, B. Z., Menkir, A. and Cleveland, T. E. (2009). Identifying aflatoxin resistance-related proteins/genes through proteomics and RNAi gene silencing. *Peanut Sci.* 36: 35–41. Doi: 10.3146/AT07-005.1.

Chen, Z., Wang, M. L., Barkley, N. A. and Pittman, R. N. (2010). A simple allele-specific PCR assay for detecting *FAD2* alleles in both A and B genomes of the cultivated peanut for high-oleate trait selection. *Plant Mol. Biol. Rep.* 28: 542–8. Doi: 10.1007/s11105-010-0181-5.

Chen, S., Lei, Y., Xu, X., Huang, J., Jiang, H., Wang, J., Cheng, Z., Zhang, J., Song, Y., Liao, B. and Li, Y. (2015). The peanut (*Arachis hypogaea* L.) gene *AhLPAT2* increases the lipid content of transgenic Arabidopsis seeds. *PLoS One* 10: e0136170. Doi: 10.1371/journal.pone.0136170.

Chen, W., Jiao, Y., Cheng, L., Huang, L., Liao, B., Tang, M., Ren, X., Zhou, X., Chen, Y. and Jiang, H. (2016). Quantitative trait locus analysis for pod- and kernel-relayed traits in the cultivated peanut (*Arachis hypogaea* L.). *BMC Genetics* 17: 25. Doi: 10.1186/s12863–016-0337-x.

Chu, Y., Holbrooke, C. C. and Ozias-Akins P. (2009). Two alleles of *ahFAD2B* control the high oleic acid trait in cultivated peanut. *Crop Sci.* 49: 2029–36. Doi: 10.2135/cropsci2009.01.0021.

Cipollini, D. and Heil M. (2009). Costs and benefits of induced resistance to herbivores and pathogens in plants. *CAB Reviews: perspectives in Agriculture, Veterinary Science, Nutrition and Natural Resources*. 5(005): 1–25. Doi: 10.1079/PAVSNNR20105005.

Cipollini, D., Walter, D. and Voelckel C. (2014). Cost of resistance in plants: from theory to evidence. In Voelckel, C. and Jander, G. (Eds), *Annual Plant Reviews Volume 47: Insect-Plant Interactions*. Wiley: Chichester, UK. Doi: 10.1002/9781118829783.ch8.

Collard, B. C. Y. and Mackill, D. J. (2008). Marker-assisted selection: an approach for precision plant breeding in the twenty-first century. *Philos. Trans. R. Soc. Lond. B. Biol. Sci.* 363: 557–72. Doi: 10.1098/rstb.2007.2170.

Craft, B. D., Kosinska, A., Amarowicz, R. and Pegg, R. B. (2010). Antioxidants properties of extracts obtained from raw, dry-roasted, and oil-roasted US peanuts of commercial importance. *Plant Foods Hum. Nutr.* 65:311–18. Doi:10.1007/s11130-010-0160-x.

Cui, H., Tsuda, K. and Parker, J. E. (2015). Effector-triggered immunity: from pathogen perception to robust defense. *Annu. Rev. Plant Biol.* 66: 487–511. Doi: 10.1146/annurev-arplant-050213-040012.

Culbreath, A. K., Gorbet, D. W., Martinez-Ochoa, N., Holbrook, C. C., Todd, J. W., Isleib, T. G. and Tillman, B. (2005). High Levels of Field Resistance to Tomato spotted wilt virus in Peanut Breeding Lines Derived from *hypogaea* and *hirsuta* Botanical Varieties. *Peanut Sci.* 32: 20–4 https://doi.org/10.3146/0095-3679(2005)32[20:hlofrt]2.0.co;2.

Culbreath, A. K., Todd, J. W. and Brown, S. L. (2003). Epidemiology and management of Tomato spotted wilt in peanut. *Annu. Rev. Phytopathol.* 41: 53–75. Doi: 10.1146/annurev.phyto.41.052002.095522.

Deom, C. M., Kapewa, T., Busolo-Bulafu, C. M., Naidu, R. A., Chitembekeza, A. J., Kimmins, F. M., Subrahmanyam, P. and van der Merwe, P. J. A. (2006). Registration of ICG 12991 peanut germplasm line. *Crop Sci.* 46: 481. Doi: 10.2135/cropsci2005.010.

Diao, E., Dong, H., Hou, H., Zhang, Z., Ji, N. and Ma, W. (2015). Factors influencing aflatoxin contamination in before and after harvest peanuts: A review. *J. Food Res.* 4: 148–54. Doi: 10.5539/jfr.v4n1p148.

Dwivedi, S. L., Nigam, S. N., Reddy, D. V. R., Reddy, A. S. and Ranga Rao, G. V. (1995). Progress in breeding groundnut varieties resistant to peanut bud necrosis virus and its vector. In Buiel, A. A. M., Parlevliet, J. E. and Lenne, J. M. (Eds), *Recent Studies on Peanut Bud Necrosis Disease*. Patancheru, India: ICRISAT, pp. 35–40.

Dwivedi, S. L., Crouch, J. H., Nigam, S. N., Ferguson, M. E. and Paterson, A. H. (2003). Molecular breeding of groundnut for enhanced productivity and food security in the semi-arid tropics: opportunities and challenges. *Adv. Agron.* 80: 153–221. Doi: 10.1016/S0065-2113(03)80004-4.

Fonceka, D., Hodo-Abalo, T., Rivallan, R., Faye, I., Sall, M. N., Ndoye, O., Favero, A. P., Bertioli, D. J., Glaszmann, J. C., Courtois, B. and Rami, J. F. (2009). Genetic mapping of wild introgressions into cultivated peanut: a way toward enlarging the genetic basis of a recent allotetraploid. *BMC Plant Biol.* 9: 103. Doi: 10.1186/1471-2229-9-103.

Food and Agricultural Organization of the United Nations (2016). FAOSTAT. http://faostat3.fao.org/browse/Q/QC/E.

Fountain, J. C., Khera, P., Yang, L., Nayak, S. N., Scully, B. T., Lee, R. D., Chen, Z.-Y., Kemerait, R. C., Varshney, R. K. and Guo, B. (2015). Resistance to *Aspergillus flavus* in maize and peanut: Molecular biology, breeding, environmental stress, and future perspectives. *Crop J.* 3: 229–37. Doi: 10.1016/j.cj.2015.02.003.

Frankel, O. H. (1984). Genetic perspective of germplasm conservation. In Arber, W., Llimensee, K., Peacock, W. J. and Starlinger, P. (Eds), *Genetic Manipulation: Impact on Man and Society.* Cambridge: Cambridge University Press, pp. 161–70. ISBN: 9780521142892.

Gan, S. and Amasino, R. M. (1995). Inhibition of leaf senescence by autoregulated production of cytokinin. *Science* 270: 1986–8. Doi: 10.1126/science.270.5244.1986.

Garcia, G. M., Tallury, S. P., Stalker, H. T. and Kockert, G. (2006). Molecular analysis of *Arachis* interspecific hybrids. *Theor. Appl. Genet.* 112: 1342–8. Doi: 10.1007/s00122-006-0236-z.

Gerland, P., Raftery, A. E., Sevcikova, H., Li, N., Gu, D., Spoorenberg, T., Alkema, L., Fosdick, B. K., Chunn, J., Lalic, N., Bay, G., Buettner, T., Heilig, G. K. and Wilmoth, J. (2104). World population stabilization unlikely this century. *Science* 346: 234–7. Doi: 10.1126/science.1257469.

Gorbet, D. W. and Tillman B. L. (2009). Registration of 'Florida-07' peanut. *J. Plant Reg.* 3: 14–18. Doi: 10.3198/jpr2008.05.0276crc.

Guimon, J. and Guimon, P. (2012). How ready-to use therapeutic food shapes a new technological regime to treat child malnutrition. *Technol. Forecast. Soc.* 79: 1319–27. Doi:10.1016/j.techfore2012.04.011.

Guo, B., Fedorova, N. D., Chen, X., Wan, C.-H., Wang, W., Nierman, W. C., Bhatnagar, D. and Yu, J. (2011). Gene expression profiling and identification of resistance genes to *Aspergillus flavus* infection in peanut through EST and microarray strategies. *Toxins* 3: 737–53. Doi: 10.3390/toxins3070737.

Guo, B., Khera, P., Wang, H., Peng, Z., Sudini, H., Wang, X., Osiru, M., Chen, J., Vadez, V., Yuan, M., Wang, C. T., Zhang, X., Waliyar, F., Wang, J. and Varshney, R. K. (2016). Annotation of trait loci on integrated genetic maps of *Arachis* species. In Stalker, H. T. and Wilson, R. F. (Eds), *Peanuts: Genetics, Processing and Utilization.* Waltham, MA: Academic Press and AOCS Press, pp. 163–96. Doi: 10.1016/B978-1-63067-038-2.00003-4.

Hamidou, F., Ratnakumar, P., Halilou, O., Mponda, O., Kapewa, T., Monyo, E., Faye, I., Ntare, B. R., Nigam, S. N., Upadhyaya, H. D. and Vadez, V. (2012). Selection of intermittent drought tolerant lines across years and locations in the reference collection of groundnut (*Arachis hypogaea* L.). *Field Crop Res.* 126: 189–99. Doi: 10.1016/j.fcr.2011.10.009.

Herselman, L., Thwaites, R., Kimmins, F. M., Courtois, B., van der Merwe, P. J. A. and Seal, S. E. (2004). Identification and mapping of AFLP markers linked to peanut (*Arachis hypogaea* L.) resistance to the aphid vector of groundnut rosette disease. *Theor. Appl. Genet.* 109: 1426–33. Doi: 10.1007/s00122-004-1756-z.

Holbrooke, C. C., Anderson, W. F. and Pittman, R. N. (1993). Selection of a core collection from the U.S. germplasm collection of peanuts. *Crop Sci.* 33: 859–61.

Holbrooke, C. C. and Dong, W. (2005). Development and evaluation of a mini-core collection for the U.S. peanut germplasm collection. *Crop Sci.* 45: 1540–4. Doi: 10.2135/cropsci2004.0368.

Holbrooke, C. C., Guo, B. Z., Wilson, D. M. and Timper, P. (2009). The U.S. breeding program to develop peanut with drought tolerance and reduced aflatoxin contamination. *Peanut Sci.* 36: 50–3. Doi: 10.3146/AT07-009.1.

Holbrooke, C. C., Burow, M. D., Chen, C. Y., Pandey, M. K., Liu, L., Chagoya, J. C., Chu, Y. and Ozias-Akins, P. (2016). Recent advances in peanut breeding and genetics. In Stalker, H. T. and Wilson, R. F. (Eds), *Peanuts: Genetics, Processing and Utilization*. Waltham, MA: Academic Press and AOCS Press, pp. 111–45. Doi: 10.1016/B978-1-63067-038-2.00004-6.

Hollis, P. (22 February 2015). A look at 13 improved peanut varieties available now or coming soon. Retrieved from: http://southeastfarmpress.com/peanuts/look-13-improved-peanut-varieties-available-now-or-coming-soon.

Hu, H. and Xiong, L. (2014). Genetic engineering and breeding of drought-resistant crops. *Annu. Rev. Plant Biol.* 65: 715–41. Doi: 10.1146/annurev-arplant-050213-040000.

Janila, P., Nigam, S. N., Pandey, M. K., Nagesh, P. and Varshney, R. K. (2013). Groundnut improvement: use of genetic and genomic tools. *Front. Plant Sci.* 4: 1–16. Doi: 10.3389/fpls.2013.00023.

Janila, P., Variath, M. T., Pandey, M. K., Desmae, H., Motagi, B. N., Okori, P., Manohar, S. S., Rathnakumar, A. L., Radhakrishnan, T., Liao, B. and Varshney, R. K. (2016a). Genomic tools in groundnut breeding program: status and perspective. *Front. Plant Sci.* 7: 1–14. Doi: 10.3389/fpls.2016.00289.

Janila, P., Pandey, M. K., Shasidhar, Y., Variath, M. T., Sriwathi, M., Khera, P., Manohar, S. S., Nagesh, P., Vishwakarma, M. K., Mishra, G. P., Radhakrishnan, T., Manivannan, N., Dobariya, K. L., Vasanthi, R. P. and Varshney, R. K. (2016b). Molecular breeding for introgression of fatty acid desaturase mutant alleles (*ahFAD2A* and *ahFAD2b*) enhances oil quality in high and low oil containing peanut genotypes. *Plant Sci.* 242: 203–13. Doi: 10.1016/j.plantsci.2015.08.013.

Jannink, J.-L., Lorenz, A. J. and Iwata, H. (2010). Genomic selection in plant breeding: from theory to practice. *Brief Funct. Genomics* 9: 166–77. Doi: 10.1093/bfgp/elq001.

Jiang, H.-F., Ren, X.-P., Wang, S. Y. and Liao, B.-S. (2006). Durability of resistance to *Aspergillus flavus* infection and effect of intact testa without injury on aflatoxin production in peanuts. *Acta. Agron. Sin.* 32: 851–5.

Jiang, H.-F., Ren, X.-P., Liao, J. Q., Huang, J. Q., Lei, Y., Chen, B. Y., Guo, B. Z., Holbrooke, C. C. and Upadhyaya, H. D. (2008). Peanut core collection established in China and compared with ICRISAT mini core collection. *Acta Agron. Sin.* 34: 25–30.

Jiang, H.-F., Ren, X.-P., Zhang, X.-J., Huang, J.-Q., Lei, Y., Yan, L.-Y., Liao B.-S., Upadhyaya, H. D. and Holbrooke, C. C. (2010). Comparison of genetic diversity based on SSR markers between peanut mini core collections from China and ICRISAT. *Acta Agron. Sin.* 36: 1084–91. Doi: 10.1016/S1875-2780(09)60059-6.

Jolly, P. E., Jiang, Y., Ellis, W. O., Wang, S., Afriyie-Gyawu, E., Phillips, T. D. and Williams, J. H. (2008). Modulation of the human immune system by aflatoxin. In Leslie, J. F., Bandyopadhyay, R. and Visconti, A. (Eds), *Mycotoxins: Detection Methods, Management, Public Health and Agricultural Trade*. Oxfordshire, UK: CAB International, pp. 41–52.

Kemerait, R., Culbreath, A. K., Prostko, E., Brenneman, T., Smith, A., Tubbs, S., Srinivasan, R., Abney, M., Monfort, S., Tillman, B., Dufault, N., Rowland, D., Mulvaney, M., Hagan, A., Sarver, J., Anco, D. and Smith, N. (2016). Peanut Rx: Minimizing diseases of peanut in the southeastern United States. In Monfort, W. S. (Ed.), *2016 Peanut Production Update*. Athens, GA: University of Georgia Cooperative Extension Publication CSS-12-0130, pp. 47–62.

Kessie, M., Shiferaw, B. and Muricho, G. (2011). Agricultural technology, crop income, and poverty alleviation in Uganda. *World Dev.* 39:1784–95. Doi: 10.1016/j.worlddev.2011.04.023.

Khedikar, Y. P., Gowda, M. V. C., Sarvamangala, C., Patgar, K. V., Upadhyaya, H. D. and Varshney, R. K. (2010). A QTL study on late leaf spot and rust revealed one major QLT for molecular breeding for rust resistance in groundnut (*Arachis hypogaea* L.). *Theor. Appl. Genet.* 121: 971–84. Doi: 10.1007/s00122-010-1366-X.

Knauft, D. A. and Wynne, J. C. (1995). Peanut Breeding and Genetics. *Adv. Agron.* 55: 393-445. Doi: 10.1016/S0065-2113(08)60544-1.

Kolekar, R. M., Sujay, V., Shirasawa, K., Sukruth, M., Khedikar, Y. P., Gowda, M. V. C., Pandey, M. K., Varshney, R. K. and Bhat, R. S. (2016). QTL mapping for late leaf spot and rust resistance using an improved genetic map and extensive phenotypic data on a recombinant inbred line population in peanut (*Arachis hypogaea* L.). *Euphytica* 209: 147–56. Doi: 10.1007/s10681-016-1651-0.

Krishna, G., Singh, B. K., Kim, E.-K., Morya, V. K. and Ramteke, P. W. (2015). Progress in genetic engineering of peanut (*Arachis hypogaea* L.) – a review. *Plant Biotech.* J. 13: 147–62. Doi: 10.1111/pbj.12339.

Lal, C., Hariprasanna, K., Chikani, B. M. and Gor, H. K. (2014). Inter-allelic interactions in the inheritance of physical-quality traits in peanut (*Arachis hypogaea* L.). *Aust. J. Crop Sci.* 8: 1049–55.

Lata, C. and Prasad, M. (2011). Role of DREBs in regulation of abiotic stress responses in plants. *J. Exp. Bot.* 62: 4731–48. Doi: 10.1093/jxb/err210.

Leal-Bertioli, S. C. M., Jose, A. C. V. F., Alves-Freitas, D. M. T., Moretzsohn, M. C., Guimaraes, P. M., Nielen, S., ViDigal, B. S., Pereira, R. W., Pike, J., Favero, A. P., Parniske, M., Varshney, R. K. and Bertioli, D. J. (2009). Identification of candidate genome regions controlling disease resistance in *Arachis*. *BMC Plant Biol.* 9: 112. Doi: 10.1186/1471-2229-9-112.

Leal-Bertioli, S. C. M., Shirasawa, K., Abernathy, B., Moretzsohn, M., Chavarro, C., Clevenger, J., Ozias-Akins, P., Jackson, S. and David Bertioli, D. J. 2015. Tetrasomic Recombination Is Surprisingly Frequent in Allotetraploid Arachis. *Genetics*, 199: 1093–105. Doi: 10.1534/genetics.115.174607.

Liang, X., Zhou, G., and Pan, R. (2003). Wax and cuticle of peanut seed coat in relation to infection and aflatoxin production by *Aspergillus flavus*. *J. Topic. Subtropic. Bot.* 11: 11–14.

Liang, X., Zhou, G., Hong, Y., Chen, X., Liu, H. and Li, S. (2009). Overview of research progress on peanut (Arachis hypogaea L.) host resistance to aflatoxin contamination and genomics at the Guangdong Academy of Agricultural Sciences. *Peanut Sci.* 36: 29–34. Doi: 10.3146/AT07-003.1.

Liao, B. S. and Holbrooke, C. C. (2007). Groundnut. In Singh, R. J. (Ed.), *Genetic Resources, Chromosomal Engineering and Crop Improvement*. Boca Raton, Fl, USA: CRC Press, pp. 51–87.

Liao, B. S. (2014). Peanut breeding. In Mallikarjuna, N. and Varshney, R. V. (Eds), *Genetics, Genomics and Breeding of Peanuts*. Boca Raton: CRC Press, pp. 61–78.

Lobell, D. B. and Field, C. B. (2007). Global scale climate-crop yield relationships and the impacts of recent warming. *Environ. Res. Lett.* 2: 014002 (pp. 1–7). Doi: 10.1088/1748-9326/2/1/014002.

Lyerly, J. H., Stalker, H. T., Moyer, J. W. and Hoffman, K. (2002). Evaluation of Arachis species for resistance to Tomato spotted wilt virus. *Peanut Sci.* 29:79–84.

MacQueen, A. and Bergelson, J. (2016). Modulation of *R*-gene expression across environments. *J. Expt. Bot.* 67: 2093–105. Doi: 10.1093/jxb/erv530.

Maestri, E., Klueva, N., Perrotta, C., Gulli, M., Nguyen, H. T. and Marmiroli, N. (2002). Molecular genetics of heat tolerance and heat shock proteins in cereals. *Plant Mol. Biol.* 48: 667–81. Doi: 10.1023/A:1014826730024.

Mallikarjuna, G., Rao, T. S. R. B. and Kirti, P. B. (2016). Genetic engineering for peanut improvement: current status and prospects. *Plant Cell Tiss. Organ Cult.* 125: 399–416. Doi: 10.1007/s11240-016-0966-9.

Mauricio, R. (1998). Costs of resistance to natural enemies in field populations of the annual plant *Arabidopsis thaliana*. *Am. Nat.* 151: 20–28. Doi: 10.1086/286099.

Monyo, E. S., Njoroge, S. M. C., Coe, R., Osiru, M., Madinda, F., Waliyar, F., Thakur, R. P., Chilunjika, T. and Anitha, S. (2012). Occurrence and distribution of aflatoxin contamination in groundnuts (arachis hypogaea) and population densities of aflatoxigenic aspergilli. *Crop Prot.* 42: 149–155.

Moose, S. P. and Mumm, R. H. (2008). Molecular plant breeding as the foundation for 21st century crop improvement. *Plant Physiol.* 147: 969–77. Doi: 10.1104/pp.108.118232.

Moyo, S., Norton, G., Alwang, J., Rhinehart, I. and Deom, C. M. (2007). Peanut research and poverty reduction: Impacts of variety improvement to control peanut viruses in Uganda. *Amer. J. Agr. Econ.* 89: 448–60. Doi:10.1111/j.1467-8276.2007.00991.x.

Murant, A. F., Rajeshwari, R., Robinson, D. J. and Raschke, J. H. (1988). A satellite RNA of Groundnut rosette virus that is largely responsible for symptoms of groundnut rosette disease. *J. Gen. Virol.* 69: 1479–86. Doi: 10.1099/0022-1317-69-7-1479.

Murant, A. F. (1990). Dependence of groundnut rosette virus on its satellite RNA as well as on groundnut rosette assistor luteovirus for transmission by *Aphis craccivora*. *J. Gen. Virol.* 71: 2163–6. Doi: 10.1099/0022-1317-71-9-2163.

Murant, A. F. and Kumar, I. K. (1990). Different variants of the satellite RNA of groundnut rosette virus are responsible for the chlorotic and green forms of groundnut rosette disease. *Ann. Appl. Biol.* 117: 85–92. Doi: 10.1111/j.1744-7348.1990.tb04197.x.

Nagaraji, S. (21 February 2014). New variety of groundnut released for the dry savanna regions of Nigeria. Retrieved from http://grainlegumes.cgiar.org/new-variety-of-groundnut-released-for-the-dry-savanna-regions-of-nigeria/.

Naidu, R. A. and Kimmins, F. M. (2007). The effect of *Groundnut rosette assistor virus* on the agronomic performance of four groundnut (*Arachis hypoaea* L.) genotypes. *J. Phytopathol.* 155: 350–6. Doi: 10.1111/j.1439-0434.2007.01243x.

Naidu, R. A., Kimmins, F. M., Deom, C. M., Subrahmanyam, P., Chiyembekeza, A. J. and van der Merwe, P. J. A. (1999). Groundnut rosette: A virus disease affecting groundnut production in sub-Saharan Africa. *Plant Dis.* 8:700–9. Doi: 10.1094/PDIS.1999.83.8.700.

Nautiyal, P. C., Rajgopal, K., Zala, P. V., Pujari, D. S., Basu, M. S. and Dhadhal, B. A. 2008. Evaluation of wild Arachis species for abiotic stress tolerance: thermal stress and leaf water relations. *Euphytica* 159: 43–57. Doi: 10.1007/s10681-007-9455-x.

Nigam, S. N., Dwivedi, S. L. and Gibbons, R. W. (1991). Groundnut breeding: constraints, achievements and future possibilities. *Plant Breed. Abstr.* 61: 1127–36.

Nigam, S. N. and Bock, K. R. (1998). Inheritance of resistance to groundnut rosette virus in groundnut Arachis hypogaea L). *Ann. Appl. Biol.* 117: 553–60. Doi: 10.1111/j.1744-7348.1990.tb04821.x.

Nigam, S. N., Waliyar, F., Aruna, R., Reddy, S. V., Lava Kumar, P, Craufurd, P. Q., Diallo, A. T., Ntare, B. R. and Upadhyaya, H. D. (2009). Breeding peanut for resistance to aflatoxin contamination at ICRISAT. *Peanut Sci.* 36: 42–9. Doi: 10.3146/AT07-008.1.

Nigam, S. N., Prasada Rao, R. D. V. J., Bhatnagar-Mathur, P. and Sharma, K. K. (2012). Genetic management of virus diseases in peanut. *Plant Breed. Rev.* 36: 293–356.

Norden, A. J., Gorbet, D. W., Knauft, D. A. and Young, C. T. (1987). Variability in oil quality among peanut genotypes in the Florida breeding program. *Peanut Sci.* 14: 7–11. Doi: 10.3146/i0095-3679-14-3.

Nowakowski, K. (5 June 2015). Aflatoxin, a silent threat to Africa's food supply. Retrieved from http://theplate.nationalgeographic.com/2015/06/05/Aflatoxin-the-silent-threat-to-africas-food-supply/.

Oerke, E. C. (2006). Crop losses to pests. *J Agri. Sci.* 144: 31–43. Doi: 10.1017/S0021859605005708.

Okello, D. K., Biruma, M. and Deom, C. M. (2010a). Overview of groundnut research in Uganda: Past, present and future. *Afri. J. Biotech.* 9: 6448–59. Doi: 10.5897/AJB09.013.

Okello, D. K., Kaaya, A. N., Bisikwa, J., Were, M. and Oloka, H. K. (2010b). *Management of Aflatoxins in Groundnut: A Manual for Farmers, Processors, Traders, and Consumers in Uganda*. Entebbe: National Agricultural Research Organization. http://pmil.caes.uga.edu/documents/UGA136/ManagementOfAflatoxinsInGroundnutGuideUganda2010.pdf.

Okello, D. K., Monyo, E., Deom, C. M., Ininda, J. and Oloka, H. K. (2013). *Groundnut Production Guide for Uganda: Recommended Practices for Farmers*. Entebbe: National Agricultural research Organization. ISBN: 978-9970-401-06-2. http://oar.icrisat.org/7385/1/GroundnutProuction_1-42_2013.pdf.

Okello, D. K., Okori, P., Puppala, N., Bravo-Ureta, B., Deom, C. M., Ininda, J., Anguria, P., Biruma, P. and Asekenye, C. (2015). *Groundnuts Seed Production Manual for Uganda*. Entebbe, Uganda: National Agricultural Research Organization. http://pmil.caes.uga.edu/documents/UGA203/Publication_UGA203_Deom_FY2015_GroundnutSeedProductionManualforUganda-Okello.pdf.

Olorunju, P. E., Ntare, B. R., Pande, S. and Reddy, S. V. (2001). Additional sources of resistance to groundnut rosette disease in groundnut germplasm and breeding lines. *Ann. Appl. Biol.* 139: 259–68. Doi: 10.1111/j.1744-7348.2001.tb00402.x.

Pandey, M. K., Guo, B., Holbrook, C. C., Janila, P., Zhang, X., Bertioli, D. J., Isobe, S., Liang, X. and Varshney, R. K. (2014). Molecular markers, genetic maps, and QLTs for molecular breeding in peanut. In Mallikarjuna, N. and Varshney, R. V. (Eds), *Genetics, Genomics and Breeding of Peanuts*. Boca Raton: CRC Press, pp. 79–113.

Peng, Z., Liu, F., Wang, L., Zhou, H., Paudel, D., Tan, L., Maku, J., Gallo, M. and Wang, J. 2017. Transcriptome profiles reveal gene regulation of peanut (*Arachis hypogaea* L.) nodulation. *Sci. Rep.* 7: 40066. Doi: 10.1038/srep40066.

Pensuk, V., Jogloy, S., Wongkaew, S. and Patanothai, A. (2004). Generation means analysis of resistance to peanut bud necrosis caused by peanut bud necrosis tospovirus in peanut. *Plant Breed.* 123: 90–2. Doi: 10.1046/j.0179-9541.2003.00928.x.

Pinstrup-Anderson, P. (2000). The future world food situation and the role of plant diseases. *Can. J. Plant Pathol.* 22: 321-331 Doi: 10.1080/07060660009500451-0425-01.

Prasad, K., Bhatnagar-Mathur, P., Waliyar, F. and Sharma, K. K. (2013). Overexpression of a chintinase gene in transgenic peanut confers enhanced resistance to major soil borne and foliar fungal pathogens. *J. Plant Biochem. Biotechnol.* 22: 222–233. Doi: 10.1007/s13562-012-0155-9.

Pruthvi, V., Narasimhan, R. and Nataraja, K. N. (2014). Simultaneous expression of abiotic stress responsive transcription factors, *AtDREB2A, AtHB7*, and *AtABF3* improves salinity and drought tolerance in peanut (*Arachis hypogaea* L.). *PLoS One* 9: e111152. Doi: 10.1371/journal.pone.0111152.

Qin, H., Gu, Q., Zhang, J., Sun, L., Kuupa, S., Zhang, Y., Burow, M., Payton, P., Blumwald, E. and Zhang, H. (2011). Regulated expression of an isopentenyltransferase gene (*IPT*) in peanut significantly improves drought tolerance and increases yield under field conditions. *Plant Cell Physiol.* 52: 1904–14. Doi: 10.1093/pcp/pcr125.

Rathnakumar, A. L., Hariprasanna, K. and Lalwani, H. B. (2012). Genetic improvement in Spanish type groundnut, *Arachis hypogaea* L. varieties in India over the years. *J. Oilseeds Res.* 27: 1–7.

Ray, D. K., Gerber, J. S., MacDonald, G. K. and West, P. C. (2015). Climate variation explains a third of global crop yield variability. *Nat. Commun.* 6: 5989. Doi: 10.1038/ncomms6989.

Reddy, D. V. R., Buiel, A. A. M., Satyanarayana, T., Dwivedi, S. L., Reddy, A. S., Ratna, A. S., Lakshmi, K. V., Ranga Rao, G. V., Naidu, R. A., and Wightman, J. A. (1995). Peanut bud necrosis disease: An overview. In Buiel, A. A. M., Parlevliet, J. E., and Lenne, J. M. (Eds), *Recent Studies on Peanut Bud Necrosis Disease*. Patancheru, India: ICRISAT, pp. 3–7.

Riley, D. G., Joseph, S. V., Srinivasan, R. and Diffie, S. (2011), Thrips vectors of tospoviruses. *J. Integ. Pest Mngmt.* 2: I1–I10. Doi: 10.1603/IPM10020.

Rivero, R. M., Kojima, M., Gepstein, A., Sakakibara, H., Mittler, R., Gepstein, S. and Blumwald, E. (2007). Delayed senescence induces extreme drought tolerance in flowering plants. *PNAS* 104: 19631–6. Doi: 10.1073/pnas.0709453104.

Robinson, D. J., Ryabov, E. V., Raj, S. K., Roberts, I. M. and Taliansky, M. E. (1999). Satellite RNA is essential for encapsidation of groundnut rosette umbravirus RNA by groundnut rosette assistor luteovirus coat protein. *Virology* 254: 104–14. PMID: 9927578.

Ros, E. (2010). Health benefits of nut consumption. *Nutrients* 2: 652–82. doi:10.3390/nu2070652.

Sarkar, T., Thankappan, R., Kumar, A., Mishra, G. P. and Dobaria, J. R. (2014). Heterologous expression of *AtDREB2A* gene in transgenic peanut conferred tolerance to drought and salinity stresses. *PLoS One* 9: e110507. DOI:10.1371/journal.pone.0110507.

Sarkar, T., Radhakrishnan, T., Kumar, A., Mishra, G. P. and Dobaria, J. R. (2016). Stress inducible expression of *AtDREB2A* transcription factor in transgenic peanut (*Arachis hypogaea* L.) conferred tolerance to soil-moisture deficit stress. *Front. Plant Sci.* 7: 935. Doi: 103389/fpls.2016.00935.

Schlenker, W. and Roberts, M. J. (2009). Nonlinear temperature effects indicate severe damages to U.S. crop yields under climate change. *PNAS* 106: 15594–8. Doi: 10.1073/pnas.0906865106.

Seijo, G., Lavia, G. I., Fernandez, A., Krapovickas, A., Ducasse, D., Bertioli, D. J. and Moscone, E. A. 2007. Genomic relationships between the cultivated peanut (*Arachis hypogaea, Leguminosae*) and its close relatives revealed by double GISH. *Am. J. Bot.* 94: 1963–71. Doi: 10.3732/ajb.94.12.1063.

Selvaraj, M. G., Belamkar, V., Ayers, J. L., Puppala, N. and Burow, M. D. (2010). Variability for drought resistant traits in the U.S. mini-core collection of peanut. In: Peanut breeding for drought adaption symposium. Colorado State University, CO.

Selvaraj, M. G., Burow, G., Burke, J. J., Belamkar, V., Puppala, N. and Burow, M. D. (2011). Heat stress screening of peanut (*Arachis hypogaea* L.) for acquired thermotolerance. *Plant Growth Regul.* 65: 83–91. Doi: 10.1007/s10725-011-9577-y.

Senthil-Kumar, M., Kumar, G., Srikanthbabu, V. and Udayakumar, M. (2007). Assessment of variability in acquired thermotolerance: potential option to study genotypic response and the relevance of stress genes. *J. Plant Physiol.* 164: 111–25. Doi: 10.1016/j.jplph.2006.09.009.

Settaluri, V. S., Kandala, C. V. K., Puppala, N. and Sundaram, J. (2012). Peanuts and their nutritional aspects – A review. *Food and Nutr. Sci.* 3:1644–50. doi:10.4236/fns.2012.312215.

Singh, A. K., Mehan, V. K. and Nigam, S. N. (1997). *Sources of Resistance to Groundnut Fungal and Bacterial Diseases: An Update and Appraisal.* Information Bulletin No 50. Patancheru, Andhra Pradesh, India: International crops research institute for the semi-arid tropics.

Singh, A. K. and Nigam, S. N. (2016). *Arachis* gene pools and genetic improvement in groundnut. In Rajpal, V. R., Rama Rao, S. and Raina, S. N. (Eds), *Gene Pool Diversity and Crop Improvement.* Volume 1. Switzerland: Springer International Publishing, pp. 17–75. Doi: 10.1007/978-3-319-27096-8_2.

Smart, C. M., Scofield, S. R., Bevan, M. W. and Dyer, T. A. (1991). Delayed leaf senescence in tobacco plants transformed with tmr, a gene for cytokinin production in Agrobacterium. *Plant Cell* 3: 647–656. Doi: 10.1105/tpc.3.7.647.

Smartt, J. (1994). The groundnut in farming systems and the rural economy – a global view. In Smartt J. (Ed.), *The Groundnut Crop: A Scientific Basis for Improvement.* London: Chapman and Hall, pp. 664–99.

Srinivasaraghavan, A., Sunkad, G., Bera, S. K. and Revadi, M. (2013). Sources of peanut bud necrosis disease in groundnut with desirable yield parameters and pod features. *Indian Phytopath.* 66: 215–16.

Stalker, H. T., Tallury, S. P., Seijo, G. R. and Leal-Bertioli, S. C. (2016). Biology, speciation, and utilization of peanut species. In Stalker, H. T. and Wilson, R. F. (Eds), *Peanuts: Genetics, Processing and Utilization.* Waltham, MA: Academic Press and AOCS Press, pp. 27–66. Doi:10.1016/B978-1-63067-038-2.00004-6.

Strange, R. N. and Scott, P. R. (2005). Plant disease: A threat to global food security. *Annu. Rev. Phytopathol.* 43: 83–116. Doi: 10.1146/annurev.phyto.43.113004.133839.

Subrahmanyam, P., Ramanatha Rao, V., McDonald, D., Moss, J. P. and Gibbons, R. W. (1989). Origins of resistance to rust and late leaf spot in peanut (*Arachis hypogaea*, Fabaceae). *Econ. Bot.* 43:444–55. Doi: 10.1007/BF02935917.

Subrahmanyam, P., McDonald, D., Waliyar, F., Reddy, L. J., Nigam, S. N., Gibbons, R. W., Rao, V. R., Singh, A. K., Pande, S., Reddy, P. M. and Subba Rao, P. V. (1995). *Screening Methods and Sources of Resistance to Rust and Late Leaf Spot of Groundnut.* Information Bulletin No 47. Patancheru, Andhra Pradesh, India: International crops research institute for the semi-arid tropics.

Subrahmanyam, P., Hildebrand, G. L., Naidu, R. A., Reddy, L. J. and Singh, A. K. (1998). Sources of resistance to groundnut rosette disease in global groundnut germplasm. *Ann. Appl. Biol.* 132: 473–85. Doi: 10.1111/j.1744-7348.1998.tb05223.x.

Subrahmanyam, P., Naidu, R. A., Reddy, L. J., Kumar, P. L. and Ferguson, M. (2001). Resistance to groundnut rosette disease in wild *Arachis* species. *Ann Appl. Biol.* 139:45–50. Doi: 10.1111/j.1744-7348.2001.tb00129.x.

Sukruth, M., Paratwagh, S. A., Sujay, V., Kumari, V., Gowda, M. V. C., Nadaf, H. L., Motagi, B. N., Lingaraju, S., Pandey, M. K., Varshney, R. K. and Bhat, R. S. (2015). Validation of markers linked to late leaf spot and rust resistance, and selection of superior genotypes among diverse recombinant inbred lines and backcross lines in peanut (*Arachis hypogaea* L.). *Euphytica* 204: 343–51. Doi: 10.1007/s10681-014-1339-2.

Sun, L., Hu, R., Shen, G. and Zhang, H. (2013). Genetic engineering peanut for higher drought- and salt-tolerance. *Food Nutr Sci.* 4: 1–7. Doi: 10.4236/fns.2013.46A001.

Sundaraj, S., Srinivasan, R., Culbreath, A. K., Riley, D. G. and Pappu, H. R. (2014). Host plant resistance against *Tomato spotted wilt virus* in peanut (*Arachis hypogaea* L.) and its impact on susceptibility to the virus, virus population genetics, and vector feeding behavior and survival. *Phytopathol.* 104: 202–10. Doi: 10.1094/PHYTO-04-13-0107-R.

Sundaresha, S., Kumar, M. A., Rohini, S., Math, S. A., Keshamma, E., Chandrashekar, S. C. and Udayakumar, M. (2010). Enhanced protection against two major fungal pathogens of peanut, *Cercospora arachidicola* and *Aspergillus flavus* in transgenic peanut over-expressing a tobacco β 1–3 glucanase. *Eur. J. Plant Pathol.* 126: 497–508. Doi: 10.1007/s10658-009-9556-6.

Sundaresha, S., Rohini, S., Appanna, V. K., Arthikala, M.-K., Shanmugam, N. B., Shashibhushan, N. B., Kishore, C. M. H., Pannerselvam, R., Kirti, P. B. and Udayakumar, M. (2016). Co-overexpression of *Brassica juncea NPR1* (*BjNPR1*) and *Trigonella foenum-graecum defensin* (*Tfgd*) in transgenic peanut provides comprehensive but varied protection against *Aspergillus flavus* and *Cerospora arachidicola*. *Plant Cell Rep.* 35: 1189–203. 10.1007/s00299-016-1945-7.

Taliansky, M. E., Robinson, D. J. and Murant, A. F. (1996). Complete nucleotide sequence and organization of the RNA genome of groundnut rosette umbravirus. *J. Gen. Virol.* 77: 2335–45. PMID: 8811035.

Taliansky, M. E. and Robinson, D. J. (1997). Trans-acting untranslated elements of groundnut rosette virus satellite RNA are involved in symptom production. *J. Gen. Virol.* 78: 1277–85. DOI:10.1099/0022-1317-78-6-1277.

Tian, D., Traw, M. B., Chen, J. Q., Kreitman, M. and Bergelson, J. (2003). Fitness costs of R-gene-mediated resistance in *Arabidopsis thaliana*. *Nature* 423: 74–77. Doi: 10.1038/nature01588.

Tillman, B. L. and Gorbet, D. W. (2015). Registration of 'FloRun 107' peanut. *J. Plant Reg.* 9: 162–7. Doi: 10.3198/jpr2014.12.0086crc.

Todesco, M., Balasubramanian, S., Hu, T. T., Traw, M. B., Horton, M., Epple, P., Kuhns, C., Sureshkumar, S., Schwartz, C., Lanz, C., Laitinen, R. A. E., Huang, Y., Chory, J., Lipka, V., Borevitz, J. O., Dangl, J. L., Bergelson, J., Nordborg, M. and Weigel, D. (2010). Natural allelic variation underlying a major fitness trade-off in *Arabidopsis thaliana*. *Nature* 465: 632–6. Doi: 10.1038/nature09083.

UNEP (2016). UNEP frontiers 2016 report: Emerging issues of environmental concern. United Nations Environment Programme, Nairobi. Retrieved from: http://web.unep.org/frontiers/sites/unep.org.frontiers/files/documents/unep_frontiers_2016.pdf.

Upadhyaya, H. D. and Nigam, S. N. (1994). Inheritance of two components of early maturity in groundnut (*Arachis hypogaea* L.). *Euphytica* 78: 59–67. Doi: 10.1007/BF00021398.

Upadhyaya, H. D., Nigam, S. N. and Thakur, R. P. (2002). Genetic enhancement for resistance to aflatoxin contamination in groundnut In: Summary Proc of the seventh ICRISAT regional groundnut meeting for Western and Central Africa, 6–8 December 2000, Cotonou, Benin. International Crops Research Institute for the Semi-Arid Tropics, Patancheru, AP, India, pp. 29–36.

Upadhyaya, H. D. and Ortiz, R. (2001). A mini core subset for capturing diversity and promoting utilization of chickpea genetic resources in crop improvement. *Theor. Appl. Genet.* 102: 1292–1298. Doi: 10.1007/s00122-001-0556-y.

Upadhyaya, H. D., Ortiz, R., Bramel, P. J. and Singh, S. (2003). Development of a groundnut core collection using taxonomical, geographical and morphological descriptors. *Genet. Res. Crop Evol.* 50: 139–48.

Upadhyaya, H. D. (2005). Variability for drought resistance related traits in the mini core collection of peanut. *Crop Sci.* 45: 1432–40. Doi: 10.2135/cropsci2004.0389.

Upadhyaya, H. D., Reddy, L. J., Gowda, C. L. L. and Singh, S. (2006). Identification of diverse groundnut germplasm: Sources of early maturity in a core collection. *Field Crops Res.* 97: 261–71. Doi: 10.1016/j.fcr.2005.10.010.

Upadhyaya, H. D., Mukri, G., Nadaf, H. L. and Singh, S. (2012). Variability and stability analysis for nutritional traits in the mini core collection of peanut. *Crop Sci.* 52: 168–78. Doi: 10.2135/cropsci2011.05.0248.

Upadhyaya, H. D., Swivedi, S. L., Vadez, V., Hamidou, S., Singh, D. S., Varshney, R. K. and Liao, B. (2014). Multiple resistant and nutritionally dense germplasm identified from mini core collection in peanut. *Crop Sci.* 54: 679–93. Doi: 10.2135/cropsci2013.07.0493.

Vadez, V., Kholova, J., Medina, S., Kakkera, A. and Anderberg, H. (2014). Transpiration efficiency: new insights into an old story. *J. Exp. Bot.* 65: 6141–53. Doi: 10.1093/jxb/eru040.

Valentine, H. (2016). The role of peanut in global food security. In Stalker, H. T. and Wilson, R. F. (Eds), *Peanuts: Genetics, Processing and Utilization*. Waltham, MA: Academic Press and AOCS Press, pp. 67–109. Doi:10.1016/B978-1-63067-038-2.00003-4.

Varshney, R. K., Pandey, M. K., Janila, P., Nigam, S. N., Sudini, M. V., Gowda, M. V., Sriswathi, M., Radhakrishnan, T., Manohar, S. S. and Nagesh, P. (2014). Marker-assisted introgression of a QTL region to improve rust resistance in three elite and popular varieties of peanut (*Arachis hypogaea* L.). *Theor. Appl. Genet.* 127: 1771–81. Doi: 10.1007/s00122-014-2338-3.

Vasavirama, K. and Kirti, P. B. (2012). Increased resistance to late leaf spot disease in transgenic peanut using a combination of PR genes. *Funct. Integr. Genomics* 12: 625–34. Doi: 10.1007/s10142-012-0298-8.

Vos, I. A., Pieterse, C. M. J. and van Wees, S. C. M. (2013). Costs and benefits of hormone-regulated plant defences. *Plant Pathol.* 62: 43–55. Doi: 10.1111/ppa.12105.

Waliyar, F., Kumar, P. L., Traore, A., Ntare, B. R., Diarra, B. and Kodio, O. (2008). Pre- and postharvest management of aflatoxin contamination in peanuts. In Leslie, J. F., Bandyopadhyay, R., and Visconti, A. (Eds), *Mycotoxins: Detection Methods, Management, Public Health and Agricultural Trade*. Oxfordshire, UK: CABI, pp. 209–18. Doi: 10.1079/9781845930820.0209.

Waliyar, F., Osiru, M., Ntare, B. R., Vijay Krishna Kumar, K., Sudini, H., Traore, A. and Diarra, B. (2015). Post-harvest management of aflatoxin contamination in groundnut. *World Mycotoxin J.* 8: 245–52. Doi: 10.3920/WMJ2014.1766.

Wang, H., Lei, Y., Yan, L., Wan, L., Ren, X., Chen, S., Dai, X., Guo, W., Jiang, H. and Liao, B. (2016). Functional genomic analysis of *Aspergillus flavus* interacting with resistant and susceptible peanut. *Toxins* 8: 46. Doi: 10.3390/toxins8020046.

Willekens, J. (2003). Mechanism of vector resistance in groundnut to control *Groundnut rosette virus* disease in sub-Saharan Africa. Ph.D. Thesis, p. 199. The University of Greenwich, Natural Resources Institute, UK.

Zhang, X., Han, S., Tang, F., Xu, J., Lui, H., Yan, M., Dong, W., Huang, B. and Zhu, S. (2011). Genetic analysis of yield in peanut (*Arachis hypogaea* L.) using mixed model of major gene plus polygene. *Afr. J. Biotechnol.* 10: 7126–30. Doi: 10.5897/AJB10.1871.

Zipfel, C. and Robatzek, S. (2010). Pathogen-associated molecular pattern-triggered immunity: veni, vidi…? *Plant Physiol.* 154: 551–4. Doi: 10.1104/pp.110.161547.

Chapter 8

Improving cultivation of groundnuts

S. N. Nigam, International Crops Research Institute for the Semi-Arid Tropics (ICRISAT), India; D. L. Jordan, North Carolina State University, USA; and P. Janila, International Crops Research Institute for the Semi-Arid Tropics (ICRISAT), India

1 Introduction
2 Limitations of present agronomic recommendations
3 Choice of variety/cultivar
4 Field preparation and soil resources management
5 Seed preparation, planting and weed and water management
6 Plant protection practices
7 Harvesting, drying, curing and storage
8 Precision cultivation
9 Seed systems
10 Conclusion
11 Where to look for further information
12 References

1 Introduction

Groundnut (also known as peanut) (*Arachis hypogaea* L.), a native of South America, has often been referred to as an unpredictable legume in the past (Gregory and Gregory, 1979; Hammons, 1994). The genus *Arachis* contains 81 described species, categorized into nine taxonomic sections, and includes both diploids and tetraploids belonging to either annual or perennial type. The classification is based on morphology, geographical distribution and cross-compatibility among the species (Valls and Simpson, 2005). The only cultivated groundnut, *Arachis hypogaea* L., is further divided into two sub-species '*hypogaea*' and '*fastigata*' based on the branching pattern and the distribution of vegetative and reproductive axes. Although it has been known to humankind for many centuries, its commercial cultivation started only in early 1900, when it began receiving research attention.

Groundnut is cultivated on 26.5 million ha globally with a total production of 43.9 million tons and an average productivity of 1660 kg/ha (FAO, 2014). Its global area, production and productivity increased at an annual growth rate of 1.0%, 2.7% and 1.4%, respectively, in the last two decades, from 1995 to 2014. Least developing countries account for 41.24% of the global area and 22.47% of global production (FAO, 2014). Groundnut yield in these countries

http://dx.doi.org/10.19103/AS.2017.0023.27

varies from 330 kg/ha in Namibia to 3200 kg/ha in Egypt with an average of 900 kg/ha. Most countries in Africa and some in Asia and South America led by India, Nigeria, Sudan, Sierra Leone, Tanzania and Myanmar produce yields below the world average. The yields are low largely because the crop is grown under rainfed conditions in marginal and eroded soils with very little or no input. There is limited opportunity for increasing area under the crop in Asia. Any increase in groundnut production in Asia has to come through gains in productivity, whereas in Africa both area expansion and gains in yield are possible to have sustained and enhanced groundnut production in the continent. In Australia, Argentina, China and the United States, groundnut production during 2014 was 0.016, 1.16, 16.5 and 2.35 million tons, respectively (FAO, 2014). Greater yields in these industrialized nations are most likely the result of access to seeds of new cultivars, inputs, technologies and mechanization that increase overall production and efficiency (Diop, 2004).

1.1 Groundnut ecologies

The crop is grown as a sole crop, intercrop or mixed crop. In rainfed ecologies, all three cropping systems are prevalent and only a single crop in a year is taken. In bimodal rainfall regions like Haiti and Uganda, two crops per year can be taken. Under irrigated and residual soil moisture conditions, the crop is largely grown as a sole crop. Under residual soil moisture conditions, mostly in rice fallows, only a short-duration single groundnut crop is possible, and in the post-rainy season and sometimes where soil moisture recedes towards later stages of crop growth, supplementary irrigation is provided (Sarkar and Pal, 2004). Being a legume, including groundnut as rotation crop contributes to soil fertility through biological nitrogen (N) fixation.

1.2 Groundnut farming systems

At least, two distinct production scenarios of groundnut cultivation exist at the global level, one is mechanized cultivation on large farms under input intensive conditions, and the other is non-mechanized or partially mechanized cultivation on small farms under limited resources. Subsistence groundnut farming is predominant in Asia and Africa and is characterized by low inputs, rainfed cultivation, smallholdings, manual operations using conventional tools, sole crop, mixed or intercrop (Tadele, 2017) and low productivity (700–1000 kg/ha). It covers more than 76% of the global groundnut area under cultivation in subsistence farming, where groundnut haulms are valued as fodder for livestock and contribute to farmers' income. Resistance to foliar fungal diseases is a must-have trait for varieties in subsistence farming to ensure good quality and quantity of haulms besides pod yield. The commercial farming is practised largely in the United States, Australia, Argentina, Brazil, China and South Africa and in small pockets in other countries. It is characterized by high inputs, availability of irrigation in some areas, medium to large holdings, mechanized or semi-mechanized cultivation, sole crop or rotated with other crops and high productivity (3000–4000 kg/ha).

1.3 Yield gap

Groundnut is an efficient crop. A pod yield of 10500 kg/ha over a small area under intensive cultivation in Shandong province in China (Yanhao, 1996), 9600 kg/ha in large plots in Zimbabwe (Hildebrand, 1996) and 9400 kg/ha in a 0.2 ha plot in summer season in

Maharashtra and 9500 kg/ha in a small area in Andhra Pradesh in India (Nigam, 2000) have been reported. However, these reflect the upper end of the yield spectrum. At the other end of the spectrum, the crop may face complete failure if drought is severe. Groundnut productivity largely varies not only among the countries but also among the regions within a country. There exists a large yield gap across all agroecologies. Simulated rainfed yields in India indicated a much higher potential for groundnut productivity than what is realized at present in the country (Bhatia, 2006).

Groundnut pod yield and quality have to be improved in the large majority of groundnut-growing areas. Under different farming systems the extent of improvement that can be achieved is variable. Genetic and management options are available to achieve and/or sustain the pod yield and quality; some of the options reduce the input cost and are key to improve the profitability of groundnut cultivation. In this chapter, the options available to improve and/or sustain groundnut pod yield and quality under different groundnut farming systems that include subsistence to precision cultivation are discussed.

2 Limitations of present agronomic recommendations

In developing countries, most of the farmers are resource-poor with little or very low risk-taking capacity. With the majority of the groundnut crop being grown under rainfed conditions, it remains vulnerable to vagaries of weather, which discourages farmers to adopt recommended crop management practices to keep the cost of cultivation low. Such farmers require low-cost technologies based on locally available resources. On the other hand, under irrigated conditions, farmers, being resourceful, are able to follow intensive agriculture following all the recommended practices to obtain higher yields where options to reduce inputs cost and precision cultivation can be useful.

In most developing countries, agronomic recommendations made are of a general nature and they fail to take into account local variations in agroclimatic conditions, soils, cropping systems and other factors across regions/groundnut fields, which are often significant. Blanket fertilizer application, in the long run, often results in nutrient imbalance in the soil, affecting nutrient-use-efficiency and productivity of the crop adversely. Moreover, diverse growing conditions of the crop necessitate development of different crop management modules meeting the requirements of different agroecologies and strata of the farmers.

The recommendations often are not adopted by farmers, particularly in developing countries as access and usage of inputs and technologies are determined by the several policies of the governments such as input and credit policy and minimum support price for the produce. Besides, prevailing market prices of the commodity, cost of technologies and business opportunities determine the extent of inputs used by farmers in developing countries.

Peanut yield is the result of genotype, environment, management and their interactions as follows:

$$\text{Yield} = \text{Cultivar/Variety (V)} + \text{Management (M)} + \text{Environment (E)} + V \times M + V \times E + M \times E + V \times M \times E$$

For higher net profits and stable yields, all the components and interactions in the above equation need to be fully exploited.

Regardless of country, farmers also must consider the ramifications of government legislation (trade policies and financial support available to the agricultural sector), regulations (pesticide registrations and availability and food safety requirements) and other incentives that promote environmental stewardship such as conservation tillage practices and adoption of integrated pest management (IPM) when developing sustainable production systems. Government policies play a critical role in availability, cost and use of technologies such as new varieties, quality seed and other inputs. Market prices often determine the extent of cultivation of groundnut, a higher market price in the current year might result in larger areas being planted in the next season/year under groundnut that may bring down the market price. The input use in Asia and Africa is largely determined by inputs supplied by the government agencies, subsidies on the inputs and credit policies of the government as farmers often depend on credit for making inputs investments. In the absence of cultivar replacement policy and seed systems, an obsolete cultivar, TMV 2 continued to occupy over 50% area under groundnut cultivation in India (Deb, 2005). On the other hand, business opportunities contribute to fast variety turnover in the United States, Argentina and Australia. Therefore, the equation above can be further expanded to include government and business; however, in this chapter, we discuss the components of the above-mentioned equation.

3 Choice of variety/cultivar

Cultivar is a critical component of any crop production module to ensure consistent sustainable production of the crop. Genetic components of a cultivar/variety include agronomic traits that determine suitability of a cultivar to a farming system, host-resistance or -tolerance to various biotic and abiotic stress factors, and both physical and nutritional quality parameters such as high oleic acid profile, for enhanced shelf-life (Janila, 2016a). Resistance/tolerance to prevailing diseases and insect pests is important for cultivars; otherwise, chemical control measures will add to the cost of cultivation besides raising environmental concerns. Several disease-resistant varieties have been released for cultivation and are summarized by Janila (2016b). The cultivar choice thus has to meet the needs of the producer, processor and consumer.

The choice of cultivar at a location/in the region is guided by the following factors: i) length of soil moisture availability period, ii) type of product preferred by the processors and market, iii) prevailing diseases and insect pests' spectrum and iv) prevailing environmental conditions including rainfall pattern. Groundnut cultivars with crop duration varying from 90/100 to 160/180 days are available. There are three distinct plant types in groundnut – erect, semi-erect/semi-spreading and runner/spreading (Stalker, 1997). In North America, where only a single crop is possible due to temperature limitations and where the crop is grown with intensive inputs, the long duration cultivars with spreading growth habit adapted to local agroecologies are grown. On the other hand, in rainfed environment in India or West Africa, in multiple cropping situations (common in East and Southeast Asia) and in residual soil moisture situation (common in South Asia), the shorter duration (90/100 days) erect type cultivars are needed. Under rainfed conditions, soil moisture stress becomes an important issue in the choice of a cultivar. Depending upon the long-term pattern of rainfall, cultivars having tolerance to mid-season or end-of-season drought can be selected.

The end-use of the product will influence the choice of cultivar to be grown. Groundnut is grown for food uses (direct consumption after roasting and boiling, in confections and peanut butter, etc.) and for extraction of high-quality oil for edible and industrial use.

4 Field preparation and soil resources management

4.1 Selection of field

The ideal groundnut soil is well drained, light coloured, with either sand, loamy sand or sandy loam texture and pH ranging between 6.0 and 6.3 (Jordan, 2016). The right soil for groundnut should not ribbon out but fall apart easily when the moist soil is rubbed between index finger and thumb. The produce from such soils is clean and bright. Groundnut should regrow in the same field only after two seasons/years.

4.2 Land preparation and conservation tillage

In conventional tillage, the field is cleared off all stubbles and plant residues of the previous crop before ploughing to obtain a fine tilth to a depth of 15–20 cm. Very deep ploughing encourages pod formation in deeper layers of soil rendering harvest more difficult and is hence not desirable. Groundnut is sown on flat beds or ridges or on raised beds separated by furrows. Raised beds with a slope of 0.4–0.8% are beneficial as they allow easy drainage of excess water, avoid compaction of seedbeds and facilitate field operations as all traffic is restricted to furrows (Kuotsu, 2014).

Conservation agriculture (continuous no-till, maintaining soil cover with plant residues) is now widely recognized as a viable concept for practising sustainable agriculture with benefits in production, economic and environmental aspects of the farming. In groundnut, conservation tillage practices can impact pest complexes and yield compared with conventional tillage, and in some cases, efficiency of digging pods is an important factor in selecting the appropriate tillage system (Johnson, 2001; Jordan, 2017). Preliminary results at ICRISAT on Alfisols suggest that conservation tillage (no-tillage and minimum tillage) produced pod yield equal to the conventional tillage. In North Carolina, conservation tillage peanut increased from 10% in 1998 to 23% and 41% in 2004 and 2009, respectively, and later decreased to 20% in 2014 (Jordan, 2017). The no-tillage system resulted in higher population of various biological agents in soil than under reduced tillage condition, thus showing the lowest incidence of peanut blight (*Sclerotinia minor*) and root rot (strains of *Fusarium solani*) in peanut (Gil, 2008). Porter and Wright (1991) reported a greater incidence of early leaf spot and defoliation in conventionally tilled plots than in those that employed the conservational tillage system (band-tilled or in-row plots). Compared with conventional tillage, conservation tillage showed a reduced incidence of tomato spotted wilt disease of peanut in North Carolina (Hurt, 2006). The nodule size is larger in conservation tillage compared to conventional tillage (Rowland, 2015). In Georgia, the conservation production system increased water-holding capacity of the soil and enhanced soil water conservation (Hawkins, 2016).

4.3 Fertilizers and amendments

Groundnut responds better to residual soil fertility than to the direct application of fertilizers. Crop(s) preceding groundnut crop should be well fertilized to build up soil fertility, particularly for phosphorus (P) and potassium (K). Basics of the fertilization strategy include nutrient application based on soil test, building up soil P and K to a medium or a high level and providing calcium (Ca) to the pegging zone. Based on the soil test results and the expected pod yield from the crop, fertilizer doses for both macro- and

Table 1 Estimated nutrients required to produce selected pod yields of groundnut

	Quantity (Kg/ha)[1]									
Pod yield (Kg/ha)	N	P	K	Ca	Mg	S	Fe	Mn	Zn	B
1000	58	5	18	11	9	4	2	0.09	0.08	0.05
2000	117	10	36	23	18	9	4	0.19	0.16	0.11
3000	174	15	54	34	27	13	6	0.29	0.24	0.16
4000	232	20	73	45	36	18	8	0.38	0.32	0.22
5000	290	25	91	56	45	22	10	0.48	0.41	0.27
6000	348	30	109	68	54	26	12	0.58	0.49	0.33
7000	406	35	126	77	63	30	14	0.68	0.56	0.38
8000	464	40	144	88	72	34	16	0.78	0.64	0.44
9000	522	45	162	99	81	38	18	0.88	0.72	0.49
10 000	580	50	180	110	90	42	20	0.98	0.80	0.54

[1] Calculation based on Sahrawat, Srinivas Rao, and Nambiar. 1988. *Plant and Soil* 109:291–293.

micronutrients should be determined. Table 1 gives an idea of the requirements of the nutrients at different levels of targeted yields. Whatever is extracted from the soil, it has to be replenished from regeneration in the soil, external fertilizers and other sources to maintain soil fertility status.

4.4 Macronutrients

Under most conditions, enough N is fixed through symbiotic relations with the native *Bradyrhizobium* ssp. However, N deficiency can occur in poor and eroded soils; in new soils brought under cultivation and rice fallows; and in soils having inefficient *Bradyrhizobium*, drought and high temperatures, etc. In addition to seed inoculation with efficient *Bradyrhizobium* strain (there is specificity between a cultivar and *Bradyrhizobium* strain in groundnut), a basal dose will be useful before nodules appear and become effective (Elsheikh and Mohamedzein, 1998).

Phosphorus deficiency in groundnut is widespread (Kumar, 2015) and can be corrected by application of P fertilizers. In calcareous sandy soils, P fixation can lead to P deficiency in spite of P fertilization. Soil P levels required for groundnut are often lower than those required for other crops.

Groundnut responds to application of K when available K in the soil is very low. There is an antagonistic effect on the uptake of K, Ca and Magnesium (Mg). The proportion of K:Ca:Mg is more important than the total amount of any one of them. A high level of K in the pod zone is harmful because it affects pod quality, especially at low levels of Ca. The suggested proportion in the literature is 4:4:2 (Nigam, 2006). The latter was reported from studies on sandy loam soil under rainfed and irrigated conditions in Andhra Pradesh, India. The P and K fertilizers should be turned in deep before sowing to keep them out of the pegging zone, where they can interfere with Ca uptake (Singh, 1999).

Ca is essential for pod and seed development. Adequate quantities of Ca should be present in the top 8–10 cm of soil to enable pods to directly absorb it (Arnold, 2017). If soil pH, based on soil test, is to be raised, lime may be used both to adjust pH and to provide Ca to the pegging zone. However, lime (calcium oxide) should be applied and incorporated to a shallow depth immediately after deep ploughing to allow enough time for the Ca in lime to dissolve and get into soil solution. The rate of application of lime depends on the degree of pH to be raised, type of lime, soil type and depth of application. Instead of lime, calcium sulphate (gypsum or land plaster) can be applied when needed if soil pH is not an issue (Nigam, 2006). Regardless of the test Ca level, gypsum should be applied to all large-seeded groundnuts at the peak flowering stage and should be mixed thoroughly with the soil rather than leaving the added gypsum on top soil (Singh, 1999). Adequate Ca supply in the podding zone reduces infection from *Aspergillus flavus* and other pod-rot-causing fungi (Fernandez, 1997).

Little response is reported for application of Mg to groundnut. Responses occur at very low soil test levels most likely on deep, excessively drained sandy soils. The deep root pattern of groundnut allows plant to forage deeply for Mg and other nutrients. Application of gypsum provides adequate sulphur (S) to the crop. S deficiency is most likely to occur in very sandy soils, which possess little anion exchange capacity (Singh, 1999).

4.5 Micronutrients

Most micronutrients are more likely to be deficient at high pH, particularly in calcareous soils. The deficiency symptoms of different micronutrients can be observed in any standard book on groundnut. Boron (B) deficiency can be corrected by application of borax to the soil at a rate of 3–4 kg/ha or by spraying 0.1% borax early in the season to assure uptake before flowering. Iron (Fe) deficiency can be a serious problem in calcareous soils and it can be alleviated by soil application of 10 kg/ha ferrous sulphate or spraying of 0.5% ferrous sulphate + 0.2% urea. Zinc (Zn) deficiency can be corrected by basal application of 10–20 kg/ha zinc sulphate once every three years to the soil where groundnut is continuously grown. Application of Zn and other micronutrients should be based on soil analysis report; otherwise, it may lead to toxicity. Manganese (Mn) deficiency occurs in soils which are inherently low in this nutrient, and foliar Mn application can correct Mn deficiency more rapidly than soil Mn application (Singh, 1999).

4.6 Organic and other amendments

Most soils in developing countries are low in soil carbon content. Application of organic matter to the soil improves its structure and reduces its compaction and crusting (Hawkins, 2016). Farmers in Asia commonly apply 8000–10 000 kg/ha of farmyard manure (FYM) and other organic manures (compost and cattle and sheep penning). It also improves nodulation and enhances the availability of P to the crop. However, with dwindling cattle populations, application of FYM to the soils has reduced drastically, leading to poor soil health and an increased incidence of soilborne diseases. Application of tank silt and green manuring crop in crop rotation are some of the options to improve soil organic matter content.

4.7 Biofertilizers

Biofertilizers help enhance the efficiency of externally applied nutrients and transform, mobilize and make available fixed or unavailable form of nutrients (Jeffries, 2003). They are low cost, reduce external application of N and P by 25% and may increase crop yield up to 15%. Some of the biofertilizers used in groundnut are *Bradyrhizobium*, plant-growth-promoting rhizobacteria (PGPR), phosphorus-solubilizing microorganisms (PSM) and arbuscular mycorrhizal (AM) fungi. In addition to fixing atmospheric N, *Bradyrhizobium* also secretes siderophores, which help in promoting Fe uptake in soils where lime-induced Fe chlorosis is observed. PGPR, a mixture of beneficial rhizosphere microbes, provides benefits to plants by mobilizing the majority of macro- and micro-nutrients besides producing hormones which lead to enhanced plant yield. It also suppresses soilborne pathogens such as *Aspergillus niger* and *Sclerotium rolfsii* (Dey, 2004). Phosphate-solubilizing bacteria and fungi play an important role in converting insoluble phosphatic compounds such as rock phosphate, bone meal and basic slag and the chemically fixed soil P into available forms. AM fungi possess special structures known as vesicles and arbuscules, with the latter helping in the transfer of nutrients from soil into the root system. Mycorrhizal plants increase the surface area of the root system for better absorption of nutrients from soil, especially when the soil is deficient in P and Zn (Bell, 1989).

Several types of formulations of biofertilizers in the form of dry products and suspensions are available in the market. They can be applied as seed treatment, as a mix with FYM and through irrigation water.

5 Seed preparation, planting and weed and water management

Groundnut differs from other legumes in that its pod develops under ground and therefore soil health and water management are very critical in groundnut cultivation. Groundnut pods directly absorb Ca from soil for their development and growth (Skelton and Sheer, 1971). The crop requires moisture both in the root zone and in the pegging zone to give sustainable high yields (Hawkins, 2016).

5.1 Sowing

Seeds should be tested for germination and postharvest seed dormancy before sowing. If required, seeds should be sprayed 2–3 times with etherel (5 mL in 1L of water) to break postharvest dormancy and air-dried (Nigam, 2006). In groundnut, seed cost is one of the major components of the cost of cultivation; in developing countries of Asia and Africa the seed cost constitutes one-third of the total cultivation cost. Every effort should be made to ensure germination and growth of each planted seed. Seed size, spacing and germination per cent of seed should determine seed rate to obtain optimal plant stand. Maintaining optimal plant stand is key to realising improved.

5.2 Seed treatment and seed priming

Plant losses of 50% or more are possible with untreated seeds (Ruark and Shew, 2010), which directly impact pod yield. Seed treatment ensures their safety, uniform and healthy seedling, fast emergence and optimum plant population to achieve high yields. Seed treatment varies from basic seed dressing to coating and pelleting. Seed treatment is resorted to in order to protect the seed from soilborne diseases and insect pests, as well as providing nutrients and growth stimulus to growing seedlings and plants. By using inert material, seed pelleting also allows modification of seed shape and size for precision planting.

5.2.1 Management of soilborne plant pathogens

In the United States, commercial seeds are almost always treated with fungicide, and insecticides are frequently applied in the seed furrow at planting to control thrips (*Frankliniella* spp.) (Jordan, 2017). In Asia and Africa, seed treatment using fungicides, pesticides or their combination is done by farmers before planting. The seed should be treated first with liquid chemicals followed by powder/dust. Recently, biocontrol agents have also found a place in seed treatment (Rojo et al., 2007; Ruark and Shew, 2010) as fungicides are not effective enough to control soilborne diseases. Now, *Trichoderma viride* and *T. harzianum* are widely used to manage soilborne diseases such as collar rot, stem rot, brown root rot, etc. Commercial preparation of *Trichoderma* spp. and *Bacillus* spp. is available in developed countries.

Evaluations conducted by Ruark and Shew (2010) in North Carolina showed that seed treatment with Kodiak, a commercial preparation of *Bacillus subtilis*, increased seedling emergence and survival over untreated seeds. The effect of seed treatment with *B. subtilis* was more pronounced when groundnuts were subjected to stresses such as limited water availability, poor rotational practices or cool soil caused by early plantings (Turner and Backman, 1991). It resulted in improved germination and emergence, increased nodulation by *Bradyrhizobium* spp., enhanced plant nutrition, reduced levels of root cankers caused by *Rhizoctonia solani* and increased root growth. Seed treatment with VOTiVOTM, a commercial product of *B. firmus* from Bayer Crop Science, is used to reduce the impact of nematodes.

Enhanced resistance to abiotic stresses: *Pseudomonas fluorescens* strain TDK 1 possessing ACC deaminase activity enhanced salinity resistance and improved yield parameters under salinity stress (Saravanakumar and Samiyappan, 2007). Pelleting groundnut seed with Ca enhanced plant growth and reduced seedling mortality in acid soils (Murata, 2008).

5.2.2 Nutrient supply

Seed treatment with *Bradyrhizobium* to makes N available to growing seedlings and plants. It is a common practice in legumes cultivation. Although commercial preparations of *Bradyrhizobium* are available, their quality and effectiveness remain doubtful particularly in developing countries. In groundnut, cultivar and *Bradyrhizobium* strain specificity is observed, and therefore, the strain of *Bradyrhizobium* that matches with cultivar should be used for better performance. If the strain of *Bradyrhizobium* is not compatible with other chemicals used in seed treatment (see manufacturer's instructions), the culture can be applied in sowing rows following the slurry method. The success and effectiveness of seed coating and pelleting with micronutrients depend on the nutrient used, coating material, soil type, soil moisture and fertility status and the nutrient:seed ratio (Halmer, 2008).

5.2.3 Seed priming

Seed priming techniques include hydro-priming (soaking the seed in water before sowing and may or may not be followed by air-drying of the seeds), halo-priming (soaking of seeds in solution of inorganic salts), osmo-priming (soaking of seeds for a certain period in solution of sugar, PEG, etc.) and hormonal-priming (seed treatment with different hormones) (Singh, 2015). Generally, in seed priming, seeds are partially hydrated to allow metabolic events to occur without actual germination, and are then re-dried (near to their original weight) to permit routine handling (Bradford, 1986). Limited studies were done on seed priming in groundnut. Seed priming of groundnut seeds with $CaCl_2$ 1% resulted in higher percentage of field emergence, plant height, number of pods per plant, seed yield per plant and seed yield per ha in addition to early flowering and early maturity than the control (Bhingarde, 2015). Seed treatment with hot water (50°C for 20 min) or seed priming with hot aqueous solutions of $Ca(OH)_2$ enhanced the rate of germination and increased final plant stand in groundnut. The rapid rate of emergence gave a chance to seedlings to escape diseases assisted by antifungal properties of $Ca(OH)_2$ (El-Rehim, 1981). Seed priming with $CaSO_4$ significantly reduced seedling mortality and resulted in the early emergence of groundnut in acid soils (Murata, 2008).

5.2.4 Addressing micronutrient deficiency

Although micronutrient deficiencies are traditionally addressed by soil or foliar application, micronutrient seed treatment is an alternative cost-effective method to supply micronutrients and in most cases, it improves stand establishment, advances phenological events and increases yield and micronutrient contents of the grains (Farooq, 2012). Seeds may be treated with micronutrients either by soaking in nutrient solution of a specific concentration for a specific duration (seed priming or nutripriming) or by coating with micronutrients. Oxygen supply, temperature of the priming media and solution concentration (water potential) are important factors affecting seed priming (Corbineau and Cõme, 2006). Seed priming has been used to address Zn, B, Mo, Mn, Cu and Co deficiencies (Farooq, 2012).

5.3 Sowing time

The temperature profile of a location and season of cultivation decide the optimum time of sowing of the crop at that location. The optimum air temperature for growth and development of groundnut is between 25°C and 30°C. Temperatures above 35°C are detrimental to groundnut production. Groundnut is more sensitive to heat stress in the reproductive phase than in the vegetative phase. Studies have shown that responses of groundnut genotypes to high temperatures are variable, and genotypes with superior performance under high temperature are identified (Akbar, 2017). Under lower temperatures, germination is delayed and crop duration is prolonged (Prasad, 2000). In places where the crop is sown after winter season, groundnut should not be sown until the average soil temperature at 10 cm depth is 18°C or greater for a minimum of three consecutive days and favourable air temperature are forecast. Sowing time can also impact pest damage and is a key component of production strategies where inputs are limited (Singh, 1986).

5.4 Planting density, sowing depth and sowing methods

Plant density in groundnut varies from region to region in a country and also from country to country. The growth habit of the cultivar, the sowing method and the type of farming have a bearing on the plant density to be adopted in a region. The plant spacing adopted should ensure groundnut cover as quickly as possible. A wider row spacing leaves ground uncovered, which attracts thrips and aphids, the vectors of viruses. Row spacing varies from as low as 20 cm to as high as 100 cm and within row spacing from 7.5 to 15 cm in different countries (Lanier, 2004). A closer spacing for Spanish/Valencia cultivars and a wider spacing for Virginia/Runner cultivars are recommended. In some places farmers use criss-cross sowing and paired row sowing, particularly in rice fallows. The optimum sowing depth is 5 cm. Deeper sowing delays germination and results in poor root and shoot development and poor nodulation, thereby adversely affecting the yield. The seed rate would be dependent on the row-to-row and plant-to-plant spacing, seed weight and adjustment to be made for germination percentage.

The sowing methods include manual (dibbling and behind the plough), animal-drawn seed drill, tractor-drawn seed drills and vacuum planters. For sowing, the use of seed drill is recommended as it results in faster sowing, quick emergence and uniform plant stand.

5.5 Intercropping/crop rotation

Intercropping/mixed cropping is often practised in rainfed semi-commercial and subsistence agriculture as an assurance against crop failures and to meet diverse food needs of the family. Groundnut is generally intercropped with cereals such as maize, sorghum and pearl millet. In some provinces in India, groundnut and pigeonpea (both legumes) intercropping is also popular (Ong, 1986; Sarkar and Pal, 2004). Recently, intercropping of two legumes is referred to as double-up legumes technology, which results in doubling grain yield and soil fertility. Pigeonpea and groundnut intercropping is gaining popularity as double-up technology in Malawi. Good crop rotation of the crops helps to maintain soil fertility, improves organic matter and physical structure of the soil and reduces diseases inoculum and insect pest population in the soil (Bationo and Ntare, 2000). As groundnut responds well to residual fertility, it is desirable to rotate it with other well-fertilized crops, particularly cereals. In Southeast Asia, groundnut–rice and groundnut–maize rotations are common.

Groundnut should be rotated with a well-fertilized cereal crop (Bationo and Ntare, 2000) as it can make effective use of residual fertilizers from the previous crop(s). A proper crop rotation can result in higher yields and in substantial saving in disease control and fertilizer requirements. The crop planted prior to groundnut can impact yield of groundnut and cost of production required to maintain yield. For example, in the United States, maize (*Zea mays* L.) and cotton (*Gossypium hirsutum* L.) are more effective rotation crops than soybean (*Glycine max* (L.) Merr.) because these crops more effectively break cycles associated with soilborne pathogens and plant parasitic nematodes (Jordan, 2008).

5.6 Weed management

It is essential to keep groundnut fields weed-free up to 45 days after crop emergence. In mechanized farming, weeds interfere with digging and inverting of pods. Weeds can harbour insect pests which may be the vector of viruses. Recommendation on chemical control of weeds is available in each country, and should be followed accordingly. Lastly, hand-weeding or intercultural operation can be done along with gypsum application so as to incorporate it into the soil. Once pegs have entered the soil, the plant should not be disturbed. Interculture in a rainfed crop helps to reduce weeds and also encourages infiltration of rainwater. Many farmers in South Asia practise earthing up (mounting soil around the plant) to allow pegs from higher nodes to enter soil. This practice may promote growth of stem-rot-causing fungus (*Sclerotium rolfsii*) and may also deteriorate the quality of the earlier set of mature pods while waiting for the later set of pods to mature.

5.7 Water management

Groundnut is very sensitive to moisture stress at flowering, pegging and pod and seed development stages. Similarly, it cannot tolerate water stagnation in its field. Thus, water management becomes very important in groundnut. In an irrigated crop in the post-rainy season, it is easy to regulate water supply to ensure optimal crop growth. Depending upon the soil type and climatic conditions at a given location, the water requirement of groundnut varies from 420 to 820 mm in India (Ong, 1986; Deb, 2005). Groundnut can withstand 2- to 3-week moisture stress at seedling stage. After providing full irrigation for the seeds to germinate, water supply can be withheld for the next 2–3 weeks. This forces the roots to go deeper in search of water, and when the moisture stress is released after 2–3 weeks, a sudden burst of profuse flowering would occur, resulting in uniform maturity and high yield. At pegging and pod and seed development stages, the crop requires frequent but light irrigation. The preferred method of irrigation is sprinkler method or any other system of overhead irrigation. Flood or furrow irrigation results in water wastage and does not distribute water uniformly (Naveen, 1992).

In rainfed agriculture, it is essential that the duration of the selected cultivar matches the duration of plant-extractable water available in the soil. In the case of groundnut, it is important that the topsoil remains moist at the time of pegging to facilitate entry into the soil and during the pod and seed development (Reddy, 2003; Hawkins, 2016). Besides increasing yield, the absence of moisture stress at the time of pod and seed development will also discourage pod and seed invasion by *A. flavus* and subsequent aflatoxin production (Nigam, 2009). Groundnut fields should be well drained. Standing water even for 4–6 h in the field can damage the crop. Planting on raised beds or ridges facilitates drainage. Quality of irrigation water can affect groundnut productivity. The limits of saline water for groundnut are EC (electrical conductivity) <4.0 mmhos/cm and RSC (residual sodium carbonate) <2 meq/L.

5.8 Groundnut cultivation under polythene mulch

Groundnut cultivation under polythene mulch is very popular in China and Vietnam. It is described in detail in a publication by Wenguang (1996). Mulching can be done either before or after sowing. In sowing before mulching, normal sowing with two seeds per hill on bed surface is followed by mulching. As soil cracks due to emerging seedlings, holes of 4–5 cm diameter are made over the crack in the film and the hole is covered with moist soil to facilitate emergence of seedlings through the holes in polythene film. In sowing after mulching, after spreading the polythene mulch over beds, holes, 3–4 cm deep, are made at the desired spacing with a hole-maker. Two seeds are placed in each hole and covered with soil. If soil moisture is low, some water can be poured into the hole and the seeds can be covered with moist soil. Some additional soil is placed over the hole to cover the edges of polythene film. Cultivation under polythene mulch can increase groundnut yield by 20–50% (Nigam, 2006; Jain, 2017). The advantages of this practice include an increase in pod yield, proportion of well-filled pods and oil and protein contents in seeds. It results in early emergence and early maturity of the crop (heat entrapment due to sunlight transmittance raises soil temperature), giving opportunities to increase cropping intensity; requires fewer irrigations; suffers less insect pest damage as compared to non-mulched crop; and provides better weed control with pre-emergence herbicide application. The crop can be sown under lower temperatures and thus helps in extending the cultivation of the crop in non-traditional areas. Polymers are also used for conserving water and enhancing crop productivity and thereby increasing water-use efficiency of crops. Use of mulching with polymer and integrated nutrient management enhanced pod, haulm, kernel, and oil yields and net economic returns (Jain, 2017).

5.9 Use of growth regulators

The use of chemical growth regulators on groundnut to suppress vegetative growth, achieve higher yield and improve pod quality has met with varying degrees of success. But their use in groundnut is very common in China and Vietnam. Some of the growth regulators are Dinocap (DPC), Fosamine, 2,3,6-trichlorobenzoic acid (TCBA) and Paclobutrazol (P 333). DPC inhibits stem growth, enhances root development and branching and increases pod yield by increasing pod number per plant. The early flowering stage is the best time to apply 80 ppm DPC as foliar spray. Foliar application of P 333 at 60 ppm around 25–30 days after the first flowering arrests excessive growth of the plant by reducing stem growth and increasing pod number per plant, pod and seed mass and pod yield. Fosamine inhibits the growth of aerial parts and flowering of groundnut. Foliar application of 500 ppm is recommended at the late pod-forming stage to reduce late-forming flowers. However, the seeds from fosamine-treated groundnut have poor germination and produce abnormal seedlings (Chengrong, 1996). Therefore, they should not be used as seeds to grow the next crop. TCBA inhibits growth of the aerial parts and late ineffective flowers. It should be sprayed (250 ppm) at the peak or the late peak flowering stage. Once the effect of the chemical has waned, the plant may grow more rapidly. Response to TCBA application (increase in yield) varies with the genotypes. It gives good results with large-seeded cultivars of medium- and long-duration. Its effect on Spanish cultivars is very little. The plant growth regulator prohexadione calcium, which is used in some parts of the United States, prevents internode elongation and increases pod retention (Beam, 2002; Jordan, 2009).

6 Plant protection practices

Groundnut yield and quality is constrained by insect pests and diseases caused by wind and soilborne pathogens and viruses; insects that consume foliage and pods; soil-parasitic nematodes; interference by weeds; and storage pests (Janila, 2013). Some of these are widespread and others are localized. Since pods develop and grow underground they are also exposed to different kinds of soilborne fungal pathogens and insect pests.

Genetic and preventive and management options and their combinations are desirable to reduce the losses to diseases and insect pests. Resistance or tolerance to diseases and insect pests, if available, is the best-bet option. Other options include cultural, chemical and biological/botanical control measures. No single option alone can give effective control of biotic constraints. An integrated insect pest and disease management (IPM and IDM) approach is both effective and safe to contain the damage caused by insect pests and diseases. Chemical sprays should be need-based, and appropriate safety measures should be taken while applying them.

6.1 Foliar fungal diseases

Cultivars resistant/tolerant to rust and late and early leaf spots are available. They should be the first choice for growing. If needed, these varieties should be supported by 1–2 chemical sprays at the critical stages to obtain the maximum quality yield from the crop. Carbendazim 50WP 500 g/ha or Mancozeb 50WP 1 kg/ha for both leaf spots, Calixin 250 mL/ha for rust alone and Chlorothalonil 75WP 1 kg/ha for both rust and leaf spots can be effectively used. Cultural management practices such as removal of debris from fields and volunteer plants in the vicinity, intercropping with cereals, deep ploughing and crop rotation can reduce inoculum load, particularly of leaf spots (Nigam, 2006).

6.2 Soilborne fungal diseases

There are several soilborne diseases which can infect groundnut seed and seedlings. Some of these are collar rot, white mould/stem rot, *Aspergillus flavus*, rhizoctonia damping-off, etc. Soilborne diseases, once appeared, are difficult to control. Therefore, it is essential to maintain good soil health following good agricultural practices. Available genetic resistance against soilborne diseases is weak and cannot offer protection against them in disease-endemic areas. These diseases are best managed by appropriate chemical seed treatment (see Section 5.2), following crop rotation and other field sanitation practices including summer ploughing. Soilborne pathogens such as stem rot are controlled using 2–5 sprays applied bi-weekly when pathogens are active (Jordan, 2017).

6.3 Aflatoxin contamination

Aflatoxins, which are carcinogenic, are secondary metabolites produced in groundnut seeds by *A. flavus* and *A. parasiticus*. These fungi (soilborne and airborne) infect groundnut seeds through microscopic cracks in pod wall and seed coat, mechanical injury to pods during intercultural operations and harvesting and insect and nematode damage to the pods. Moisture stress at the pod and seed development stage predisposes them to *A. flavus*/*A. parasiticus* infection in the field (Navya, 2015). The rainfed crop is more vulnerable to fungal infection in the field. Postharvest infection is serious under wet and humid conditions. In addition to the seed treatment, the following steps minimize the extent of aflatoxin contamination.

1. Grow tolerant cultivars, if available.
2. Apply farm yard manure/compost @5–10 t/ha to the soil.
3. Apply gypsum @400–500 kg/ha at peak flowering.
4. Provide light but frequent irrigation, if available, during pod and seed development stages.
5. Avoid mechanical damage to pods during weeding, harvesting, curing, threshing and storage.
6. Harvest the crop as soon as it matures (in case of severe drought, the crop should be harvested early).
7. Dry pods to lower moisture content below 8%.
8. Do not mix gleanings (pods left into soil at the time of harvesting) with main produce.
9. Remove damaged and underdeveloped pods from the produce.

10 Store groundnut in-shell and at low temperature, low humidity and in moisture-free conditions.
11 Protect the stored produce from damage of storage insect pests.
12 Remove immature pods from haulm before feeding it to the animals.

6.4 Virus diseases

Important virus diseases include peanut bud necrosis disease (PBND) in South Asia and Southeast Asia, tomato spotted wilt virus (TSWV) in the United States, peanut stem necrosis diseases (PSND) in some parts of India, peanut stripe disease (PStV) in East and Southeast Asia and groundnut rosette disease (GRD) in Africa (Janila et al., 2013). Except for PStV disease, other virus diseases are not seedborne. Cultivars resistant/tolerant to viruses causing these diseases and to their vectors are available, and they should be grown in virus endemic areas (Nigam, 2012). Cultural practices such as timely sowing, optimum plant population, effective control of vectors and intercropping and border cropping with tall growing cereals, removal of alternate weed hosts of the virus and the vector during off-season should be adopted to manage virus diseases. Chemical seed treatment helps in reducing virus diseases through the control of their vectors. In the United States, risk indices have been developed to suppress TSWV and include cultivar selection, plant population and pattern, insecticides to control thrips (the vector of TSWV), the sowing date, and the tillage system (http://tomatospottedwiltinfo.caes.uga.edu/peanut/risk.html).

Efficacy to manage virus diseases through chemical control of vector is also dependent on the nature of the virus (persistent or nonpersistent), its acquisition period, its transmission period, virus retention in the insect and insect knock-down time, etc. It is easy to manage GRD by chemical control as the acquisition period of the virus complex by aphid vector is long thereby giving enough opportunity to chemicals to kill the vector. On the other hand, management of PBNV through chemical control of thrips vector is not effective as the virus is nonpersistent and its transmission period is very short before chemicals could kill the vector. In such situations, it is likely that agitated vectors could spread virus inoculum further before they die.

6.5 Field insect pests

Major insect pests include thrips, jassids and aphids as sucking pests; leaf miner, tobacco caterpillar, gram pod borer and hairy caterpillar as defoliators; and white grubs, termites, earwig and jewel beetle as soil-inhabiting insect pests. Readers may note that the chapter presents some broad recommendations for control of insect pests, which may differ from specific recommendations in the growing ecology, and the availability of the chemicals, their formulations and regulations of their use may vary in different countries.

IPM is desirable for effective control of insect pests in an eco-friendly manner. For sucking insect pests, grow resistant cultivars, treat seeds with insecticide to protect the crop in the early seedling stage, and spray systemic insecticides at economic threshold levels. For leaf miner, chemical control should be adopted when the larval population reaches 61–70 per 100 leaflets. Other options include installing of pheromone traps and light traps for destroying the moths.

For tobacco and hairy caterpillars and gram pod borer, provide perches, grow sunflower and castor bean trap crops on borders and inside the groundnut field (1 plant per 20 m^2), and destroy egg masses on trap crops and groundnut plants manually. Biological control measures like nuclear polyhedrosis virus (NPV) and neem seed kernel extract can also be used. Insecticide spray is recommended at economic threshold levels (Ghorpade and Thakur, 1989). For red hairy caterpillars, digging 15–20 cm deep trench all around the field or short barricade of polythene fence (10 cm high) across the migrating route can prevent their entry in the field and they can be collected manually and destroyed.

6.6 Control of soil insect pests

As a prophylactic measure, soil application of granular insecticides is recommended, and seed treatment with insecticides offers protection against white grubs in the initial stages. Feeding trees of adult white grubs can be sprayed with insecticides as a community approach. Destroying the termite mounds in the vicinity of fields, removal of plant residues and debris from fields and timely harvesting can help to minimize the termite damage.

6.7 Major storage insect pests

Groundnut bruchids damage both pods and kernels. Eggs are laid on the pods and kernels and young larvae cut through the shell and burrow into the kernel and feed. Cocoons are papery white and tough, and pupation takes place on the pods or kernels. Red flour beetle and rice moth are considered secondary pests as they are not able to infest intact pods. However, with mechanical damage caused to pods due to poor intercultivation and harvesting and curing practices, their damage in storage is quite common. Signs of damage include webbing in the case of rice moth and powdery remnants without webbing in the case of red flour beetle. Red flour beetles are oblong in shape and brown in colour and rice moth has greyish brown forewings. Development of storage insect pests can be arrested if seed moisture content is kept low (not more than 5%). Storage sanitary measures are recommended.

7 Harvesting, drying, curing and storage

Because of indeterminate growth habit, the pod maturity in groundnut is not uniform. It is important to harvest the crop at optimum maturity for maximum economic returns. Several biochemical (optical density of oil, arginine maturity index, methanol extract and maturity protein marker) and physical (kernel density, internal pericarp colour, seed hull maturity index and hull scrape) methods to estimate maturity in groundnut are available. The hull scrape method is commonly used in large farms, while the easiest method of assessing optimum time of harvesting on small farms in Asia and Africa is by evaluating internal pericarp colour of the pods collected from a few representative plants in the field. If more than 75–80% pods in case of Spanish/Valencia cultivars and 70–75% pods in case of Virginia cultivars show internal pericarp darkening, the crop is ready for harvest. If sprouting of seeds is observed in Spanish/Valencia cultivars (due to rains at harvest time in cultivars lacking fresh seed dormancy), the crop should be harvested as soon as the conditions permit without waiting for 75–80% pods to mature. In the case of end-of-season drought or under severe disease pressure, sometimes, forced premature harvesting may be required to salvage the crop. However, premature/forced harvesting may result in low shelling turnout. Over maturity or delay in harvesting can result in greater pod loss in the soil and deterioration in pod quality. In the absence of dormancy, *in-situ* germination of kernels in overmature pods can be a serious quality concern.

Although commercial groundnut combines are available, the crop in Asia and Africa is harvested by uprooting the plants manually. In some cases, an animal- or tractor-drawn digger, which cuts the roots of the plants below the podding zone, is also used and is followed by manual lifting and shaking of the plants.

Proper drying (removal of moisture from the produce to a point at which the moisture content of the produce comes into equilibrium with the moisture of the surrounding air) and curing (the total process of moisture removal and flavour and texture development in bringing the produce into storable condition) of the harvested produce ensure good quality of the produce. Harvested plants are kept inverted with pods facing upwards in windrows for about 2–3 days in the field for initial curing of plants and pods under ambient temperatures. If day temperatures are high, the plants should be dried in circular heaps avoiding direct exposure of pods to the sun. Exposure to high temperatures adversely affects the seed quality and germination. Groundnut pods at harvest contain moisture ranging between 35% and 60%. Plants should be dried till the moisture content in pods is brought down to 18–20% for mechanical threshing and 15% for hand threshing. Pods should be further dried, preferably under shade, to <10%.

7.1 Storage

Groundnut stores better in pods than in seeds. It is important to remove all damaged, discoloured, rotted, immature and sprouted seeds/pods, other plant materials and soil from the produce before storage. Well-dried pods with about 5% moisture content should be stored to avoid fungal and insect pests attack in storage. Pods should be stored in polythene-lined gunny bags or in some other safe storage structure in a well-ventilated and rodent free room which is not in general use and out of bound for children. The use of triple-layer 'Purdue Improved Crop Storage (PICS)' bags, was found to be safe for short-term groundnut pod storage. These hermetic triple-layer bags consist of three bags, one inside the other, made up of an outer woven polypropylene layer for strength, and two inner bags composed of 80-micron thick high-density polyethylene. They have been used with success for storing several crops including cowpea, maize and groundnut. PICS bags support retention of seed weight, germinability and oil content significantly better than cloth bags along with safeguarding the groundnuts from bruchids and retarding aflatoxin accumulation (Sudini, 2015). Bags should be placed on wooden planks (not more than five in a stack) and away from walls to avoid damage from dampness and should be protected from damage by storage pests by dusting the bags with insecticides. In case of pest outbreak during storage, the seed bags, after covering with polythene sheet, should be fumigated with insecticide. Storage temperature should be low; temperatures below 13°C inactivate most insects, arresting insect growth and the influence of other quality-deteriorating factors. Seed quality can be maintained for at least a year at temperatures ranging from 1°C to 5°C and moisture content of 7% or lower. Relative humidity in storage should be between 65% and 70%. Under unfavourable conditions, groundnut seeds lose viability quickly. High kernel moisture causes more deterioration than any other single factor.

8 Precision cultivation

In Argentina, Australia, and the United States farmers have greater access to synthetic fertilizers and crop protection materials including fungicides, herbicides, and insecticides than farmers in many other countries in the world. These farmers also have greater access to mechanization and technology that enables greater efficiency and flexibility in production systems and often results in higher yields and market grades than realized in resource-poor countries. Drying and storing facilities also contribute to minimize postharvest loss and improve and sustain quality over a longer period of time than in some regions of the world. Regulations through State and Federal inspection agencies contribute to greater food safety and quality for the consumer. An established and maintained seed system in both private and public sectors that insures known origin, quality, purity, and germination of varieties is a major contributor to consistency of production. While these resources, technologies and regulations add cost to farmers and the supporting industry, they are often cost-effective to the farmer and increase the likelihood of meeting consumer demands in both domestic and international markets.

Land preparation and precision planting are important steps in establishing adequate and uniform plant populations that optimize yield potential. Implements associated with both conventional and conservation tillage allows seed placement that results in rapid and uniform emergence across the field. Fertilizers and lime are often applied using equipment that utilizes global positioning systems (GPS) which enable variable application to increase uniformity of soil resources and in some instances to decrease produciton costs. Air and vacuum planters with plates that match seed size (runner market types versus Virginia market types, for example) allow precision seed placement. Modern twin-row planters also allow staggered seed placement which optimizes plant arrangement and minimizes competition between groundnut plants. Modern planters are also equipped to deliver insecticides, fungicides and *Bradyrhizobium* inoculant in the furrow, along with seed drop. These materials are applied either in water as a carrier or as granular formulations and in some instances as seed treatments. In Argentina, where the length of season and rainfall is limited, the seed is treated with a propriety polymer that allows successful stand establishment in conservation tillage systems when groundnut is planted in soil with sub-optimum temperatures.

During the growing season fungicides, herbicides, insecticides, micronutrients and plant growth regulators are applied to water solutions using ground equipment. Sprayers are calibrated frequently to ensure proper rates are applied to obtain adequate pest control, fertilizer delivery, and decrease the likelihood of crop injury and negative environmental impact. When equipped with GPS, sprayers have capacity to deliver pesticides and fertilizers in a manner that prevents overlap and subsequent decreases in efficiency (Battea and Eshanib, 2006; Luck, 2010). In current production systems pesticides are applied frequently (Jordan, 2016) and savings can result in a positive economic impact. Although not widely used, some modern sprayers have capacity to apply variable rates depending upon pest incidence, severity and soil characteristics. Research efforts are currently underway to determine if multi-spectral and hyperspectral imaging can be used to define incidence and severity of disease and other biotic stresses as well as abiotic stresses. Weather-based models have currently been used by some farmers to initiate fungicide sprays to protect groundnut from leaf spot disease (caused by *Cercospora arachidicola* or *Cercosporidium personatum*) and *Sclerotinia minor* (Jordan, 2016). These respective models use algorithms that use a combination of air temperature and soil temperature in concert with air humidity to define when the pathogen is active and the likelihood of epidemics developing. Predictive models are being developed to determine the likelihood of aflatoxin contamination in peanut based on temperature and rainfall (Chauhan, 2010). Understanding temperature and moisture thresholds for aflatoxin contamination will help to improve crop management to minimize levels of contamination at harvest (Bowen and Hagan, 2015).

Irrigation is an important component of efficient peanut production where water is available for this purpose. Several methods are used to initiate and schedule overhead or sprinkler irrigation and many of these systems are becoming sophisticated and automated. Although limited, surface and sub-surface drip irrigation has proven successful in some fields (Jordan, 2014; Sorensen, 2005).

Timely, efficient and precise digging of pods and inversion of vines is important to optimize pod yield and market-grade characteristics and to ensure desirable fatty acid composition and economic return. Heat unit accumulation using growing degree day calculations with minimum and maximum temperatures of 56°F (13.3°C) and 95°F (35°C) are used to approximate digging. Using thermal data to target the digging date assumes limited or no biotic stress, especially from inadequate soil moisture, minimal damage from insects and abiotic stress from pesticide damage. Farmers often use thermal data as the estimate for the first possible digging date and then sample fields several times to fine-tune the actual digging date. Commercial pressure washers with rotating nozzles are used to remove the exocarp, which reveals the mesocarp. Darker mesocarp colour is correlated with greater pod and kernel maturation. This approach enables farmers to determine when to initiate digging and vine inversion and establish a timeline for each field. Some attempts have been made to use multi-spectral imaging of the peanut canopy to determine when to initiate digging but this approach has had limited success (Colvin, 2014; Rowland, 2006; Rowland et al., 2008).

Many farmers use GPS to dig pods and invert groundnut in a precise manner regardless of morphological characteristics of groundnut, planting pattern, and row configuration. Roberson and Jordan (2014) reported that yield was higher when peanut was dug using GPS than when done manually. The increase in yield was greater with GPS than when the plant growth regulator prohexadione calcium was applied to minimize excessive vine growth later in the season. Pod loss in the digging process can occur when changes in soil texture and other characteristics occur in the same field and the digger is set at the same setting on the tool bar across the entire field. Prototype diggers have been developed and are currently being evaluated that allow digger blades to be maintained at the same depth across the entire field regardless of the changes in soil conditions (Warner, 2014). Yield monitors for peanut have the potential to enable farmers to define areas in fields that are less productive and can be managed more specifically based on biotic and/or abiotic stresses in those areas (Vellidis et al., 2001). Commercial availability and cost of drones and other platforms have the capacity to generate multi-spectral and hyperspectral images to assist growers in managing groundnut. Efforts are currently underway to correlate imaging with biotic and abiotic stresses so that the information derived from field scouting using drones could be used to increase efficiency of management.

Success of production systems is enabled by the availability of accurate and timely information that can be used by farmers. In addition to traditional methods of outreach and delivery of information, risk management through development of indices for specific pests and composite risk of multiple pests can help farmers more effectively implement practices with less risk (Lassiter, 2016; Lassiter et al., 2017). In future developing applications that can use complicated information to present effective options for pests in real time will be valuable in managing groundnut in a sustainable manner.

9 Seed systems

In developed countries, an established and well maintained seed system in both private and public sectors ensures production and supply of quality seed of improved varieties. On the contrary, the adoption rate of improved groundnut cultivars in most developing countries remains low primarily because of non-availability of their seeds in required quantities and at affordable price. Old cultivars, which have outlived their utility, continue to dominate groundnut cultivation affecting productivity in many countries. In Asia and Africa, ICRISAT together with national partners promotes formal and informal seed systems to improve the availability of quality seed of new cultivars where groundnut seed production has failed to attract participation of private seed sector due to absence of hybrids, low seed multiplication ratio (1: 8–10), large seed mass/volume resulting in high storage and transportation costs, and quick loss of seed viability. Most of the formal seed production remains under the domain of public sector seed-producing agencies, which often fail to meet their expected obligations. A majority of the groundnut seed production in developing countries occurs in the informal seed sector by well-off farmers often passing off the commercial produce after grading it as seed. Local traders also indulge in groundnut seed trade by procuring local commercial produce and selling it as seed in local markets after cleaning and grading the produce. Consequently, the seed sold in the local markets is often a mixture of several cultivars. A majority of the farmers either use saved seed or buy it from local traders. In the United States, the farmers source seed from a seed dealer or can save their own seed; sometimes, a farmer may bring saved seed into the co-op to be shelled, cleaned, treated, and bagged and can take back the seed for planting next season crop (Jordan, 2017). In China, some seed companies have emerged in partnership with public research institutes that produce and sell the seeds of improved varieties. In the absence of formal seed systems, informal and semi-formal seed systems have been contributing to seed supply to some extent in several countries of Africa and Asia.

A seed production crop requires more care and attention than commercial crop to meet the standards of genetic purity and physical quality. The field selected for seed production crop should not have had groundnut crop in the past two seasons to avoid contamination from volunteer groundnut plants. An isolation distance of 3 m is required between cultivars in all classes of seeds in the field. In addition to cultural practices described above, the following additional measures should be taken for the seed crop. Only one seed per hill should be sown to facilitate roguing. Diseased plants in the field, if any, should be removed and destroyed. Seed crop should be weed-free till harvest as weeds interfere with inspection and monitoring of the field and with harvest. A minimum two (preferably three) times of roguing should be carried out before harvest to remove off-type groundnut plants in the seed production field.

10 Conclusion

Reducing the input cost is critical to increasing the profitability of groundnut cultivation across different farming systems, from subsistence to precision farming. Host resistance to biotic and abiotic factors improves the pod yields and reduces the input cost, thus making cultivars/varieties critical component of improved groundnut cultivation. Achieving optimal plant population is central to optimize pod yields; thus, seed treatment and priming are cost-effective methods to achieve a good crop establishment. Groundnut farming in Asia and Africa is highly labour intensive and urgently requires mechanization that is suitable to operate on small farms. Awareness among producers, traders and consumers in Asia and Africa on food safety concerns is needed along with low-cost drying and storage techniques to reduce aflatoxin contamination in the absence of reliable host-resistance. Environmentally sustainable management of weeds may be a viable option to reduce the losses caused by weeds in Asia and Africa where herbicides are not readily available or affordable. Understanding the nutrient-use efficiency under water-limiting conditions may be useful to better manage soil and water resources. For precision farming, yield monitors, drones and applications will be valuable in managing groundnut farming. Studies on conservation tillage practices may provide options for sustainable management of groundnut on small farms in Asia and Africa.

11 Where to look for further information

A great deal of information exists on groundnut production and pest management in many parts of the world. Universities that support agriculture are also critical resources and may have solutions that address farmers' questions. However, depending on the country or regions of the world, universities and extension services may lack the infrastructure and resources to conduct research and deliver these findings to farmers. In these cases nongovernmental organizations can play a major role in developing and promoting production practices. These organizations often partner with local groups not only to foster improved production through education and technology (improved varieties, for example) but also to work toward developing markets and supply chains for groundnut. ICRISAT together with national agriculture research and extension partners provides information on production packages and supports technology delivery. Private industry also plays a major role in developing new technology and delivery systems.

ICRISAT's collaborative work with partners in Asia and Africa provides good information, and institutes like Oil Crops Research Institute in Wuhan, China, and Directorate of Groundnut Research in India, and Universities in the United States (e.g. North Carolina State University, University of Georgia and New Mexico State University) are sources of additional information.

12 References

Akbar, A., Manohar, S. S., Variath, M. T., Kurapati, K. and Janila, P. (2017), 'Efficient partitioning of assimilates in stress-tolerant groundnut genotypes under high-temperature stress', *Agron.*, 7, 30.

Arnold, J. A., Beasley Jr., J. P., Harris, G., Timothy, L. G. and Cabrera, M. (2017), 'Effect of gypsum application rate, soil type, and soil calcium on yield, grade and seed quality of runner peanut cultivars', *Peanut Sci.*, 44(1), 13–18.

Bationo, A. and Ntare, B. (2000), 'Rotation and nitrogen fertilizer effects on pearl millet, cowpea and groundnut yield and soil chemical properties in a sandy soil in the semi-arid tropics, West Africa', *J. Agric. Sci.*, 134, 277–84.

Battea, M. T. and Ehsanib, M. R. (2006), 'The economics of precision guidance with auto-boom control for farmer-owned agricultural sprayers', *Comput. Electron. Agric.*, 53, 28–44.

Beam, J. B., Jordan, D. L., York, A. C., Isleib, T. G., Bailey, J. E., McKemie, T. E., Spears, J. F. and Johnson, P. D. (2002), 'Influence of prohexadione calcium on pod yield and pod loss of peanut', *Agron. J.*, 94(2), 331–6.

Bell, M. J., Middleton, K. J. and Thompson, J. P. (1989), 'Effects of vesicular-arbuscular mycorrhizae on growth and phosphorus and zinc nutrition of peanut (*Arachis hypogaea* L.) in an oxisol from subtropical Australia', *Plant Soil.*, 117, 49–57.

Bhatia, V. S., Singh, P., Wani, S. P., Kesava Rao, A. V. R. and Srinivas, K. (2006), Yield gap analysis of soybean, groundnut, pigeonpea and chickpea in India using simulation modelling. Global Theme on Agroecosystems Report no. 31. Patancheru 502 324, Andhra Pradesh, India: International Crops Research Institute for the Semi-Arid Tropics (ICRISAT), 156p.

Bhingarde, M. T., Kadam, R. S. and Tagad, L. N. (2015), 'Effect of seed priming on seed yield and seed quality of groundnut (*Arachis hypogaea* L.)'. *Life Sci. Int. Res. J.*, 2(2), 25–9.

Bowen, K. L. and Hagan, A. K. (2015), 'Temperature and moisture conditions that affect aflatoxin contamination of peanuts', *Peanut Sci.*, 42, 121–7.

Bradford, K. J. (1986), 'Manipulation of seed water relations via osmotic priming to improve germination under stress conditions'. *Hort. Sci.*, 21, 1105–12.

Chauhan, Y. S., Graeme, W., Rachaputi, R. C. N., Holzworth, D., Broom, A., Krosch, A. and Robertson, M. S. (2010), 'Application of a model to assess aflatoxin risk in peanuts'. *J. Agric. Sci.*, 148(03), 341–51.

Chengrong, N. (1996), 'Effect of fosamine and nitrogen manure on nitrogen metabolism and productivity of peanut', *China Oil*, 18(2), 31–3.

Colvin, B. C., Rowland, D. L., Ferrell, J. A. and Faircloth, W. H. (2014), 'Development of a digital analysis system to evaluate peanut maturity', *Peanut Sci.*, 41, 8–16.

Corbineau, F. and Cõme, D. (2006), 'Priming: A technique for improving seed quality', *Seed Test Int.*, 132, 38–40.

Deb, U. K., Bantilan, M. C. S. and Nigam, S. N. (2005), *Impact of Agricultural Research: Post-Green Revolution Evidence from India*. International Crops Research Institute for the Semi-Arid Tropics, Patancheru, Hyderabad, India.

Dey, R., Pal, K. K., Bhatt, D. M. and Chauhan, S. M. (2004), 'Growth promotion and yield enhancement of peanut (*Arachis hypogaea* L.) by application of plant growth-promoting rhizobacteria', *Microbiol. Res.*, 159, 371–94.

Diop, N., Beghin, J. and Sewadeh, M. (2004), Groundnut policies, global trade dynamics, and the impact of trade liberalization, World Bank Policy Research Working Paper No. 3226. SSRN, Washington DC, USA, pp. 1–36. https://ssrn.com/abstract=610267.

El-Rehim, M. A. Abd Backman, P. A., Rodriguez-Kabana, R. and Crawford, M. A. (1981), 'Peanut seed treatment with hot calcium hydroxide solutions', *Peanut Sci.*, 8, 36–9.

Elsheikh, E. A. E. and Mohamedzein, E. M. M. (1998), 'Effect of Bradyrhizobium, VA mycorrhiza and fertilisers on seed composition of groundnut', *Ann. Appl. Biol.*, 132, 325–30.

FAO. (2014), Food and Agriculture Organization. Faostat.fao.org/site/567/DesktopDefault.aspx?PageID=567#ancor.

Farooq, M., Wahid, A. and Siddique Kadambot, H. M. (2012), 'Micronutrient application through seed treatments – Review', *J. Plant Sci. Plant Nutr.*, 12(1), 125–42.

Fernandez, E. M., Rosolem, C. A., Maringoni, A. C. and Oliveira, D. M. T. (1997), 'Fungus incidence on peanut grains as affected by drying method and Ca nutrition', *Field Crop Res.*, 52, 9–15.

Ghorpade, S. A. and Thakur, S. G. (1989), 'Assessment of need based monocrotophos application for the control of major pest of groundnut', *Pesticides*, 23(11), 19–20.

Gil, S. V., Meriles, J. M., Haro, R., Casini, C. and March, G. J. (2008), 'Crop rotation and tillage systems as a proactive strategy in the control of peanut fungal soilborne diseases', *BioControl*, 53(3), 685–98.

Gregory, M. P. and Gregory, W. C. (1979), 'Exotic germplasm of Arachis L. interspecific hybrids', *J. Hered.*, 70, 185–93.

Halmer, P. (2008), 'Seed technology and seed enhancement'. *Acta Hort.*, 771, 17–26. http://www.actahort.org/books/771/771_1.htm.

Hammons, H. O. (1994), *The Origin and History of the Groundnut*. Springer, Dordrecht, pp. 24–42. ISBN: 978-94-011-0733-4.

Hawkins, G. L., Kelton, J., Smith, N. and Balkcom, K. (2016), 'A note on comparing rate of soil moisture loss for conventional and conservation tillage production methods for peanut (*Arachis hypogaea*)', *Peanut Sci.*, 43(2), 168–72.

Hildebrand, G. L. (1996), 'The status of technologies used to achieve high groundnut yields in Zimbabwe', In: *Achieving High Groundnut Yields: Proceedings of an International Workshop* (Renard, C., Gowda, C. L. L., Nigam, S. N. and Johansen, C. (Eds)), 25–29 August 1995, Laixi City, Shandong, China, pp. 101–14.

Hurt, C. A., Brandenburg, R. L., Jordan, D. L., Royals, B. M. and Johnson, P. D. (2006), 'Interactions of tillage with management practices designed to minimize tomato spotted wilt of peanut (*Arachis hypogaea* L.)', *Peanut Sci.*, 33(2), 83–9.

Jain, N. K., Meena, H. N. and Bhaduri, D. (2017), 'Improvement in productivity, water use efficiency and soil nutrient dynamics of summer peanut (*Arachis hypogaea* L.) through use of polythene mulch, hydrogel and nutrient management', *Comm. Soil Sci. Plant Anal.*, 48(5), 549–64.

Janila, P. and Nigam, S. N. (2013), *Phenotyping for Plant Breeding*. Springer, Heidelberg, Germany, pp. 129–67.

Janila, P., Nigam, S. N., Pandey, M. K., Nagesh, P. and Varshney, R. K. (2013), 'Groundnut improvement: Use of genetic and genomic tools', *Front. Plant Sci.*, 4, 23.

Janila, P., Pandey, M. K., Shasidhar, Y., Variath, M. T., Sriswathi, M., Khera, P., Manohar, S. S., Nagesh, P., Vishwakarma, M. K., Mishra, G. P., Radhakrishnan, T., Manivannan, N., Dobariya, K. L., Vasanthi, R. P. and Varshney, R. K. (2016a), 'Molecular breeding for introgression of fatty acid desaturase mutant alleles (ahFAD2A and ahFAD2B) enhances oil quality in high and low oil containing peanut genotypes', *Plant Sci.*, 242, 203–13.

Janila, P., Murali, T. V., Pandey, Desmae, H., Motagi, B. N., Okori, P., Manohar, S. S., Rathnakumar, A. L., Radhakrishnan, T., Liao, B. and Varshney, R. K. (2016b), 'Genomic tools in groundnut breeding program: Status and perspectives', *Front. Plant Sci.*, 7, 289.

Jeffries, P., Gianinazzi, S., Perotto, S., Turnau, K. and Barea, J. M. (2003), 'The contribution of arbuscular mycorrhizal fungi in sustainable maintenance of plant health and soil fertility', *Biol. Fertil. Soils*, 37, 10–16.

Johnson, W. C., Brenneman, T. B., Baker, S. H., Johnson, A. W., Sumner, D. R. and Mullinix Jr., B. G. (2001), 'Tillage and pest management considerations in a peanut–cotton rotation in the southeastern Coastal Plain', *Agron. J.*, 93, 570–6.

Jordan, D. (2017), 'Peanut production practices', pp. 21–46. In: *Peanut Information 2017* (Jordan, D. L. (Coord.)). North Carolina Cooperative Extension Serv. AG-331. Raleigh, NC, USA, 172p.

Jordan, D. L., Barnes, J. S., Corbett, T., Bogle, C. R., Johnson, P. D., Shew, B. B., Koenning, S. R., Ye, W. and Brandenburg, R. L. (2008), 'Crop response to rotation and tillage in peanut based cropping systems'. *Agron. J.*, 100, 1580–6.

Jordan, D. L., Johnson, P. D., Grabow, G. L. and Corbett, T. (2014), 'Corn, cotton, and peanut response to tillage and sub-surface drip irrigation in North Carolina', *Agron. J.*, 106, 962–7.

Jordan, L. D., Brandenburg, R. L., Brown, A. B., Bullen, S. G., Roberson, G. T. and Shew, B. (2016), *Peanut Information*. North Carolina Cooperative Extension, Raleigh, NC, USA. https://content.ces.ncsu.edu/peanut-information/peanut-seed.

Jordan, L. D., Nuti, R. C., Beam, J. B., Lancaster, S. H., Lanier, J. E. and Dewayne Johnson, P. (2009), 'Influence of application variables on peanut (*Arachis hypogaea* L.) response to prohexadione calcium', *Peanut Sci.*, 36, 96–103.

Kumar, A., Pandey, A. and Gowda, M. V. C. (2015), 'Evaluation of groundnut genotypes for tolerance to phosphorus deficiency', *J. Plant Nutr.*, 38(5), 687–99.

Kuotsu, K., Das Lal, A. R., Munda, G., Ghosh, P. and Ngachan, S. (2014), 'Land forming and tillage effects on soil properties and productivity of rainfed groundnut (*Arachis hypogaea* L.)–rapeseed (*Brassica campestris* L.) cropping system in northeastern India', *Soil Till. Res.*, 142, 15–24.

Lanier, J. E., Jordan, D. L., Spears, J. F., Wells, R., Johnson, P. D., Barnes, J. S., Hurt, C. A., Brandenburg, R. L. and Bailey, J. E. (2004), 'Peanut response to planting pattern, row spacing, and irrigation', *Agron. J.*, 96, 1066–72.

Lassiter, B. R., Wilkerson, G. G., Buol, G. S., Jordan, D. L., Shew, B., Brandenburg, R., Herbert Jr. A., Phipps, P. and Chapin, J. (2016), 'Peanut production in Virginia and the Carolinas: Development of a website and program', *J. Crop Forage Turfgrass Manag.*, 2.

Lassiter, B. R., Wilkerson, G. G., Buol G. S., Jordan, D. L., Shew, B., Brandenburg, R., Herbert Jr., A., Phipps, P. and Chapin, J. (2017), 'Development and deployment of a composite pest risk assessment decision support system for peanut (*Arachis hypogaea* L.) in the Virginia-Carolinas Region', *J. Crop Forage Turfgrass Manag*, 3 (accepted).

Luck, J. D., Pitla, S. K., Shearer, S. A., Mueller, T. G., Dillon, C. R., Fulton, J. P. and Higgins, S. F. (2010), 'Potential for pesticide and nutrient savings via map-based automatic boom section control of spray nozzles', *Comput. Electron. Agric.*, 70, 19–26,

Murata, M. R., Zharare, G. E. and Hammes, P. S. (2008), 'Pelleting or priming seed with calcium improves groundnut seedling survival in acid soils', *J. Plant Nutr.*, 31(10), 1736–45.

Naveen, P., Daniel, K. V., Subramanian, P. and Kumar, P. S. (1992), 'Response of irrigated groundnut (*Arachis hypogaea* L.) to moisture stress and its management', *Ind. J. Agron.*, 37, 82–5.

Navya, H. M., Naveen, J., Hariprasad, P. and Niranjana, S. R. (2015), 'Beneficial rhizospheric microorganisms mediated plant growth promotion and suppression of aflatoxigenic fungal and aflatoxin contamination in groundnut seeds', *Ann. Appl. Biol.*, 167, 225–35.

Nigam, S. N. (2000), 'Some strategic issues in breeding for high and stable yield in groundnut in India', *J. Oilseeds Res.*, 17, 1–10.

Nigam, S. N., Aruna, R., Giri, D. Y., Ranga Rao, G. V. and Reddy, A. G. S. (2006), *Obtaining Sustainable Higher Groundnut Yields: Principles and Practices of Cultivation*. Patancheru 502 324, Andhra Pradesh, India: International Crops Research Institute for the Semi-Arid Tropics, 48p. ISBN: 92-9066-484-3. http://oar.icrisat.org/id/eprint/3217.

Nigam, S. N., Waliyar, F., Aruna, R., Reddy, S. V., Lava Kumar, P. L., Craufurd, P. Q., Diallo, A. T., Ntare, B. R. and Upadhyaya, H. D. (2009), 'Breeding peanut for resistance to aflatoxin contamination at ICRISAT', *Peanut Sci.*, 36 (1), 42–9.

Nigam, S. N., Prasada Rao, R. D. V. J., Bhatnagar-Mathur, P. and Sharma, K. K. (2012), 'Genetic management of virus diseases in peanut'. *Plant Breeding Reviews*, 36, 293–356.

Ong, C. K. (1986), 'Agroclimatological factors affecting phenology of groundnut' In: *International Symposium: Agrometeorology of Groundnut*, 21–26 August 1985, Niamey, Niger, pp. 115–24.

Porter, D. M. and Wright, F. S. (1991), 'Early leafspot of peanuts: Effect of conservational tillage practices on disease development', *Peanut Sci.*, 18(2), 76–9.

Prasad, P. V. V., Craufurd, P. Q., Summerfield, R. J. and Wheeler, T. R. (2000), 'Effects of short episodes of heat stress on flower production and fruit-set of groundnut *Arachis hypogaea* L', *J. Exp. Bot.*, 51(345), 777–84.

Reddy, T. Y., Reddy, V. R. and Anbumozhi, V. (2003), 'Physiological responses of groundnut (Arachis hypogea L.) to drought stress and its amelioration: a critical review', *Plant Growth Regul.*, 41, 75–88.

Roberson, G. T. and Jordan, D. L. (2014), 'RTK GPS and automatic steering for peanut digging', *Am. Soc. Agric. Biol. Eng.*, 30(3), 405–9.

Rojo, F. G., Reynoso, M. M., Ferez, M., Chulze, S. N. and Torres, A. M. (2007), 'Biological control by Trichoderma species of Fusarium solani causing peanut brown root rot under field conditions', *Crop Prot.*, 26(4), 549–55.

Rowland, D. L., Sorensen, R. B., Butts, C. L. and Faircloth, W. H. (2006), 'Determination of maturity and degree day indices and their success in predicting peanut maturity', *Peanut Sci.*, 33, 125–36.

Rowland, D. L., Sorensen, R. B., Butts, C. L., Faircloth, W. H. and Sullivan, D. G. (2008), 'Canopy characteristics and their ability to predict peanut maturity', *Peanut Sci.*, 35, 43–54.

Rowland, D. L., Smith, C., Cook, A., Mason, A., Schreffler, A. and Bennett, J. (2015), 'Visualization of peanut nodules and seasonal nodulation pattern in different tillage systems using a minirhizotron system', *Peanut Sci.*, 42(1), 1–10.

Ruark, S. J. and Shew, B. B. (2010), 'Evaluation of microbial, botanical and organic treatments for control of peanut seedling diseases', *Plant Dis.* 94, 445–54.

Saravanakumar, D. and Samiyappan, R. (2007), 'ACC deaminase from Pseudomonas fluorescens mediated saline resistance in groundnut (*Arachis hypogaea*) plants', *J. Appl. Microbiol.*, 102(5), 1283–92.

Sarkar, R. K. and Pal, P. K. (2004), 'Effect of intercropping rice (*Oryza sativa*) with groundnut (*Arachis hypogaea*) and pigeonpea (*Cajanus cajan*) under different row orientations on rainfed uplands', *Ind. J. Agron.*, 49(3), 147–50.

Singh, A. L. (1999), 'Mineral nutrition of groundnut', *Adv. Plant Physiol.*, 2, 161–200.

Singh, B., Sandhu, B. S., Khera, K. L. and Aujla, T. S., (1986), 'Groundnut response to irrigation and sowing time on a deep loamy sand in a subtropical monsoon region', *Field Crops Res.*, 13, 355–66.

Singh, H., Jassal, R. Kaur, Kang, J. S., Sandhu, S. S., Kang, H. and Grewal, K. (2015), 'Seed priming techniques in field crops - A review'. *Agri. Rev.*, 36(4), 251–64.

Skelton, B. J. and Shear, G. M. (1971), 'Calcium translocation in the peanut (*Arachis hypogaea* L.)', *Agron. J.*, 63(3), 409–12.

Sorensen, R. B., Butts, C. L. and D. L. Rowland. (2005), 'Five years of subsurface drip irrigation on peanut: What have we learned?', *Peanut Sci.*, 32, 14–19.

Stalker, H. T. (1997), 'Peanut (*Arachis hypogaea* L.)', *Field Crops Res.*, 53, 205–17.

Sudini, H., Ranga Rao, G. V., Gowda, C. L. L., Chandrika, R., Margam, V., Rathore, A. and Murdock, L. L. (2015), 'Purdue Improved Crop Storage (PICS) bags for safe storage of groundnuts', *J. Stored Prod. Res.*, 64, 133–8.

Tadele, Z. (2017), 'Raising crop productivity in Africa through intensification', *Agron.*, 7(1), 22.

Turner, J. T. and Backman, P. A. (1991), 'Factors relating to peanut yield increases after seed treatment with Bacillus subtilis', *Plant Dis.*, 75, 347–53.

Valls, J. F. M. and Simpson, C. E. (2005), 'New species of *Arachis* (Leguminosae) from Brazil, Paraguay and Bolivia', *Bonplandia*, 14, 35–63. http://www.jstor.org/stable/41941266

Vellidis, G., Perry, C. D., Durrence, J. S., Thomas, D. L., Hill, L. W., Kvien, C. K., Hamrita, T. K., and Rains, G. (2001), 'The peanut yield monitoring system', *Trans. ASAE*, 44, 775–85.

Warner, A. C., Kirk, K. R., Thomas, J. S., Monfort, W. S., White, J. W., Brantley, S. A., Peele, J. S., Han, J. and Massey, H. F. (2014), Variable depth Peanut digger: Part II – digging loss analysis. Paper Number 141914272, American Society of Agricultural and Biological Engineers, Saint Joseph, MI, USA.

Wenguang, H., Shufen, D. and Qingwei, S. (1996), 'High yield technology for groundnut', *Int. Arachis News.*, 15, 22.

Yanhao, S., Shouxiang, T. and Caibin, W. (1996), 'Theoretical foundations for high yield of groundnut in China', In: *Achieving High Groundnut Yields: Proceedings of an International Workshop* (Renard, C., Gowda, C. L. L., Nigam, S. N. and Johansen, C. (Eds)), 25–29 August 1995, Laixi City, Shandong, China. ICRISAT, Patancheru 502 324, Andhra Pradesh, India, pp. 129–39.

Chapter 9

Preventing mycotoxin contamination in groundnut cultivation

David Jordan, Rick Brandenburg and Gary Payne, North Carolina State University, USA; David Hoisington, Nick Magnan and James Rhoads, The University of Georgia, USA; Mumuni Abudulai, Savanna Agricultural Research Institute, Ghana; Koushik Adhikari and Jinru Chen, The University of Georgia, USA; Richard Akromah, William Appaw and William Ellis, Kwame Nkrumah University of Science and Technology, Ghana; Maria Balota and Kumar Mallikarjunan, Virginia Polytechnic Institute and State University, USA; Kenneth Boote and Greg MacDonald, University of Florida, USA; Kira Bowen, Auburn University, USA; Boris Bravo-Ureta and Jeremy Jelliffe, University of Connecticut, USA; Agnes Budu, University of Ghana, Ghana; Hendrix Chalwe, Alice Mweetwa and Munsanda Ngulube, University of Zambia, Zambia; Awere Dankyi and Brandford Mochia, Crops Research Institute, Ghana; Vivian Hoffmann, International Food Policy Research Institute, USA; Amade Muitia, Mozambique Institute of Agricultural Research, Mozambique; Agnes Mwangwela, Lilongwe University of Agriculture and Natural Resources, Malawi; Sam Njoroge, International Crops Research Institute for the Semi-Arid Tropics (ICRISAT) (ICRISAT), Malawi; David Okello, National Semi-Arid Resources Research Institute (NaSARRI), Uganda; and Nelson Opoku, University for Development Studies, Ghana

1 Introduction
2 Factors impacting aflatoxin development
3 Prediction models for aflatoxin development
4 Minimizing aflatoxin contamination
5 Aflatoxin contamination during drying, storage and processing
6 Challenges in mycotoxin research
7 Value chain projects
8 Conclusion
9 Where to look for further information
10 References

1 Introduction

Mycotoxins have a substantial negative impact on human and livestock health around the world. *Aspergillus flavus* and *Aspergillus parasiticus*, both capable of producing aflatoxins, are distributed widely in crops and can contaminate produce in the field and

http://dx.doi.org/10.19103/AS.2017.0023.28

during storage and transport, as well as food products consumed directly by humans (Wu and Guclu, 2012). Secondary metabolites of toxigenic *A. flavus* and *A. parasiticus* are carcinogenic and adversely affect liver function (CAST, 2003; IARC, 1993; Wild et al., 2015). The impact of consumption of aflatoxins is especially severe in infants and children (CAST, 2003; Gong et al., 2002; Wild and Gong, 2010; Williams et al., 2004). It has been estimated that at least 25% of agricultural commodities in the world are contaminated by mycotoxins (Charmly et al., 1994; Payne, 2016). As many as 5 billion people in developing countries are estimated to be exposed on a consistent basis to aflatoxin levels that are considered dangerous to human health (Jolly et al., 2008; Strosnider et al., 2006; Williams et al., 2004). More recently, exposure to aflatoxins and other mycotoxins prior to and during pregnancy has been linked to stunting at birth (IARC, 2015). Minimizing the exposure of humans to aflatoxin is the goal of a number of health and agricultural initiatives and partnerships among public and private sectors around the world. This particular contaminant, derived from biological organisms, requires interventions across components of crop value chains and research and extension expertise across multiple disciplines.

In-depth reviews related to mycotoxins and their impact on human health (Jelinek et al., 1989; Kensler et al., 2011; Payne, 2016; Rai and Varma, 2010; Shepard, 2008; Wild and Gong, 2010; Williams et al., 2004), management in the field and in storage (Torres et al., 2014; Waliyar, 1997; Waliyar et al., 2013), and approaches that incentivize all segments of food systems from production of raw agricultural products through processing and marketing (Pitt et al., 2013) are available elsewhere. The goal of this chapter is to describe factors that affect growth of *A. flavus* and *A. parasiticus*, models predicting contamination, cultural and biological control measures designed to minimize contamination, challenges associated with research and quantification of aflatoxin, and three case studies from current value chain projects in Ghana, Haiti and Malawi with peanut (groundnut, *Arachis hypogaea* L.). This chapter will demonstrate the need for a more focused and concerted effort to address the issue of aflatoxin contamination in peanut.

2 Factors impacting aflatoxin development

Pre-harvest aflatoxin contamination of peanut is associated with drought conditions (Fig. 1) and high soil temperatures (Blankenship et al., 1984; Bowen and Hagan, 2015; Cole et al., 1985; Cotty and Jamie-Garcia, 2007; Hill et al., 1983; Holbrook et al., 2000, 2009; Kambiranda et al., 2011; Nigam et al., 2009; Parmar et al., 1997; Sanders et al., 1985; Umeh et al., 2000). The primary organisms that produce aflatoxin, *A. flavus* and *A. parisiticus*, are ubiquitous in soil and can readily infect seeds and other plant materials (Accinelli and Abbas, 2011; Diener and Davis, 1977; Ehrlich, 2014; Okoth et al., 2012). Non-damaged peanut pods grown in soil temperatures ranging from 29°C to 30.5°C can have higher levels of aflatoxin contamination than peanuts exposed to either lower or higher temperatures (Blankenship et al., 1983, 1984; Cole et al., 1985). Drought is necessary for aflatoxin contamination to occur in peanuts, and late-season drought in particular favours infection (Cole et al., 1985; Cole, 1989; Hill et al., 1983; Sanders et al., 1981, 1984, 1985; Wilson and Stansell, 1983). *Aspergillus flavus* and *A. parasiticus* are thermophilic and xerotolerant fungi that grow and develop in drier conditions and at higher temperatures than many other fungi or microorganisms that infect crop plants (Abdel–Azim and Khalil,

Figure 1 Peanut experiencing drought stress in the United States within 4 weeks of harvest. Photo credit, David Jordan.

1979; Hell et al., 2010; Magan et al., 1984; Pitt et al., 1994; Reddy et al., 2009; Taligoola et al., 2011; Wareing, 1997; Wu et al., 2011).

In addition to high soil temperatures, low exchangeable calcium associated with soil acidity, low soil organic matter content, drought stress, high ambient temperatures and relative humidity above 75% influence the health of peanut as well as the growth and development of *A. flavus* and *A. parasiticus* and subsequent infection (Achar and Sanchez, 2006; CAST, 2003; Hell and Mutegi, 2011). Under severe and prolonged drought conditions, the soil surface near the growing plants may be exposed to direct sunlight, resulting in increased temperatures in the geocarposphere (i.e. the region under the immediate influence of the peanut pod). While these environmental conditions force many soil microorganisms to enter dormancy, *A. flavus* and *A. parasiticus* can continue to grow and develop. Prolonged moisture deficits and elevated soil temperatures promote the presence of these organisms and increase the risk of contamination in the harvested crop (Torres et al., 2014). Drought-stressed plants usually have a reduced natural defence against infection due to reduced metabolism and promotion of adapted organisms including *A. flavus* and *A. parasiticus* (Pitt et al., 2013; Wang et al., 2013). Expression of genes with possible resistance to the production of aflatoxin by *A. flavus* and *A. parasiticus* has been shown to be lower in peanuts grown under drought conditions (Guo et al., 2008; Wang et al., 2013).

Phytoalexins, a family of compounds known to have antimicrobial activity to several fungi including *A. flavus*, have been shown to be lower in drought-stressed plants compared to non-stressed plants (Ahuja et al., 2012; Diao et al., 2015; Dorner et al., 1989; Hammerschmidt, 1999; Philippe, 2015; VanEtten et al., 1994; Vidhyasekaran et al., 1972; Wotton and Strange, 1987). Dorner et al. (1989) reported that lower kernel moisture was associated with decreased phytoalexin production in kernels and increased aflatoxin accumulation. Wotton and Strange (1987) showed decreased resistance to aflatoxin contamination in peanut as a result of low kernel water activity, most likely as a result of lower production of phytoalexin in kernels.

Irrespective of these environmental factors that impact aflatoxin accumulation, research has also shown that the nutrient composition and content of the crop can affect the risk of aflatoxin contamination. Ma et al. (2015) reported variations in aflatoxin contamination in crops and that high-lipid substrates such as those in peanut had a higher concentration of aflatoxin relative to low-lipid substrates produced by other crops. The oil content of a substrate is one of the factors affecting aflatoxin production by mycotoxic organisms. Liu et al. (2016) reported that the removal of lipids from ground substrates reduced the potential for aflatoxin production by *A. flavus*. They also suggested that arginine and stachyose in grains stimulate aflatoxin B1 production and the growth of *A. flavus*, and offer a possible new dimension to the relationship of cultivar development and presence of aflatoxin.

The following section will address models that predict the risk of *A. flavus* and *A. parasiticus* development and possible contamination by aflatoxin and implications for management. These are based on the observation that the key factors that influence the presence of *A. flavus* and *A. parasiticus* at levels that pose a risk for aflatoxin production are hot and dry soil conditions for a prolonged period of time.

3 Prediction models for aflatoxin development

Prediction models for aflatoxin contamination have been developed using soil moisture or water-deficit and soil temperature data (Cole, 1989; Thai et al. 1990; Craufurd et al., 2006). Thai et al. (1990) developed a model that predicted aflatoxin contamination using soil temperature and daily accumulation of aflatoxin during a drought period. Parmar et al. (1997) developed an artificial neural network model using soil temperature, number of drought days and crop age at harvest; results obtained from this model correlated well with observed and predicted aflatoxin contamination. Thai et al. (1990) developed a model using data from automated rain-out systems in the United States. Henderson et al. (2000) added a genetic component to the neural network model. DSSAT-CROPGRO-Peanut is a crop growth simulation model that calculates soil temperature and plant water deficits from weather inputs, soil information, plant growth data and management (Craufurd et al., 2006). Craufurd et al. (2006) used this prediction model successfully in Niger. Chauhan et al. (2010) also used automated constructs to develop algorithms for another simulation model, the Agricultural Production Systems Simulator (APSIM) peanut module. Results of this model correlated well with aflatoxin contamination in cultivation fields in Australia (Chauhan et al., 2010).

Bowen and Hagan (2015) noted that these existing models may be too complex for widespread adaption and showed how maximum daily temperature and '3-day dry periods' could be used to evaluate the risk for high levels of aflatoxin contamination. They suggested that inverting vines and digging pods 2 to 6 days earlier than the optimum harvest date could decrease the risk of aflatoxin contamination when soil temperature and moisture algorithms from the preceding 42 days were used in the prediction model. Rachaputi et al. (2002) reported that digging 2 weeks prior to optimum pod maturity could reduce aflatoxin contamination, although digging prior to optimum pod maturity negatively impacts pod yield and market-grade characteristics (Mixon and Branch, 1985).

Models have also been derived to predict the risk of aflatoxin development after harvest. Modifying the gaseous environment, storage temperature and water activity using a synthetic media, and a process optimization technique, referred to as response surface methodology, prediction models were developed which allowed for evaluating several variables simultaneously (Ellis et al., 1993). This model was tested on peanut samples together with the absorbent technology and showed positive outcomes (Ellis et al., 1994). While the focus of these efforts was on *A. flavus*, the model also showed potential for controlling aflatoxin production in packaged peanut products.

The utility of current prediction models for managing *A. flavus* and *A. parasiticus* and possible aflatoxin contamination of crops and food sources is not known. Peanut buyers can use these models to focus on sources of peanut grown in regions where development and contamination by aflatoxin in peanut is less likely, although subsequent poor storage might eliminate the gains of low contamination of the crop prior to harvest. While this can have a positive impact on value after harvest, farmers in high-risk areas will be at a major disadvantage with respect to access to markets, even if the actual level of contamination is within the acceptable range. For this reason, developing a comprehensive strategy that effectively minimizes aflatoxin development and contamination is needed. Approaches to minimize aflatoxin contamination in high-risk environments will be discussed in the next section.

4 Minimizing aflatoxin contamination

Increases in *A. flavus* and subsequently the likelihood of increases in aflatoxin contamination of peanuts can occur in the field, during drying, transportation and storage. Storage, which is the culmination of both field production and drying, is a key phase in which aflatoxin contamination often increases. While aflatoxigenic fungi may be present in bags or other storage facilities, the initial contamination is a reflection of contamination from the field and during drying. Poor drying exacerbates the magnitude of field aflatoxin contamination and poor storage further increases the problem. Identifying and addressing the initial source of contamination, in this case peanut in the field, is a critical first step. Reducing the pathogen from entering the drying, transport and storage steps will decrease the magnitude of contamination in storage. In this section management practices that can be employed to minimize *A. flavus* development and subsequent aflatoxin contamination will be discussed. A section on drying, storing and processing is also provided as these steps can minimize further development of *A. flavus* even if contamination from the production field is higher than desired.

Integrated pest management is an important concept in production and post-harvest handling of crops, especially peanut with a goal of increasing yield, quality and profit. The integration of preventative and avoidance strategies to preserve plant health as well as consistent monitoring and suppression can be used to minimize the impact of pests in the field and in storage (SRIPMC, 2016). Several strategies can be employed to minimize *A. flavus* and reduce aflatoxin contamination of peanut. A discussion of each of the major management practices will follow.

4.1 Cropping sequence

Diversification of cropping systems through crop rotation and sequence with rotations can reduce *A. flavus* populations in soil and minimize the risk of aflatoxin presence in pods and kernels. A rotation of crops that is not affected by *A. flavus* infection can decrease populations of the pathogen (Mutegi et al., 2009; Torres, et al., 2014; Waddington and Karigwindi, 2001). Selection of crops and the sequence of plantings are often dictated by economic pressures, acceptance of crops as food sources and availability of alternatives. For example, in Malawi, corn (*Zea mays* L.) and peanut are preferred crops for human consumption, with both crops being susceptible to contamination by *A. flavus* (Hell and Mutegi, 2011). Although a rotation of corn and peanut can provide benefits to soil fertility and improvement of yield of both crops (Hagan et al., 2003; Jeranyama et al., 2007), reproduction of *A. flavus* continues because both crops are suitable hosts (Prasad et al., 2008; Waliyar et al., 2013). Monocropping these or other crops can build inoculum levels that readily impact adjacent crops or subsequent crops. However, the magnitude of benefits of crop rotation and sequence in minimizing aflatoxin contamination is inconclusive (Munkvold, 2003) and requires focused studies and attention.

4.2 Resistant cultivars

Cultivar selection is an important management decision for farmers relative to many pests in peanut and has been proposed as a key strategy to minimize aflatoxin contamination (Bhatnagar-Mathur et al., 2015; Wang et al., 2015). Early attempts to define relations of cultivars and aflatoxin included research in India (Suryanarayana Rao and Tulpule, 1967), Niger (McDonald, 1964) and the United States (Doupnik, 1969). Efforts by Mixon (1983, 1986), Mehan et al. (1987) and Blankenship et al. (1985) resulted in the deployment of cultivars with known resistance to *A. flavus* infection, but the combined interactions of genetics, environment and management rendered these efforts unsatisfactory. Incorporating resistance traits into cultivars for deployment has proven challenging, in part because the variation between infection by the pathogen and production of aflatoxin and heritability of traits influence this interaction in peanut (Arunyanark et al., 2010; Bhatnagar-Mathur et al., 2015; Girdthai et al., 2010a,b; Hamidou et al., 2014; Olwari et al., 2013; Rajasekaran et al., 2009; Wang et al., 2010). Recent advances in molecular biology and genomics of peanut could make important contributions to improved resistance to *A. flavus* and other biotic and abiotic stresses in the future (Bhatnagar-Mathur et al., 2015; Dash et al., 2016; Guo et al., 2011; Wang et al., 2013, 2015).

Mechanisms of reduced aflatoxin contamination in peanut have been associated with delayed expression of aflatoxin biosynthesis pathway genes in *A. flavus* when infecting resistant cultivars (Arias et al., 2014; Zhang et al., 2015). In terms of phenotypic expression, Dieckert and Dieckert (1977) demonstrated that cultivars resistant to *A. parasiticus* had thinner and tightly compacted testa compared with susceptible cultivars. Seed coat tannin reduced the growth of *A. parasiticus* and aflatoxin production, a trait that is genotype dependent (Azaizeh and Pettit, 1987; Azaizeh et al., 1990). Lower aflatoxin contamination was associated with high protein and oleic acid content while higher levels of contamination were associated with total fat and linoleic fatty acid (Arias et al., 2014; Ebrahimi et al., 2009). Phytoalexin production in peanut seed has been reported in response to pathogenic infection and drought, and the genotype had a determinant role in the amount of phytoalexins produced (Arias et al., 2014; Sobolev 2008, 2013; Wotton and Strange, 1987). Phytoalexins are secondary metabolites induced most often by biotic stress and possess antimicrobial activity against pathogens (Arias et al., 2014; Ahuja et al., 2012; Holland and O'Keef, 2010).

The linkage between drought tolerance in cultivars and aflatoxin contamination has been investigated (Reddy et al., 2003; Sobolev et al., 2007; Sudhakar et al., 2007; Sullivan and Holbrook, 2007), and is logical, given that more drought-tolerant cultivars that have access to more soil water and better shade soil are likely to have less aflatoxin. However, a major complication to the development of cultivars resistant to *A. flavus* and *A. parasiticus* is the often highly significant interaction of genotype with environment observed in peanut breeding programmes and the natural existence of aflatoxigenic fungi in soils and their infection (Snigdha et al., 2015). Because the growth of *A. flavus* and *A. parasiticus* is mainly influenced by temperature and relative humidity, which are difficult to control in field soils, new strategies may be necessary to improve cultivar resistance (Bhatnagar-Mathur, 2015; Nakai et al., 2008). Guo et al. (2011) recently demonstrated that targeted marker-assisted breeding can be employed for the development of cultivars with resistance to infection or aflatoxin production. Huang et al. (2012) identified a total of 16 loci associated with *A. flavus* infection and aflatoxin contamination.

Research leading to the development of peanut cultivars with improved drought tolerance and lower aflatoxin contamination was initiated in the 1990s in Niger, Senegal and Burkina Faso (Schilling and Misari, 1993) and is being continued by the International Crops Research Institute for the Semi-Arid Tropics (ICRISAT, 2016), in China and the United States (Ferguson et al., 2001; Holbrook et al., 2008; Reddy et al., 2011). Using the worldwide peanut core collections, identification of *A. flavus*-resistant and drought-tolerant sources and incorporation of these traits into new cultivars is underway (Damicone et al., 2010; Chamberlin et al., 2010; Holbrook and Dong, 2005; Jiang et al., 2010; Upadhyaya et al., 2002). Additionally, other valuable traits such as pest resistance can provide indirect benefits relative to aflatoxin contamination through improved plant health. Traits related to plant development and pod and kernel maturation may also be valuable as early maturing cultivars may be harvested prior to higher levels of aflatoxin contamination.

4.3 Soil fertility and plant nutrition

Calcium deficiencies (Fig. 2) have been linked to higher infection of kernels by *A. flavus* and subsequently to higher aflatoxin levels (Adams and Bowen, 1993; Bowen et al., 1996; Wilson and Walker, 1988). Calcium plays an important role in cellular integrity. Deficiencies can inhibit pod fill (Smith et al., 1993) and compromise pods such that they are more readily damaged (Chari et al., 1986). Calcium deficiency is more likely to occur in times of drought since peanut plants growing in conditions of inadequate moisture may be unable to absorb adequate calcium (Cox et al., 1976; Skelton and Shear, 1971). During periods of drought, poorly developed and weak pods may crack and provide an entry point for toxigenic fungi and thereby increase the chances of aflatoxin contamination (Alva et al., 1989).

Calcium applied at the base of the plant 45 to 60 days after planting strengthens the pod shell and may indirectly reduce aflatoxin contamination by reducing fungal infection. The use of calcium amendments has been shown to increase pod yield depending on soil characteristics (Cox et al., 1976). Chari et al. (1986) observed increased drought tolerance in peanut resulting from sufficient calcium fertilization. Because of its diverse roles in plant growth and pod development, low levels of calcium in the soil have been associated with high rates of fungal infection and subsequent aflatoxin contamination.

4.4 Soil organic matter and texture

Soil organic matter has shown potential to reduce the level of pre-harvest aflatoxins in peanut kernels. Waliyar et al. (2013) observed a 42% reduction in aflatoxin contamination due to application of farmyard manure. This may be attributed to improved water-holding

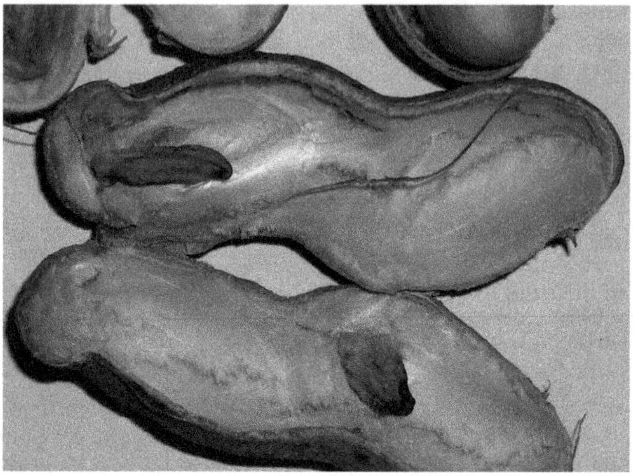

Figure 2 Results of drought and calcium deficiency in peanut. Photo credit, David Jordan.

capacity and nutrient status due to the addition of manure. A recent study investigating the effect of organic matter application using partially decomposed poultry manure showed a general decrease in pre-harvest aflatoxin concentrations in peanut kernels with an increase in the quantity of poultry manure applied to the soil at planting. Soil analysis from the same experiment showed an improved soil water-holding capacity. Higher rates of organic matter application increased the overall microbial activity in soil which can suppress the dominance exhibited by *A. flavus*. Soil texture (sand, silt and clay) can influence pre-harvest infection of peanut by *A. flavus* and *A. parasiticus*. Torres et al. (2014) reported that infection of peanut pods was higher when grown on coarse-textured soils prone to drought due to low water-holding capacity. Soils with higher water-holding capacities due to higher levels of organic matter and lower sand content are at lower risk because higher soil water encourages competition among soil fauna including *A. flavus* and *A. parasiticus* (Bowen, 2009).

4.5 Sowing date and harvest date

The influence of sowing date on infection by *A. flavus* and *A. parasiticus* and subsequent risk of aflatoxin contamination of peanut is associated with the interaction of pod maturity, harvest date, soil moisture and temperature during the final month of the season. In some regions of peanut production, soil temperatures during drought conditions may not be sufficiently high to promote *A. flavus* development and colonization compared with other fungi (Azaizeh et al., 1989; Bowen and Hagan, 2015; CAST, 2003). When the final stages of reproductive growth and pod formation and fill occur under dry conditions with elevated temperatures, the risk of aflatoxin contamination is high. In regions without irrigation and a distinct rainy season, for example in many areas of sub-Saharan Africa, rainfall is limited late in the growing season, thus placing peanut at a higher risk of contamination by aflatoxin. In contrast, in temperate production regions such as the mid-Atlantic region of the United States, ambient and soil temperatures often decrease towards the end of the growing season and thus minimize *A. flavus* development, even though peanut is exposed to drought on coarse-textured soils (Azaizeh et al., 1989; Bowen and Hagan, 2015).

Table 1 Aflatoxin contamination of peanut in Zambia as influenced by seeding date in relation to drought stress during the latter part of the growing season*

	Rainfall during final 30 days of the season	Mean aflatoxin B_1 concentration in farmer stock peanut
Planting date	cm	μg/kg
31 December 2014	9.90	4 c
8 January 2015	2.73	37 b
15 January 2015	0.00	89 a

*Means for aflatoxin contamination followed by the same letter are not significantly different according to Fisher's protected LSD at $P \leq 0.05$.

Shorter-duration cultivars, early planting and practices that minimize plant stresses that promote pod infection or delay pod and kernel maturation can decrease the risk of aflatoxin contamination by decreasing the duration of exposure to conditions favourable for *A. flavus* infection (Bowen and Hagan, 2015). However, unpredictable rainfall patterns complicate crop management decisions, including both planting and harvest, especially with respect to aflatoxin contamination. For example, bimodal rainfall patterns in Ghana have been less predictable in recent years, making decisions related to sowing date complicated. Ultimately, this challenge may translate into greater risk of aflatoxin contamination in the field.

A recent example from research in Zambia demonstrates the risk of increased aflatoxin contamination when planting is delayed and peanut experiences a longer period of drought and higher soil temperatures during pod maturation within 30 days prior to harvest (Table 1). In addition to synchronizing sowing and rainfall to optimize yield, farmers can minimize aflatoxin contamination by sowing in a manner that minimizes exposure to drought close to harvest.

Immature or overmature peanut pods have increased susceptibility to aflatoxin contamination (Arias et al., 2014; Blankenship et al., 1983; Cole et al., 1985; Diao et al., 2015; Dorner et al., 1989; Hell et al., 2003; Sanders et al., 1993; Waliyar et al., 2013). This dynamic is influenced by the ability of *A. flavus* and *A. parasiticus* to infect pods that are not fully developed and are susceptible to insect damage or pods that have a greater propensity for breakage because they are fully developed. Furthermore, immature peanut loses the capacity to produce phytoalexins faster than that of mature peanut, leading to more rapid contamination (Dorner et al., 1989). In both instances the avenue to aflatoxin-producing mould infection and colonization is greater than that of pods that have reached optimum maturity for harvest but do not remain in soil after that point in time (Diao et al., 2015).

4.6 Irrigation

Yield and quality of peanut is affected by timing and duration of drought. Clearly, the ability to supplement rainfall with timely and adequate irrigation not only will increase peanut yield and kernel quality but can create a soil environment where *A. flavus* and *A. parasiticus* are not the dominant soil organisms (Holbrook et al., 2009). Reddy et al. (2003) found that aflatoxin contamination was widespread where peanut was grown under rain-fed conditions compared with irrigated peanut. Of the strategies discussed in this chapter to address pre-harvest contamination of peanut by aflatoxin, irrigation might have the greatest impact. However, many farmers worldwide do not have access to irrigation because of the quantity or quality of water. Resources to establish and maintain irrigation systems are also limited.

4.7 Minimizing pod and kernel damage

Contamination by aflatoxin in peanut occurs more frequently when pods or kernels are damaged compared with sound, non-damaged pods and kernels (McDonald and Harkness, 1967). Physical damage during handling, drying or storage and damage caused by insects and nematodes affect infection by increasing entry of soil containing *A. flavus* and *A. parasiticus* into pods (Waliyar et al., 2013). Lesser cornstalk borer (*Elasmopalpus lignosellus*) (Bowen and Mack, 1993; Lynch and Wilson, 1991), burrower bug (*Cydnus aterrimus*) (Chapin et al., 2004) and termites (Microlermes, Isoptera: Termitidae) (Umeh et al., 2000) can damage peanut pods and promote contamination by *A. flavus* and increase aflatoxin contamination. Root-knot nematodes have also been associated with higher aflatoxin content in peanut (Minton et al., 1969; Timper et al. 2004, 2013). Waliyar et al. (2013) reported that mechanical damage to pods by soil-borne insects, nematodes or cultural operations can allow infection and subsequently result in the formation of aflatoxins even under less-severe temperature and moisture conditions. Once kernels are opened they often dry down to moisture content level that promotes fungal growth (Wagacha and Muthoni, 2008).

4.8 Biological control

Employing biological controls or enhancing natural controls of organisms are fundamental approaches to pest management. Pre-harvest management practices affect natural controls that can minimize growth and development of *A. flavus* and *A. parasiticus* or promote competition by soil fauna that do not produce toxins with those that do. Additionally, biological controls to minimize *A. flavus* and *A. parasiticus* have been introduced (Dorner et al., 2003). Application of the strains of *A. flavus* and *A. parasiticus* that are non-aflatoxigenic, such as the *A. flavus* NRRL 21882 strain, has been successful in reducing aflatoxin contamination in peanut (Dorner, 2009). For example, Dorner (2009) reported reductions in aflatoxin concentrations following application of Afla-Guard® compared to non-treated peanut (Table 2). Use of the non-aflatoxigenic biological control is being evaluated worldwide in the public and private sectors. In the coming years the value of this potentially long-term strategy to minimizing aflatoxin contamination in peanut will become apparent. However, the cost-effectiveness and sustainability of this approach still need further evaluation.

5 Aflatoxin contamination during drying, storage and processing

While cultural practices in the field can impact infection of peanut pods and kernels, the eventual impact of infection can be exacerbated by drying and storing techniques. Drying peanut within 7 days of harvest to below 10% moisture prior to storage and processing

Table 2 Mean aflatoxin concentrations in farmer stock peanut from fields treated with the nonaflatoxogenic product Alfa-Guard® compared with non-treated peanut at 7 locations in the south-eastern United States during 2004

Location		Mean aflatoxin concentration	
		Control	Treated
Town	State	µg/kg	
Harford	Alabama	16.6	5.5
Newton	Alabama	319.7	49.0*
Ft. Gaines	Georgia	96.6	0.2*
Sasser	Georgia	0.0	0.0
Smithfield	Georgia	0.0	0.1
Unadilla 1	Georgia	37.4	0.0
Unadilla 2	Georgia	2.6	1.0*
Pooled	–	78.9	11.7*

*indicates significance at $P \leq 0.01$ within a location and when pooled over all locations using nonparametric Mann-Whitney rank sum test. Adapted from Dorner (2009).

can limit the development and growth of *A. flavus* and *A. parasiticus* and the subsequent levels of aflatoxin contamination (Amoako-Attah et al., 2007). In most developing countries, peanuts are dried on the farm in windrows, drying cocks, staked poles (e.g. Mandela corks) or on plastic sheeting off the farm. In some cases, drying is on bare soil or concrete slab, both of which create situations where *A. flavus* infection can easily occur, leading to higher levels of aflatoxin contamination.

Prior to marketing of peanut and peanut products, processing, either through sorting of unshelled and shelled peanuts or through blanching (removal of skins that are potentially infected by mould), can decrease aflatoxin in the consumed product and decrease the risk to human health. Sorting can occur in three steps, the first step involves sorting of in-shell peanuts to remove peanut pods that are broken, insect damaged and mouldy including loose shelled kernels (Fig. 3). The second level of sorting occurs on shelled nuts where discoloured, rotten, mouldy, broken, shrivelled and de-skinned kernels are removed. The third level of sorting occurs after blanching where discoloured kernels are removed. These sorting practices have been demonstrated to significantly reduce aflatoxin levels in peanuts (Whitaker et al., 2005). However, these steps in processing can be expensive and time-consuming and are not employed equally. In addition, use of sorting at the processor level can result in considerable loss of raw material that might have no economic value for the processor. This is a key component for smaller processors who aim at optimizing the output from the raw materials procured.

Pitt et al. (2013) reported that maintaining relative humidity below 85% with temperatures less than 10°C will prevent *A. flavus* and *A. parasiticus* growth and aflatoxin production in storage. However, one of the main challenges worldwide is the ability of farmers to grow and store peanut and other commodities below these thresholds. Kernel moisture content of 7.5%, ambient temperature of 10°C and relative humidity of 65% are optimal bulk storage conditions for peanut, allowing storage of up to 1 year (Pattee et al., 1982).

Figure 3 Examples of peanut at different stages of sorting prior to marketing in Malawi. Photo credit, David Jordan.

6 Challenges in mycotoxin research

A number of challenges exist to compare currently available practices that minimize the risk of aflatoxin contamination in peanuts. Inherent variation in *A. flavus* distribution in soil and in harvested peanuts contributes to the experimental challenge. Implementing an adequate sampling procedure that is representative of the compared interventions is critical. Minimizing contamination across treatments that may be introduced while handling and processing peanuts is necessary to reduce error and enable effective comparisons. Increasing the sample size and providing adequate replication can improve precision but require resources to accomplish. For example, standard methodology for agronomic research (3 to 6 replications of a treatment) may be inadequate to characterize experimental error with respect to treatment effects in the analysis of variance. Preliminary data mentioned later in this chapter were the result of 12 replications in the village and not the standard 3 to 6 used in many agronomic trials. Sampling error for aflatoxin determination can be significant since aflatoxin distribution is not uniform within a lot (Whitaker, 2003).

Proper storage of samples prior to analysis can be a limiting factor in terms of the scope of work that can be accomplished while maintaining the desired quality and applicability of results. Placing peanut samples below 3°C is necessary to prevent continued growth of *A. flavus*. When stored at higher temperatures, results may reflect higher concentrations than those that were present at the completion of a specific intervention. Resources and access to proper storage can be a challenge, especially in some regions where electricity is limited or sporadic or when samples cannot be subject to analytic procedures in a timely manner. Devices that permit immediate and rapid assessment of samples either in the field or, soon after arrival, in the laboratory would help minimize problems associated with poor storage.

Finally, a major challenge associated with comparing aflatoxin interventions is the cost of analysing samples. On the basis of study objectives, researchers may have to limit observations, especially if analysis is through high-performance liquid chromatography (HPLC) and if the precision in concentration is needed to compare treatments. HPLC is the most expensive procedure currently employed in aflatoxin testing. Colorimetric assays are less expensive than HPLC, but can be qualitative rather than quantitative in some cases and at some concentrations. Improvements in devices used to analyse samples and store data continue to increase and have the potential to increase test efficiency and lower expense to levels that facilitate a greater number of treatment comparisons which could lead to development of higher-quality data sets that accurately represent the impact of interventions on aflatoxin contamination in developing countries.

7 Value chain projects

Aflatoxin contamination is a problem throughout the peanut value chain whether the perspective is in a local village or when considering peanut entering regional, national and international markets. In this section, the current understanding of aflatoxin contamination in peanut value chains will be discussed with three case studies addressing aflatoxin contamination in Ghana, Haiti and Malawi. These case studies reflect ongoing research through the US Feed the Future Innovation Lab for Collaborative Research on Peanut Productivity and Mycotoxin Control (PMIL, 2016).

7.1 Current status of value chain projects

Efforts to develop and increase efficiency and sustainability of value chains for safe, high-quality peanut and other crops have been documented in numerous regions and countries within regions (Afri-Nut, 2016; Humphrey and Navas-Alemán, 2010). Availability, quality and expertise associated with technologies for mycotoxin control, as well as financial resources and market forces influence the decisions of value chain actors at each step.

7.2 Attitudes to aflatoxins in Ghana, Mozambique and Uganda

Knowledge of the conditions that favour development and contamination of peanut and other crops by aflatoxin is important in developing mitigation strategies. At a basic level, proper identification of the problem and implementation of effective interventions are essential components. In the United States, farmers are well aware of financial losses that could occur if farmer stock peanuts are contaminated by aflatoxin. The potential of this direct economic loss has resulted in concerted efforts in research and education programmes that focus on steps within both pre-harvest and post-harvest components of the value chain. However, in countries with limited research and education programmes where inspection programmes at the point of sale are not in place, knowledge of aflatoxin and its management are less well established.

Recent surveys of farmers in Uganda (2014), Ghana (2015) and Mozambique (2016) demonstrate the variation in knowledge of aflatoxin in peanut. Farmers in Uganda had the greatest existing knowledge of aflatoxin (93%) compared with the farmers in Ghana (11%)

Table 3 Results from surveys of peanut farmers in Ghana, Mozambique and Uganda with respect to knowledge of aflatoxin and management strategies to mitigate aflatoxin in peanut

Questionnaire country (Survey year)	Uganda (2014)	Ghana (2015)	Mozambique (2016)
% Respondents with knowledge of aflatoxin prior to the survey	93%	11%	15%
Most common source of aflatoxin knowledge: *other farmers*	50%	32%	38%
Farmers that experienced aflatoxin problem in peanuts	66%	83%	75%
Aflatoxin incidence during the last 5 years:			
Mean number of years	1.4	2.4	2.0
% Respondents with no aflatoxin	34%	31%	28%
1–2 years	50%	24%	40%
3–5 years	16%	44%	32%
Most common resource for aflatoxin information exchange:			
Farmer group members	14%	6%	3%
Neighbours	12%	10%	15%
Relatives	6%	14%	8%
Most important influence of aflatoxin information to farmers practice:	Drying	Variety choice	Harvest timing
	18%	8%	10%
Most important causes of aflatoxin given by farmers:			
Don't know	41%	25%	50%
Poor drying	38%	12%	14%
Too much rain	12%	10%	24%
Pests and disease	5%	11%	4%
Poor storage	1%	5%	8%
Removal of visibly rotten (aflatoxin) peanuts before:			
Eating	61%	80%	81%
Storage	25%	76%	52%
Experienced buyers cross-checking peanuts for aflatoxin	42%	88%	59%
Most common uses for visibly mouldy or rotten peanuts:			
Same as other peanuts (no difference)	34%	9%	26%
Throw away	24%	81%	69%
Burn	36%	0%	2%
Bury	3%	1%	1%
Use for soup or stew	2%	1%	1%
Sell	0%	8%	1%
Animal feed	2%	1%	1%

and Mozambique (15%) (Table 3). This variability may be due to prior interventions and the presence of farmer groups. In all cases, the most common source of aflatoxin knowledge is other farmers. Of the three countries, Ghana has the greatest proportion of farmers who have experienced problems with aflatoxin contamination (83%), followed by Mozambique (75%) and Uganda (66%). Farmers in Ghana also experienced greater persistence with aflatoxin contamination on average over the last 5 years (2.4 years) followed by the farmers in Mozambique (2.0 years) and Uganda (1.4 years). The most common resources for aflatoxin information exchange are among farmer group members, neighbours and relatives. Based on the information exchanged, farmers were most likely to change drying practices (Uganda), cultivar choice (Ghana) or harvest timing (Mozambique).

Most farmers did not know the cause of aflatoxin contamination, but among those who did, the most important causes mentioned include poor drying, too much rain, pests and disease, and poor storage. Visibly mouldy and rotten peanuts were removed before consumption by 61%, 80% and 81% of farmers in Uganda, Ghana and Mozambique, respectively, while removal before storage was done by 25%, 76% and 52% of farmers in these respective countries. Cross-checking peanuts for aflatoxin by buyers varied across the countries from 42% in Uganda to 88% in Ghana. The most common uses of visibly mouldy or rotten peanuts were disposal or the same use as other peanut, presumably consumption. Results from these surveys reveal the need for education programmes involving farmers on the negative impact of aflatoxin on human health and strategies that reduce aflatoxin contamination. While solutions that minimize aflatoxin contamination must be tailored for countries and regions based on resources, the fundamental message of the impact of aflatoxin exposure on human health is clear and does not vary and should be delivered to all populations.

7.3 Controlling aflatoxin contamination at village level: a case study

The range of cultural and management practices discussed above are essential components of value chains at local, regional, national and international levels. Differences in mechanisms or drivers and efficiencies at each step in the value chain can be hindered or enhanced depending on geographical level and regulation.

Preliminary results from a value chain research project in the rural village of Dagomba in the Ashanti region of southern Ghana will be used to describe the interactions at pre-harvest and post-harvest that affect volume, quality and safety of peanuts leaving the village for external markets, larger-scale processing or staying within the village cluster for trade and consumption. This case study is a component of a larger research project conducted with farmers in five villages in Ghana from 2014 to 2016 to explore the level of aflatoxin in peanut following harvest, drying and storage. This case study is presented to demonstrate the importance of each step of the value chain.

In the Ghana case study, peanuts in the field were subjected to improved management of aphids, the vector of rosette virus, and fertility compared with the traditional farmer practice. Although the improved practices resulted in increased peanut yield, quality and economic return (774 vs 978 USD/ha, $P = 0.0002$) when pooled over 12 farmers in one village, only a minor difference in aflatoxin concentration in peanut farmer stock was noted when sampling occurred immediately after harvest but prior to drying when comparing management treatments (1.0 vs 0.5 μg/kg, $P = 0.0015$). Aflatoxin levels subsequently increased during drying over levels documented immediately after harvest. The impact of

Table 4 Aflatoxin concentration in farmer stock peanut from the village Dagomba in the Ashanti region of Ghana after storage following various practices in the field, during drying and in storage*

Management practices				Peanut response			
				Kernel status after storage for 4 months			
			Aflatoxin after	Good	Shrivelled	Mouldy	Discoloured
Field	Drying	Storage	storage (μg/kg)	g/500 g sample			
Farmer	Ground	Poly bag	1407 a	223 f	124 ab	20 a	134 a
Farmer	Ground	Sealed bag	297 bc	264 e	129 a	2 c	106 bc
Farmer	Tarp	Poly bag	298 bcd	306 cd	110 abc	1 c	82 c
Farmer	Tarp	Sealed bag	100 e	339 bc	103 c	0 c	59 d
Improved	Ground	Poly bag	548 b	263 e	118 abc	8 b	111 ab
Improved[†]	Ground	Sealed bag	177 cde	299 d	106 bc	0 c	95 bc
Improved	Tarp	Poly bag	149 de	366 ab	80 d	0 c	57 d
Improved	Tarp	Sealed bag	53 f	393 a	80 d	0 c	27 e

*Means followed by the same letter are not significantly different according to Fisher's protected LSD test at $P \leq 0.0001$. Actual data are presented with mean separation performed on log transformed data. Data are pooled over 12 farmers in a village cluster.

†Improved practice included one additional weeding, application of local soaps to aphids, the vector of rosette virus, and calcium applied at flowering.

minor differences at harvest was manifested after drying, but approaches to drying affected aflatoxin concentration more than management prior to harvest (data not shown). The importance of effective drying on plastic tarps (29–80 μg/kg aflatoxin) became apparent compared with ground drying (153–226 μg/kg) regardless of the level of aflatoxin coming out of the field at harvest. As these peanuts continued through the value chain, following relatively low input in the field, drying on the ground and storage in readily available poly sacks with limited protection from the weather resulted in an aflatoxin concentration of 1407 μg/kg (Table 4). Use of the most effective approaches in field production, drying with tarps and storage in hermetically sealed bags resulted in the lowest aflatoxin concentration (53 μg/kg). Adopting a single improved practice, or two of the three possible improved practices, resulted in aflatoxin concentrations between 100 and 548 μg/kg.

As buyers become more discriminating in their purchases, the amount of harvested peanut that is marketable could be lowered if quality is compromised. In the Ghana example, after shelling, variation in categories of good kernels, shrivelled kernels, kernels with visible mould and discoloured kernels were observed based on practices in the field and following drying and storing (Table 4). When correlated with aflatoxin concentration in peanut after storing, coefficients were highly significant ($P \leq 0.0001$) for good kernels ($R = -0.84$), shrivelled kernels ($R = 0.59$), kernels with visible mould ($R = 0.64$) and discoloured kernels ($R = 0.83$). These results were not surprising given the strong correlation between damaged kernels (Fig. 4 and 5) and aflatoxin contamination (Cole et al., 1985; Dorner, 2008). When improved practices were included in the field and during drying and storage, 79% of the 500 g sample were considered good and would be of high value in the market and safe for consumption. In contrast, the current practices used by many farmers resulted

Figure 4 Poor-quality peanut kernels caused by poor drying and storing conditions. Photo credit, William Appaw.

Figure 5 Contamination of shelled peanut by *Aspergillus* spp. Photo credit, William Appaw.

in less than half of the kernels (49%) in the good category (Fig. 6). The lower yield of farmer stock peanut under current farmer practices in the field would be compounded by the lower percentage of high-quality peanuts available for the market after storage. This would be the case even if aflatoxin contamination was not a deciding factor for peanut sales. Removal of good peanut kernels for household consumption or removal for seed used for sowing during the next season would further reduce income.

Figure 6 High-quality peanut kernels following adequate drying prior to storage. Photo credit, William Appaw.

These results from this village during the major season in southern Ghana are not conclusive, they do provide an example of importance of each step in the value chain in terms of impact the concentration of aflatoxin has on the final product. Peanut after storage will be consumed directly by individuals in the household or will enter the market in some form. Higher yield and economic return captured immediately after harvest could be lower or higher after storage, depending on quality and seasonal price dynamics, especially if buyers consider aflatoxin contamination in their decision making. The final price of peanut the farmer receives compared with selling soon after harvest could be reflected in their own decision to process peanut prior to marketing or to accept the price offered by the buyer.

The impact of aflatoxin contamination on human health is the key issue, although market pressures often dictate decisions on acceptance of risk of aflatoxin at village and broader levels. While pre-storage segregation to remove damaged or mouldy peanuts is effective in reducing the risk of aflatoxin contamination (Cole et al., 1985; Dorner, 2008), the decision by the farmer to remove damaged farmer stock prior to selling or saving higher-quality farmer stock for seed for the following season's planting can impact the farming household. Failure to save high-quality seed for planting will most likely affect stand establishment, seedling vigour and yield, the next season. Poor-quality peanuts that are removed from the marketed product could be used for household consumption or as livestock feed, increasing exposure to aflatoxin. From the buyer's perspective, the cost of processing purchased farmer stock prior to the next step in the value chain will impact the price that can be offered to the farmer, and this could impact financial risk to purchasers in the market. Market price based on quality of pods and kernels could increase exposure of individuals purchasing lower-quality peanut to aflatoxin. Efficiencies in the market are important at each step in the value chain, and finding a use for each product in processing, whether be the initial focus or by-products of steps, is essential in competitive markets. Minimizing aflatoxin contamination at earlier steps, going all the way back to the field level, can make the value chain more efficient and profitable with less risk of exposure to aflatoxin.

Presenting incentives at each step in the value chain to deliver a high-quality and safe peanut is critical if effective and sustainable value chains are to exist. In the example above, developing incentives for farmers to adopt greater management and inputs in the field that increase cost during drying and in storage increases the likelihood that the entire value chain delivers higher-quality peanuts with less risk of aflatoxin to the consumer. While on the surface the difference in aflatoxin immediately after harvest was low even after adoption of greater inputs that improve plant health and subsequent yield, small differences in aflatoxin contamination were translated into larger differences after drying.

The investment in improved practices in the field most likely would be readily adopted by many farmers because the increase in yield following those inputs more than offset the increase in cost associated with them. Certainly, the limited local availability of fertilizer and other inputs increased time and financial cost of pest management and fertilizer application, and limited money or credit for initial purchase of these can negatively impact adoption. However, in this case increased economic return would justify the improved field practices and would continue to decrease aflatoxin contamination during drying and storage. Use of tarps in the drying step compared with ground drying may be limited because of the expense of the tarp relative to overall budget. Whether or not the farmer would adopt tarps could also be determined by availability, especially in rural villages, and what other uses the farmer may have for the tarp once harvest begins.

In an ongoing randomized trial in the northern and upper east regions of Ghana, farmers were educated on the health consequences of aflatoxin exposure and the benefits of tarps relative to ground drying to prevent contamination. Farmers were then offered the opportunity to purchase inexpensive plastic sheeting, procured from the regional commercial hub, at cost. Preliminary results indicate a significant increase in reported tarp drying relative to a control group (Gajate-Garrido et al., 2016, unpublished data). However, the level of tarp usage remained below 15%. Two-thirds of farmers adopted tarps when these were provided free of charge together with the educational intervention. These results indicate that most but not all farmers are willing to use a simple technology for mycotoxin control when this does not entail a cash outlay, but that financial constraints constitute a major hurdle to adoption.

Farmers often receive the lowest price for peanut during the season if they are forced to sell immediately after harvest, either because of financial commitments incurred earlier that must be addressed or because storage quality, storage capacity or a combination of these are limited. In the Ghana example, it is estimated that use of hermetically sealed bags could increase the farmer's total production cost by as much as 10%. In addition to the issue of local availability of such bags, whether the farmer can afford to increase production and storage costs by 10–15% is an open question. Based on the above-mentioned tarp demand trial, demand for aflatoxin control technologies in this price range appears quite modest.

Will the farmer receive a dividend for investing in tarps and hermetic bags? There are several dimensions to this issue, and additional work is needed to address the topic. Should the farmer receive the same price at some point after storage that would have been obtained around the timing of harvest without lengthy storage, there would be no financial incentive for investment in hermetic bags, at least on the surface of the issue. Certainly, an increase in price over time could at the very least offset the input cost associated with storing in hermetic bags, provided hermetic storage reduces the extent of either or both physical losses or the deterioration of qualities that are valued by farm households or the market.

Peanut prices in Ghana tend to increase steadily after harvest and peak immediately prior to the following harvest as stocks run low, implying that to the extent that farmers perceive storage losses as a problem, high-quality storage could be attractive. However, only 7.6% of farmers enrolled in the tarp demand trial reported experiencing any damage to stored nuts. Among these, the primary reason for damage, reported by 5.4% of farmers, was rodent infestation, against which hermetic bags are ineffective. Only 2% of farmers reported insect damage during storage and 1.6% reported problems with mould. However, based on the results reported in Table 4, the actual extent of damage due to moulds is far higher than that perceived by farmers.

The invisibility of fungal contamination presents a formidable challenge to increasing farmer adoption of practices and technologies for mycotoxin control. As the costs of mycotoxin screening and testing technologies fall (ICRISAT, 2016), and awareness of the health consequences of mycotoxin exposure grows and drives demand for safer food, increased screening and testing for these contaminants in peanut value chains will be important and can be expected. The further upstream these occur, the more likely incentives are passed on to farmers, and the more likely these translate into improvements in pre- and post-harvest practices. Screening for mycotoxins further down the value chain will result in directing the most contaminated nuts to the lowest value markets (and thus the poorest consumers), without much effect on the overall level in the food supply. However, when there are many small producers selling small quantities of peanut, upstream testing is extremely costly and potentially infeasible. This implies that aggregation of smallholder produce for collective testing and sale will be critical for improving food safety in this value chain. The challenge for buyers demanding safe peanut will be to establish a price that not only incentivizes the farmer to adopt management practices that increase quality and lower aflatoxin but also maintains an adequate margin for their financial interests.

The case study presented above presents considerations experienced by many farmers, buyers and processors around the world in a scenario where effective regulation is not in place. In contrast, for the value chain in the US at the farmer level, peanut is dried and stored either using facilities on the farm or at regional buying points. Federal standards for quality and safety of peanut have been developed and adopted through state agencies and include restrictions on *A. flavus* and aflatoxin entering the commercial trade. The incentive for farmers to deliver peanuts with aflatoxin levels that meet regulatory standards is market driven. Farmer stock peanuts with aflatoxin levels exceeding 15 μg/kg cannot enter the domestic trade, as these peanuts lower the price substantially. Farmer stock at this aflatoxin level or higher is sent to the oil market for crushing at a substantially lower price, often at a price resulting in a financial loss to the farmer when considering the overall budgetary constraints. While farmers can try to reduce contamination in loads exceeding 15 μg/kg, these approaches are limited in effectiveness and are time and resource consuming.

In this value chain, the final consumer is considered with respect to risk associated with exposure to aflatoxin in peanut-based products. In many parts of the world the ability to monitor peanut at the first point of sale does not exist with respect to a third-party regulatory system. Lack of a regulatory system at this step in the value chain is unlikely to change in the near future in many countries, and this continues the challenges associated with producing, processing and trading in many parts of the world. The regulatory component of the US peanut value chain and strong consumer demand for safe peanut products creates incentives and pressures throughout the value chain with respect to management that minimizes aflatoxin levels at multiple steps. The importance of the regulatory component cannot be over-emphasized at the point of sale. While consumer demand is a strong driver in the US, efficiency of the value chain is enhanced because of the regulatory component at the point of farmer delivery to the purchaser. Less *A. flavus* entering the value chain from the farmer stock can serve to reduce the contamination at steps later in the value chain. In the absence of an effective regulatory component in other parts of the world, altruistic concerns to the consumer and market pressures will make efficiencies at managing aflatoxin within the value chain challenging. However, regulatory components associated with agriculture and food safety are not without their own set of challenges to develop, implement and sustain, and the US system is just one example of the possible contributions of third-party regulation.

7.4 Interface between nongovernmental organizations and the private sector: a case study

In locations with limited public institutions, protecting consumers from aflatoxin contamination can be difficult. Without regulatory bodies to set and enforce standards, there is limited incentive for producers or processors to self-regulate. Collaboration between nongovernmental organizations (NGOs), non-profits and private sector operators can be effective in reducing aflatoxin risk, raising critical awareness of the public health concern and building capacity towards improved monitoring and reducing of aflatoxin contamination.

One example of this has been the collaboration in Haiti between Meds & Food for Kids (MFK), a US-based NGO operating a food processing facility in Cap-Haïtian, Haiti, and the Acceso Peanut Enterprise Corporation, a for-profit Haitian supply chain company with philanthropic investment by the Clinton Guistra Enterprise Partnership. MFK began producing peanut-based ready-to-use therapeutic food (RUTF) in 2006 and quickly learned that controlling aflatoxin would be a major limitation to the use of locally sourced peanuts. Beginning in 2007, in collaboration with the USAID-funded Peanut Collaborative Research Support Program, MFK began working with farmers and implementing technology that would help meet the strict standards for RUTF required by their international customers, primarily UNICEF. As MFK's production capacity increased, their demand for locally sourced peanuts created an opportunity for farmers. However, the coordination of aggregation and quality control management at the farm level was beyond the capacity of either MFK or local cooperatives. Eyeing the potential to sustainably improve farmer livelihood by helping them both increase yields through access to inputs, technology, and training and connect with this growing market demand, Acceso was formed in collaboration with MFK and other non-profit and private sector actors.

Acceso manages a growing network of small depots that act as both distribution points for inputs, such as seed, fertilizer, fungicide and training, and buying points where farmers pay off these input credits and sell their surplus crop. As of the Fall 2016, Acceso manages 13 depots in the two major peanut-producing regions, with over 2000 farmers participating and coordinating the sale of over 500 metric tons of low-aflatoxin (<20 μg/kg) farmer stock peanuts per season. Acceso's depot managers train the farmers in farm business management and application of best management practices for both pre- and post-harvest, based on the expert input from the US Feed the Future Innovation Lab for Collaborative Research on Peanut Productivity and Mycotoxin Control that is continuously evaluating these and new technologies to increase yield and quality and reduce costs. During the season, the depot managers visit farmer fields to assure compliance of best practices and evaluate the future crop. All the data are collected using tablet-based software to monitor loans, survey farmers and record observations. In the future, weather data can be included with historic aflatoxin data to predict the level of risk by location.

After the peanuts are harvested, the farmers dry the peanuts on tarps provided by Acceso. They then bring the crop to the depot for preliminary sorting, moisture testing and quantitative aflatoxin test using ELISA strip tests and a tablet reader. The crop is then bagged and entered into a traceability system, also managed by the tablet software. This testing step, prior to aggregation, prevents mixing of contaminated product and incentivizes farmers to follow the recommended post-harvest practices. Peanut lots with positive aflatoxin tests are further dried and resorted, or Acceso will purchase at a reduced price and further process the lot internally by blanching and sorting. The early segregation of lots and immediate purchase after harvest to prevent storage contamination at the farm level are important for maintaining quality.

While only in their third season, Acceso has expanded their purchasing to meet a growing demand for low-aflatoxin peanuts in the urban markets. Partially due to publicity around these joint efforts, public awareness of the health risks of aflatoxin has increased and local peanut product manufacturers have increased their investments. Beyond aflatoxin control, many medium- and larger-scale peanut processors prefer to deal with a reliable supplier of high-quality products, rather than contest with the variable markets and disorganized aggregators. Likewise, Acceso has diversified beyond peanuts to other commodities where processors are seeking regular, high-quality products. MFK, Acceso and their clients hope that through investments in improved production that benefit the farmer, they can stabilize their supply chain and potentially reduce the cost of raw materials over time while still helping the farmer.

Although there are factors of this arrangement unique to Haiti, similar supply chain arrangements that align NGOs and private sector are being scaled elsewhere. For example, Exagris Africa Ltd., a commercial farming company in Malawi, is working with smallholders in a similar way along with their own anchor farms to supply Afri-Nut Ltd., an affiliated commercial peanut processing company, and Valid Nutrition, another RUTF producer, in Malawi.

7.5 Supplying regional peanut markets: a case study

The opportunities to focus efforts to reduce aflatoxin contamination in the peanut value chain in Malawi have many similarities with the programme in Ghana. However, unlike the southern region of Ghana which experiences bimodal rainfall, Malawi has only a single rainy season and this season is critical for the majority of all food production for this land-locked country. There has not been as long a history of collaboration as we have had in Ghana, so there is much less actual village-level evaluations and technology transfer. However, there have been ongoing projects associated with aflatoxin contamination in peanuts through ICRISAT-Malawi, Chitedze Agricultural Research Station and Lilongwe University of Agriculture and Natural Resources, and linkages with corporate enterprises including Exagris Africa Ltd. and Afri-Nut Ltd., a processor in which Exagris has invested.

The value chain process begins with the utilization of well-adapted varieties provided by the ICRISAT and Chitedze Agricultural Research Station peanut breeding programmes. The most promising germplasm along with popular local land races are evaluated in production practices that promote higher yields and economic returns. These same practices are also evaluated for the impact on in-field contamination with *A. flavus*. The general pre-harvest strategies are to employ strategies that encourage good plant health and thus higher yields while simultaneously reducing plant stress to biotic and abiotic stresses that may encourage aflatoxin contamination. The production practices that are evaluated and integrated into optimal production systems include sowing date; planting density; row spacing; pest management; and soil amendments such as gypsum, crop rotations and harvest dates. In Malawi, corn is an important crop that is planted extensively. As in many areas of Africa, timely planting of corn take precedence over planting of other crops. Corn is often rotated with many other crops including peanut, and since both are hosts for the *Aspergillus* fungi, practical rotations and adjustments in planting dates once the rains start each year are often not possible.

Recent research at a number of institutions in Malawi has demonstrated significant value in specific practices that not only increase yield and profit, but also reduce aflatoxin contamination of the crop while in the field. Early planting, increased plant populations, calcium, management of insects and diseases, crop rotations with non-host crops and timely harvest have all shown great benefits individually and in more comprehensive cropping studies to not only increase yield but also reduce aflatoxin contamination to more acceptable levels. These studies have been supported by collaborative studies in the adjacent countries of Zambia and Mozambique.

Exagris Africa Ltd., a corporate farm which supplies peanuts for Afri-Nut Ltd., utilizes these research findings on their farms throughout Malawi and is rapidly moving forward with cost-effective strategies to produce profitable, higher-yielding and higher-quality peanuts with lower aflatoxin content than is typically produced at the village level where these practices are yet to be adopted. In addition, Exagris provides training through their out grower programme for local farmers and is involved in seed multiplication and distribution of improved varieties to local farmers who also seek to sell high-quality peanuts to Afri-Nut.

Post-harvest strategies include improved drying technologies and training as well as research focused on food processing. Information derived from recent surveys reveal the need for education programmes associated with the inadequate knowledge about practices in handling of peanuts from procurement, preparation and processing of peanuts, the raw material in peanut butter processing to control aflatoxins. Targeted educational programmes for small- and medium-scale producers of peanut butter are resulting in changes in all phases of the industry.

8 Conclusion

This chapter has provided a comprehensive discussion of strategies to minimize aflatoxin contamination in peanut at various steps in the value chain. The case studies in Ghana, Haiti and Malawi provide insight into the challenges associated with implementation of these strategies. The complexity of aflatoxin mitigation in peanut and other crops will require interactions among public and private sector entities across many disciplines in order to effectively decrease risk of aflatoxin and improve human health.

9 Where to look for further information

Developing strategies to address sources and health risks associated with aflatoxin will continue to be important components of development projects around the world. Additional information related to aflatoxin will emerge from public and private sector institutions and the clientele they serve. Peer-reviewed literature, book chapters similar to this chapter and recommendations from public and private sector resources are available and will continue being developed into comprehensive approaches to mitigating aflatoxin contamination in peanuts and other foods.

10 References

Abdel–Azim, O. Z. and Khalil, I. I. (1979), 'Studies on storage moulds in Egypt. III. Deterioration and discoloration of rice grains', *Agricultural Research Review*, 59, 79–94.

Accinelli, C. and Abbas, H. K. (2011), 'New perspectives for the application of bioplastic materials in the biocontrol of *Aspergillus flavus* in corn', *Toxin Reviews*, 30(2–3), 71–8.

Achar, P. N. and Sanchez, A. (2006), 'Effects of substrate and temperature on growth of *Aspergillus flavus* in peanuts from Georgia', *Georgia Journal of Science*, 64, 76–80.

Adams, J. F. and Bowen, K. L. (1993), Peanut quality as affected by gypsum and soil calcium, *Agronomy Abstracts*, 265.

Afri-Nut.(2016),'AfriNut,pro-poorpeanutprocessinginMalawi',Accessed13September2016at:http://www.inclusivebusinesshub.org/project/aproject-profile-afrinut-pro-poor-peanut-processing-in-malawi/

Ahuja, I., Kissen, L. and Bones, A. M. (2012), 'Phytoalexins in defense against pathogens', *Trends in Plant Science*, 17(2), 73–90.

Alva, A. K., Gascho, G. J. and Guang, Y. (1989), 'Gypsum material effects on peanut and soil calcium', *Communications in Soil Science and Plant Analysis*, 20(17–18), 1727–44.

Amoako-Attah, I., Awuah, R. T., Kpodo, K. A., Fialor, S. C. and Jolly, C. M. (2007), 'Cost effectiveness of selected post harvest pod handling techniques against damage, mouldness and aflatoxin contamination of shelled groundnut in Ghana', *Journal of Science and Technology*, 27, 17–27.

Arias, R. S., Sobolev, V. S., Orner, V. A., Dang, P. M. and Lamb, M. C. (2014), 'Potential involvement of *Aspergillus flavus* laccases in peanut invasion at low water potential', *Plant Pathology*, 63, 354–64.

Arunyanark, A., Jogloy, S., Wongkaew, S., Akkasaeng, C., Vorasoot, N., Kesmala, T. and Patanothai, A. (2010), 'Heritability of aflatoxin resistance traits and correlation with drought tolerance traits in peanut', *Field Crops Research*, 117, 258–64.

Azaizeh, A. H. and Pettit, R. E. (1987), 'Influence of tannin-related compounds from peanut seed coats and cotyledons on *Aspergillus parasiticus* growth and aflatoxin production', *Phytopathology*, 77, 1703.

Azaizeh, H. A., Pettit, R. E., Sarr, B. A. and Phillips, T. D. (1990), 'Effect of tannin extracts on growth of *Aspergillus parasiticus* and aflatoxin production', *Mycopathologia*, 110(3), 125–32.

Azaizeh, H. A., Pettit, R. E., Smith, O. D. and Taber, R. A. (1989), Reaction of peanut genotypes under drought stress to *Aspergillus flavus* and *A. parasiticus*', *Peanut Science*, 16, 109–13.

Bhatnagar-Mathur, P., Sunkara, S., Bhatnagar-Panwar, M., Waliyar, F. and Sharma, K. K. (2015), 'Biotechnological advances for combating *Aspergillus flavus* and aflatoxin contamination in crops', *Plant Science*, 234, 119–32.

Blankenship, P. D., Cole, R. J. and Sanders, T. H. (1985), 'Comparative susceptibility of four experimental peanut lines and the cultivar Florunner to preharvest aflatoxin contamination,' *Peanut Science*, 12(2), 70–2.

Blankenship, P. D., Cole, R. J., Sanders, T. H. and Hill, R. A. (1983), Environmental control plot facility with manipulable soil temperature', *Oleagineux*, 38, 615–20.

Blankenship, P. D., Cole, R. J., Sanders, T. H. and Hill, R. A. (1984), 'Effect of geocarposphere temperature on preharvest colonization of drought-stressed peanuts by *Aspergillus flavus* and subsequent aflatoxin contamination', *Mycopathologia*, 85, 69–74.

Bowen, K. L. (2009), 'Risk factors for pre-harvest aflatoxin contamination of peanuts'. *Proceedings American Peanut Research and Education Society*, 41, 75–6.

Bowen, K. L. and Hagan, A. K. (2015), Temperature and moisture conditions that affect aflatoxin contamination of peanuts', *Peanut Science*, 42, 121–7.

Bowen, K. L. and Mack, T. P. (1993), 'Relationship of damage from the lesser cornstalk borer to *Aspergillus flavus* contamination in peanuts', *Journal Entomological Science*, 28, 29–42.

Bowen, K. L., Adams, J. F. and Bahaminyakamwe, L. (1996), 'Calcium and pH effects on *Aspergillus flavus* invasion and aflatoxin contamination of peanut', *Proceedings American Peanut Research and Education Society*, 28, 29.

[CAST] Council for Agricultural Science and Technology. (2003), 'Mycotoxins – risks in plant, animal and human systems', *Task Force Report*, 139, 1–191.

Chamberlin, K. D., Melouk, H. A. and Payton, M. E. (2010), 'Evaluation of the U. S. peanut mini-core collection using a molecular marker for resistance to *Sclerotinia minor* Jagger', *Euphytica*, 172, 109–15.

Chapin, J. W., Dorner, J. W. and Thomas, J. S. (2004), Association of a burrower bug (Heteroptera: Cydnidae) with aflatoxin contamination of peanut kernels', *Journal of Entomological Science*, 39, 71–83.

Chari, M., Gupta, K., Prasad, T. G., Sastry, K. S. and Udaya, K. M. (1986), 'Enhancement of water status by calcium pretreatment in groundnut and cowpea plants subjected to moisture stress', *Plant and Soil*, 91, 109–14.

Charmley, L. L., Rosenberg, A. and Trenholm, H. L. (1994), 'Factors responsible for economic losses due to fusarium mycotoxins of grains, foods, and feedstuffs', *Mycotoxins in Grain: Compounds other than Aflatoxin*. The American Association of Cereal Chemists, Inc., St. Paul, 471–86.

Chauhan, Y. S., Wright, G. C., Rachaputi, R. C. N., Holzworth, D., Broome, A., Krosch, S. and Robertson, M. J. (2010), 'Application of a model to assess aflatoxin risk in peanuts', *Journal of Agricultural Science*, 148, 341–51.

Cole, R. J. (1989), 'Technology of aflatoxin decontamination', *Mycotoxin and Phycotoxin*, 177–84.

Cole, R. J., Sanders, T. H., Hill, R. A. and Blankenship, P. D. (1985), 'Mean geocarposphere temperatures that induce preharvest aflatoxin contamination of peanuts under drought stress', *Mycopathologia*, 91, 41–6.

Cotty, P. J. and Jaime-Garcia, R. (2007), 'Influences of climate on aflatoxin producing fungi and aflatoxin contamination', *International Journal of Food Microbiology*, 119(1–2), 109–15.

Cox, F. R., Sullivan, G. A. and Martin, C. K. (1976), 'Effect of calcium and irrigation treatments on peanut yield, grade and seed quality', *Peanut Science*, 3, 81–5.

Craufurd, P. Q., Prasad, P. V. V., Waliyar, F. and Taheri, A. (2006), 'Drought, pod yield, pre-harvest *Aspergillus* infection and aflatoxin contamination on peanut in Niger', *Field Crops Research*, 98, 20–9.

Damicone, J. P., Holbrook, C. C., Smith, D. L., Melouk, H. A. and Chamberlin, K. D. (2010), 'Reaction of the core collection of peanut germplasm to Sclerotinia blight and pepper spot', *Peanut Science*, 37, 1–11.

Dash, S., Cannon, E. K. S., Kalberer, S. R., Farmer, A. D., and Cannon, S. B. (2016), 'PeanutBase and other bioinformatic resources for peanut', In: Stalker, H. T. and Wilson, R. F. (eds), *Penauts: Genetics, Processing, and Utilization, AOCS Monograph on Oilseeds*, pp. 253–66.

Diao, E., Dong, H., Hou, H., Zhang, Z., Ji, N. and Ma, W. (2015), 'Factors influencing aflatoxin contamination in before and after harvest peanut: A review', *Journal of Food Research*, 4, 148–53.

Dieckert, M. C. and Dieckert, J. W. (1977), 'Genetically determined structural parameters of the seed coat affecting the colonization of peanut seeds by aflatoxin producing *Aspergilli*', *Annales de Technologie Agricole*, 26(3), 353–66.

Diener, U. L. and Davis, N. D. (1977), Aflatoxin formation in peanut by *Aspergillus flavus*', Alabama Agricultural Experiment Station Bulletin No. 93.

Dorner, J. W., Cole, R. J., Sanders, T. H. and Blankenship, P. D. (1989), 'Interrelationship of kernel water activity, soil temperature, maturity, and phytoalexin production in preharvest aflatoxin contamination of drought-stressed peanuts', *Mycopathologia*, 105, 117–28.

Dorner, J. W. (2008), 'Management and prevention of mycotoxins in peanuts', *Food Additives and Contamination*, 25, 203–8.

Dorner, J. W. (2009), 'Development of biocontrol technology to manage aflatoxin contamination in peanuts', *Peanut Science*, 36, 60–7.

Dorner, J. W., Cole, R. J., Connick, W. J., Daigle, D. J., McGuire, M. R. and Shasha, B. S. (2003), 'Evaluation of biological control formulations to reduce aflatoxin contamination in peanuts', *Biological Control*, 26, 318–24.

Doupnik, B. (1969), 'Aflatoxins produced on peanut varieties previously reported to inhibit production', *Phytopathology*, 59(10), 1554.

Ebrahimi, T., Khomeiri, M., Maghsoudlou, Y. and Ahmadi, G. M. (2009), 'Relation between chemical compositions of peanut with its resistance to *Aspergillus flavus*', *Journal Agricultural Science*, http://en.journals.sid.ir/ViewPaper.aspx?ID=176745.

Ellis, W. O., Smith, J. P., Simpson, B. K., Khanizadeh, S. and Oldham, J. H. (1993), 'Control of growth and aflatoxin production of *Aspergillus flavus* under modified atmosphere packaging (MAP) conditions', *Food Microbiology*, 10, 9–21.

Ellis, W. O., Smith, J. P., Simpson, B. K., Ramaswamy, H. and Doyon, G. (1994), 'Growth of and aflatoxin production by Aspergillus flavus in peanuts stored under modified atmosphere packaging (MAP) conditions', *International Journal of Food Microbiology*, 22, 173–87.

Ehrlich, K. C. (2014), 'Non-aflatoxigenic Aspergillus flavus to prevent aflatoxin contamination in crops: advantages and limitations', *Frontiers in Microbiology*, 5, 50.

Ferguson, M. E., Upadhyaya, H. D. and Bramel, P. J. (2001), 'Status of the *Arachis* germplasm collection at ICRISAT', *Peanut Science*, 28(2), 89–96.

Gajate-Garrido, G., Hoffmann, V., Kanyam, D., Magnan, N. and Opoku, N. (2016), 'Technological and market interventions to reduce aflatoxin in Ghanaian groundnuts', unpublished data.

Girdithai, J. S., Vorasoot, N., Akkasaeng, C., Wongkaew, S., Holbrook, C. C. and Patanothai, A. (2010a), 'Associations between physiological traits for drought tolerance and aflatoxin contamination in peanut genotypes under terminal drought', *Plant Breeding*, 129, 693–9.

Girdithai, J. S., Vorasoot, N., Akkasaeng, C., Wongkaew, S., Holbrook, C. C., Patanothai, A. (2010b), 'Heritability of, and genotypic correlations between, aflatoxin traits and physiological traits for drought tolerance under end of season drought in peanut (*Arachis hypogaea* L.)', *Field Crops Research*, 118, 169–76.

Gong, Y. Y., Cardwell, K., Hounsa, A., Egal, S., Turner, P. C., Hall, A. J. and Wild, C. P. (2002), 'Dietary aflatoxin exposure and impaired growth in young children from Benin and Togo: cross sectional study', *British Medical Journal*, 325 (7354), 20–1.

Guo, B., Chen, C. Y., Lee, R. D. and Scully, B. T. (2008), Drought stress and preharvest aflatoxin contamination in agricultural commodity: genetics, genomics and proteomics', *Journal of Integrated Plant Biology*, 50, 1281–91.

Guo, B. Z., Fedorova, N. D., Chen, X. P., Wan, C. H., Wang, W., Nierman, W. C., Bhatnagar, D. and Yu, J. J. (2011), 'Gene expression profiling and identification of resistance genes to *Aspergillus flavus* infection in peanut through EST and microarray strategies', *Toxins*, 3(7), 737–53.

Hagan, A. K., Campbell, H. L., Weeks, J. R., Rivas-Davila, M. E., and Gamble, B. (2003), 'Impact of bahiagrass, cotton, and corn cropping frequency on the severity of diseases of peanut', Proceedings Sod Based Cropping Systems Conference, North Florida Research and Extension Center, http://citeseerx.ist.psu.edu/viewdoc/download?doi=10.1.1.503.9424&rep=rep1&type=pdf

Hamidou, F., Rathore, A., Waliyar, F. and Vadez, V. (2014), 'Although drought intensity increases aflatoxin contamination, drought does not lead to less aflatoxin contamination', *Field Crops Research*, 156, 103–10.

Hammerschmidt, R. (1999), 'Phytoalexins: what have we learned after 60 years?', *Annual Review of Phytopathology*, 37, 285–306.

Hell, K. and Mutegi, C. (2011), 'Aflatoxin control and prevention strategies in key crops of Sub-Saharan Africa', *African Journal of Microbiology Research*, 5(5), 459–66.

Hell, K., Cardwell, K. F. and Poehling, H. M. (2003), 'Relationship between management practices, fungal infection and aflatoxin for stored groundnuts in Benin', *Journal Phytopatology*, 151, 690–8.

Hell, K., Mutegi, C. and Fandohan, P. (2010), 'Aflatoxin control and prevention strategies in maize for sub-Saharan Africa', In *10th International Working Conference on Stored Product Protection*, pp. 534–41.

Henderson, C. E., Porter, W. D., McClendon, R. W. and Hoogenboom, G. (2000), 'Predicting aflatoxin contamination in peanuts: a genetic algorithm/neural network approach', *Applied Intelligence*, 12(3), 183–92.

Hill, R. A., Blankenship, P. D., Cole, R. J. and Sanders, T. H. (1983), 'Effects of soil moisture and temperature on preharvest invasion of peanuts by the *Aspergillus flavus* group and subsequent aflatoxin development', *Applied Environmental Microbiology*, 45, 628–33.

Holbrook, C. C., Kvien, C. K., Rucker, K. S., Wilson, D. M., Hook, J. E. and Matheron, M. E. (2000), 'Preharvest aflatoxin contamination in drought-tolerant and drought-intolerant peanut genotypes', *Peanut Science*, 27, 45–8.

Holbrook, C. C., Guo, B. Z., Wilson, D. M. and Timper, P. (2009). 'The U. S. breeding program to develop peanut with drought tolerance and reduced aflatoxin contamination', *Peanut Science*, 36, 50–3.

Holbrook, C. C. and Dong, W. (2005), 'Development and evaluation of a mini core collection for the U. S. peanut germplasm collection', *Crop Science*, 45(4), 1540–4.

Holbrook, C. C., Ozias-Akins, P., Timper, P., Wilson, D. M., Cantonwine, E., Gou, B. Z., Sullivan, D. G. and Dong, W. (2008), Research from the Coastal Plain Experiment Station, Tifton, Georgia, to minimize aflatoxin contamination in peanut', *Toxin Reviews*, 27, 391–410.

Holland, K. W. and O'Keef, S. F. (2010), 'Recent applications of peanut phytoalexins', *Recent Patents on Food Nutrition*, 2(3), 221–32.

Huang, L., Ren, X. P., Zhang, X. J., Chen, Y. N. and Jiang, H. F. (2012), 'Association analysis of agronomic traits and resistance to *Aspergillus flavus* in the ICRISAT peanut mini-core collection', *Acta Agronomica Sinica*, 38(6), 935–46.

Humphrey, J. and Navas-Alemán, L. (2010), 'Value chains, donor interventions and poverty reduction: A review of donor practice', *IDS Research Reports*, 63, 1–106.

[IARC] International Agency for Research on Cancer. (1993), 'Some naturally occurring substances: Food items and constituents, heterocyclic aromatic amines and mycotoxins', IRAC Monographs on the Evaluation of Carcinogenic Risks to Humans, 56, 599pp.

[IARC] International Agency for Research on Cancer. (2015), 'Human exposure to aflatoxins and fumonisins', IRAC Monographs on the Evaluation of Carcinogenic Risks to Humans.

[ICRISAT] International Crops Research Institute for the Semi-Arid Tropics. (2016), 'First mobile low-cost device developed for rapid detection of aflatoxins', Hyderabad, India and Lilongwe, Malawi: ICRISAT, 26 July 2016.

Jelinek, C. F., Pohland, A. E. and Wood, G. E. (1989), 'Worldwide occurrence of mycotoxins in foods and feeds, an update', *Journal of the Association of Analytical Chemistry*, 72, 223–30.

Jeranyama, P., Waddington, S. R., Oran, B., Hesterman, O. B. and Harwood, R. R. (2007), 'Nitrogen effects on maize yield following groundnut in rotation on smallholder farms in sub-humid Zimbabwe', *African Journal of Biotechnology*, 6(13), 1503–8.

Jiang, H. F., Ren, X. P., Wang, S. Y., Zhang, X. J., Huang, J. Q., Liao, B. S., Holbrook, C. C. and Upadhyaya, H. D. (2010), 'Development and evaluation of peanut germplasm with resistance to *Aspergillus flavus* from core collection', *Acta Agronomica Sinica*, 36(3), 428–34.

Jolly, P. E., Jiang, Y., Ellis, W. O., Sheng-Wang, J., Afriyie-Gyawu, E., Phillips, T. D. and Williams, J. H. (2008), Modulation of the human immune system by aflatoxin, *Mycotoxins: Detection Methods, Management*, Public Health and Agricultural Trade, CAB International, UK, pp. 41–52.

Kambiranda, D. M., Vasanthaiah, H. K. N., Ananga, R. K. A., Basha, S. M. and Naik, K. (2011), 'Impact of drought stress on peanut (*Arachis hypogaea* L .) productivity and food safety', In: Vasanthaiah, H. and Kambiranda, D. M. (eds), *InTech, Plants and Environment, Rijeka, Croatia*, pp. 249–72.

Kensler, T. W., Roebuck, B. D., Wogan, G. N. and Groopmant, J. D. (2011), 'Aflatoxin: A 50-year odyssey of mechanic and tranlational', *Toxicological Sciences*, 120, S28–48.

Liu, J., Sun, L., Zhang, N., Zhang, J., Guo, J., Li, C., Rajput, S. A. and Qi, D. (2016), 'Effects of nutrients in substrates of different grains on aflatoxin B1 production by *Aspergillus flavus*', Biological Medical Research Institute, https://www.hindawi.com/journals/bmri/2016/7232858/abs/.

Lynch, R. E. and Wilson, D. M. (1991), 'Enhanced infection of peanut, *Arachis hypogaea* L., seed with *Aspergillus flavus* group fungi due to external scarification of peanut pods by the lesser cornstalk borer, *Elasmopalpus lignosellus* (Zeller)', *Peanut Science*, 18, 110–16.

Ma, H., Zhang, N., Sun, L. and Qi, D. (2015), 'Effects of different substrates and oils on aflatoxin B_1 production by *Aspergillus parasiticus*', *European Food Research and Technology*, 240(3), 627–34.

Magan, N., Cayley, G. R. and Lacey, J. (1984), 'Effect of water activity and temperature on mycotoxin production by *Alternaria alternata* in culture and wheat grain', *Applied and Environmental Microbiology*, 47, 1113–17.

McDonald, D. (1964). 'Effects of drought stress on root growth characteristics of peanut during mid-to-late growth stages', *Samaru Agricultural Newsletter of Nigeria*, 6(71), 153–7.

McDonald, D. and Harkness, C. (1967), 'Aflatoxin in the groundnut crop at harvest in Northern Nigeria', Tropical Science, 9, 148–61.

Mehan, V. K., McDonald, D. and Rajagopalan, K. (1987), 'Resistance of peanut genotypes to seed infection by *Aspergillus flavus* in field trials in India', *Peanut Science*, 14(1), 17–21.

Mixon, A. C. (1983), 'Two peanut germplasm lines, GFA-1 and GFA-2', *Crop Science*, 23(5), 1020–1.

Mixon, A. C. (1986), 'Reducing *Aspergillus* species infection of peanut seed using resistant genotypes', *Journal of Environmental Quality*, 15(2), 101–3.

Minton, N. A., Bell, D. K. and Doupnik, Jr., D. (1969), 'Peanut pod invasion by *Aspergillus flavus* in the presence of *Meloidogyne hapla*', *Journal of Nematology*, 1, 318–20.

Mixon, A. C. and Branch, W. D. (1985), 'Agronomic performance of a Spanish and runner cultivar harvested at six different digging intervals', *Peanut Science*, 12, 50–4.

Munkvold, G. P. (2003), 'Cultural and genetic approaches to managing mycotoxins in maize', *Annual Review Phytopathology*, 41(10), 99–116.

Mutegi, C. K., Ngugi, H. K., Hendriks, S. L. and Jones, R. B. (2009), 'Prevalence and factors associated with aflatoxin contamination of peanuts from western Kenya', *International Journal of Food Microbiology*, 130, 27–34.

Nakai, V. K., de O. Rocha, L., Goncalez, E., Fonseca, H., Ortega, E. M. M. and Correa, B. (2008), 'Distribution of fungi and aflatoxins in a stored peanut variety', *Food Chemistry*, 106(1), 285–90.

Nigam, S. N., Waliyar, F., Aruna, R., Reddy, S. V., Kumar, P. L., Crauford, P. Q., Diallo, A. T., Ntare, B. R. and Upadhyaya, H. D. (2009), 'Breeding peanut for resistance to aflatoxin contamination at ICRISAT', *Peanut Science*, 36, 42–9.

Olwari, F., Bisikwa, J., Kaaya, A. N. and Okello, D. K. (2013), 'Tolerance levels of peanut varieties against *Aspergillus flavus* infection', *Journal of Plant Pathology and Microbiology*, doi:10.4172/2157–7471.1000195.

Okoth, S., Nyongesa, B., Ayugi, V., Kang'ethe, E., Korhonen, H. and Joutsjoki, V. (2012), 'Toxigenic potential of Aspergillus species occurring on maize kernels from two Agro-ecological zones in Kenya', *Toxins*, 4, 991–1007.

Parmar, R. S., McClendon, R. W., Hoogenboom, G., Blankenship, P. D., Cole, R. J. and Dorner, J. W. (1997), 'Estimation of aflatoxin contamination in preharvest peanuts using neural networks', *Transactions of the ASAE*, 40(3), 809–13.

Patee, H. E., Young, J. L., Pearson, J. L., Singleton, J. A. and Giesbrecht, F. G. (1982), 'Storage and moisture effects on peanut composition and roasted flavor', *Peanut Science*, 9, 98–101.

Payne, G. A. (2016), 'Mycotoxins and product safety, peanuts: genetics, processing, and utilization', *AOCS Monograph Series on Oil Seeds*, 347–61.

Philippe, J. (2015), 'Phytoalexins: current progress and future prospects', *Molecules*, 20, 2770–4.

Pitt, J. I., Hocking, A. D., Budhasamai, K., Miscamble, B. F, Wheeler, K. A. and Tanboon, P. (1994), 'The normal mycoflora of commodities from Thailand: 2- Beans, rice, small grains and other commodities', *International Journal of Food Microbiology*, 23, 35–53.

Pitt, J. I., Taniwaki, M. H. and Cole, M. B. (2013), 'Mycotoxin production in major crops as influenced by growing, harvesting, storage and processing, with emphasis on the achievement of Food Safety Objectives', *Food Control*, 32, 205–15.

[PMIL] Peanut and Mycotoxin Innovation Lab. (2016), http://pmil.caes.uga.edu/documents/UGA136/ManagementOfAflatoxinsInGroundnutGuideUganda2010.pdf.

Prasad, P. V. V., Staggenborg, S. A. and Ristic, Z. (2008), 'Impacts of drought and/or heat stress on physiological, developmental, growth, and yield processes of crop plants', *Advances in Agricultural Systems Modeling Series* 1, 301–56.

Rachaputi, R., Krosch, S. and Wright, G. (2002), 'Management practices to minimize pre-harvest aflatoxin contamination in Australia peanuts', *Animal Production Science*, 42(5), 595–605.

Rai, M. and Varma, A. (2010), 'Mycotoxins in food, feed and bioweapons', DOI: 10.1007/978-3-642-00725-5_2, Springer, NY.

Rajasekaran, K., de Lucca, A. J. and Cary, J. W. (2009), 'Aflatoxin control through transgenic approaches', *Toxin Reviews*, 28, 89–101.

Reddy, K. R. N., Reddy, C. S. and Muralidharan, K. (2009), 'Detection of *Aspergillus* spp. and aflatoxin B1 in rice in India', *Food Microbiology*, 26, 27–31.

Reddy, E. C. S., Sudhakar, C., Reddy, N. P. E., Reddy, B. R., Yellareddyygari, S. K. R., Reddy, M. S., Kumar, K. V. K., Zhou, X. G., Yuan, Z. L., Yang, Z. L., Qi, W. and Du, B. H. (2011), 'Studies on pre-harvest aflatoxin contamination in peanut and its integrated management using host plant resistance', 2nd Asian Plant Growth-Promoting Rhizobacteria Conference, 2, 556.

Reddy, T. Y., Sulochanamma, B. N., Subramanyam, A. and Balaguravaiah, D. (2003), 'Influence of weather, dryspells and management practices on aflatoxin contamination in groundnut', *Indian Phytopathology*, 56(3), 262–5.

Sanders, T. H., Blankenship, P. D., Cole, R. J. and Hill, R. A. (1984), 'Effect of soil temperature and drought on peanut pod and stem temperatures relative to *Aspergillus flavus* invasion and aflatoxin contamination', *Mycopathologia*, 86, 51–4.

Sanders, T. H., Cole, R. J., Blankenship, P. D. and Hill, R. A. (1985), 'Relation of environmental stress duration to *Aspergillus flavus* invasion and aflatoxin production in preharvest peanuts', *Peanut Science*, 12, 90–3.

Sanders, T. H., Hill, R. A., Cole, A. J. and Blankenship, P. D. (1981), 'Effect of drought on occurrence of *Aspergillus flavus* in maturing peanuts', *Journal American Oil Chemistry Society*, 58, 966–70.

Sanders, T. H., Cole, R. J., Blankeship, P. D. and Dorner,J. W. (1993), 'Aflatoxin contamination of peanuts from plants drought stressed in pod or root zones', *Peanut Science*, 20, 5–8.

Schilling, R. and Misari, S. M. (1993), 'Assessment of groundnut research achievements in the savannah regions of West Africa', *Oleagineux*, 48(7), 323–32.

Shepard, G. S. (2008), 'Impact of mycotoxins on human heath in developing countries', *Food Additive Contamination*, 25, 146–51.

Skelton, B. J. and Shear, G. M. (1971), 'Calcium translocation in the peanut (*Arachis hypogaea* L.)', *Agronomy Journal*, 63, 409–12.

Smith, D. H., Wells, M. A., Porter, D. M. and Cox, F. R. (1993). Nutrient deficiencies and toxicities in crop plants', In: Peanuts, APS Press, pp. 193–247.

Snigdha, M., Hariprasad, P. and Venkateswaran, G. (2015), 'Transport via xylem and accumulation of aflatoxin in seeds of groundnut plant', *Chemoshere*, 119, 524–9.

Sobolev, V. S. (2008), 'Localized production of phytoalexins by peanut (*Arachis hypogaea*) kernels in response to invasion by *Aspergillus* species', *Journal Agriculture and Food Chemistry*, 56(6), 1949–54.

Sobolev, V. S. (2013), 'Production of phytoalexins in peanut (*Arachis hypogaea*) seed elicited by selected microorganisms', *Journal Agriculture and Food Chemistry*, 61(8), 1850–8.

Sobolev, V. S., Guo, B. Z., Holbrook, C. C. and Lynch, R. E. (2007), 'Interrelationship of phytoalexin production and disease resistance in selected peanut genotypes', *Journal of Agricultural and Food Chemistry*, 55, 2195–2000.

[SRIPMC] Southern Region IPM Center. (2016), 'What is IPM?', http://www.sripmc.org/index.cfm/ipm-in-the-south/what-is-ipm/

Strosnider, H., Azziz-Baumgarthner, E., Banziger, M., Bhat, R. V., Breiman, R., Brune, M. N., DeCock, K., Dilley, A., Groopman, J., Hell, K., et al. (2006), 'Workgroup report: public health strategies for reducing aflatoxin exposure in developing countries', *Environmental Health Perspectives*, 114, 1898–1903.

Sudhakar, P., Latha, P., Babitha, M., Reddy, P. V. and Naidu, P. H. (2007), 'Relationship of drought tolerance traits with aflatoxin contamination in groundnut', *Indian Journal of Plant Physiology*, 12(3), 261–5.

Sullivan, D. G. and Holbrook, C. C. (2007), 'Using ground-based reflectance measurements as selection criteria for drought- and aflatoxin-resistant peanut genotypes', *Crop Science*, 47(3), 1040–50.

Suryanarayana Rao, K. and Tulpule, P. G. (1967), 'Varietal differences of groundnut in the production of aflatoxin', *Nature*, 214(5089), 738–9.

Taligoola, H. K., Ismail, M. A. and Chebon, S. K. (2011), 'Incidence of mycobiota and aflatoxins during storage of paddy and milled rice grown in Uganda', *Journal of Basic and Applied Mycology*, 2, 37–53.

Thai, C. N., Blankenship, P. D., Cole, R. J., Sanders, T. H. and Dorner, J. W. (1990), 'Relationship between aflatoxin production and soil temperature for peanuts under drought stress', *Transactions of the ASAE*, 33, 324–9.

Timper, P., Wilson, D. M., Holbrook, C. C. and Maw, B. W. (2004), 'Relationship between *Meloidogyne arenaria* and aflatoxin contamination in peanut', *Journal of Nematology*, 36, 167–70.

Timper, P., Wilson, D. M. and Holbrook, C. C. (2013), 'Contribution of root-knot nematodes to aflatoxin contamination in peanut (*Arachis hypogaea*)', *Peanut Science*, 40, 31–9.

Torres, A. M., Barros, G. G., Palacios, S. A., Chulze, S. N. and Battilani, P. (2014), 'Review on pre- and post-harvest management of peanuts to minimize aflatoxin contamination', *Food Research International*, 62, 11–19.

Umeh, V. C., Waliyar, F. and Traoré, A. (2000), '*Aspergillus* species colonization of termite-damaged peanuts in parts of West Africa and its control prospects', *Peanut Science*, 27, 1–6.

Upadhyaya, H. D., Bramel, P. J., Ortiz, R. and Singh, S. (2002), 'Developing a mini core of peanut for utilization of genetic resources', *Crop Science*, 42, 2150–6.

VanEtten, H. D., Mansfield, J. W., Bailey, J. A. and Farmer, E. E. (1994), Two classes of plant antibiotics: phytoalexins versus 'phytoanticipins', *Plant Cell*, 6, 1191–2.

Vidhyasekaran, P., Lalithakumari, D. and Covindaswamy, C. V. (1972), 'Production of a phytoalexin in groundnut due to storage fungi', *Indian Phytopathology*, 25, 240–9.

Waddington, S. R. and Karigwindi, J. (2001), 'Productivity and profitability of maize + groundnut rotations compared with continuous maize on smallholder farms in Zimbabwe', *Experimental Agriculture*, 37, 83–98.

Wagacha, J. M. and Muthomi, J. W. (2008), 'Mycotoxin problem in Africa: current status, implications to food safety and health and possible management strategies', *International Journal of Food Microbiology*, 124(1), 1–12.

Waliyar, F. (1997), An overview of research on the management of aflatoxin contamination of groundnut', Working Group Meeting Ministry of Agriculture and Rural Development, Hanoi, Vietnam, ICRISAT, http://oar.icrisat.org/2979/1/An_overview_of_research_on_the_management_of_aflatoxin_contamination_of_groundnut.pdf

Waliyar, F., Osiru, M., Sudini, H. K. and Njoroge, S. (2013), 'Aflatoxins: finding solutions for improved food safety-reducing aflatoxins in groundnuts through Integrated Management and Biocontrol', *International Food Policy Research Institute*, 18.

Wang, H. O., Chen, X.-P., Hu, Z. C. and Xie, H. X. (2015), 'The contamination and control of aflatoxin in peanut', *Jiangsu Agricultural Science*, 43(1), 270–3.

Wang, T., Zhang, E. E., Chen, Z. P., Li, L. and Liang, X. Q. (2010), 'Identification of seed proteins associated with resistance to pre-harvest aflatoxin contamination in peanut (*Arachis hypogaea* L.)', *BMC Plant Biology*, 10(267), 267.

Wang, T., Chen, X. P., Liu, H. Y., Hong, Y. B., Yang, Q. L., Chi, X. Y., Yang, Z., Yu, S. L., Li, L. and Liang. X. Q. (2013), 'Transcriptome identification of the resistance-associated genes (RAGs) to *Aspergillus flavus* infection in pre-harvested peanut (*Arachis hypogaea*)', *Functional Plant Biology*, 40(3), 292–303.

Wareing, P. W. (1997), 'Incidence and detection of thermotolerant and thermophilic fungi from maize with particular reference to *Thermoascus* spp.', *International Journal of Food Microbiology*, 35(2), 135–47.

Whitaker, T. B., Dorner, J. W., Lamb, M. and Slate, A. B. (2005), 'The effect of sorting farmers' stock peanuts by size and color on partitioning aflatoxin into various shelled peanut grade sizes', *Peanut Science*, 32, 103–18.

Whitaker, T. B. (2003), 'Standardisation of mycotoxin sampling procedure: An urgent necessity', *Food Control*, 14, 233–7.

Wild, C. P. and Gong, Y. Y. (2010), 'Mycotoxins and human disease: a largely ignored global health issue', *Carcinogenesis*, 31, 71–82.

Wild, C. P., Miller, J. D. and Groopman, J. D. (2015), Mycotoxin control in low- and middle income countries', IARC Working Group Report No. 9., International Agency for Research on Cancer, France.

Williams, T. H., Phillips, T. D., Jolley, P. E., Stiles, J. K., Jolly, C. M. and Aggarwal, S. (2004), 'Human aflatoxin developing countries: a review of toxicology, exposure, potential health consequences, and interventions', *American Journal of Clinical Nutrition*, 80, 1106–22.

Wilson, D. M. and Stansell, J. R. (1983), 'Effect of irrigation regimes on aflatoxin contamination of peanut pods', *Peanut Science*, 10, 54–6.

Wilson, D. M. and M. E. Walker. (1988), 'Effects of gypsum and irrigation on *Aspergillus flavus* group colonization of peanuts', *Proceedings of the American Peanut Research and Education Society*, 20, 24.

Wotton, H. R. and Strange, R. N. (1987), 'Increased susceptibility and reduced phytoalexin accumulation in drought-stressed peanut kernels challenged with *Aspergillus flavus*', *Applied and Environmental Microbiology*, 53(2), 270–3.

Wu, F. and Guclu, H. (2012), 'Aflatoxin regulations in a network of global maize trade', *PLoS ONE*, 7, 9.

Wu, F., Bhatnagar, D., Bui-Klimke, T., Carbone, I., Helmich, R. L., Munkvold, G. P., Paul, P. Payne, G. and Takel, E. S. (2011), 'Climate change impacts on mycotoxin risks in US maize', *World Mycotoxin Journal*, 4, 79–93.

Zhang, H. L., Scharfenstein, L., Zhang, D. H., Chang, P. K., Montalbano, B. G., Guo, B. Z., Meng, X. J. and Yu, J. J. (2015), 'Peanut resistance gene expression in response to *Aspergillus flavus* infection during seed germination', *Journal of Phytopathology*, 163(3), 212–21.

Part 2

Cultivation of cowpea, faba beans and pigeonpea

Chapter 10

Breeding improved varieties of cowpea

B. B. Singh, G.B. Pant University of Agriculture and Technology, India

1. Introduction
2. Production constraints
3. Cowpea breeding programmes and past challenges
4. Cowpea breeding at IITA
5. Highlights of progress made
6. Cowpea international trials
7. Future trends and conclusion
8. Where to look for further information
9. References

1 Introduction

Cowpea (*Vigna unguiculata* (L.) Walp) is a versatile food legume widely cultivated in the tropics and subtropics covering over 65 countries. With up to 30% protein in its grain, tender leaves and immature pods, cowpea is consumed in a variety of ways and it has been given indigenous names such as 'lobia', 'chowlee' and many other names in India; 'kunde' in East Africa; 'beans' and 'wake' in Nigeria; 'niebe' in francophone Africa; 'southern pea', 'crowder pea' and 'black eye pea' in the United States; 'feijão caupe' in Brazil; and a host of other names in local languages in different countries around the world. Current estimates indicate that it is grown in about 14.5 million hectares with an annual production of over seven million tons on a worldwide basis (Singh, 2014). The important cowpea-growing countries are Nigeria, Niger Republic, Mali, Burkina Faso, Senegal, Ghana, Togo, Benin, Cameroon and Chad in the West and Central Africa; Sudan, Somalia, Kenya, Malawi, Uganda, Tanzania, Zambia, Zimbabwe, Botswana and Mozambique in East and Southern Africa; India, Bangladesh, Nepal, Myanmar, Sri Lanka, Indonesia, China and Philippines in Asia, and Brazil, Cuba, Haiti and the West Indies in Central and South America. However, the bulk of cowpea production comes from the drier regions of northern Nigeria (five million ha and 2.9 million tons), Niger Republic (four million ha and 1.5 million tons) and North East Brazil (about 1.9 million ha and 0.8 million tons).

http://dx.doi.org/10.19103/AS.2017.0023.29

1.1 Origin

The distribution of wild cowpea species and genetic diversity indicate that cowpea originated in Southern African region and first domesticated as crop in Africa about 1700–1500 BCE. Cowpea was subsequently taken from Africa to Asia and Europe by the early travellers and explorers (Padulosi, 1993 and Padulosi and NG, 1997). It has been suggested that cowpea probably moved from eastern Africa to India before 150 BCE (Steel and Mehra, 1980), and to West Asia and Europe about 300 BCE. Since the climatic conditions in South Asia and South East Asia were highly suitable for cowpea growth and development, a great deal of genetic variability and selection occurred over time making this region as the secondary centre of cowpea diversity. Cowpea came to the Americas from Africa through Jamaica about 1675 by the slave traders carried as food for their captive slaves on the way to Americas. Due to its adaptability to the warm weather of the tropics, cowpea cultivation spread in the West Indies. From there it was brought to Florida about 1700 and subsequently spread to several southern and central states in the United States primarily as a fodder, cover and food crop. The name 'cowpea' is of American origin and it appeared in print in 1798.

1.2 Versatility

Cowpea is tolerant to drought and heat and it has the ability to fix nitrogen even in very poor soils with a pH as low as 4.5, organic matter below 0.2% and sand content of over 85% (Kolawale et al., 2000, Sanginga et al., 2000). Also, it is shade-tolerant and, therefore, very compatible as an intercrop with a number of cereals and root crops, as well as with cotton, sugarcane and several plantations tree crops (Singh and Emechebe, 1998). Coupled with these attributes, its quick growth and rapid ground cover have made cowpea an essential component of sustainable subsistence agriculture in marginal drier regions of the tropics where rainfall is erratic and scanty and soils are sandy with little organic matter (Carsky et al., 2001, Mortimore et al., 1997).

1.3 Regional preferences

Unlike other crops, cowpea has diverse plant types, pod types, seed types, growth habit, and time to maturity and it is grown and used in many different ways. These attributes have been amply exploited by farmers in selecting suitable varieties for different regions. For example, in West Africa the preferred varieties are photosensitive spreading types with large white and brown seeds and a rough seed coat. In contrast, countries in Central America and the Caribbean prefer non-photosensitive varieties with red, black or white seeds and a smooth coat. The preference in Mexico, Guatemala, Nicaragua, Costa Rica and Cuba is for black seeds, while in Honduras, El Salvador, Venezuela and Jamaica it is for red. In East Africa and Asia, any colour other than black is acceptable, but tan and red are preferred in East Africa and white and cream in Asia. Varieties in these regions are erect, semi-determinate and non-photosensitive. The preference for a rough coat in West Africa is related to the plant's uses here. About 50% of the region's crop is used for snack foods, for which the seed coat must be removed before making a paste of the cotyledons. In the absence of mechanical devices to remove seed coats, the West African women soak the seeds in water for a few minutes. The rough seed coat absorbs water faster than the cotyledons, such that a gently rubbing by hand easily removes it. This is not possible if the seed coat is smooth. However, smooth seed coat is desirable if cowpeas are eaten as boiled beans, as is the case in East and Southern Africa and parts of Asia and Latin America.

2 Production constraints

In spite of its wide cultivation and importance, the overall productivity of cowpea has been very low with average yield particularly in Africa ranging from 100 to 400 kg/ha. This is due to several biotic and abiotic constraints as well as due to cultivation of cowpea as an intercrop with cereals in marginal environments without inputs (Henriet et al., 1997, Singh, 2005). Also, the local varieties have low yield potential because of their prostrate growth habit and late maturity.

2.1 Biotic constraints

Cowpea is attacked by several diseases, insect pests and parasitic plants (Singh and Allen, 1979). The major diseases are anthracnose, web blight, brown blotch, *Cercospora* leaf spot, *Septoria*, scab and *Macrophomina* caused by fungi; bacterial pustule and bacterial blight caused by bacteria; and cowpea yellow mosaic, cowpea aphid-borne mosaic, blackeye cowpea mosaic, cowpea severe mosaic and southern bean mosaic caused by viruses. Nematodes are important in some areas and parasitic weeds such as *Striga gesnerioides* and *Alectra vogelii* are important in Africa (Singh and Emechebe, 1990, Lane et al., 1994). *Striga* causes severe damage to cowpeas in the Sudan savanna and Sahel of West Africa whereas *Alectra* is more prevalent in the Guinea and Sudan savannas of West and Central Africa. *Alectra* is also widespread in Eastern and Southern Africa but *Striga* is not a problem there.

The major insect pests of cowpea are aphid (*Aphis craccivora*), thrips (*Megalurothrips sjostedti,*) Maruca pod borer (*Maruca vitrata*), a complex of pod-sucking bugs (*Clavigralla* spp., *Acanthomia* spp., *Riptortus* spp.) and the storage weevil (*Callosobruchus maculatus*). Of these, thrips and *Maruca* cause major damage in sub-Saharan Africa. There are some location-specific insect pests such as *Lygus* in Americas, bean fly in Asia and East Africa and Ootheca beetles in wetter regions of the tropics.

2.2 Abiotic constraints

Cowpea suffers considerable yield reduction from erratic rainfall in the beginning and towards the end of the rainy season which are common in the semi-arid tropics. Early maturing cultivars escape terminal drought but if exposed to intermittent moisture stress during the vegetative or reproductive stages, they perform very poorly. Cowpea is inherently more drought tolerant than other crops but it still suffers considerable damage due to frequent drought in the Sahelian region where rainfall is scanty and irregular. Low soil fertility due to low organic matter and low phosphorus in the savanna soils is also a major constraint for cowpea production.

3 Cowpea breeding programmes and past challenges

Limited programmes for cowpea improvement began in many countries from early 1900s which resulted in identification of promising land races and varieties which became popular in varied cropping systems. However, systematic cowpea breeding programmes and sustained efforts continued in only a few countries including India, Nigeria, Senegal and the United States where greater resources were allocated from

1960 onwards. Cowpea breeding received a big push and international attention from 1967 onwards after establishment of the International Institute of Tropical Agriculture (IITA) which was tasked with a global mandate for cowpea research and development. A comprehensive cowpea breeding programme was initiated at IITA from 1970 when a critical mass of scientists involving breeders, agronomists, microbiologists, soil scientists, biochemists, food scientists and economists began working as a team in collaboration with national cowpea scientists in Asia, Africa, and Central and South America and IITA emerged as the centre of excellence for cowpea research. Cowpea research received another push from 1980 onwards when the USAID-funded Bean/ Cowpea CRSP (now 'Feed the Future Innovation Lab for Collaborative Research on Grain Legumes') became operational and complemented IITA's efforts in strengthening cowpea research and development in West Africa and East and southern Africa. Since then, the cowpea breeding programme progressed over the years in a logical succession. Initially, from 1970 to 1980, the scientists concentrated on collection and evaluation of cowpea germplasm for desirable plant types and disease resistance. From 1981 to 1990, the focus was on breeding for disease and insect resistance combined with 60–75 day maturity, erect and semi-erect plant types and desirable seed types for different regions. The major thrust in 1990s was to incorporate resistance to nematodes, and parasitic plants, *Striga gesneriodes* and *Alectra vogelii* combined with drought and heat tolerance, high biological nitrogen fixation and efficient acquisition, use of phosphorus from low-P soils and exploring the possibility of biotechnological interventions for cowpea improvement. Concerted efforts in the first decade of the twenty-first century have been to develop improved and intensive cowpea-based cropping systems to maximize the food production per unit area and per unit time by using short-duration cowpeas in intensive and multiple cropping systems with cereals and root crops. Simultaneously studies were initiated on cowpea genomics to identify molecular makers to facilitate marker-assisted selection for *Striga* resistance and to develop a Bt-cowpea for resistance to Maruca pod borer.

4 Cowpea breeding at IITA

The global mandate for cowpea breeding presented a very challenging task to the scientists at IITA because the variety requirements for cowpea differed from region to region in respect of the seed colour preference, use patterns, maturity and growth habit and no single variety could be suitable for all countries. Therefore, IITA located additional scientists in Philippines, Nigeria, Niger, Burkina Faso, Cameroon, Congo and Brazil in order to address the regional constraints in cowpea production. IITA scientists began with the identification of major production constraints in different regions in collaboration with the national partners and defined breeding objectives to address the common constraints across several regions which included late maturity of local varieties, their spreading plant type and susceptibility to many diseases and insects. This was followed by systematic collection, evaluation and screening of cowpea germplasm to identify sources of desirable genes for breeding programme. Cheap, reliable and fast screening methods were developed to screen for resistance to major biotic and abiotic constraints and quality traits. Essential field and green house facilities were developed for advancing 3–4 generations each year for fast-track breeding and multi-location testing of the selected breeding lines to ensure stable performance over diverse environments. Annual workshops were held with national partners and collaborators and a global network for research, testing and release of the improved varieties was developed. Recent reviews by Singh et al. (1997), Hall et al. (1997), Singh et al. (2002, 2005) and Singh (2014) have described cowpea breeding programme and their progress in different regions of the world. IITA continues to be the centre of excellence for cowpea research and the Bean/Cowpea CRSP supported by USAID is also catalysing and supporting research on cowpea improvement. Recently cowpea improvement programmes at the University of California, Riverside and EMBRAPA, Brazil have been strengthened and expanded. Significant research on various aspects of cowpea improvement is also being done in Nigeria, Burkina Faso, Senegal, Mali, and India and to a lesser extent in a number of other countries.

4.1 Breeding objectives

Since cowpea is cultivated in a range of environments and cropping systems and used for grain, leaves, pods, and fodder, no single variety can be suitable for all conditions. Therefore, cowpea breeding is more challenging. Fortunately cowpea germplasm has maximum genetic diversity for plant type, growth habit, maturity and seed type, and it offers an unique opportunity to cowpea breeders to design and develop specific plant types with desired maturity to suit different cropping systems and agro-ecological niches (Singh and Sharma, 1996). Therefore, IITA cowpea breeding programme in collaboration with national programmes has focused on developing a range of high-yielding cowpea varieties with diverse maturity, varied plant types and seed types combined with high protein and resistance to major biotic and abiotic stresses.

4.2 Germplasm resources

Prior to 1967, a few national agricultural research programmes such as Nigeria, Senegal, Tanzania and Uganda in Africa; India in Asia; and the United States in Americas were maintaining some collection of cowpea germplasm. However, with the establishment of the IITA in 1967 and its global mandate for the improvement of cowpea, IITA has collected cowpea germplasm exceeding 15 000 accessions of cultivated varieties drawn from over 100 countries including over 500 accessions of wild cowpeas. These have been characterized and evaluated for desirable traits and being preserved and used in the breeding programme (Ng and Singh, 1997). Extreme cowpea genotypes have been observed with respect to many traits and genetic studies have identified over 207 genes (Ferry, 1985, Ferry and Singh, 1997 and Singh, 2002b) which control plant pigmentation, plant type, plant height, leaf type, growth habit, photosensitivity and maturity, nitrogen fixation, fodder quality, heat and drought tolerances, root architecture, resistance to major bacterial, fungal and viral diseases, resistance to root-knot nematodes, resistance to aphid, bruchid and thrips, and resistance to parasitic weeds such as *Striga gesnerioides* and *Alectra vogelii* pod traits, seed traits and grain quality (Singh and Emechebe, 1990, Singh, 2002a, 2005). Over the years, many of these desirable traits have been bred into improved breeding lines. Limited studies have also been conducted on inter-specific crosses and genetic maps including classical as well as DNA markers (Fatokun and Singh, 1987, Fatokun, 1991, Fatokun, 1997, Fatokun et al., 1993, Ouedraogo et al., 2001, 2002 and Boukar et al., 2004). Quantitative genetic studies on heritability of yield and yield components and protein content, and heterosis for important traits and improved breeding methods have facilitated selection of parents and increased efficiency in breeding programmes (Singh, 2005).

4.3 Variety development and testing

In order to meet the varied requirements of different countries and regions, the general strategy is to develop new cowpea breeding lines with a range of maturity, plant type, seed quality, etc., combining resistance to major diseases, insect pests, *Striga*, *Alectra* and broad adaptability. The selected parents are crossed to obtain segregating populations. Sometimes, partial back crossing is done to bring appropriate traits of the local varieties into new breeding lines. Using a combination of greenhouse and field facilities, three to four generations of breeding materials are advanced each year so that within two years, F_6/F_7 lines combining major attributes become available. The segregating materials are handled by pedigree method and while advancing the generations, these are exposed to major diseases, insect pests and diverse environments. The selected advanced breeding lines are bulked at F_6–F_7 stage are then tested for yield potential and also used as parents in another cycle of crossing programme. The bulked progenies from F_6 to F_7 plants are grouped according to maturity, plant type and seed type and evaluated first in initial evaluation trials (IET) at three locations with single replication in augmented design. Promising lines are then tested in succession through preliminary variety trials (PVT) and advanced variety trials (AVT) at four locations and 3–4 replications. These locations are on a north-south axis covering Nigeria and Niger and represent different agro-ecological zones: i) Ibadan (7 25′N, 3 37E) with relatively high and bimodal rainfall (1500 mm) in the forest/savanna transition zone; ii) Samaru (11 10′N, 7 38E) in the northern Guinea savanna with about 1000 mm rainfall; iii) Minjibir (12 08′N, 8 40E) in the Sudan savanna with about 700 mm rainfall and Toumnia (13 58′N, 9 01E) representing the Sahel with about 350 mm rainfall. The lines with best performance for individual locations as well as across locations are then selected for distribution to national programmes.

5 Highlights of progress made

5.1 Direct selections from the germplasm and early crosses

The initial focus from 1970 to 1980 was to screen germplasm lines and select the most based on multiple disease resistance and agronomic traits. These were directly distributed to international collaborators for broad-based testing. The results from these trials indicated four lines, TVu-201, TVu- 1190, TVu-1977 and TVu-4577 to be resistant to many diseases and had very high yield potential. These were described as VITA-1, VITA-3, VITA-4 and VITA-5 (Vigna IITA-1, 3, 4, 5) respectively and subsequently released in many countries. These VITA lines were also extensively used as parents for the initial crossing programmes and development of segregating populations. The focus was primarily to develop multiple disease-resistant breeding lines with high yield potential. Based on the good performance across many countries, 5 new lines were described as VITA numbers and released in many countries. These were TVx 1193-7D as VITA-6, TVx 289-4G as VITA-7, TVx 66-2H as VITA-8, TVx 1948-01F as VITA-9, and TVx 1836-013J as VITA-10. However, all these VITA lines were primarily released in Asia and South America and not in West Africa where bulk of the cowpea is grown. This was because all these lines (except for VITA-5) had smooth seeds that are not acceptable in West Africa. The desired types in West Africa are brown and white seeds with rough seed coat texture.

5.2 Hybridization and selection for diverse plant type and maturity

Beginning from 1981, the breeding objectives were broadened and a systematic hybridization and selection programme was initiated to develop a diverse set of improved cowpea varieties differing in plant type, growth habit, maturity and seed type to suit the regional preferences and cropping systems.

5.3 Development of '60-day' erect type cowpea varieties

The idea of developing 60-day cowpea varieties came at the Agricultural Research Institute (ARI), Ilonga near Morogoro in Tanzania, where the author was first posted in 1979 as a Grain Legume Breeder by IITA in the IITA/USAID/TANZANIA Project. The region had a bimodal rainfall with a long rainy season from March to June and a short rainy season from October to December. Maize being the major staple of the country, farmers planted maize in both the seasons. The maize crop in the main rainy season performed well but suffered frequent terminal drought during the short rainy season. It was thought that an extra-early (60–70 day) grain legume crop would be ideal for the short rainy season from food security as well as from cereal-legume rotation stand points. He evaluated all the available germplasm lines of cowpea, pigeon pea, soybean, *Phaseolus* bean and mung bean at station and selected some lines of cowpea and mung bean which matured between 60 and 65 days. He tested the selected extra-early cowpea and mung bean lines in the short rainy season in 1980 and found that 60-day cowpeas had had higher yield potential (1.5–2 t/ha) compared to mung beans (0.7–1.2 t/ha). This marked the concept of 60-day cowpeas and a full-fledged breeding programme was launched to develop a range of high-yielding 60-day cowpea varieties with resistance to major diseases and wide adaptability. The author was transferred to IITA Headquarters at Ibadan, Nigeria w.e.f. 1 October 1981. Discussion with international collaborators confirmed the need for developing extra-early cowpea varieties to be grown in areas with short rainy seasons as well as a niche crop in multiple cropping systems in many countries. Based on the genetic variability for maturity in the breeding populations, the earliest lines with good yield potential were between 60 and 69 days in maturity and these were collectively called as extra-early or '60-day' cowpea varieties (Fig. 1). The year-round tropical weather at Ibadan (Nigeria) permitted growing of four crops of early cowpeas each year enabling extremely fast-track progress. A number of promising progenies were selected from the existing segregating populations and planned crosses among selected lines were made. The main focus was early maturity and multiple disease resistance with diverse seed types. Beginning from 1982 early season, a large number of '60-day' cowpea varieties have been developed and distributed to national collaborators. Some of the prominent varieties of this group that have been released and become popular in many countries are IT82E-9 (black), IT 82E-60 (white blackeye), IT82E-16 (red), IT82E-18 (tan), IT82E-32 (red), IT82D-752 (tan), IT82D-789 (light brown), IT82D-889 (red), IT83S-818 (white blackeye), IT85F-867-5, IT86D-1010 (white blackeye), IT93K-452-1(white blackeye), IT97K-1042-3 (red), IT98K-1111-1(white blackeye) and IT98K-205-8 (white small eye). Of these, IT82D-889, IT83S-818, IT85F-867-5 and IT86D-1010 are resistant to all the major viruses

Figure 1 Short-duration (60–70 day) cowpea varieties.

and IT 98K-205-8 is resistant to major viruses as well as to aphid, thrips, bruchid, *Striga* and *Alectra*.

5.4 Medium maturing semi-erect type varieties

In addition to the extra-early varieties, a number of medium maturing varieties (75–80 days) with semi-erect plat type combining multiple pest resistance and diverse seed types were also developed. The traditional cowpea varieties in West Africa are photosensitive late maturing types which are commonly grown as they fit well in relay intercrop in 'millet-sorghum-cowpea, systems in many countries and serve as dual-purpose varieties providing grain as well as fodder. However, these varieties are too late (120–150 days) and often suffer serious yield loss due to terminal drought. Therefore, some of the existing photosensitive varieties were used as parents and crossed and backcrossed to improved early types to select a new set of improved medium maturing varieties (75–100 days) were developed which retained the acceptable seed types and other desirable traits of the traditional varieties. Some of these had combined resistance to major diseases and insects such as aphid, thrips and bruchid. Some of the prominent varieties of this group that have been widely tested, released and become popular in many countries are VITA-1, VITA-3, VITA-4, VITA-5, IT81D-985, IT81D-994, IT84S-2163, IT84S-2246-4, IT84D-449, IT84D-666, IT85F-2020, IT86D-368, IT86D-719, IT87D-697, IT87D-1627, IT88S-574-3, IT89KD-374, IT90K-277-2, IT90K-372-1-2, IT97K-368-18 and IT98K-506-1 IT89KD-245, IT89KD-288, IT89KD-391. Yield potential of selected lines is presented in Table 1.

Table 1 Performance of improved cowpea varieties compared to local varieties and other legumes in Africa

Variety	Grain yield kg/ha	Fodder yield kg/ha
IT97K-568-19	2582	5998
IT90K-277-2	2544	3415
IT97K-568-18	2510	2166
IT98K-205-8	2143	2332
IT95K-627-34	2143	2083
IT86D-719	1788	1208
Dan Ila (local)	1753	2249
Bambaranut	334	1000
Groundnut	421	5165
SED	312	420

5.5 Breeding for resistance to major diseases

Using a combination of field and laboratory screening, a number of new of sources of resistance to major diseases like cercospora, smut, rust, septoria, scab, ascochyta blight and bacterial blight, macrophomina and anthracnose, have been identified and used in the breeding programmes (Abadassi et al., 1987). Several cowpea breeding lines have been developed with combined resistance to cowpea yellow mosaic, blackeye cowpea mosaic and many strains of cowpea aphid-borne mosaic. Among these, IT82D-889, IT83S-818, IT86D-880, IT86D-1010, IT86F 2089-5, IT87D-611-3, IT90K-76, IT90K-277-2 and IT90K-284-2, have been tested in many countries and found resistant (van Boxtel et al., 2000). Good progress has also been made in breeding for resistance to nematodes. Some of the improved breeding lines with nematode resistance are IT849-2049, IT89KD-288, IT86D-634, IT87D-1463, IT95K-398-14, IT96D-772, IT96D-748, IT95K-222-5, IT96D-610, IT87K-818-18 and IT97K-556-4. Among these, IT89KD-288 is a high-yielding variety with high level of resistance to nematodes in Nigeria (Mohamed et al., 2000) and it is the most popular variety in the dry season in northern Nigeria. IT89KD-288 was also found to be resistant to four strains of *Meloidogyne incognita* in the United States (Ehlers et al., 2000).

5.6 *Striga* and *Alectra* resistance

Good progress has been made in genetics and breeding for combined resistance to *Striga* and *Alectra* (Singh and Emechebe 1990, Singh et al., 1993, Atokple et al., 1995, Singh, 2002a), which have recently become important if Africa. A local landrace, B 301 from Botswana, confers complete resistance to *Striga* and *Alectra* in Burkina Faso, Mali, Cameroon, Niger and Nigeria, but only moderate level of resistance to the strain from Benin Republic. Other lines such as IT81D-994, IT89KD-288, 58-57 and Gorom local confer complete resistance to strains from Benin Republic and Burkina Faso. Therefore, by using these complementary parents in crosses, a number of new varieties have been developed with combined resistance to *Alectra* as well as to all the five strains of *Striga*. The most promising new cowpea varieties are IT93K-693-2, IT95K-1090-12, IT97K-499-35, IT97K-497-2, IT97K-819-118, IT03K-325-11, IT03K-378-4, IT03K-364-1, IT04K-217-5 and

IT04K-218-26, which have combined resistance to *Striga* and *Alectra*. Some of these lines are also resistant to bacterial blight, aphid, bruchid, thrips, and viruses with much higher yield potential than the local varieties (Singh, 2002a, Carsky et al., 2003). Most of these lines also serve as a false host for *S. hermonthica*, reducing its seed bank in the soil when grown as intercrop or in rotation with cereals.

5.7 Breeding for resistance to major insects

Insect pests are the major constraint in cowpea production. Several improved cowpea varieties with combined resistance to aphid; thrips and bruchid have been developed (Adjadi et al., 1985, Bata et al., 1987, Singh et al., 1985, Singh et al., 2002). Of these, IT90K-76, IT90K-59, IT 89KD-288 and IT90K 277-2 are already popular varieties in several countries. Among the new varieties IT97K-207-15, IT95K-398-14 and 98K-506-1 have high level of bruchid resistance. Yunes et al. (1998) observed that the 7s-storage protein, 'vicillin', is responsible for bruchid resistance in cowpea lines related to TVu 2027.

Only low level of resistance has been bred for Maruca pod borer and pod bugs. This is because none of the cultivated cowpea germplasm lines and cross-compatible wild cowpeas are resistant to Maruca pod borer. A distant wild relative of cowpea *Vigna vexillata* has shown high level of resistance to Maruca pod borer and bruchid but all the efforts made at IITA to transfer Maruca resistance genes from *Vigna vexillata* to cowpea has not been successful (Fatokun, 1997). However, by evaluating new breeding lines without insecticide sprays and using the best lines in successive cycles of crossing and selection, some progress has been made for field resistance to Maruca and pod bugs. The most promising varieties with some level of field resistance to major insects are IT90K-277-2, IT93K-452-1, IT95K-321-1, IT95K-193-12, IT97K-837 and IT97K-499-38 which yield up to 500 kg/ha without any chemical protection compared to less than 50 kg/ha yields of the local varieties in the same trials. These lines are also resistant to major foliar diseases, aphid, thrips and bruchid with pods at wide angle and suffer less damage due to Maruca. IT94K-437-1 and IT97K-499-38 also have combined resistance to *Striga* and *Alectra*. Developed through conventional breeding approaches, the new field resistant lines require only 1 or 2 sprays of insecticide for normal yield of 1.5–2.5 tons compared to 5–6 sprays needed for the susceptible varieties.

5.8 Pyramiding genes for resistance in improved varieties

Over the years, efforts have been made to systematically incorporate genes for resistance in improved breeding lines as well as in some selected varieties as recurrent parents. Starting with a susceptible but high-yielding variety, Ife Brown, as a base in 1973, successive improved but related varieties like TVx 3236, IT82D-716, IT84S-2246-4, IT90K-59, IT90K-76 and IT97K-499-35 have been developed with successive addition of new genes for resistance such that the newest varieties are resistant to all the major pests except for Maruca and pod bugs. In addition to these, there are some other cowpea breeding lines such as IT86D-880, IT86D-1010, IT90K-277-2 and IT98K-205-8 which are resistant to all the three isolates of blackeye cowpea mosaic and five strains of cowpea aphid-borne mosaic.

5.9 Breeding for tolerance to drought and heat

Since cowpea is grown in varied environments it encounters different types of stresses including drought, heat and cold temperatures. Early maturing varieties escape

Figure 2 Breeding for tolerance to drought (top left), heat (top right), deep roots (bottom left) and low-P (bottom right).

terminal drought (Singh, 1987), but if exposed to intermittent moisture stress during the vegetative or reproductive stages, they perform equally poorly. Also, cowpea suffers due to high temperatures in the Sahelian region. Using simple screening methods for tolerance to drought, heat, and root architecture (Fig. 2), major gene inheritance have been identified for all these traits and incorporated into improved lines (Singh et al., 1999, Mai-Kodomi et al., 1999a,b, Singh and Matsui, 2002, Lucas et al., 2013, Verbree et al., 2014, 2015, Angira et al., 2016). The best drought-tolerant varieties are IT89KD-374-57, IT88DM-867-11, IT89KD-288, IT98D-1399, IT98K-131-1, IT97K-568-19, IT98K-452-1, IT98K-241-2; the best heat-tolerant lines are IT93K-452-1, IT98K-205-8, IT98K-1111-1, IT93K-693-2, IT97K-472-12, IT97K-472-25, IT97K-819-43, IT97K-499-35 and TX08-49-3.

5.10 Breeding for enhanced N fixation and tolerance to low soil phosphorus

Significant variation in cowpea rhizobium strains has been observed for nodulation in cowpea (Mandal et al., 1999), but the local *Rhizobia* invariably outpopulates the introduced strains. Therefore, in recent years, major efforts are concentrated to exploit genetic variability in cowpea as a host for effective nodulation and nitrogen fixation. Sanginga et al. (2000) screened 94 cowpea lines and observed major varietal differences in cowpea for growth, nodulation and arbuscular mycorrhizal fungi root infection as well as for performance under low and high phosphorus. The improved cowpea variety IT86D-715 showed equally good growth under low as well as high phosphorus levels. It also showed better N fixation than others. Based on its adaptability to grow in low-P soils and overall positive N balance, they recommended cultivation of IT86D-715 cowpea variety in soils with low fertility. Kolawale et al. (2000) screened 15 cowpea varieties for tolerance to aluminium and to determine the effect of phosphorus addition on the performance of Al-tolerant lines. The results indicated IT91K-93-10, IT93K-2046-1 and IT90K-277-2 cowpea varieties to be tolerant to aluminium and they gave higher response to phosphorus fertilization when grown in soils with aluminium toxicity problems. Recent work (Saidu et al., 2011, Allexander et al., 2011) have shown major varietal differences in cowpea for growth, nodulation and performance under low phosphorus (Fig. 1) and under rock phosphate application. Some of the promising lines under low-P condition were IT90K-372-1-2, TN5-78, IT98D-1399, TN27-80, IT99K-1060, IT89KD-374-57, TN 256-80, IT97K-1069-6 and IT98K-476-8.

5.11 Breeding for improved nutritional quality

Initially, cowpea breeding programme did not include systematic improvement for nutritional traits as an objective but the advanced breeding lines were analysed for protein content and cooking time and the best ones were selected for distribution to national programmes (Nielsen et al., 1993). However, recently a systematic breeding programme for improved protein and micronutrient contents in cowpea has been initiated through a bio-fortification project funded by the Gates Foundation and World Bank.

The project started in 2003 and considerable progress has been made. A total of 50 selected cowpea varieties and breeding lines grown in 2003 were analysed at Waite Analytical Services, Australia, in 2004. The mean and range values in the 50 cowpea lines were 24.4% and 21–31% for protein, 59.4 ppm and 46–79 ppm for iron, 829 ppm and 545–1330 ppm for calcium, 38 ppm and 23–48 ppm for zinc and 14721 ppm and 12750–16250 ppm for potassium. The best variety in respect of high protein and high iron and zinc was IT97K-1042-3 (Table 2).

Considerable differences have also been observed for physical quality of cowpeas. The relative density of cowpea seed ranged from 1.01 to 1.09, hardness (crushing weight) ranged from 3.96 kg for IT89KD-288 to 8.4 kg for Aloka local. The seed coat content ranged from 5.7% in IT95K-207-15 to 13.8% in TVu 12349 and cooking time ranged from 27.5 minutes for IT90K-277-2 to 57.5 minutes for Aloka local. The seed hardness was positively correlated with cooking time (Singh, 2001).

Progress has also been made on taste and acceptability. Akara prepared from high-yielding new cowpea varieties Ayiyi (IT83S-728-13) and Bengpla (IT83S-818) were the best. Similar results were observed in Nigeria by the women in agriculture (WIA) section of the Kano Agricultural and Rural Development Authority KNARDA. They evaluated three improved cowpea varieties, IT98D-867-11, IT89KD-288, and IT90K-277-2 and one local variety Danlla for four popular local dishes – 'kosai', 'danwake', 'alale' and 'dafaduka'. These were subjected to an independent taste panel of over 50 persons of different economic status and background. The improved variety IT90K-277-2 was rated as the best and others were as good as the local variety. None of the varieties were rated as unacceptable. IT90K-277-2 has already become very popular in Nigeria and Cameroon as a high-yielding variety.

Table 2 Protein, iron and zinc contents in improved cowpea varieties

Variety	Protein (%)	Iron (ppm)	Zinc (ppm)
IT97K-1042-3	30.7	77	46
IT99K-216-48	27.1	64	46
IT98K-205-8	27.1	60	42
IT98K-503-1	26.0	64	43
Aloka local	23.0	48	38
CD	0.8	2.5	1.7

6 Cowpea international trials

The top-performing breeding lines from the advanced trials are multiplied, and grouped into seven different cowpea international trials (CITs) and distributed to various national programmes. CIT-1 has extra-early maturity lines, CIT-2 has medium-maturity lines, CIT-3 has *Striga*-resistant lines, CIT-4 has insect-resistant lines (no spray), CIT-5 has dual-purpose lines, CIT-6 has vegetable type lines and CIT-7 has virus-resistant lines. After the trials have been formulated, a circular is sent to the collaborators in 65 countries informing them about the nature and composition of the trials. According to their requirements, individual collaborators request one or more sets of cowpea international trials. IITA then provides these sets free to the requesting collaborators along with the experimental field books containing the detailed protocols. Based on their results the scientists in national programmes select the best lines from the cowpea international trials and formulate their own multi-location national trials in the following years. The most promising varieties are then multiplied by them and either released to farmers or used as breeding stock.

7 Future trends and conclusion

7.1 IITA cowpea varieties released by national programmes

The collaborative interactions between the IITA, Bean/Cowpea CRSP and the national programme scientists have been very effective. A total of 65 countries have identified over 40 improved cowpea varieties and released for general cultivation. Many countries, where new cowpea varieties are making a difference, have given specific names to the new varieties and, in some areas, farmers themselves have given names and facilitated farmer to farmer diffusion of seeds. A few examples are 'Big Buff' in Australia; 'BRS-Mazagao' in Brazil; 'Nijwe' in Burkina Faso; 'BR-1' in Cameroon; 'Bira' in Angola; 'Titan and Cubinata' in Cuba; 'Asontem', 'Ayiyi' and 'Bengpla' in Ghana; 'GLM 93' in Camroon; 'Vamban', 'Pusa Reshmi', 'Pant Lobia,1,2,3,4,5' in India; 'Akash' (sky) and 'Prakash' (light) in Nepal; 'Sosokoyo' in Gambia; 'Pkoko Togboi' in Guinea Conakry; 'Korobalen', 'Sangaraka', 'Jiguia' and 'Fakson' in Mali; Dan IITA (son of IITA), 'Dan Bunkure', 'SAMPEA10', 'SAMPEA11', 'SAMPEA12' in Nigeria; 'IT 1', 'IT 2', 'IT 3' and 'IT4' in Niger; 'Pannar 3' in South Africa; 'Vuli-1', and 'Vuli-2' in Tanzania; 'Dahal Elgoz' (gold from the sand) in Sudan (Elawad and Singh 2003); 'Umtilane' in Swaziland; and 'Bubebe' in Zambia.

7.2 Impact of improved cowpea varieties and increased global cowpea production

Through international cowpea trials and multi-location national trials, a number of improved cowpea varieties have been released in many countries and being widely cultivated in varied cropping systems. This is catalysing a rapid increase in cowpea cultivation including its extension in non-traditional areas. The new improved 60-day cowpea varieties have enabled farmers in Nigeria to produce two crops in the same season and use these varieties in improved intercropping systems for maximizing cereals, root crops and legume productivity (Singh and Ajeigbe, 2002). Recent experiments conducted using 60-day cowpea varieties in northern Nigeria and in India have clearly demonstrated that an intensive cropping system involving 'wheat-cowpea-rice' can be successfully grown each year using supplementary irrigation for wheat and cowpea with a total output of about 7–11 tons of food/ha comprising about 3–5 tons of wheat, 3.5–6 tons of rice and about 1–1.5 tons of cowpea (Singh, 2005, Singh et al., 2010). The additional cowpea crop in the summer season after wheat harvest (April–June) not only provides extra employment but also provides nutritious grain and fodder which fetch higher prices at that time. Including cowpea in systems also makes the rotation more sustainable and productive because it fixes nitrogen and leaves nutritious residues in the field. Similarly there has been significant increase in cowpea cultivation in Brazil with soybean-cowpea double cropping in the Cerrado region. As a result, the world cowpea production has increased from about 1.1 million tons in 1974 to over 7.5 million tons in 2014 – over 8 fold increase. Also, the cowpea production has increased about 73% from 2001 to 2013, when there was little or no increase in other pulses. This is because the new '60–70 day' cowpea varieties are being grown as a niche crop in the cereals and root crop systems increasing the area under cowpea cultivation without competing with other crops. Therefore, cowpea is being considered as the 'Food legume of the twenty-first century' (Singh, 2014).

7.3 Biotechnology for insect resistance

The level of resistance to Maruca pod borer in the existing cowpea germplasm is very low and it causes 80–100% yield losses if not protected. Therefore, all the improved varieties still require at least two sprays of insecticides to protect against this insect. Unfortunately, most of the farmers in Africa do not have access to insecticides and often adulterated and expensive when available. Therefore, efforts were initiated in 2004 onwards to introduce Bt (*Bacillus thuringiensis)* gene in cowpea which imparts very high level of protection against cotton bollworm and maize stalk borer. Laboratory studies using the Bt endotoxins were highly effective against Maruca pod borer larvae in cowpea. Therefore, joint efforts were made by many international partners including IITA, USAID, Rockefeller Foundation, advanced laboratories in the United States and Australia, African Agricultural Technology Foundation (AATF), Network for Genetic Improvement of Cowpea for Africa (NGICA) and Monsanto Corporation kindly agreed to donate Bt gene free of royalty to develop Maruca-resistant cowpea varieties for Africa. The group then requested Dr. T.J. Higgins of Council of Scientific and Industrial Research Organization (CSIRO), Australia, and provided him with sufficient funds to introduce the Bt gene in cowpea. This work has progressed very well and Bt gene was successfully introduced in cowpea in 2008 and confined tests were initiated from 2009 onwards. Controlled greenhouse and field tests of Bt-cowpea have been conducted in Nigeria, Ghana and Burkina Faso for many years with excellent results in controlling Maruca and it is expected that some Bt-cowpea varieties would be released to farmers in these countries in 2018. This is expected to bring a quantum jump in cowpea production in West Africa. Efforts are also underway in Brazil (Singh et al., 2014) to develop transgenic cowpeas with virus resistance, herbicide tolerance and increased sulphur-containing amino acids.

7.4 Genomic resources for cowpea improvement

Even though studies on cowpea genomic resources started much later than other crops, there has been considerable progress during the last decade. Since the first attempts on cowpea genetic mapping (Fatokun et al., 1993; Menendez et al., 1997), there has been a progression of increasingly informative maps with a greater number of traits analysed and greater depth of marker coverage (Ouedraego et al., 2002; Timko et al., 2007; Timko and Singh 2008; Li et al., 2009). With the advent of new and improved technologies such as high-throughput DNA and cDNA sequencing and single-nucleotide polymorphism (SNP) detection, there was a rapid and significant improvements in the cowpea genetic map. Using SNP assays, Muchero et al. (2009) were able to map 928 expressed sequence tag (EST)-derived SNPs. This map represented a substantial improvement over previously available genetic maps because it was not population specific and surveyed polymorphism at 1536 identical loci in six recombinant inbred line (RIL) populations. Building upon this work, a new consensus map containing 1107 EST-derived SNP markers was reported by Lucas et al. (2011). This new map was developed by integrating 13 population-specific maps and contained 1107 markers which brought 20% increase in marker density compared to the earlier consensus map created by Muchero et al. (2009) and the average distance between markers was reduced from 1.05 to 0.79 cM. The SNP-based maps of Muchero et al. (2009) and Lucas et al. (2011) also have an additional value in that the polymorphic loci mapped are associated with expressed genes. As a consequence of being associated with a known coding region rather than a random or repetitive sequence, it is possible to examine synteny of these gene positions among closely and more distally related legume species. The use of SNPs in syntenic comparisons with both closely related species and subspecies (such as *V. unguiculata* subsp. *sesquipedalis)* and more distally related genera such as *Glycine, Medicago* and *Phaseolus* has been discussed by Varshney et al. (2009) and Lucas et al. (2011). In a very comprehensive recent study, Monoz-Amatriain et al. (2016) have described foundational genome resources and their application to analysis of germplasm currently in use in West African breeding programmes. They used resources developed from the African cultivar IT97K-499-35 which included a whole-genome shotgun (WGS) assembly, a bacterial artificial chromosome (BAC) physical map, and assembled sequences from 4355 BACs. These resources and WGS sequences of an additional 36 diverse cowpea accessions supported the development of a genotyping assay for 51 128 SNPs, which was then applied to five biparental RIL populations to produce a consensus genetic map containing 37 372 SNPs. This genetic map enabled the anchoring of 100 Mb of WGS and 420 Mb of BAC sequences, an exploration of genetic diversity along each linkage group, and clarification of macrosynteny between cowpea and common bean. The SNP assay enabled a diversity analysis of materials from West African breeding programmes. Two major subpopulations exist within those materials, one of which has significant parentage from South and East Africa and more diversity. Abaidu et al. (2012) and Boukar et al. (2016) have presented a detailed review of the recent progress made in cowpea genomics along with available information on mapping of several important traits in cowpea. They strongly feel that the emerging genomic resources and a combination of conventional and molecular breeding strategies offer a unique opportunity to cowpea breeders for the development of climate-resilient improved varieties for increased cowpea production in the varied agro-climatic zones.

8 Where to look for further information

The most comprehensive details about various aspects of cowpea research and development programmes can be found in a recent book, *Cowpea: The Food Legume of the 21st Century* by B.B. Singh and published by the Crop Science Society of America, Madison, Wisconsin, USA, 2014. Also, the annual reports of IITA at iita.org/cowpea provide progress about ongoing research on cowpea.

9 References

Abadassi, J. A., Singh, B. B., Ladiende, T. A. O., Shoyinka, S. A. and Emechebe, A. M. 1987. Inheritance of resistance to brown blotch, Septoria leaf spot and scab in cowpea. *Indian J. Genet. Plant Breed.* 47:299–303.

Abaidoo, R., Boukar, O., Boahen, S., Coulibaly, O., Bandyopadhyay, R., Fatokun, C., Kumar, L., Kamara, A., Muranaka, S., Touré, M., Adamou, M., Ishiyaku, M., Mligo, J. 2012. Enhancing cowpea productivity and production in drought-prone areas of Sub-Saharan Africa. In Abate, T. (Ed.), *Four Seasons of Learning and Engaging Smallholder Farmers: Progress of Phase 1.* Nairobi: International Crops Research Institute for the Semi-Arid Tropics, pp. 81–112.

Adjadi, O., Singh, B. B. and Singh, S. R. 1985. Inheritance of bruchid resistance in cowpea. *Crop Sci.* 25:740–2.

Allexander, T., Rothe, J., Singh, B. B., Mahamane, S., Hays, D. and Loeppert, R. H. 2011. Breeding cowpeas for efficient acquisition and use of phosphorus from low-P soils and rock phosphate. American Society of Agronomy Abs. 2011. Paper presented at ASA Meetings, San Antonio Texas. 16–20 October 2011.

Angira, B., Masor, L., Zhang, Y., Scheuring, C. F., Zhang, Y., Singh, B. B., Zhang, H.-B., Hays, D. B. and Zhang, M. 2016. Molecular Mapping of Heat Tolerance in Cowpea. Paper presented at the World Cowpea Conference 28 February to 4 March 2016, Livingstone, Zambia.

Atokple, I. D. K., Singh, B. B. and Emechebe, A. M. 1995. Genetics of resistance to *Striga* and *Alectra* in cowpea. *J. Hered.* 86:45–9.

Bata, H. D., Singh, B. B., Singh, S. R., and Landeinde, T. A. O. 1987. Inheritance of resistance to aphid in cowpea. *Crop Sci.* 27:892–4.

Boukar, O., Kong, L., Singh, B. B., Murdock, L. and Ohm, H. W. 2004. AFLP and AFLP-derived SCAR markers associated with *Striga gesnerioides* resistance in cowpea (*Vigna unguiculat* (L.) Walp.). *Crop Sci.* 44:1259–64.

Boukar, O., Fatokun, C., Huynh, B., Roberts, P. A. and Close, T. J. 2016. Genomic tools in cowpea breeding programs: Status and perspectives. *Front. Plant Sci.*, 7:1–13.

Carsky, R. J., Singh, B. B. and Oyewole, B. 2001. Contribution of early season cowpea to late season maize in the savanna zone of West Africa. *Biol. Agric. Hortic.* 18:303–16.

Carsky, R. J., Akakpo, C., Singh, B. B. and Detongnon, J. 2003. Cowpea yield gain from resistance to Striga gesnerioides parasitism in southern Benin. *Exp. Agri.* 39:327–33.

Ehlers, J. D., Matthews, W. C., Hall A. E. and Roberts, P. A. 2000. Inheritance of a broad-based form of nematode resistance in cowpea. *Crop Sci.* 40:611–18.

Fatokun, C. A. and Singh, B. B. 1987. Interspecific hybridization between *Vigna pubescens* and *V. unguiculata* (L.) Walp. through embryo rescue. *Plant Cell Tissue Organ Cult.* 9:229–33.

Fatokun, C. A. 1991. Wide hybridization in cowpea: Problems and prospects. *Euphytica* 54:137–40.

Fatokun, C. A., Menancio-Hautea, D. D. I. and Young, N. D. 1993. A linkage map for cowpea (*Vigna unguiculata*) [L.] Walp.) based on DNA markersin Genetic maps. In O'Brien, S. J. (Ed.), *Locus Maps of Complex Genomes*. VI Edition. USA: Cold Spring Harbor Laboratory Press, p. 256.

Fatokun, C. A. 1997. Breeding cowpea for resistance to insect pests: Attempted crosses between cowpea and V. vexillata. In Singh B. B., Mohan Raj, D. R., Dashiell, K. and Jackai, L. E. N. (Eds), *Advances in Cowpea Research*. Copublication of International Institute of Tropical Agriculture (IITA) and Japan International Center for Agricultural Sciences (JIRCAS), Ibadan, Nigeria: IITA, pp. 52–61.

Fery, R. L. 1985. The genetics of cowpea. A review of the World Literature. In Singh, S. R. and Rachie, K. O. (Eds), *Cowpea Research, Production and Utilization*. Chichester, U.K.: John Wiley and Sons, pp. 25–62.

Fery, R. L. and Singh, B. B. 1997. Cowpea genetics a review of recent literative. In Singh, B. B., Mohan Raj, D. R., Dashiell, K. E. and Jackai, L. E. N. (Eds), *Advances in Cowpea Research*. Co publication of IITA and JIRCAS, Ibadan, Nigeria: IITA, pp. 13–29.

Hall, A. E., Singh, B. B. and Ehlers, J. D. 1997. Cowpea breeding. *Plant Breed. Rev.* 15:215–74.

Henriet J., van Ek, G. A., Blade, S. F. and Singh, B. B. 1997. Quantitative assessment of traditional cropping systems in the Sudan savanna of northern Nigeria: I. Rapid survey of prevalent cropping systems. *Samaru J. Agr. Res.* 14:37–45.

Kolawale, G. O., Tian, G. and Singh, B. B., 2000. Differential response of cowpea varieties to Aluminum and phosphorus application. *J. Plant Nutr.* 23:731–40.

Lane, J. A., Moore, I. H. M., Child, D. V., Cardwell, K. F., Singh, B. B., Singh Bailey, J. A. 1994. Virulence characteristics of a new race of the parasitic angiosperm *Striga gesnerioides*, from southern Benin of on cowpea, (*Vigna unguiculata*). *Euphytica* 72(3):183–8.

Li, J., Lis, K. E. and Timko, M. P. (2009). Molecular genetics of race specific resistance of cowpea to *Striga gesnerioides*. *Pest Manag. Sci.* 65:520–7. doi:10.1002/ps. 1722.

Lucas, M. R., Diop, N.-N., Wanamaker, S., Ehlers, J. D., Roberts, P. A. and Close, T. J. (2011). Cowpea-soybean synteny clarified through an improved genetic map. *Plant Genome* 4:218–25.

Mai-Kodomi, Y., Singh, B. B., Myers Jr. O., Yopp, J. H., Gibson, P. J. and Terao, T. 1999a. Two mechanisms of drought tolerance in cowpea. *Indian J. Genet.*, 59:309–16.

Mai-Kodomi, Y., Singh, B. B., Terao, T., Myers Jr. O. , Yopp, J. H. and Gibson, P. J. 1999b. Inheritance of drought tolerance in cowpea. *Indian J. Genet.*, 59:317–32.

Mandal, J., Chattpadhyaya, A., Hizra, P., Dasgupta, T. and Som, M. G. 1999. Genetic variability for three biological nitrogen fixation components in cowpea cultivars. *Crop Res. (Hissar)* 18:222–5.

Menéndez, C. M., Hall, A. E. and Gepts, P. 1997. A genetic linkage map of cowpea (*Vigna unguiculata*) developed from a cross between two inbred, domesticated lines. *Theor. Appl. Genet.* 95:1210–17.

Mohammed, S. G., Singh, B. B. and Mustapha, O. T. 2000. An evaluation of variation for root-knot nematode resistance in cowpea (*Vigna unguiculata*). *Nig. J. Botany* 13:31–7.

Mortimore, M. J., Singh, B. B., Harris, F. and Blade, S. B. 1997. Cowpea in traditional cropping systems. In Singh, B. B., Mohan Raj, D. R., Dashiell, K. and Jackai, L. E. N. (Eds), *Advances in Cowpea Research*. Copublication of International Institute of Tropical Agriculture (IITA) and Japan International Center for Agricultural Sciences (JIRCAS), Ibadan, Nigeria: IITA, pp. 99–113.

Muchero, W., Diop, N. N., Bhat, P. R., Fenton, R. D., Wanamaker, S., Pottorff, M., Hearne, S., Cisse, N., Fatokun, C., Ehlers, J. D., Roberts, P. A. and Close, T. J. (2009). A consensus genetic map of cowpea [Vigna unguiculata (L) Walp.] and synteny based on EST-derived SNPs. *Proc. Natl. Acad. Sci. U.S.A.* 106:18159–64.

Muñoz-Amatriaín, M., Mirebrahim, H., Xu, P., Wanamaker, S. I., Luo, M., Alhakami, H., Alpert, M., Atokple, I., Batieno, B. J., Boukar, O., Bozdag, S., Cisse, N., Drabo, I., Ehlers, J. D., Farmer, A., Fatokun, C., Gu, Y. Q., Guo, Y.-N., Huynh, B.-L., Jackson, S. A., Kusi, F., Lawley, C. T., Lucas, M. R., Ma, Y., Timko, M. P., Wu, J., You, F., Roberts, P. A., Lonardi, S. and Close, T. J. (2016). Genome resources for climate-resilient cowpea, an essential crop for food security. *Plant J.* 89:1042–54. Accepted Author Manuscript. doi:10.1111/tpj.13404.

Ng, N. Q. and Singh, B. B. 1997. Cowpea. In Fuccillo, D., Sears, L. and Stapleton, P. (Eds), *Biodiversity and Trust*. Cambridge, U.K.: Cambridge University Press, pp. 82–99.

Nielsen, S. S., Brandt, W. E. and Singh, B. B. 1993. Genetic variability for nutritional composition and cooking time of improved cowpea lines. *Crop. Sci.* 33:469–72.

Ouedraogo, J. T., Maheshwari, V., Berner, D. K., St-Pierre, C. A., Belize, F. and Timko, M. P. 2001. Identification of AFLP markers linked to resistance of cowpea to parasitism by *Striga gesnerioides. Theor. Appl. Genet.* 102:1029–36.

Ouedraogo, J. T., Gowda, B. S., Jean, M., Close, T. J., Ehlers, J. D., Hall, A. E., Gillaspie, A., Roberts, P. A., Ismail, A. M., Bruining, G., Gepts, P., Timko, M. P. and Belzile, F. J. 2002. An improved genetic linkage map for cowpea (*V.unguiculata*) combining AFLP, RFLP, RAPD, biochemical markers and biological resistance traits. *Genome* 45:175–88.

Padulosi, S. 1993. Genetic diversity, taxonomy and ecogeographic survey of the wild relatives of cowpea (V. unguicullata). Ph.D. Thesis. University Catholique Lovain-la-Neuve, Belgique.

Padulosi, S. and Ng, N. Q. 1997. Origin, taxonomy, and morphology of *Vigna unguiculata* (L.)Walp. In Singh B. B., Mohan Raj, D. R., Dashiell, K. E. and Jackai, L. E. N. (Eds), *Advances in Cowpea Research*. Copublication Intl Inst Tropical Agric (IITA) and Japan Intl Res Center Agric Sci (JIRCAS). Sayce, Devon, UK, pp. 1–12.

Saidu, A. K., Abaidu, R. C., Ewuafor, E. N. O., Sanginga, N. and Singh, B. B. 2011. Genotypic variation in cowpea for rock phosphate use in low phosphorus soils of dry Sudan and Sahel savannas of Niger and Nigeria, West Africa. *American-Eurasian J. Agric. & Environ. Sci.* 11:62–70.

Sanginga, N., Lyasse, O., and Singh, B. B. 2000. Phosphorus use efficiency and nitrogen balance of cowpea breeding lines in a low P soil of the derived savanna zone in West Africa. *Plant Soil* 220:119–28.

Singh, S. R. and Allen, D. J. 1979. *Cowpea Pests and Diseases*. Booklet published by the International Institute of Tropical Agriculture., Ibadan, Nigeria, 113 pages.

Singh, B. B., Adjadi, O. and Singh, S. R., 1985. Bruchid resistance in cowpea. *Crop Sci.* 25:736–9.

Singh, B. B. 1987. Breeding cowpea varieties for drought escape. In Menyonga, J. M., Bezuneh, T. and Youdeowei, A. (Eds), *Food Grain Production in Semi-Arid Africa*. OUA/STRC-SAFGRAD, pp. 299–306.

Singh, B. B. and Emechebe, A. M. 1990. Genetics of *Striga* resistance in cowpea genotype B 301. *Crop Sci.* 30:879–81.

Singh, B. B., Emechebe, A. M. and Atokple, I. D. K. 1993. Inheritance of *Alectra* resistance in cowpea genotype B 301. *Crop. Sci.* 33:70–2.

Singh, B. B. and Sharma, B. 1996. Restructuring cowpea for higher yield. *Indian J. Genet.* 56:389–405.

Singh, B. B., Cambliss, O. I. and Sharma, B. 1997. Recent advances in cowpea breeding. In Singh B. B., Mohan Raj, D. R., Dashiell, K. and Jackai, L. E. N. (Eds), *Advances in Cowpea Research*. Copublication of International Institute of Tropical Agriculture (IITA) and Japan International Centre for Agricultural Sciences (JIRCSA), Ibadan, Nigeria: IITA, pp. 30–49.

Singh, B. B. and Emechebe, A. M. 1998. Increasing productivity of millet-cowpea intercropping systems. In Emechebe, A. M., Ikwelle, M. C., Ajayi, O., Aminu Kano, M. and Anaso, A. B. (Eds), *Pearl Millet in Nigeria Agriculture: Production, Utilization and Research Priorities*. Proceedings of the Pre-season Planning Meeting for the Nationally Coordinated Research Programme for Pearl Millet, Maiduguri, 21–24 April 1997. P.M.B. 1239, Maiduguri, Nigeria: Lake Chad Research Institute, pp. 68–75.

Singh, B. B., Mai-Kodomi, Y. and Terao, T. 1999. A simple screening method for drought tolerance in cowpea. *Indian J. Genet.*, 59:211–20.

Singh, B. B. 2001. Genetic variability for physical properties of cowpea seeds and their effect on cooking quality. *African Crop Sci. Proc.* 5:43–6.

Singh, B. B., Ehlers, J. D., Sharma, B. and Freire-Filho, F. R. 2002. Recent progress in cowpea breeding. In Fatokun, C. A., Tarawali, S. A., Singh, B. B., Kormawa, P. M. and Tamo, M. (Eds),

Challenges and Opportunities for Enhancing Sustainable Cowpea Production. Ibadan, Nigeria: IITA, pp. 22–40.

Singh, B. B. 2002a. Breeding cowpea varieties for resistance to *Striga gesnerioides* and *Alectra vogelii*. In Fatokun, C. A., Tarawali, S. A., Singh, B. B., Kormawa, P. M. and Tamo, M. (Eds). *Challenges and opportunities for enhancing sustainable cowpea production*. Ibadan, Nigeria: IITA. pp. 154–66.

Singh, B. B. 2002b. Recent genetic studies in cowpea. In Fatokun, C. A., Tarawali, S. A., Singh, B. B., Kormawa, P. M. and Tamo, M. (Eds). *Challenges and opportunities for enhancing sustainable cowpea production*. Ibadan, Nigeria: IITA, pp. 3–13.

Singh, B. B. and Matsui, T. 2002. Breeding cowpea varieties for drought tolerance. In Fatokun, C. A., Tarawali, S. A., Singh, B. B., Kormawa, P. M. and Tamo, M. (Eds), *Challenges and Opportunities for Enhancing Sustainable Cowpea Production*. Ibadan, Nigeria: IITA. pp. 287–300.

Singh, B. B. 2005. Cowpea [*Vigna unguiculata* (L.) Walp. In Singh, R. J. and Jauhar, P. P. (Eds), *Genetic Resources, Chromosome Engineering and Crop Improvement. Volume 1*. Boca Raton, Florida, USA: CRC Press, pp. 117–62.

Singh, B. B., Timko, M. P. and Aragao, F. L. 2014. Advances in cowpea improvement and genomics. In Gupta, S., Nadarajan, N. and Gupta, D. S. (Eds), *Legumes in the Omic Era*. N.Y.: Springer, pp. 131–54.

Singh, B. B. 2014. *Cowpea: The Food Legume of the 21st Century*. Madison, Wisconsin, USA: Crop Science Society of America.

Singh, Y. V., Nautiyal, M. K., Singh, B. B., Pandey, B. and Sharma, C. L. 2010. Performance of short duration cowpea varieties for enhanced pulses production in different seasons under Tarai conditions of Uttarakhand. *Pantnagar J. Res.* 8:198–201.

Steele, W. M. and K. L. Mehra. 1980. Structure, evolution and adaptation to farming system and environment in Vigna. In Summerfield, R. J. and Bunting, A. H. (Eds). *Advances in Legume Science*. London, UK: HMSO, pp. 393–404.

Timko, M. P., Gowda, B. S., Ouedraogo, J. and Ousmane, B. (2007). Molecular markers for analysis ofresistance to *Striga gesnerioides* in cowpea. In Ejeta, G. and Gressell, J. (Eds), *Integrating New Technologies for Striga Control: Towards Ending the Witch-hunt*. Singapore: World Scientific Publishing Co. Pte Ltd, pp. 115–28.

Timko, M. P. and Singh, B. B. (2008). Cowpea, a multifunctional legume. In Moore, P. H. and Ming, R. (Eds). *Genomics of Tropical Crop Plants*. New York, NY: Springer Science + Business Media, LLC, pp. 227–58.

van Boxtel, J., Singh, B. B., Thottappilly, G. and Maule, A. J. 2000. Resistance of (*Vigna unguiculata* (L.) Walp.) breeding lines to blackeye cowpea mosaic and cowpea aphid borne mosaic poty virus isolates under experimental conditions. *J. Plant Dis. Protect.* 107:197–204.

Varshney, R. K., Close, T. J., Singh, N. K., Hoisington, D. A. and Cook, D. R. (2009). Orphan legume crops enter the genomics era. *Curr. Opin. Plant Biol.*, 12:202–10.

Verbree, D. A., Singh, B. B. and Payne, W. A. 2014. Role of gas exchange, leaf water status and carbohydrate partitioning during the early vegetative stage on drought tolerance in cowpea. *Crop Sci.* 54:770–8.

Verbree, D. A., Singh, B. B. and Payne, W. A. 2015. Genetics and heritability of shoot drought tolerance in cowpea. *Crop Sci.* 55:146–53.

Yunes, A. A., de Sales, M. T., Andrade, M. P., Morais, R. A., Fernandes, K. V. S., Gomes, V. M., Xavier-Filho, J. and De Andrade, M. T. 1998. Legume seed cicilins (7S storage proteins) interfere with the development of the cowpea weevil (Callosobruchus maculates (F). *J. Sci. Food Agr.* 76(1):111–16.

Chapter 11

Improving cultivation of cowpea in West Africa

Alpha Y. Kamara, Lucky O. Omoigui and Nkeki Kamai, International Institute of Tropical Agriculture (IITA), Nigeria; Sylvester U. Ewansiha, University of Benin, Nigeria; and Hakeem A. Ajeigbe, International Crops Research Institute for the Semi-Arid Tropics (ICRISAT), Nigeria

1 Introduction
2 Optimal plant population
3 Plant configuration in intercropping systems in West Africa
4 Manipulating planting dates to improve cowpea productivity
5 Nutrient management for increased cowpea productivity
6 Integrated pest management (IPM) in cowpea production
7 Future trends and conclusion
8 Where to look for further information
9 References

1 Introduction

Cowpea [*Vigna unguiculata* (L.) Walp.] is a legume crop of vital importance to the livelihoods of millions of people in West and Central Africa (WCA). It provides a nutritious grain and a less expensive source of protein for both rural and urban poor consumers (Inaizumi et al., 1999). It can be grown and harvested in as little as 60–80 days. This enables households to harvest leaves and grains for consumption or sale during the 'hungry season' when grain reserves from the previous cereal harvests have been depleted and current crops are not ready for harvest. Most of the world's cowpea (>90) is grown in sub-Saharan Africa, most of which is in West Africa particularly in Nigeria and Burkina Faso. Over 12.61 million ha are grown to cowpea worldwide, with an annual grain production of about 5.59 million tons (FAO, 2014). Of this amount, Africa accounts for 94% of grain production. Nigeria is the largest cowpea producer in the world and accounts for over 2.5 million tons grain production from an estimated 4.9 million ha (FAO, 2014). Other major producers in West Africa are Mali, Niger and Senegal. Cowpea cultivation is mainly under traditional systems and cowpea grain yields in farmers' fields are low especially in the West African sub-region (0.025–0.3 t ha^{-1}). This is caused by severe attacks of pest complexes, diseases, low soil fertility, drought, inadequate planting systems, inappropriate cultivars and lack of inputs (Ajeigbe et al., 2010a). In addition

http://dx.doi.org/10.19103/AS.2017.0023.30

to biotic and abiotic stresses, existing planting practices limit crop yields. Despite the availability of *Striga* and disease-resistant cowpea cultivars, grain yields on farmers' fields are still low. However, on-station and researcher-managed plot yields are high and encouraging. Grain yields ranging from 0.5 to 2.76 t ha^{-1} have been reported in sole crop (Ajeigbe et al., 2005, 2008), whereas grain yields ranging from 0.37 to 1.27 t ha^{-1} have been reported in intercrop in the savannahs of Africa (Ajeigbe et al., 2005, 2010b). Yield potential assumes unconstrained crop growth and adequate management that avoids limitations from nutrient deficiencies; inadequate planting systems and water stress and reductions from weeds, pests and diseases (Evans and Fisher, 1999). Considering the large differences between farmers' yields (0.3 t ha^{-1}) and experimental station yields (1.5–2.5 t ha^{-1}), potential for on-farm yield increase in the region is high. This has stimulated interest in agronomic practices that could enhance crop yields. Some of the agronomic practices that may increase cowpea productivity are optimal plant population, appropriate planting date, nutrient management, integrated pest management and suitable cropping system.

2 Optimal plant population

Cowpea is generally cultivated on rows of ridges in the cowpea growing region of West Africa. The ridges are spaced 75 cm apart because equipment used in the region for ridging is the same as for the other grain crops such as maize, soybean, sorghum and millet. This general row spacing does not consider individual crop and varietal requirements. The use of ridges spaced 75 cm apart with the recommended plant spacing of 20 cm corresponds to plant population of 133 333 plants ha^{-1}, which may not be sufficient for optimal cowpea yield. The low plant density resulting from wide row spacing usually leads to low yields in some grain legume crops, such as cowpea (Kamara et al., 2016). Grain yields of the widely available stress-tolerant cowpea cultivars hardly go above 1.7 t ha^{-1} on farmers' fields, despite the enormous gain in genetic improvement over the past three decades (Kamara et al., 2010). In Nigeria, cowpea planting density recommendations ranged from 33 000 plants ha^{-1} in the more spreading and traditional variety to 66 000 plants ha^{-1} in the improved erect varieties (Dugje et al., 2009; Utoh et al., 2008). Plant density is an important component of yield in grain crops such as cowpea and soybean. Adjusting planting density is an important method to optimize crop growth and the time required for canopy closure and to achieve maximum biomass and grain yield (Liu et al., 2008). Crop cultivars respond differently to high plant density because of differences in growth habit. Some cultivars record high grain yield when grown at high densities (Liu et al., 2008). High plant density increases light interception, radiation use efficiency, dry matter and yield components (pods and seeds) by both decreasing row spacing and increasing plant density (Bruns, 2011). Ezedinma (1974) reported that close spacing between and within rows increased biological and grain yields of cowpea, while Jallow and Fergusson (1985) reported a linear response of grain yield to plant density between 40 000 and 250 000 plants ha^{-1}. The response of cowpeas to changes in density depends on the morphology of the cultivars and the growing environment. Kwapata and Hall (1990) found that cowpea grain yield for some cultivars was significantly greater at 400 000 plants ha^{-1} than at 100 000 plants ha^{-1} under irrigated conditions in California. Semi-dwarf lines produce relatively greater yield than standard lines at narrower row spacing (Ismaila and Hall, 2000). Breeders in West Africa have produced several cowpea varieties with varying growth habits. While some are erect and early maturing, others are semi-erect and have a prostrate growth habit which allows them to spread and close canopy faster. These varieties are usually spaced 30 cm apart to avoid overcrowding stress, while the erect types are spaced 20 cm apart to compensate for lower branching habit. To ensure that the optimum population is achieved on the widely spaced ridges, Kamara et al. (2016) recommended planting two rows of cowpeas placed equidistant on the ridges. In an experiment in northern Nigeria, they planted four cowpea varieties with contrasting maturity duration in single, double and triple rows on ridges spaced 75 cm apart to achieve corresponding densities of 133 333, 266 666 and 400 000 plants ha^{-1}, respectively. Plant densities of 266 666 and 400 000 plants ha^{-1} gave higher crop performance in terms of light interception, biomass production, yield and yield components for all cowpea varieties. Yield increases were related largely to increased pod number and grain production, but the effect of grain size on yield was relatively minor. Their results provide evidence that the current density of 133 333 plants ha^{-1} used by farmers is not optimum for cowpea production. Smallholder farmers can increase cowpea grain and fodder yields if they use a density of 266 666 plants ha^{-1} in cowpea cultivation. Further yield increases when cowpea is planted at 400 000 plants ha^{-1} may not be sufficient to offset the cost of seed.

3 Plant configuration in intercropping systems in West Africa

In West Africa, cowpea is traditionally intercropped with other food crops such as maize, pearl millet, sorghum and cassava. Although sole cropping of cowpea is profitable, farmers continue to grow cowpea as an intercrop with these crops. This is because it fits well into the low input labour-intensive tradition of growing crops in the region. Several advantages are accruable when crops are intercropped. Lithourgidis et al. (2011) have summarized these advantages to include: the production of greater yield on a given piece of land by making more efficient use of the available growth resources; using a mixture of crops of different rooting ability, canopy structure, height and nutrient requirements based on the complementary utilization of growth resources by the component crops; improving soil fertility through biological nitrogen fixation with the use of legumes and increasing soil conservation with greater ground cover than sole cropping as well as providing better lodging resistance for crops that are susceptible to lodging when grown as sole crops. Others include reduction in pest incidence, improvement in forage quality by increasing crude protein yield of forage, provision of insurance against crop failure or against unstable market prices for a given commodity. This provide greater financial stability than sole cropping which makes the system particularly suitable for labour-intensive small farms and allows for lower inputs use through reduced fertilizer and pesticide requirements.

However, intercropping cowpea with maize has a major weakness of very low cowpea yield (Olufajo and Singh, 2002). Cowpea farmers in the dry savannah areas of sub-Saharan Africa obtain low yields, estimated at about 350 kg ha^{-1} (Olufajo and Singh, 2002; IITA, 2009). A major reason for the low productivity of cowpea in intercropping systems is shading from the taller cereal plants (Olufajo and Singh, 2002). Performance of cowpea intercropped with cereals is dependent on the growth habit of the cowpea crop. Indeterminate cowpea varieties with spreading growth habit normally performed better than the erect cowpea varieties because they are tolerant to shade (Ewansiha et al., 2014a). Ecological studies carried out by Terao et al. (1997) showed that at least 40% of incident light is necessary to grow healthy cowpea plants. In the traditional system, intercropped cowpea receives from <30 to >75% of incident light. In these light-limited conditions, cowpea varieties with a spreading growth habit get more light than those with an erect growth habit by producing more leaves as well as expanding their leaf area. N'tare and Williams (1992) and Terao et al. (1997) concluded that spreading cowpea type with a well-developed root system and high transpiration efficiency is best adapted to intercropping. However, intercropping would become more productive if the effect of shading were reduced.

Improved intercropping systems may be achieved by the plant configuration adopted by the farmer. Plant configuration refers to the spatial arrangement of the component crops as the intimacy of the crop mixture has important effects on the interactions between the crop species. Three plant configurations may be described: same (within) row intercropping – where the component crops are planted within the same row (Fig. 1); alternate row intercropping – where two different crops are cultivated in separate alternate rows (Fig. 2) and strip intercropping – where few–several rows of a crop are alternated with few–several rows of another crop (Fig. 3). The success of intercropping systems depends much on the interactions between the adapted crop cultivars that form the component crops, available management practices which include plant population and planting date of component crops, and the environmental conditions (Lithourgidis et al., 2011; Ewansiha et al., 2014a,b, 2015a,b). The practice of a given plant configuration or intercropping system is related to one or more cultural practices.

In a same row intercropping system trial that involved adapted and improved cultivars of cowpea and maize in the northern Guinea savannah (NGS) of Nigeria, indeterminate cowpea cultivar intercepted more light and produced higher fodder and grain yields compared to semi-determinate and erect cowpea (Ewansiha et al., 2014a). Light interception and fodder and grain production by cowpea was higher when intercropped

Figure 1 Same row intercropping system.

Figure 2 Alternate row intercropping system.

Figure 3 Strip intercropping system.

with extra early and early maize cultivars than with late-maturing cultivar. The late-maturing and indeterminate cowpea was able to grow and use more light to produce higher fodder and grain yields after harvesting of the earlier maturing maize cultivars. On the other hand, the medium-maturing and semi-determinate cowpea completed its life cycle soon after harvesting of maize, making it unable to compensate for the reduced growth suffered when growing in the companion maize crop. In a same row intercropping system which involved intercropping of semi-determinate and indeterminate cowpea cultivars with early-maturing maize having populations of 17 777, 26 666 and 53 333 plants ha^{-1}, cowpea grain and fodder yields decreased with increase in maize population (Ewansiha et al., 2013, 2015a). The combined mean fodder yield of cowpea and maize was 6.3, 4.9 and 5.1 t ha^{-1} when cowpea was intercropped with maize having populations of 53 333, 26 666 and 17 777 plants ha^{-1}, respectively. Similarly, at these plant populations, the combined mean grain yield of cowpea and maize intercrop was 5.8, 4.5 and 3.8 t ha^{-1}, respectively. However, maize population of 0–26 666 plants ha^{-1} favoured better cowpea performance compared with 53 333 plants ha^{-1} because at these lower plant populations, maize plants had lower leaf area indices which allowed maize canopy to transmit more light into the understorey cowpea. However, in this system, the negative effects of shade were more pronounced in the semi-determinate cowpea than in the indeterminate at full maize population. Therefore, in high maize populations, indeterminate spreading cowpeas should be grown, while semi-determinate cowpeas should be planted in low to moderate maize populations because of their intolerance to severe shade. For same row intercropping systems, interactions among maize cultivar, cowpea cultivar and planting date were studied in the Sudan savannahs (SS) of Nigeria by Kamara et al. (2011) and Ewansiha et al. (2014b). The authors reported higher fodder and grain yield of cowpea intercropped with extra early maize cultivar compared with early maize cultivar. Fodder and grain yields were also higher for cowpea intercropped at four weeks after sowing maize than for cowpea intercropped at six weeks after sowing maize. Also, intercropping cowpea at six weeks after sowing maize gave a higher yield than intercropping at eight weeks after maize was sown. Furthermore, indeterminate cowpea cultivars produced higher yields than semi-determinate cowpea cultivars when intercropped early with maize (four weeks > six weeks > eight weeks). In another study comprising interactions among cowpea cultivar, plant population and planting date in the SS of Nigeria (Ewansiha et al., 2015b), the best grain yield potential for intercropped cowpea was achieved by sowing early in low to moderate maize plant populations, whereas indeterminate cowpea had higher grain yield at full maize population. Fodder yields were always higher for indeterminate cowpea across maize populations. Early sowing was more conducive to achieving a higher number of branches, higher number of peduncles, higher number of pods and higher fodder and grain yields. Cowpea performance reduced progressively with increase in maize plant population because of increased shading from maize plants. Growing cowpea under high maize population was more favourable for indeterminate cowpea cultivar, whereas growing under zero to moderate maize populations favoured semi-determinate cowpea cultivar in terms of grain production. Thus, when planning to grow cowpea with maize at full maize crop, farmers may need to sow indeterminate cowpea cultivars early under earlier maturing maize cultivars. However, at reduced maize plant populations, growing maize with semi-determinate cowpea cultivar will be preferable. The choice of maize plant population to use may depend on the income, food nutrition and feed needs of the farmer.

Studies have shown that by intercropping millet with cowpea in the semi-arid zone, millet yield is only reduced if cowpea is planted simultaneously with millet (N'tare and Williams, 1992). The relative planting dates of component crops can contribute significantly to the yield of intercrop systems as it modifies the relative periods of complementarity and competitiveness of component crops (Midmore, 1993). Terao et al. (1997) stated that cowpea planted 2–3 weeks later than millet grew to only 20% of sole-cropped cowpea planted at the same time, but millet yields are reduced when millet and cowpea are simultaneously intercropped, especially in the arid zones. It is therefore very important to strike a balance where there would be minimal reductions in yield of intercrops. This is especially important in the arid and semi-arid zones where root competition for moisture is more severe. Planting cowpea later than millet is one way to reduce millet yield loss. However, planting cowpea late will drastically reduce cowpea yield. Therefore, modifying the planting arrangement and the development of a cropping system that use the root zone complementarily is important.

There are several reports on the effect of spatial arrangement on the productivity of millet-cowpea intercrop in the Sudan savannah. The influence of component crop densities and manipulation of spacing between component crops, such as row arrangement and inter-row spacing on yield and production efficiency of intercrops, were evaluated by Ofori and Stern (1987). They suggested that the cereal components were less affected by these manipulations, whereas the legume yield usually decreased significantly depending on proximity of the cereal, perhaps due to the top of the legume canopy being shaded. Planting arrangement is an important determinant of how effectively the available resources are used, especially soil moisture. Clark and Myers (1994) also noted that cowpea in narrow strips (2:2) yielded average of 46% less than in wide strips (2:4) or in sole crop. They attributed the reduction in yield to the narrow strips; both of the cowpea rows were bordered by non-legume, and therefore, competition was greater than in the wide strips. Strip intercropping is advantageous in terms of ease of crop management, fertilizer and insecticide application, weeding and reduction of the shading effect of cereal on cowpea (Olufajo and Singh, 2002). Therefore, to boost cowpea productivity under intercropping, a 2cereal:4cowpea row to row planting pattern was recommended (Ajeigbe et al., 2005, 2006, 2010b). Singh and Ajeigbe (2002) noted that this system might also be more suitable and help maintain soil fertility because two-thirds of the area is legume and only one-third is cereal. In a similar trial involving different row to row arrangements, Mohammed et al. (2008) reported that 2:4 row arrangement relative to other arrangements had highest gross monetary return. Odion et al. (1994) observed that intercropping two rows of millet with four rows of cowpea was superior to intercropping both crops in alternate stands in the same row. Singh et al. (1997) reported superior yield of cowpea and higher intercrop productivity at 1:4 and 2:4 compared with 1:1cereal:cowpea row arrangement, but millet yield in 1:1 row arrangement was lower than in 1:4 and 2:4 row arrangements. These authors reported that strip cropping with two rows cereal:four rows cowpea offers an opportunity for selective input application and better economic advantage than the traditional one row cereal:one row cowpea spatial arrangement. Furthermore, alternating three rows of cowpea with two or three rows of sorghum plus one to two insecticide applications gave a yield advantage of 58–69%.

4 Manipulating planting dates to improve cowpea productivity

Despite the potential of cowpea in the dry savannahs of WCA, its production and productivity is constrained by several biotic and abiotic factors. The major yield-limiting factors of cowpea in the regions are insect pests and diseases, drought stress and parasitic weeds. Rainfall in the region has also been so variable and droughts have been so extreme that local cowpea cultivars and some modern cultivars that had evolved over the years in the Sahel hardly produce significant quantities of grain and fodder in recent times due to climate change. Furthermore, poor cultural practices such as inappropriate planting dates and plant population contribute to low productivity of cowpea (Singh et al., 2002; Ishiyaku et al., 2005).

Habitually, farmers in Northern Nigeria take the risk of planting their crops with the first rains in order to achieve early food security and also capture the flush of soil N that comes with the first rains (Jagtap and Abamu, 2003). Early planting with the first rain will not only make the crop mature during the rains but also predispose the crop to insect pests and disease pressure. On the other hand, planting too late may risk the danger of an early cessation of rains which may affect the quantity and quality of cowpea seed produced (Isubikalu et al., 1999). The important criteria in planting cowpea, therefore, is to determine the onset and duration of the rains and, more importantly, the maturity period of the cowpea variety. Kamara and Godfrey-Sam-Aggrey (1979) reported that high yields of good quality seeds are obtained when cowpeas are planted late so that the crop matures in dry weather. Studies by Mbong et al. (2010) showed that the grain yield of cowpea planted early in the Guinea savannah ecology was of poor quality due to disease infection, while those from late planting were of good quality. In an earlier study in Nigeria by IITA (1982), grain yield of cowpea planted early in the season was reported to be higher than those from the late planted crops. The increased yield was attributed to low pest population levels.

Planting date also affects the use of insecticide for controlling insect pests in cowpea. Farmers in the dry savannahs manipulate cowpea planting dates to avoid insect pest and disease attack. Kamara et al. (2010) reported the most effective combination of planting dates with insecticide spraying regimes for the management of insect pests of cowpea. In their study, they found that delaying planting beyond mid-August reduced cowpea grain yield by 12.3%, while the yield of the medium-maturing variety was significantly higher when planted in mid-August and sprayed twice than when planted earlier or on later dates in the savannahs of northeastern Nigeria. In that study, they also found that the yields of the indeterminate late-maturing variety were higher when planted in early August and sprayed with insecticide three times. They concluded that early and medium maturing cowpea varieties should be planted in mid-August, while the late-maturing indeterminate varieties should be planted in early August and sprayed thrice.

In WCA, reproductive development of cowpea is determined primarily by their response to photoperiod (Craufurd et al., 1996). Some cowpea genotypes are photoperiod-sensitive, while others are photoperiod-insensitive. Photoperiod which is determined by daylength is a critical factor in determining the appropriate planting time to establish cowpea. Among all the legumes, cowpea has the maximum diversity for plant type, growth habit, maturity and seed type. Breeders in West Africa have developed a range of cowpea varieties differing in growth habits (Singh and Sharma, 1996). These varieties respond differently to different photoperiods and growing environments. Some varieties of cowpeas, such as the local and indeterminate improved varieties, are photoperiod-sensitive. Planting these varieties at the onset of the rainy season (June or early July) will delay flowering and promote excessive vegetative growth leading to low yield (Kamara et al., 2010). These varieties should, therefore, be planted in mid-July to mid-August. However, planting these varieties later than mid-August in the dry savannahs will risk crop failure because of early cessation of rain in October (Kamara et al., 2016). On the other hand, the erect, early and medium maturing varieties that are photoperiod-insensitive can be planted anytime in the year provided there is sufficient rainfall or irrigation facilities. Ewansiha and Tofa (2016) reported that photosensitive cowpea varieties planted at the end of July produced significantly higher number of pods, number of seeds and grain yield compared with other planting dates, whereas the medium maturing erect cowpea varieties such as IT99K-573-1-1 (photoperiod-insensitive) produced higher pods, number of seeds and grain yield at all sowing dates in the Sudan savannah. They concluded that for optimum yield, indeterminate and prostrate cowpea varieties should be sown at the end of July, while the erect, medium or early maturing photoperiod-insensitive varieties could be planted at any time. In another study, Asante et al. (2001) reported that photoperiod-insensitive elite cowpea lines performed better in terms of grain yield when planted between mid-June and mid-July without insecticide protection, whereas the local varieties, which are mostly photoperiod-sensitive, produced higher grain yield when planted between late July and early August. This suggests that an important strategy to maximize cowpea yield is the ability to fit the cowpea varieties into their different planting dates for optimum performance. This makes the choice of planting date very important management decision in cowpea production. Table 1 highlights the planting dates recommended in some locality in WCA. The growing evidence from scientific and local observations suggests that the rainy season no longer comes at the same time as it did in the past. Farmers reported shifts in the onset of the rainy season, which used to start in April, toward May (Sanfo et al., 2014). With the effects of climate change, prediction of planting dates of cowpea in the savannahs from climate alone is becoming more difficult. Thus, planting dates may be chosen based on the plant maturity by ensuring that the critical growth stage, such as flowering, is synchronized with the availability of sufficient moisture during the cropping period.

As one moves from the Sahelian and Sudan to the Derived savannah, rainfall increases and the date to plant cowpea is further delayed. Distinct variations have been observed in the growth and reproduction of cowpea planted at different times in this ecology. Some authors have argued that early planting (before August) will make the plant flower when

Table 1 Rainfall distribution and planting dates of cowpea in WCA

Commencement of rains	Duration	Cowpea growth habit	When to plant
May	May–October	Erect (early and extra-early maturity)	August, week 2
		Semi-erect (medium maturity)	August, week 1
		Prostrate (late)	August, week 2
June	June–October	Erect (early and extra-early maturity)	August, week 3
		Semi-erect (medium maturity)	August, week 1
		Prostrate (late)	August, week 3
June/July	July–October	Erect (early and extra-early maturity)	End of July
		Semi-erect (medium maturity)	End of July
		Prostrate (late)	August, week 1

Source Dugje et al. (2009)

rainfall is heavy and then require frequent spraying, while others argued for late planting. Studies have shown that for the medium/late maturing cowpea varieties, planting should commence from mid-August to early September in the Guinea savannahs, depending on the onset of late season rains. Akande et al. (2012) reported that cowpea planted in August and September matured early and produced significantly higher grain yield than cowpea planted in June and July in the Derived savannah of West Africa. Higher incidence of diseases was also reported for early planting in June and July. However, Ezeaku et al. (2015) reported that some cowpea varieties planted early (mid-July) produced significantly higher yield and yield components than those planted late (mid-September). They found that the mean grain yield of early planting ranged from 921 to 1220 kg ha^{-1} compared to late planting dates that produced grain yields that ranges from 326 to 723 kg ha^{-1}.

It is clear that planting date is a very critical factor in determining cowpea performance in the savannahs of WCA. Therefore, for optimization of yield, planting cowpea at the appropriate time is an important agronomic requirement for obtaining higher yield.

5 Nutrient management for increased cowpea productivity

Despite the importance of cowpea particularly in the dry savannahs of West Africa, its yields are very low due to several constraints including poor soil, insect pests and drought. One striking feature of the soils is their inherent low fertility expressed in low levels of organic carbon (generally less than 0.3%), total and available phosphorus and nitrogen and effective cation exchange capacity (ECEC) (Bationo et al., 2002). Total and available P levels are very low and P deficiency is the most limiting soil fertility factor for cowpea production. Apart from low P stocks, the low-activity nature of these soils results in a relatively low capacity to fix added phosphorus (Bationo et al., 1995). Soil, water and nutrient management practices are inadequate to sustain food production and to meet the food requirements of the fast-growing population. Although organic amendments such as crop residue, manure or compost are essential in the sustainability of the cropping systems, they are often not sufficient to prevent nutrient depletion. As a result, the use of external inputs such as inorganic plant nutrients or local sources of P such as phosphate rock is an essential requirement for soil productivity (Batieno et al., 2002). Significant cowpea responses to nitrogen applied as urea have been obtained in different agroecological zones of the West African savannahs (Batieno et al., 2002) suggesting that the crop does not fix enough nitrogen because of limitations in other crop nutrients. Cowpea has a high P requirement. P is essential for growth, pod formation and N fixation in legumes (Abaidoo et al., 2007). According to Adu-Gyamfi et al. (2009), legumes generally need a high P requirement for adequate growth, nodulation and N fixation. However, the available P content of soils in West Africa (averaging 8 mg kg^{-1}) is seldom adequate for optimal plant growth (Manu et al., 1991). Consequently, production of cowpea in West Africa and many parts of sub-Saharan Africa is hindered by low soil-available P, which is also a characteristic of about 5·7 billion ha of land worldwide (Bationo et al., 1985; Hinsinger, 2001). Furthermore, the soils of many sub-Saharan regions of Africa are characterized by low-activity clays with high P-fixing capacities, which render P less available to plants (Abaidoo et al., 2007). Smalberger et al. (2006) observed that soil P deficiency in sub-Saharan Africa was so severe that other technologies such as optimum seeding rate, improved germplasm and inoculation would not work without some form of P added as fertilizers.

One of the major benefits of cowpea production in many parts of Africa is its capacity for biological N fixation (BNF). Cowpea can fix between 20 and 100 kg N ha^{-1} with an estimated N fertilizer replacement value ranging from 10 to 80 kg N ha^{-1} (Carsky et al., 2002; Sanginga et al., 2000). The N fixed is made available to associated or succeeding cereals in the predominantly poor savannah soils. Cowpea has therefore become important in the cropping systems which allow crop rotation with cereals to supply N through BNF. However, BNF in legumes can be limited or enhanced by P availability and utilization (Vance et al., 2003; Waluyo et al., 2004). Because of its multiple effects on plant nutrition (not only on nodulation), a P fertilizer is recommended to increase yields (P_2O_5: 20–60 kg ha^{-1}). Singh et al. (2011) reported that cowpea showed a significant response to applied P on pods per plant, grain and stover yield and 100-seed weight with highest response of application to 60 kg P ha^{-1}. Despite the importance of P in the soils of cowpea growing regions in West Africa, the use of commercial P fertilizers is limited due to the high cost of imported fertilizers. Batieno et al. (2002), therefore, recommended the direct application of indigenous phosphate rocks (PR) as an alternative to the use of more expensive water-soluble phosphorus fertilizers. These phosphate rocks, however, produce variable results because of low agronomic effectiveness compared to the water-soluble fertilizers such as single super phosphate and triple super phosphate. Karikari et al. (2015) evaluated three varieties of cowpea at four P rates of 0, 20, 40 and 60 kg ha^{-1} P_2O_5 in Ghana. The rate of P fertilizer application was directly proportional to the grain yield in all three cowpea varieties. The highest grain yield of 1682 and 1476 kg ha^{-1} for major and minor seasons, respectively, was produced at 60 kg ha^{-1} P_2O_5. Ndor et al. (2012) reported significant increase in grain yield of cowpeas when P was applied at 40 kg P_2O_5 ha^{-1} in the southern Guinea savannah of Nigeria. In the predominantly cowpea growing region of the West Africa dry savannahs, the addition of some quantities of organic materials in combination with mineral fertilizers can maintain soil fertility and increase crop yields. Results from a long-term experiment at Sadore in Niger indicated that the application of small quantities of fertilizers and crop residues resulted in an increase of cowpea fodder yield from 1700 to 5300 kg ha^{-1}. In on-farm trials, pocket applications of small quantities of manure (3 t ha^{-1}) plus 4 kg ha^{-1} of P at seedling time increased cowpea yield from 180 kg ha^{-1} in the control plot to 400 kg^{-1} ha^{-1} (Batieno et al., 2002).

6 Integrated pest management (IPM) in cowpea production

Despite the importance of cowpea and its high yield potential in the West African savannahs, insect pest attack is a major constraint to production (Singh et al., 1990). Severity can vary and sometimes leads to total yield loss (Singh and Allen, 1980). Yield losses of up to 70%, from insect pests alone, have been reported (Rusoke and Rubaihayo, 1994). In some areas, the losses caused by insect pests account for a reduction in grain yield as much as 80% (ICIPE, 1980). Cowpea growers in West Africa are at risk of losing the entire crop to insect pests in most growing seasons. The most damaging of all insect pests are those that attack the crop during the flowering and podding stages (Jackai et al., 1985). Worldwide, insect pests, especially *Aphis craccivora* Koch, *Megalurothrips sjostedti* Trybom, *Maruca virata* Fab., and a complex of pod-feeding bugs cause the greatest yield reductions (Omongo et al., 1997). In a recent study, Kamara et al. (2007) reported that flower thrips, the legume pod-borer (Maruca) and a range of pod-feeding bugs were the major insect pests of cowpea in the dry savannahs of northeast Nigeria. Maruca larvae damage flower buds, flowers, green pods and seeds (Singh and Jackai, 1985). Thrips start to attack at flower initiation, causing flower bud abortion (Akingbohungbe, 1982). Adults and nymphs of pod bugs remove sap from green pods, causing abnormal pod and seed formation (Singh and Jackai, 1985). High levels of insect resistance are not available in current cultivars (Oghiakhe et al., 1995; H.A. Ajeigbe, pers. comm.). The development of integrated IPM strategies is the key for successful cowpea production. Insecticide application is the most widely known means of insect pest control method in cowpea (Matteson, 1982); it is not otherwise feasible to grow cowpea commercially (Jackai et al., 1985). Farmers can improve yield 10-fold if insecticides are used (Singh and Jackai, 1985).

In Nigeria, Alghali (1992) obtained yield increases of 50–200% following the application of insecticide, once each at flowering and podding stages. Spraying once at flowering stage increased grain yield by 75%; two sprays, once each at flowering and podding stages, significantly, reduced insect pest population levels and increased grain yield by 126% (Kamara et al., 2007). In Kenya, Kyamanywa (1996) obtained a 15-fold increase in grain yield after two sprays, once each at flowering and podding stages. Some farmers in Nigeria and elsewhere have resorted to the indiscriminate use of insecticides to reduce pest damage (Kamara et al., 2007), sometimes applying as many as 8–10 sprays per season. However, most Nigerian farmers are resource poor and require pest management strategies that are cost-effective and sustainable. The use of insecticides must be minimized because of high costs and harmful effects on human health and the environment (Giliomee, 1997). To increase effectiveness and reduce overuse, their application should be integrated with other cultural practices for insect pest management in cowpea (Kamara et al., 2010). In recent years, planting date has been identified as an important component of IPM practices (Kamara et al., 2010). It has been suggested that adjusting of planting dates could cause asynchrony between crops and insect pests (Pedigo, 1989). For example, in the Delmarva Region, USA, planting cowpea early in combination with an application of insecticide resulted in a much higher grain yield than planting it late (Javaid et al., 2005). Similarly, in Uganda, Karungi et al. (2000) reported that early planting reduced levels of infestation by aphids, thrips and pod-feeding bugs but increased levels of infestation by Maruca. There is usually a build-up of pests as the season progresses that causes most damage to late planted cowpea. Early sowing has also been reported to enable the crop to escape high temperatures during the flowering stages when the crop is sensitive to heat (Hall, 1992; Ismaila and Hall, 1998). Taylor (1978) and Akingbohungbe (1982) suggested that differences in planting dates could be explored as they might offer some scope in avoiding various insect pests. Cultural practices, when combined with insecticides, are probably effective against some pests and could be used as components of IPM (Javaid et al., 2005). In an earlier study in Nigeria, the grain yield of cowpea planted early in the season was higher than from the late planted crops (IITA, 1982). The increased yield was attributed to low pest population levels, and the crop was allowed to mature before the rains ceased. Some authors have argued that early planting in West Africa (before August) will allow the plant to flower in September when rainfall is heavy and cowpea flowering would usually require frequent spraying. Also, most of the cowpea varieties grown by the farmers are photosensitive and should be planted late in the season to enhance early flowering (H. A. Ajeigbe, pers. comm.). Farmers are therefore usually advised to plant their cowpea around late July to early August. Kamara et al. (2010) sought to establish the most effective combination of planting dates with insecticide spraying regimes for the management of insect pests of cowpea in the savannahs of northeast Nigeria. They reported that three sprays, made once each at the bud initiation, flowering and podding stages, did not differ significantly from those of two sprays, made once each at flowering and podding, in terms of reducing insect pest population and increasing grain yield. Despite the reduction in insect infestation, delaying planting beyond mid-August reduced cowpea grain yield by 12.3%, on average. The yield of the medium-maturing variety IT89KD-391 was significantly higher when planted in mid-August and sprayed twice than when planted on the earlier or later dates. The yield of the indeterminate late-maturing variety ITKD89-288 was higher when planted in early August and sprayed thrice. Early and medium maturing cowpea varieties should, therefore, be planted in mid-August and sprayed twice. Late-maturing indeterminate varieties should be planted in early August and sprayed thrice.

7 Future trends and conclusion

Cowpea production in West Africa is increasing because of the high yield of improved varieties and perhaps increases in market outlets. However, yields of cowpea in smallholders' farms are quite low due to poor crop management despite the availability of improved varieties that are tolerant to biotic and abiotic stresses. When intercropped with maize, indeterminate cowpea varieties should be sown early into extra early or early maturing maize varieties under full maize population for higher grain and fodder. IPM would increase cowpea yield by over 200%. Proper nutrients especially P fertilization would increase the productivity in Africa. Response to fertilizer is, however, dependent on good agronomic practices such as using appropriate planting systems, plant configuration, IPM and availability of water. The usage of agricultural management strategies such as tailored crop planting dates and optimal plant populations would contribute not only to reduce crop failure but also to increase cowpea production. Supplementary irrigation and deployment of *Striga* resistant, heat- and drought-tolerant cowpea varieties would help in improving cowpea productivity.

8 Where to look for further information

1 Fatokun, C. A., S. A. Tarawali, B. B. Singh, P. M. Kormawa and M. Tamo (Eds), (2002). Challenges and opportunities for enhancing sustainable cowpea production. Proceedings of the World Cowpea Conference held at the International Institute of Tropical Agriculture (IITA), Ibadan, Nigeria, 4–8 September 2000. IITA, Ibadan, Nigeria, 433p.

2 Boukar, O., O. Coulibaly, C. A. Fatokun, K. Lopez and M. Tamo (Eds), Innovative research along the cowpea value chain. Proceedings of the Fifth World Cowpea Conference on improving livelihoods in the cowpea value chain through advancement in science, held in Saly, Senegal, 27 September–1 October 2010. IITA, Nigeria, 432p.

9 References

Abaidoo, R. C., N. Sanginga, J. A. Okogun, G. O. Kolawole, B. K. Tossah and J. Diels (2007). Genotypic variation of soybean for phosphorus use efficiency and their contribution of N and P to subsequent maize crops in three ecological zones of West Africa. In: *Proceedings of the Demand-Driven Technologies for Sustainable Maize Production in West and Central Africa: Proceedings of the 5th Biennial Regional Maize Workshop, IITA-Cotonou Benin Republic, 3–6 May 2007*. Nigeria: WECAMAN/IITA, pp. 194–224.

Adu-Gyamfi, J. J, M. Aigner and D. Gludovacz (2009). Variations in phosphorus acquisition from sparingly soluble forms by maize and soybean in low- and medium-P soils using P-32. In: *The Proceedings of the International Plant Nutrition Colloquium XVI*, Publisher is Department of Plant Science, UC Davis, p. 1317.

Ajeigbe, H. A., B. B. Singh and T. O. Oseni (2005). Cowpea-cereal intercrop productivity in the Sudan Savanna Zone of Nigeria as affected by planting pattern, crop variety and pest management. *African Crop Science Journal*, 13:269–79.

Ajeigbe, H. A., T. O. Oseni and B. B. Singh (2006). Effect of planting pattern, crop variety and insecticide on the productivity of cowpea-cereal systems in Northern Guinea Savanna of Nigeria. *Journal of Food, Agriculture and Environment*, 4:101–7.

Ajeigbe, H. A, A. Y. Kamara and D. Chikoye (2008). Improved legume-cereals based cropping systems for improved productivity and natural resource management by resource poor crop-livestock farmers in West Africa. Presented at the Tropentag conference (Tropentag 2008), held at the University of Hohenheim, Germany.

Ajeigbe, H. A., F. Ekeleme and D. Chikoye (2010a). *Improved Crop– livestock System for Enhanced Food Security and Income Generation in West Africa: Final Project Report: Gatsby Improved Crop–livestock Project (Project Number: GAT2833)*. International Institute of Tropical Agriculture (IITA), Ibadan, Nigeria, 50p.

Ajeigbe, H. A., B. B. Singh, J. O. Adeosun and I. E. Ezeaku (2010b). Participatory on-farm evaluation of improved legume-cereals cropping systems for crop-livestock farmers: Maize-double cowpea in northern Guinea savannah Zone of Nigeria. *African Journal of Agricultural Research*, 5:2080–8.

Akande, S. R, S. A. Olakojo, S. A. Ajayi, O. F. Owolade, J. A. Adetumbi, O. N. Adeniyan and B. A. Ogunbodede (2012). Planting date effects on cowpea seed yield and quality at southern Guinea Savanna of Nigeria. *Seed Technology*, 34(1):879–88.

Akingbohungbe, A. E. (1982). Seasonal variation in cowpea crop performance at Ile-Ife, Nigeria, and relationship to insect damage. *Insect Science Applied*, 3:287–96.

Alghali, A. M. (1992). Insecticide application schedules to reduce grain yield losses caused by insect pests of cowpea in Nigeria. *Insect Science Applied*, 13:725–30.

Asante, S. K., M. Tamo and L. E. N. Jackal (2001). Integrated management of cowpea insect pests using elite cultivars, date of planting and minimum insecticide application. *African Crop Science Journal*, 9:655–65.

Bationo, A. B., M. Gaoh, B. C. Christianson and D. K. Friesen (1985). IFDC phosphate, nitrogen and sulfur fertilizer research in Niger: Objectives and progress. In: J. D. Axtell and J. W. Clark (Eds), *Niger Sorghum and Millet Workshop. Report of Research Collaboration in Niger. Proceedings of the Workshop INTSORMIL-INRAN-PARA. Niamey, Niger, 13–17 October, 1985*. INTSORMIL, Univ. Nebraska, Lincoln, pp. 105–15.

Bationo, A., A. Buerkert, M. P. Sedogo, B. C. Christianson and A. U. Mokwunye (1995). A critical review of crop residue use as soil amendment in the West African semi-arid tropics. In: Powell, J. M., S. Fernandez Rivera, T. O. Williams and C. Renard (Eds), *Livestock and Sustainable Nutrient Cycling in Mixed Farming Systems of sub-Saharan Africa. Proceedings of International Conference*. Addis Ababa, Ethiopia: ILCA, pp. 305–22.

Bationo, A., B. R., N'tare, S. A. Tarawali and R. Tabo (2002). Soil fertility management and cowpea production in the semiarid tropics. In: Fatokun, C. A., S. A. Tarawali, B. B. Singh, P. M. Kormawa and M. Tamò (Eds), *Challenges and Opportunities for Enhancing Sustainable Cowpea Production*. Nigeria: IITA, pp. 301–18.

Bruns, H. A. (2011). Planting date, rate, and twin-row vs. Single-row soybean in the Mid-South. *Agronomy Journal*, 103(5):1308–13

Carsky, R. J., B. Vanlauwe and O. Lyasse (2002). Cowpea rotation as a resource management technology for cereal-based systems in the savannahs of West Africa. In: Fatokun, C. A, Tarawali, S. A., Singh, B. B., Kormawa, P. M. and M. Tamo (Eds), *Challenges and Opportunities for Enhancing Sustainable Cowpea Production*. Nigeria: IITA, pp. 252–66.

Clark, K. M. and R. L. Myers (1994). Intercrop performance of pearl millet, amaranth, cowpea, soybean, and guar in response to planting pattern and nitrogen fertilization. *Agronomy Journal*, 86(6):1097–102.

Craufurd, P. Q., Q. Aiming, R. J. Summerfield, R. H. Ellis and E. H. Roberts (1996). Development in cowpea (*Vigna unguiculata*). III. Effects of temperature and photoperiod on time to flowering in photoperiod-sensitive genotypes and screening for photothermal responses. *Experimental Agriculture*, 32:29–40.

Dugje, I. Y., L. O. Omoigui, F. Ekeleme, A. Y. Kamara and H. Ajeigbe (2009). *Farmers' Guide to Cowpea Production in West Africa*. Ibadan, Nigeria: IITA, p. 3.

Evans, L. T. and R. A. Fischer (1999). Yield potential: Its definition, measurement and significance. *Crop Science*, 39:1544–51.

Ewansiha, S. U. and A. I. Tofa (2016). Yield response of cowpea varieties to sowing dates in a Sudan savannah agroecology of Nigeria. *Bayero Journal of Applied Science*, 9(1):62–7.

Ewansiha, S. U. Chiezey, U. F. Onyibe, J. E. and Kamara, A. Y. (2013) Intercropping for increased fodder and grain yields in maize and cowpea. *Nigerian Journal of Applied Science*, 31:54–65.

Ewansiha, S. U., A. Y. Kamara and J. E. Onyibe (2014a). Performance of cowpea cultivars when grown as an intercrop with maize of contrasting maturities. *Archives of Agronomy and Soil Science*, 60:597–608,

Ewansiha, S. U., A. Y. Kamara and U. F. Chiezey (2014b). Cowpea response to planting date under different maize maturity types in West African Sudan savannah. *Journal of Tropical Agriculture*, 52(2):37–42.

Ewansiha, S. U., A. Y. Kamara, U. F. Chiezey and J. E. Onyibe (2015a). Performance of cowpea grown as an intercrop with maize of different populations. *African Crop Science Journal*, 23(2):113–22.

Ewansiha, S. U., A. Y. Kamara, U. F. Chiezey and J. E. Onyibe (2015b). Response of cowpea to sowing date and maize plant population in a Sudan savannah environment. *Tropical Agriculture (Trinidad)*, 92(1):80–91.

Ezeaku, I. E., B. N. Mbah and K. P. Baiyeri (2015). Planting date and cultivar effects on growth and yield performance of cowpea (*Vigna unguiculata* (L.) Walp). *African Journal of Plant Science*, 9(11):439–48.

Ezedinma, F. O. C. (1974). Effect of close spacing on cowpea (*Vigna unguiculata* (L.)) in southern Nigeria. *Experimental Agriculture*, 10:289–98.

FAO (2014) *FAOSTAT*, http://faostat3.fao.org/download/Q/QC/E 1/04/2014.

Giliomee, J. H. (1997). The future of pesticides in crop production. *African Crop Science Conference Proceeding, Pretoria, South Africa*, 3:57–66.

Hall, A. E. (1992). Breeding for heat tolerance. *Plant Breed Review*, 10:129–68.

Hinsinger, P. (2001). Bioavailability of trace elements as related to root-induced chemical changes in the rhizosphere. In: Gobran, G. R., W. Wenzel and E. Lombi (Eds), *Trace Elements in the Rhizosphere*. Boca Raton, FL, USA: CRC Press, pp. 173–95.

ICIPE (International Centre of Insect Physiology and Ecology) (1980). *Annual Report for 1979*. Nairobi, Kenya, pp. 28–42.

IITA (International Institute of Tropical Agriculture) (1982). *Annual report for the 1981*, Ibadan, Nigeria.

IITA (International Institute of Tropical Agriculture) (2009). *Crops* (Online). Available: http://www.iita.org (13 June 2010).

Inaizumi, H., Singh, B. B., Sanginga, P. C., Manyong, V. M., Adesina, A. A. and S. Tarawali (1999). *Adoption and Impact of Dry Season Dual Purpose Cowpea in the Semi-arid Zone of Nigeria.* IITA IMPACT Series. IITA (International Institute of Tropical Agriculture), Ibadan, Nigeria, 14p.

Ishiyaku, M. F., Singh B. B and P. Q. Craufurd (2005). Inheritance of time to flowering in cowpea (*Vigna unguiculata* (L.) Walp). *Euphytica*, 142:291–300.

Ismaila, A. M. and A. E. Hall (1998). Positive and potential negative effects of heat-tolerance genes in cowpea lines. *Crop Science*, 38:7–12.

Ismaila, A. M. and A. E. Hall (2000). Semi-dwarf and standard-height cowpea responses to row spacing in different environments. *Crop Science*, 40:1618–23

Isubikalu, W. P., J. M. Erbaugh, A. R. Semana and E. Adipala (1999). Influence of farmer production goals on cowpea pest management in eastern Uganda: Implications for developing IPM programs. *African Crop Science Journal*, 7(4):539–48.

Jackai, L. E. N., S. R. Singh A. K. Raheja and F. Wiedijk (1985). Recent trends in the control of cowpea pests in Africa. In: Singh, S. R., K. O. Rachiel (Eds), *Cowpea Research, Production and Utilization.* Chichester (UK): Wiley and Sons Ltd, pp. 233–46.

Jagtap, S. S. and F. J. Abamu (2003). Matching improved maize production technologies to the resource base of farmers in a moist savanna, *Agricultural Systems*, 76(3): 1067–84.

Jallow, A. T and T. U. Fergusson(1985). Effects of plant densities and cultivar on seed yield of cowpeas (*Vigna unguiculata* (L.)) in Trinidad. *Tropical Agriculture*, 62:121–4

Javaid, J., R. B. Dadson, F. M. Hashem and L. A. Arthur (2005). Effects of insecticide spray applications, sowing dates and cultivar resistance on insect pests of cowpea in the Delmarva Region of the United States. *Journal of Sustainable Agriculture*, 26:57–68.

Kamara, A. Y., Chikoye D., Omoigui L. O and I. Y. Dugje (2007). Influence of insecticide spraying regimes and cultivar on insect pests and yield of cowpea in the dry savannahs of north-east Nigeria. *Journal of Food, Agriculture and Environment* 5(1):154–8.

Kamara, A. Y. S. U. Ewansiha, H. A. Ajeigbe, R. Okechukwu, H. Tefera, O. Boukar and L. O. Omoigui (2010a) Improvements in grain and fodder yield of cowpea (*Vigna unguiculata*) varieties developed in the Sudan savannas of Nigeria over the past four decades. In: Boukar, O., O. Coulibaly, C. A. Fatokun, K. Lopez, M. Tamo (Eds), *Innovative Research along the Cowpea Value Chain. Proceedings of the Fifth World Cowpea Conference on Improving Livelihoods in the Cowpea Value Chain Through Advancement in Science, held in Saly, Senegal, 27 September–1 October 2010.* IITA, Nigeria, pp. 179–88.

Kamara, C. S. and W. Godfrey-Sam-Aggrey (1979). Time of planting, rainfall and soil moisture effects on cowpea in Sierra Leone. *Experimental Agriculture*, 15:315–20.

Kamara, A. Y., F. Ekeleme, L. O. Omoigui, T. Abdoulaye, P. Amaza, D. Chikoye, and I. Y. Dugje (2010). Integrating planting date with insecticide spraying regimes to manage insect pests of cowpea in North-Eastern Nigeria. *International Journal Pest Management*, 56:243–53

Kamara, A. Y., L. O. Omoigui, S. U. Ewansiha, F. Ekeleme, D. Chikoye and H. Ajeigbe (2011). Performance of semi-determinate and indeterminate cowpeas intercropped with maize in Northeast Nigeria. *African Journal of Agricultural Research*, 8:1763–70.

Kamara, A. Y., T. I. Abdullahi, S. K. Boahen, R. Solomon, H. A. Ajeigbe and N. Kamai (2016). Effects of plant density on the performance of cowpea in Nigerian savannas. *Experimental Agriculture*: P1–13.

Karikari, B., Arkorful, A. and S. Addy (2015). Growth, Nodulation and yield response of cowpea to phosphorus fertilizer application in Ghana. *Journal of Agrononomy*, 14(4):234–40,

Karungi, J, E. Adipala, S. Kyamanywa, M. W. Ogenga-Latigo, N. Oyobo and L. E. N. Jackai (2000). Pest management in cowpea. Part 1. Influence of planting time and plant density on cowpea field pests infestation in eastern Uganda. *Crop Protection*, 19:231–6.

Kwapata, M. B. and A. E. Hall (1990). Determinants of cowpea (*Vigna unguiculata* (L.)) grain yield at extremely high plant density. *Field Crops Research*, 24:23–32.

Kyamanywa S. (1996). Influence of time of insecticide application on control of insect pest of cowpea and grain yield on cowpea at Mtwapa, Coastal Province of Kenya. *African Crop Science Journal*, 4:373–82.

Lithourgidis, A. S., C. A. Dordas, C. A. Damalas and D. N. Vlachostergios (2011). Annual intercrops: an alternative pathway for sustainable agriculture. *Australian Journal of Crop Science*, 5(4):396–410.

Liu, X. B., J. Jin, G. H. Wang and S. J. Herbert (2008). Soybean yield physiology and development of high-yielding practices in Northeast China. *Field Crops Research*, 105(3):157–71.

Manu, A., A. Bationo and S. C. Geiger (1991). Fertility status of selected millet producing soils of West Africa with emphasis on phosphorus. *Soil Science*, 152:315–20.

Matteson P. C (1982). The effect of intercropping with cereal and minimal permethrin application on insect pests of cowpea and their natural enemies in Nigeria. *Tropical Pest Management*, 28:374–80.

Mbong, G. A., C. N. Akem, O. Alabi, A. M. Emechebe and M. D. Alegbejo (2010). Effect of sowing date on the yield and yield components of cowpea infected with scab. *Asian Journal of Agricultural Science*, 2(2):57–62

Midmore, D. J. (1993). Agronomic modification of resource use and intercrop productivity. *Field Crops Research*, 34:357–80.

Mohammed, I. B., O. O. Olufajo, B. B. Singh, K. O. Oluwasemire and U. F. Chiezey (2008). Productivity of millet/cowpea intercrop as affected by cowpea genotype and row arrangement. *World Journal of Agricultural Sciences*, 4(5):818–24.

Ndor, E., N. Dauda, E. Abimuku, D. Azagatu and H. Anzaku (2012). Effect of phosphorus fertilizer and spacing on growth, nodulation count and yield of cowpea (*Vigna unguiculata* (L) Walp) in southern Guinea savanna agroecological zone, Nigeria. *Asian Journal of Agricultural Science*, 4:254–7.

N'tare, B. R and J. H Williams (1992). Response of cowpea cultivars to planting pattern and date of sowing intercrops with pearl millet in Niger. *Experimental Agriculture*, 28(1):41–8. Cambridge University Press.

Odion, E. C., Y. Yusuf, D. A. Labe and C. I. Amatobi (1994). Effect of crop arrangement, number and technique of insecticidal spray on cowpea and millet mixture in the Nigerian Sudan Savanna. *Samaru Journal of Agricultural Research*, 11:21–9.

Ofori, F. and W. R. Stern. (1987). Cereal-legume intercropping systems. *Advances in Agronomy*, 41:41–90.

Oghiakhe, S., L. E. N. Jackai and W. A. Makanjuola (1995). Evaluation of cowpea genotypes for field resistance to the legume pod borer, Maruca testulalis, in Nigeria. *Crop Protection*, 14(5):389–94.

Olufajo, O. O. and B. B. Singh. (2002). Advances in cowpea cropping systems research. In: Fatokun, C. A., S. A. Tarawali, B. B. Singh, P. M. Kormawa and M. Tamo (Eds), *Challenges and Opportunities for Enhancing Sustainable Cowpea Production. Proceedings of the World Cowpea Conference iii held at International Institute of Tropic Agriculture, (IITA), Ibadan, Nigeria, 4–8 September*. IITA, Ibadan, Nigeria, pp. 267–77.

Omongo, C. A., M. W. Ogenga-Latigo, S. Kyamanywa and E. Adipala (1997). The effect of seasons and cropping systems on the occurrence of cowpea pests in Uganda. *African Crop Science Conference Proceedings, Pretoria, South Africa*, 3:1111–16.

Pedigo, L. P (1989). *Entomology and Pest Management*. New York (USA): McMillan Publishing Company.

Rusoke, D. G. and P. R. Rubaihayo (1994). The influence of some crop protection management practices on yield stability of cowpeas. *African Crop Science Journal*, 2:143–8

Sanfo, S., Lamers, J. P., M. Mueller and F. M. William (Eds) (2014). Farmers' perception of climate change and climate variability versus climatic evidence in Burkina Faso, West Africa. In: *Climate change in Africa. Negotiations, Translations, and Socio-political Implications, Bonn, 10–12 September*. Kansas State University, USA.16p

Sanginga, N., O. Lyasse and B. B. Singh (2000). Phosphorus use efficiency and nitrogen balance of cowpea breeding lines in a low P soil of the derived savanna zone in West Africa. *Plant Soil*, 220:119–28.

Singh, S. R. and D. J. Allen (1980). Pests, diseases, resistance and protection of *Vigna unguiculata* (L.) Walp. In: Summerfield R. J. and A. H. Bunting (Eds), *Advances in Legume Science*. London (UK): Royal Botanical Garden, Kew, and Ministry of Agriculture, Fisheries and Food, pp. 419–23.

Singh, S. R. and L. E. N. Jackai (1985). Insect pests of cowpea in Africa: their life cycle, economic importance, and potential for control. In: Singh, S. R. and K. O. Rachie (Eds), *Cowpea Research, Production, and Utilization*. John Wiley and Sons, Chichester, UK, pp. 217–31

Singh, B. B. and H. Sharma. 1996. Restructuring cowpea for higher yield. *Indian Journal of Genetics* 56:389–405.

Singh, B. B. and Ajeigbe, H. A. (2002). Improving cowpeacereals based cropping systems in the dry savannas of West Africa. In: Fatokun, C. A., Tarawali, T. A. Singh, B. B., Kormawa, P. M. and Tamo, M. (Eds), Challenges and Opportunities for Enhancing Sustainable Cowpea Production. Proceedings of the World Cowpea Conference 111, 4–8 September 2000. IIITA, Ibadan, Nigeria, pp. 278–86.

Singh, S. R., L. E. N. Jackai, J. H. R. Dos Santos, C. B. Adalla (1990). Insect pests of cowpeas, pp. 43–90. In: Singh, S. R. (Ed.), *Insect Pests of Tropical Legumes*. Chichester, United Kingdom: John Wiley & Sons.

Singh, B. B., O. L. Chambliss and B. Sharma (1997). Recent advances in cowpea breeding. In: Singh, B. B., D. R. Mohan, K. E. Dashiell and L. E. N Jackai (Eds), *Advances in Cowpea Research*. A copublication of International Institute of Tropical Agriculture (IITA) and Japan International Research Center for Agricultural Sciences (JIRCAS), IITA, Ibadan, Nigeria, pp. 30–49.

Singh, B. B, J. D. Ehlers, B. Sharma, F. R. Freire-Filho (2002). Recent progress in cowpea breeding. In: Fatokun, C. A., Tarawali, S. A., Singh, B. B., Kormawa, P. M., Tamo, M. (Eds), *Proceedings, World Cowpea Conference III, Challenges and Opportunities for Enhancing Sustainable Cowpea Production*. IITA, Ibadan, Nigeria, pp. 22–40

Singh, A., A. L. Baoule, H. G. Ahmed, A. U. Dikko and U. Aliyu (2011). Influence of phosphorus on the performance of cowpea (Vigna unguiculata (L.) Walp.) varieties in the Sudan savanna of Nigeria. *Agricultural Science*, 2:313–17.

Smalberger, S. A., U. Singh, S. H. Chien, J. Henao and P. W. Wilkens (2006). Development and validation of a phosphate rock decision system. *Agronomy Journal* 98:471–83

Taylor, T. A (1978). Maruca testulalis, an important pest of tropical grain legumes. In: Singh, S. R., H. F. van Emden and T. A. Taylor (Eds), *Pests of Grain Legumes: Ecology and Control*. London (UK): Academic Press, pp. 193–200.

Terao, T., I. Watanabe, R. Matsunaga, S. Hakoyama and B. B. Singh (1997). Agro-physiological constraints in intercropped cowpea: An analysis. In: Singh, B. B., D. R. Mohan, K. E. Dashiell and L. E. N Jackai (Eds), *Advances in Cowpea Research*. A copublication of International Institute of Tropical Agriculture (IITA) and Japan International Research Center for Agricultural Sciences (JIRCAS), IITA, Ibadan, Nigeria, pp. 129–40

Utoh, N. O., O. O. Oyebanji and H. A. Ajeigbe (2008). A manual on cowpea seed production. Nigeria: Publication of the National Programme for Food Security, FAO/NPFS, 33p.

Vance, C. P., C. Uhde-Stone and D. Allan (2003). Phosphorus acquisition and use: critical adaptation by plants for securing non-renewable resources. *New Phytologist*, 15:423–47.

Waluyo, S. H, T. A. Lie and L. Mannetje (2004). Effect of phosphate on nodulation primordial of soybean (Glycine max. Merr) in acid soils in minirhizotron experiments. *Indonesian Journal of Agricultural Science*, 5:37–44.

Chapter 12

Developing improved varieties of faba bean

Fouad Maalouf, International Center for Agricultural Research in the Dry Areas (ICARDA), Lebanon; and Seid Ahmed and Somanagouda Patil, International Center for Agricultural Research in the Dry Areas (ICARDA), Morocco

1 Introduction
2 Heat tolerance
3 Drought tolerance
4 Herbicide tolerance
5 Breeding for resistance to foliar diseases
6 Breeding for broomrape resistance
7 Biological nitrogen fixation
8 Major achievements
9 Future trends and conclusion
10 References

1 Introduction

Faba bean (*Vicia faba* L.) is the third most important cool-season food legume crop grown under rainfed and irrigated conditions in more than 55 countries. It is grown under different cropping systems as dry grains, green pods, animal feed and green manure worldwide on 2.4 million ha, with 4.4 million tons produced in 2014 (Fig. 1). Faba bean production has been shown to contribute to reduction of carbon footprint of cereal-based agricultural production systems through its ability to fix nitrogen that can be used by the succeeding cereal crop and by breaking the cycle of biotic stresses (Jensen et al. 2010).

In North Africa, area under faba bean cultivation has remained stagnant during the last five decades, but its production has increased from 405 000 tons in 1961–65 to 605 000 tons in 2011–14. North Africa remains in short supply with an annual import of 0.44 million tons mainly from Australia and Canada (FAOSTAT 2016). In North Africa, the major faba bean-producing countries are Egypt, Morocco and Sudan, which are importing 0.34 MT with a value of 320 million USD, while West Asia imports 93 589 tons with a value of 65 million USD in 2013 (FAOSTAT 2016).

In East Africa, Ethiopia is the major consumer and producer of faba bean. The area under faba bean grain cultivation has increased from 348 400 ha in 1993–95 to 471 700 ha

http://dx.doi.org/10.19103/AS.2017.0023.31

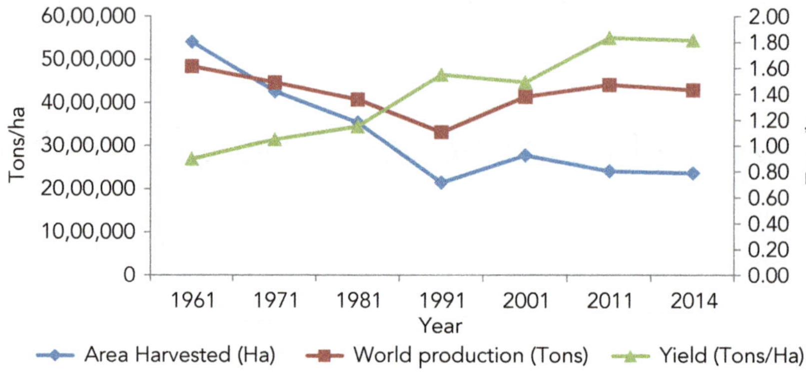

Figure 1 Trends of faba bean area, production and productivity in the world (FAOSTAT 2016).

in 2011–14 with an yield increase from 1 t/ha, to 1.9 t/ha, respectively. Productivity of improved varieties is higher (3.5 t/ha) than the country's average yield (1.8 t/ha). Although many countries in North Africa and the Nile valley region, especially Egypt, Sudan and Morocco, imported faba bean with total amount of 0.4 MT, Ethiopia exported 0.04 MT of faba bean with a value of 25 million USD in 2013 (FAOSTAT 2016).

The productivity of faba bean is decreasing due to various biotic and abiotic stresses. The major biotic stresses are parasitic weeds (*Orobanche* and *Phelipanche* spp.), foliar diseases, mainly chocolate spot (*Botrytis fabae*), rust (*Uromyces viciae fabae*), ascochyta blight (*Ascochyta fabae*), powdery mildew (*Microsphaera penicillata* var. *ludens*) and faba bean leaf gall (*Olpidium viciae* (Kusano)) and soil-borne diseases (*Fusarium* root rots), in West Asia, North Africa and East Africa (El-Sayed et al. 2011; Hanounik 1980; Maalouf et al. 2011; Tivoli et al. 2006; Abebe et al. 2013; Hailu et al. 2014). In Middle Egypt, faba bean necrotic yellow virus devastated the crop in 1992, which has led to the complete disappearance of all types of faba bean landraces and cultivars (Katul et al. 1993; Makkouk et al. 1994). In many production regions in the Mediterranean basin, the crop is seldom if ever irrigated and generally relies on rainfall for its growth and development. Variation in the amount and distribution of rainfall is generally the major reason for variability in seed yield of faba bean (Siddique et al. 2001). Heat stress is increasingly becoming a serious constraint to faba bean production in certain regions due to the shift from cooler, long-season environments to warm, short-season environments (Gaur et al. 2015). The area under faba bean cultivation in Egypt was reduced from 113 000 ha in 2001 to 46 000 ha in 2014 while in Sudan, where the weather is hotter and the season is shorter, the area increased from 50 000 ha in 2001 to 76 000 ha in 2014 (FAOSTAT 2016). This is in addition to climatic change where dry areas are becoming hotter. Heat sensitivity in grain legumes can reduce grain yield, quality and leads to restricted geographic adaptation. Soil acidity and salinity are the common problems that limit production of faba bean in Ethiopia (Asefa et al. 2010). Soil acidity can adversely affect survival, growth, nitrogen-fixation efficiency of rhizobia and formation of symbiotic association between rhizobia and their host legumes (Graham 1992; Zahran 1999).

The purpose of this review is to present major research achievements with regard to faba bean, which include identification of sources showing heat and drought tolerance, herbicide tolerance, broomrape resistance, disease resistance and high nitrogen fixation.

2 Heat tolerance

Heat, drought, waterlogging and frost are the major abiotic stresses affecting faba bean productivity in North, East Africa and West Asian countries. Heat stress is particularly harmful at the reproductive stage of faba bean plants. In fact, high sensitivity of pollen to heat stress is shown in legumes and other plant species (Prasad et al. 2006; Wang et al. 2006). The seed-filling period is also very sensitive to elevated temperatures (Egli et al. 2005). The duration from seedling emergence to flowering and maturity of faba bean has been shown to be governed by the photo-thermal duration. The optimum temperature for flowering of faba bean is 22–23°C (Patrick and Stoddard 2010). Suboptimal temperatures and photoperiods during early plant growth delay flowering. Ellis et al. (1988) noted marked differences in the rate of seedling emergence and progress to flowering of six faba bean genotypes in response to supra-optimal temperatures. Ellis et al. (1990) demonstrated that various faba bean genotypes show similar sensitivity towards the rate of progress from flowering to temperature regime but variable sensitivity to photoperiod.

Extreme heat is the major threat to faba bean production in south Egypt, Sudan and the Ethiopian lowlands. Artificially induced terminal heat stress can significantly reduce yield and the yield components of faba bean genotypes (Abdelmula and Abuanja 2007). This adverse effect could be attributed mainly to high temperatures during the vegetation period, which checked growth and led to the development of a small, short-stemmed crop with few branches and pods. Abdelmula and Abuanja (2007) concluded that the genotype C.52/1/1/1 could be used to improve heat tolerance in faba bean and to make it possible to extend production of faba bean in the non-traditional areas of Sudan. Moreover, the traits that exhibited strong and positive association with yield could be used as selection criteria for improving faba bean sensitivity under heat stress conditions.

3 Drought tolerance

Drought, an interval of water deficiency leading to a significant reduction in yield, is widely considered to be the most important environmental constraint to crop productivity (Borlaug and Dowswell 2005). Variation in the amount and distribution of rainfall is generally considered the major reason for variability in grain yield of faba bean (Bond et al. 1994; Siddique et al. 2001).

Terminal drought is one important constraint to faba bean production, particularly in Ethiopia and Morocco, where it is largely grown under rainfed conditions. Identification of early maturing genotypes adapted to drylands with a short rainfed period and use of supplemental irrigation would have a significant impact on final yield. Although genotypic variation in the response of faba bean to drought has been documented (Abdelmula et al. 1999; Amede et al. 1999), the development of drought-tolerant cultivars is essential to improve its yield stability. Plant breeders evaluate and select breeding material empirically for adaptation to drought under experiment conditions based on grain yield at drought-prone sites. Maalouf et al. (2015) reported significant differences among faba bean genotypes under different water regimes without any significant interactions. FLIP06-010FB was the highest yielding and most stable among the tested genotypes across different environments. Days to flowering and maturity were positively correlated with grain yield.

Some studies have reported physiological traits associated with drought, such as carbon isotope discrimination, leaf temperature and stomatal conductance (Khan et al. 2007). However, the physiological processes associated with adaptation to drought are not as well understood as those for some other crop species. To breed drought-tolerant cultivars, it is imperative to identify the specific physiological traits that improve adaptation to water-limited environments (Subbarao et al. 1995). So far, progress using this approach has been slow mainly due to the lack of an efficient screening technique (Wery et al. 1994; Stoddard et al. 2006). The physiological status of the canopy may be measured via spectral reflectance, which is a practical method of integrating physiological trait selection within crop improvement and research programmes (Peñuelas et al. 1995a,b; Osborne et al. 2002; Babar et al. 2006). In faba bean the spectral indices, Structure Insensitive Pigment Index and Normalized Phenophytinization Index, were found to positively correlate with grain yield and could therefore be used for selection under drought-prone environments (Maalouf et al. 2015).

4 Herbicide tolerance

Non-parasitic and parasitic weeds are the major production constraints in faba bean because of slow initial growth and being poor competitors to weeds. Mechanical or manual weeding is considered to be the only management option in faba bean because of high sensitivity to post-emergence herbicides. Pre-emergence herbicides are effective in controlling weeds at the early stage of seedling growth, but weeds germinating after crop emergence become dominant in the field and cause substantial yield losses (Solh and Pala 1990). In addition, faba bean is also heavily infested by parasitic weeds, *Orobanche* spp., in North and East Africa. Resistance to *Orobanche* in current varieties is partial and consequently there is a need to develop integrated management with herbicide tolerance such as low dosage of glyphosate, which may affect the crop in some cases. Therefore, to supplement the partial resistance to *Orobanche*, there is a need to screen faba bean lines for high dosage of glyphosate or to other effective herbicides such as Imazethapyr and Metribuzin.

Figure 2 Grouping of genotypes based on the herbicide tolerance level to Metribuzin.

At the International Center of Agricultural Research in the Dry Areas (ICARDA), 289 mutant lines of faba bean were screened against high dosage of glyphosate, resulting in the identification of two lines Mu38 and Mu418 as tolerant to 1600 g a.i./ha (Balech et al. 2016, 300). Similarly faba bean lines were screened against various post-emergence herbicides like Metribuzin, Imazethapyr and Oxyfluorfen with different modes of action (photo-synthesis inhibitor and amino acid synthesis inhibitor) during the 2014/2015 season; among the total lines, 45 genotypes were selected as tolerant to moderately tolerant to Metribuzin and Imazethapyr, respectively. In general, the sensitivity for Imazethapyr was higher than that for Metribuzin. The selected lines were retested in 2015/2016 under dry conditions, and two genotypes (Lattakia3 and Superagualdulce) were found to be the most tolerant to Metribuzin (Fig. 2).

5 Breeding for resistance to foliar diseases

Agronomical practices such as crop rotation, delayed planting date, fungicide applications and use of certified seeds may provide partial protection against the foliar diseases. However, the most effective and eco-friendly disease control method is through developing resistant cultivars. The effective resistant sources for ascochyta blight and chocolate spot were identified at ICARDA (Robertson 1984; Hanounik and Robertson 1988, 1989). These lines were used by ICARDA and National Agricultural Research Systems (NARS) to develop breeding lines with good levels of resistance and a high yield potential. As a result, varieties were released in Australia, Canada, China, Egypt, Ethiopia and Spain (Hanounik and Robertson 1989; Sillero et al. 2001; Trvoli et al. 2006; Redden et al. 2008; Villegas-Fernández et al. 2009; Sillero et al. 2010; Rubiales et al. 2012; Maalouf et al. 2013; Temesgen et al. 2015). Although most breeding programmes are focusing on developing resistant genotypes for a single disease of economic importance, efforts have recently been directed to develop faba bean lines with multiple disease resistance (Villegas-Fernández et al. 2011).

Recently, ICARDA has identified new sources for multiple disease resistance for ascochyta blight, chocolate spot and rust for further use in breeding programmes. The main area for the sources of resistance to chocolate spot and ascochyta blight (Maalouf et al. 2016a), Fertile Crescent, coincides with the centre of origin of this crop (Cubero 1973). More than 180 new sources for resistance to chocolate spot, ascochyta blight and rust were identified at ICARDA in 'sick plots' of heavy soils infested at Lattakia, Syria, by a mixture of the most virulent pathogens collected in Syria. Lines with combined resistance have been developed at ICARDA and sent to different NARS to observe the response of the resistant lines to different races in varying environments. The Ethiopian Institute of Agricultural Research (EIAR) released several varieties with a high-level resistance to chocolate spot, derived from the ICARDA breeding programme, in Ethiopia.

Through collaborative research, EIAR researchers have released several high-yielding faba bean varieties through direct selection from germplasm supplied by ICARDA, or by transferring good levels of resistance from ICARDA germplasm into locally adapted varieties. Among the faba bean varieties released with partial resistance to chocolate spot are 'Moti' (ILB 4432 X Kuse-2-27-33), 'Gebelcho' (ILB 4726 X 'Tesfa'), Obsie (ILB 4427 X CS20DK) and 'Walki' (ILB 4615 X Bulga 70). The variety 'Walki' was developed for water-logged areas, and is gaining popularity in the central highlands of Ethiopia, and more recently the varieties 'Gora' (Didea (EH01048-1: ILB2717-1x R878-1) and Didea, which have chocolate spot resistance and large seeds (EH01048-1: ILB2717-1x R878-1), have been released in Ethiopia.

6 Breeding for broomrape resistance

Faba bean cultivation is strongly hampered by the occurrence of broomrape, *Orobanche* spp., in the Middle East and North and East Africa. Faba bean is infested by various broomrape types (*Orobanche* spp.); among the most deleterious species, *Orobanche crenata* Forsk is considered indigenous to the Mediterranean basin (Rubiales et al. 2006). In Morocco, the infestation of *Orobanche* in faba bean progressed from 12% in 1981 to 51% in 2003 growing season. The estimated average yield losses due to *Orobanche* in Morocco was 37.4% in 1998. *Orobanche* infestation in Algeria spreads over 60% of the area in the middle coast and is continuing its spread from the middle zone towards the west. In Tunisia, the main *Orobanche* species include *O. crenata, O. foetida* and *O. ramosa*, affecting 7.1% area of total food legume-growing areas (Abu-Irmaileh and Labrada 2016). Faba bean was abandoned in Upper Egypt mainly due to *Orobanche* infestation, and only recently has faba bean regained some of its area, mainly due to the availability of *Orobanche*-resistant varieties. *O. crenata* is also affecting the productivity of faba bean in Ethiopia and northern Sudan.

Breeding for *Orobanche* resistance is difficult in faba bean due to limited sources of resistance, complex nature of the resistance mechanism and low heritability (Rubiales et al. 2006, 2009; Pérez de- Luque et al. 2010). Radwan et al. (1988a) found that parasitic weeds varied in their influence on host genotype and host–parasite relationship and were largely dependent on environmental conditions. The levels of resistance to *Orobanche* in faba bean can be characterized as necrosis of host cells that prevent *Orobanche* intrusion either before or just after penetration in the host root, or a barrier developing in the host root after the formation of a small tubercle (Pérez-de-Luque et al. 2007). Despite breeding difficulties, there have been significant successes in developing tolerant/resistant varieties and breeding materials in Egypt, Spain, Italy, Morocco and ICARDA (Elia 1964; Cubero 1973; Nassib et al. 1978, 1982; Fischbeck et al. 1986; Perrino et al. 1988; Radwan et al. 1988b; Cubero et al. 1992; Khalil et al. 2004) using the main source of *Orobanche* resistance line F402 identified by Egyptian researchers in early 1970s (Nassib et al. 1978) and some minor sources available in different countries.

7 Biological nitrogen fixation

The increasing demand for food to meet the growing human population put in evidence the importance of sustainable intensification of agricultural practices, which includes legume crops for their ability of biological nitrogen fixation. Abiotic stresses, such as heat, drought, salinity and low soil pH, are the major factors limiting the nitrogen-fixation ability of faba bean (Küçük and Kivanç 2008). Soil salinity is a significant problem facing agricultural production in many parts of the world and soil infertility in these areas is often due to the presence of large concentrations of salt. Most leguminous plants adapted to soils with neutral pH for their optimal growth, as they depend on the symbiotic N_2 fixation and more sensitive to salinity than their rhizobia counterparts. Consequently the symbiosis being more sensitive to salt stress than free-living rhizobia (Zahran 1999).

Soil acidity is among the common problems that limits the production of faba bean in Ethiopia (Asefa et al. 2010). Acid soil infertility is caused by toxicities of hydrogen ion, aluminium and manganese, and deficiency of calcium, molybdenum and phosphorus (Graham, 1992; Chen et al. 1993). Soil acidity can adversely affect survival, growth and nitrogen-fixation efficiency of the rhizobia, and the formation of symbiotic association between rhizobia and their host legumes (Chen et al. 1991; Graham, 1992; Zahran, 1999). Generally, *Rhizobium* strains vary markedly in their acid tolerance and ability to nodulate their host legume on acid soil (Zahran 1999). Several studies indicated that some rhizobia strains are acid tolerant when grown in acidic laboratory medium (Chen et al. 1993; Del Papa et al. 1999; Asefa et al. 2010; Jida and Assefa 2011).

The use of pink-pigmented facultative methylotrophic bacteria (PPFMs) may increase nitrogen fixation, which benefits not only a given crop but the successive crop as well. ICARDA and ARC-Egypt evaluated the response of 25 faba bean genotypes under natural conditions, treatment with improved rhizobia and with PPFMs. The study showed that plant nitrogen content increased from 238.6 to 284.2 and 360.6 mg/plant when inoculated with improved rhizobia only and combined treatment of rhizobia and PPFMs, respectively. Two faba bean accessions 45/018/F8/7307/06A and Sakha1 showed higher plant nitrogen content when compared to other genotypes (Maalouf et al. 2016b).

8 Major achievements

8.1 Partial *Orobanche* resistance varieties

ICARDA has developed faba bean lines with different levels of tolerance and resistance to *Orobanche* for the parasite population existing in Syria (Khalil et al. 2004) and have tested in different locations with NARS partners (Maalouf et al. 2011). The programme shifted to Douyet-Morocco and Sids-Egypt after the Syrian war broke out in 2012. Negative correlation between grain yield and number of *Orobanche* shoot (−0.46, $p < 0.001$) was identified across locations and season, which indicates that infection of *Orobanche* reduces yield of faba bean genotypes. Among 194 genotypes tested, 49 lines were found partially resistant to *Orobanche* adapted to diverse locations (Table 1).

Faba bean lines with resistance to *Orobanche*, disease resistance and heat tolerance were released in different countries as a result of the joint effort of ICARDA breeding programme and NARS partners in different locations: Giza843 and Misr3 in Egypt, Hashbenge (ILB4358) in Ethiopia and Najah and Chourok in Tunisia. Under the EU-IFAD project integrated management options comprising resistant varieties, Misr3 and Giza-843 and chemical control of *Orobanche* were disseminated during 2013–15. Organic fertilizers were also added to the package to improve soil fertility and carbon sink. This technology intervention has led to a 25% increase in faba bean production with enhanced

Table 1 *Orobanche*-resistant lines identified in diverse locations

Number	Pedigree	Number	Pedigree
11A15OFS-1	WRB 1-5 X F7/8983/05	11248-3	10901-1/2010/Giza - 402
11A15OFS-2	Fam 2-1-1 X F7/8984/05	11248-4	10901-1/2010/Giza - 402
10931-2	10170-2/2010/HBP/DSO/2000	11248-6	10901-1/2010/Giza - 402
10931-3	10170-2/2010/HBP/DSO/2000	11248-7	10901-1/2010/Giza - 402
10931-5	10170-2/2010/HBP/DSO/2000	11256-3	10170-2/2010/HBP/DSO/2000
10931-7	10170-2/2010/HBP/DSO/2000	11268-2	10338-2/2010/HBP/ESO/2000
10945-4	9883-3/2010/HBP/DSO/2000	11314-2	10591-3/2010/HBP/ESO/2000
10953-3	10706-3/2010/S 98013 (DC)	11389-4	10158-1/2010/HBP/DSO/2000
11072-1	9151-1/2010/HBP/DSO/2000	11392-2	9019-6/2010/HBP/DSO/2000
11248-1	10901-1/2010/Giza - 4	11414-2	9289-6/2010/HBP/DSO/2000
11418-2	10456-1/2010/HBP/ESO/2000	11543-7	10919-1/2010/Giza - 429
11419-2	9883-3/2010/HBP/DSO/2000	11544-1	9303-2/2010/HBP/DSO/2000
11420-3	9845-4/2010/HBP/DSO/2000	11544-4	9303-2/2010/HBP/DSO/2000
11426-1	9029-2/2010/HBP/DSO/2000	11549-1	9738-1/2010/HBP/DSO/2000
11432-2	10130-4/2010/HBP/DSO/2000	11549-2	9738-1/2010/HBP/DSO/2000
11432-3	10130-4/2010/HBP/DSO/2000	11555-4	10170-2/2010/HBP/DSO/2000
11510-4	9608-1/2010/HBP/DSO/2000	11567-3	9160-10/2010/HBP/DSO/2000
11535-1	10129-4/2010/HBP/DSO/2000	11599-1	9402-1/2010/HBP/DSO/2000
11539-5	10559-3/2010/HBP/ESO/2000	11626-1	9414-2/2010/HBP/DSO/2000
11543-3	10919-1/2010/Giza - 429	11719-3	9411-3/2010/HBP/DSO/2000

soil fertility (Fig. 3). It was found that the recommended technology package increased farmers' grain yield by 256 kg ha^{-1} and reduced production costs by 350 USD ha^{-1}. The result was an increase in net revenue by 550 USD ha^{-1}. From the samples of 695 surveyed farmers, 46% were using new varieties and parts of the technology package and 14% were using the full technology package.

8.2 Heat tolerance varieties

Heat stress due to temperature above 32°C affects the faba bean crop severely, especially by leading to the reduction of the percentage of germinated seeds, increase in occurrence of abnormal seedlings; early flowering; degeneration of nodules affecting the nitrogen-fixation efficiency; and reduction of plant biomass. Heat stress during the reproductive phase affects pollen viability, fertilization, pod set and seed development, leading to abscission of flowers and pods and substantial losses in grain yield. Genetic variation for heat tolerance has been identified in faba bean.

Pollen viability of faba bean tested lines varied from 0 to 50% under 32°C and from 0 to 33% under 35°C. Significant differences were found among evaluated

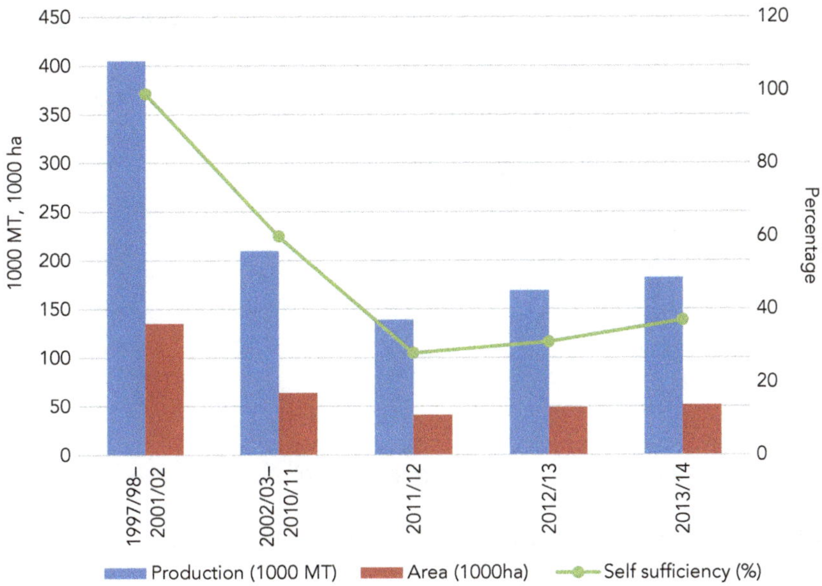

Figure 3 Trends of faba bean production and cultivation along with the self-sufficiency rate in Egypt.

Table 2 Accession with higher degree of tolerance to heat

Accession name	Pedigree
Sel TerWS11-81/2012	S 88134-14-1-4-33018-2/TR2010/WS11-81
Sel TerWS11-106/2012	BPL 5061-33005/TR2010- WS11-106
Sel TerWI11-18-2/2012	S 88094-8-1-32114-1/TR2010/WI11-18-2
Sel TerWI11-164/2012	S 88096-6-2-32219/TR2010/WI11-164
Sel TerWI11-181-1/2012	Giza 4-32087-1/TR2010/WI11-181-1
Sel TerWI11-244-1/2012	ILB445-32122-1/TR2010/WI11-244-1
Sel TerWI11-249-1/2012	S 88094-8-2-32334/TR2010/WI11-249-1
Sel TerWI11-312-2/2012	Selection from Hudaiba- Fam Hu-15/WI11-312-2

genotypes for all traits studied, indicating a wide range of genetic variability. Heat stress can cause significant reduction in pollen germination and yield components of the majority of studied accessions. Principle Component Analysis revealed that at 35°C, yield component and pollen viability explained 58% of the total variation. Based on these results, seven accessions were selected for being tolerant to heat stress. Seven accessions with higher rate of pollen viability under 35°C and higher values of pod number and seed number per plant were identified (IG11743, IG11843, VF351, VF420, VF522, VF626 and INRA1631). Improved lines for heat tolerance evaluated under Sudanese conditions were identified (Table 2) and will be further evaluated in the farmers' fields.

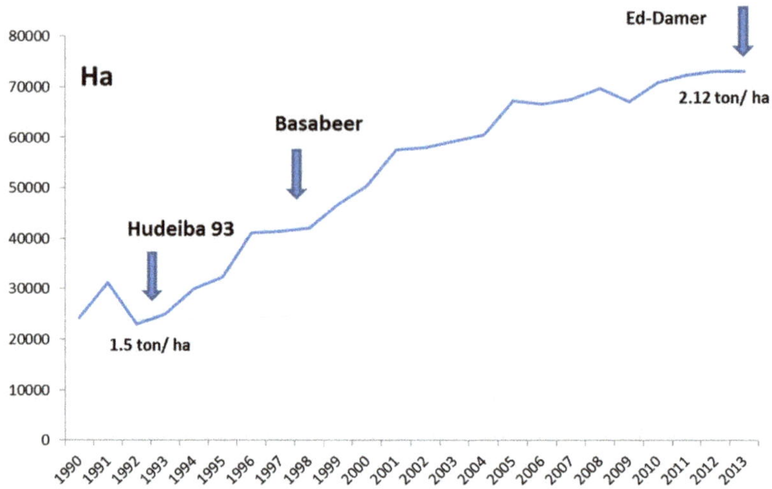

Figure 4 Trends of faba bean area and productivity in Sudan (FAOSTAT 2016).

The improved faba bean varieties (Hudeiba-93 and Basabeer) developed during the Nile Valley projects in collaboration with ICARDA in 1993 and 2000, respectively, and Eddamer in 2012 during the EU-IFAD project were brought to farmers in North part of Sudan, where heat stress is the major constraint. The area of faba bean cultivation increased from 20 000 in 1993 to 75 000 ha in 2014 and the production increased from approximately 30 000 tons in 1993 to 154 000 tons in 2014 (Fig. 4).

9 Future trends and conclusion

To enhance farm profitability through adoption of improved faba bean technologies in North Africa and Nile valley regions, there is a need to develop high-yielding nutritious varieties resistant/tolerant to multiple pests and suitable for different cropping systems prevalent in different agro-ecological zones in North and East Africa. There is a need to develop climate-smart varieties, which can tolerate stress of drought, heat, waterlogged and acid soils. There is also need to develop improved lines resistant to faba bean necrotic yellow virus in Egypt and to the new gall disease in Ethiopia, and lines with tolerance to stem borer (*Lexus*) in North Africa. There is also a need to develop herbicide-tolerant faba bean germplasm to solve the problem of weeks in North Africa, as well as water and nutrient-efficient varieties for dry areas. Acidic soils and salinity are becoming a problem for faba bean production in Ethiopia, which requires special attention. Finally Faba bean straw has added value for animal feeding which need to be explored.

Breeding programmes are relatively slow due to the nature of the mechanism of resistance for foliar diseases, *Orobanche* resistance, major abiotic stresses (heat and drought) and herbicide resistance. To accelerate breeding cycles, there is a need to use emerging biotechnological tools, such as marker-assisted selection, which have not been widely adopted despite significant achievements of quantitative trait loci studies associated with ascochyta blight, rust and chocolate spot resistance and *Orobanche* resistance (Torres et al. 2010) and the development of consensus maps in faba bean (Satovic et al. 2013; Webb et al. 2015). The knowledge generated in the model legumes *Medicago truncatula* and *Lotus japonicus* will help to overcome these gaps and speed the process (Dita et al. 2006; Rispail et al. 2010). Gene expression studies and models will help in understanding plant/pathogen interaction. In addition, progress in tissue culture and genetic modification can be used to target genes responsible for stress tolerance/resistance, especially the introduction of glyphosate tolerance in faba bean cultivars (Hanafy et al. 2013).

In conclusion, faba bean is a major food and feed legume, because of the high nutritional value of its seeds, which are rich in protein and starch and antioxidants important for human health. Faba bean is also used to improve the quality and quantity of the proteins in meals when mixed with cereals, particularly in Mediterranean countries. Many people in the Mediterranean region. The main faba bean producer countries are China, some European countries, Ethiopia, Egypt, Sudan, Morocco and Australia. The consumption is very high highest in middle East and North Africa). However, the crop suffers from diverse biotic and abiotic stresses (Rubiales et al. 2006; Sillero et al. 2010; Maalouf et al. 2013, 2016). ICARDA has a global mandate on faba bean and hosts a large number of faba bean accessions (10 036 accessions) which are being used to develop improved breeding lines and cultivars. Many faba bean cultivars have been developed and released, including some high-yielding cultivars released by NARS and ICARDA as International Public Goods (IPGs), which are improving productivity in China, Ehiopia, Sudan and Egypt. For example, high yielding, disease resistant with large seed size faba bean cultivars are being scaled out in the wheat-based cropping system in Ethiopia (Temesgen et al. 2015). Other examples include improved lines for *Orobanche* resistance developed and tested in West Asia and North Africa (Maalouf et al. 2011) and the released varieties for *Orobanche* resistance like Misr3 in Egypt, Najah and Chourouk in Tunisia, and Hashbenge in Ethiopia. Collaborative research on faba bean with China reviewed recently by Li et al. (2017) showed the impact of the faba bean varieties developed by Chinese academies.

During the international year of pulses in 2016, faba bean research was presented at various international conferences. Other sources of information about research include the International Food Legumes Research Conferences (IFLRC), held every four years, and the conferences of the European Association for Grain Legume Research held every two years.

10 References

Abdelmula, A. A. and Abuanja, I. K. (2007), Genotypic responses, yield stability, and association between characters among some of Sudanese faba bean (*Vicia faba* L.) genotypes under heat stress. In *Conference on International Agricultural Tropentag Research for Development*. University of Kassel-Witzenhausen and University of Göttingen, pp, 1–7.

Abdelmula, A. A., Link, W., Von Kittlitz, E. and Stelling, D. (1999), Heterosis and inheritance of drought tolerance in faba bean (*Vicia faba* L). *Plant Breed.* 118, 485–90.

Abebe, T., Meles, K., Nega, Y., Beyene, H. and Kebede, A. (2013), Interaction between broomrape (*orobanche crenata*) and resistance faba bean genotypes (*Vicia faba* L) in Trigray region of Ethiopia. *Can. J. Plant Prod.* 1(3), 104–9.

Abu-Irmaileh, B. E. and Labrada, R. (2016), http://www.fao.org/agriculture/crops/thematicsitemap/theme/biodiversity/weeds/ (Accessed on 15 November 2016).

Amede, T., Kittlitz, E. V. and Schubert, S. (1999), Differential drought responses of faba bean (*Vicia faba* L) inbred lines. *J. Agron. Crop Sci.* 183, 35–45.

Asefa, K., Assefa, F. and Prabu, P. C. (2010), Characterization of acid and salt tolerant rhizobial strains isolated from faba bean fields of Wollo, Northern Ethiopia. *J. Agric. Sci. Technol.* 12, 365–76.

Babar, M. A., Reynolds, M. P., Van Ginkel, M., Klatt, A. R., Raun, W. R. and Stone, M. L. (2006), Spectral reflectance indices as potential indirect selection criteria for wheat yield under irrigation. *Crop Sci.* 46, 578–88.

Balech, R., Singh, S. and Maalouf, F. (2016), Identification of faba bean lines tolerant to high dosage of Glyphosate. In: *International Conference on Pulses*, Marrakesh, Morocco, 18–20 April 2016. Poster presentation Theme 2. p. 33.

Bond, D. A., Jellis, G. J., Rowland, G. G., Le Guen, J., Robertson, L. D., Khalil, S. A. and Li-Juan, L. (1994), Present status and future strategy in breeding faba beans (*Vicia faba* L.) for resistance to biotic and abiotic stresses. *Euphytica* 73, 151–66.

Borlaug, N. and Dowswell, C. R. (2005), Feeding a world of ten billion people: a 21st century challenge. In Tuberosa, R., Phillips, R. L. and Gale, M. D. (Eds), *In the Wake of the Double Helix: From the Green Revolution to the Gene Revolution*, 27–31 May 2003. Avenue Media: Bologna, Italy.

Chen, H., Richardson, A. E., Cartner, E., Diordjevic, M. A., Roughley, R. J. and Rolfe, B. G. (1991), Construction of an acid tolerant *Rhizobium leguminosarum biovar trifolii* strain with enhanced capacity for nitrogen fixation. *Appl. Environ. Microbiol.* 57, 2005–11.

Chen, H., Cartner, E. and Rolfe, B. G. (1993). Involvement of genes on a megaplasmid in the acid tolerant phenotype of *Rhizobium leguminosarum biovar trifolii. Appl. Environ. Microbiol.* 59, 1058–64.

Cubero, J. I. (1973), Evolutionary trends in *Vicia faba. Theor. Appl. Genet.* 43(2), 59–65. https://doi.org/10.1007/bf00274958

Cubero, J. I., Moreno, M. T. and Hernandez, L. (1992), A faba bean (*Vicia faba L.*) cultivar resistant to broomrape (*Orobanche crenata* Forsk.). In Plancquaert, P. (Ed.), *Proceedings of the 1st European conference on grain legumes*, Angers, France, 1–3 June 1992 (pp. 41–2). L'association Europeenne des Proteagineux: Paris, France.

Del Papa, M. F., Balaque, L. J., Sowinski, S. C., Wegner, C., Segudo, E., Abrca, F. M., Toro N., Nichaul, K., Puhler, A., Aguilar, O. M., Martinez-Drets, G. and Lagares, A. (1999), Isolation and characterization of alfalfa nodulating rhizobia present in acidic soils of central Argentina and Uruguay. *Appl. Environ. Microbiol.* 65, 1420–7.

Dita, M. A., Rispail, N., Prats, E., Rubiales, D. and Singh, K. B. (2006), Biotechnology approaches to overcome biotic and abiotic stress constraints in legumes. *Euphytica* 147, 1–24.

Egli, D. B., TeKrony, D. M., Heitholt, J. J. and Rupe, J. (2005), Air temperature during seed filling and soybean seed germination and vigor. *Crop Sci.* 45, 1329–35.

Elia, M. (1964), Indagini preliminari sulla resistenza varietale della Fava (*Vicia faba*) all'orobanche, *Phytopathol. Mediter.* 3, 31–2. In Kasasian, L. *Orobanche spp. PANS* 17(1), 35–41.

Ellis, R. H., Roberts, E. H. and Summerfield, R. J. (1988), Variation in the optimum temperature for rates of seedling emergence and progress towards flowering amongst six genotypes of faba bean (*Vicia faba*). *Ann. Bot.* 62, 119–26.

Ellis, R. H., Summerfield, R. J. and Roberts, E. H. (1990), Flowering in faba bean: genotypic differences in photoperiod sensitivity, similarities in temperature sensitivity, and implications for screening germplasm. *Ann. Bot.* 65, 129–38.

El-Sayed, A., Sahar, R. Z., El-Shennawy, M. and Ismail, A. I. (2011), Fungicidal management of chocolate spot of faba bean and assessment of yield losses due to the disease. *Ann. Agric. Sci.* 56, 27–35.

FAOSTAT (2016), http://www.fao.org/faostat/en/#data/QC.

Fischbeck, G., Abdalla, M. M. F., Metwally, A. A. and Darwish, D. S. (1986), Variation of *Vicia faba* L. genotypes and populations of *Orobanche crenata* Forsk. In: *Proc. 2nd Agron. Conf.* vol. 2, Alex, Egypt, pp. 489–514. Fischer and Turner 1978.

Gaur, P., Saminen, S., Krishnamurthy, L., Kumar, S., Ghanem, M., Beebe, S., Rao, I., Chaturvedi, S., Basu, P., Nayyar, H., Jayalakshmi, V., Babbar A. and Varshney R. (2015), High temperature tolerance in grain legumes. *Legume Perspectives*, 7, 23–4.

Graham P. H. (1992), Stress tolerance in *Rhizobium* and *Bradyrhizobium*, and nodulation under adverse soil conditions. *Can. J. Microbiol.*, 38, 475–84.

Hailu, E., Getaneh, G., Sefera, T., Tadesse, N., Bitew, B., Boydom, A., Kassa, D. and Temesgen, T. (2014), Faba bean gall; a new threat for faba bean (*Vicia faba*) production in Ethiopia. *Adv. Crop Sci. Technol.* 2, 144. doi:10.4172/2329-8863.1000144.

Hanafy M. S., El-Banna, A., Schumacher, H. M., Jacobsen, H. J. and Hassan, F. (2013), Enhanced tolerance to drought and salt stresses in transgenic faba bean (*Vicia faba* L.) plants by heterologous expression of the PR10a gene from potato. *Plant Cell Rep.* 32, 663–74. doi:10.1007/s00299-013-1401-x.

Hanounik, S. B. (1980), Effect of chemical treatments and host genotypes on disease severity/yield relationships of Ascochyta blight in faba beans. *FABIS Newslett.* 2, 50.

Hanounik, S. B. and Robertson, L. D. (1988), New sources of resistance in *Vicia faba* to Chocolate Spot Caused by *Botrytis fabae*. *Plant Dis.* 72, 696–8.

Hanounik, S. B. and Robertson, L. D. (1989), Resistance in *Vicia faba* germplasm to blight caused by *Ascochyta fabae*. *Plant Dis.* 73, 202–5.

Jensen, E. S., Peoples, M. B. and Hauggaard-Nielsen, H. (2010), Faba bean in cropping systems. *Field Crops Res.* 115, 203–16.

Jida, M. and Assefa, F. (2011), Phenotypic and plant growth promoting characteristics of *Rhizobium leguminosarum bv. viciae* from lentil growing areas of Ethiopia. *Afr. J. Microbio. Res.* 5, 4133–42. https://doi.org/10.5897/ajmr11.400

Katul, L., Vetten, H. J., Maiss, E., Makkouk, K. M., Lesemann, D. E. and Casper, R. (1993), Characterization and serology of virus-like particles associated with faba bean necrotic yellows. *Ann. Appl. Biol.* 123, 629–47.

Khalil, S., Kharrat, M., Malhotra, R. S., Saxena, M. C. and Erskine, W. (2004), Breeding faba bean for Orobanche resistance. In: Dahan, R. and El-Mourid, M. (Eds.), International Management of Orobanche in Food Legumes in the Near East and North Africa. Proceedings of the Expert Consultation on IPM for Orobanche in Food Legume Systems in the Near East and North Africa. Rabat, Morocco, pp. 1–18.

Khan, H. R., Link, W., Hocking, T. J. and Stoddard, F. L. (2007), Evaluation of physiological traits for improving drought tolerance in faba bean (*Vicia faba* L.). *Plant Soil* 292, 205–17.

Küçük, C. and Kıvanç, M. (2008), Preliminary characterization of *Rhizobium* strains isolated from chickpea nodules. *Afr. J. Biotechnol.* 7, 772–5.

Maalouf, F., Ahmed, S., Shaaban, K., Bassam, B., Nawar, F., Singh, M. and Amri, A. (2016a), New faba bean germplasm with multiple resistances to ascochyta blight, chocolate spot and rust diseases. *Euphytica* 211, 157–67. doi:10.1007/s10681-016-1726-y.

Maalouf, F., Hamwieh, A., Abou Taleb, H. and Abdul Rehman, R. (2016b), Studies on influence of pink-pigmented facultative methylotrophic bacteria PPFMs and Rhizobium inoculation on BNF in Faba bean and Chickpea genotypes. CRP PHASE1-2012-2016-REVIEW MEETING- Reflections on four years of research aimed to improve the lives of smallholder farmers. Presentation and poster abstracts 03 to 07 October 2016, p. 33.

Maalouf, F., Khalil, S., Ahmed, S., Akintunde, A. N., Kharrat, M., Hajjar, S. and El Shama'a, K. (2011), Yield stability of faba bean lines under diverse broomrape prone production environments. *Field Crop Res.* 124, 288–94

Maalouf, F., Nawar, M., Hamwieh, A., Amri, A., Zong, X., Bao, S. and Yang, T. (2013), Faba bean. In Singh, M., Upadhyaya, H. D. and Singh Bisht I. (Eds), *Genetic and Genomic Resources of Grain Legume Improvement*. Elsevier Insight, pp. 113–36. ISBN: 978–0-12-397.

Maalouf, F., Nachit, M., Ghanem Edmond, M. and Singh, M. (2015), Evaluation of faba bean breeding lines for spectral indices, yield traits and yield stability under diverse environments. *Crop Pasture Sci.* 66, 1012–23. http://dx.doi.org/10.1071/CP14226.

Makkouk, K. M. Rizkallah, L., Madkour, M., El-Sherbeiny, M., Kumari, S. G., Amriti, A. W. and Solh, M. (1994), Survey of faba bean (*Vicia faba* L.) for viruses in Egypt. *Phytopathol. Mediterr.* 33, 207–11.

Nassib, A. M., Ibrahim, A. A. and Khalil, S. A. (1982), Breeding for resistance to Orobanche. In Hawtin, G. and Webb, C. (Eds.), *Faba Bean Improvement*. Martinus Nijhoff.: The Hague, the Netherlands, pp. 199–206.

Nassib, A. M., Ibrahim, A. A. and Saber, H. A. 1978. Broomrape (*Orobanche crenata*) resistance in broad beans. In *Proceedings of a Workshop on Seed Legumes*, ICARDA: Aleppo, Syria, pp. 133–5.

Osborne, S. L., Schepers, J. S., Francis, D. D. and Schlemmer, M. R. (2002), Use of spectral radiance to estimate in-season biomass and grain yield in nitrogen- and water-stressed corn. *Crop Sci.* 42, 165–71. doi:10.2135/cropsci2002.0165.

Patrick, J. W. and Stoddard, F. L. (2010), Physiology of flowering and grain filling in faba bean. *Field. Crop. Res.* 115, 234–42.

Peñuelas, J., Baret, F. and Filella, I. (1995a), Semi-empirical indices to assess carotenoids/chlorophyll *a* ratio from leaf spectral reflectance. *Photosynthetica* 31, 221–30.

Peñuelas, J., Filella, I., Lloret, P., Munoz, F. J. and Vilajeliu, M. (1995b) Reflectance assessment of plant mite attack on apple trees. *Int. J. Remote Sensing* 16, 2727–33.

Pérez-de-Luque, A., Eizenberg, H., Grenz, J. H., Sillero, J. C., Ávila, C. M., Sauerborn, J. and Rubiales, D. (2010), Broomrape management in faba bean. *Field Crops Res.* 115, 319–28. https://doi.org/10.1016/j.fcr.2009.02.013

Pérez-de-Luque, A., Lozano, M. D., Moreno, M. T., Testillano, P. S. and Rubiales, D. (2007), Resistance to broomrape (*Orobanche crenata*) in faba bean (*Vicia faba*): cell wall changes associated with pre-haustorial defensive mechanisms. *Ann. Appl. Biol.* 151, 89–98. https://doi.org/10.1111/j.1744-7348.2007.00164.x

Perrino, P., Pace, M. S. and Polignano, G. B. (1988), Evaluation for tolerance to broomrape in a germplasm collection of *Vicia faba*. *FABIS Newslett.* 20, 40–4.

Prasad, P., Boote, K., Allen, L., Sheehy, J. and Thomas, J. (2006), Species, ecotype and cultivar differences in spikelet fertility and harvest index of rice in response to high temperature stress. *Field Crops Res.* 95, 398–411. doi:10.1016/j.fcr.2005.04.008.

Radwan, M. S., Abdalla, M. M. F., Fishbeck, G., Metwally, A. A. and Darwish, D. S. (1988a), Variation in reaction of faba bean lines to different accessions of *Orobanche crenat* Forsk. *Plant Breed.* 101, 208–21.

Radwan, M. S., Abdalla, M. M. F., Fishbeck, G., Metwally, A. A. and Darwish, D. S. (1988b), Selection in faba bean for tolerance to broomrape (*Orobanche crenata* Forsk). *Plant Breed.* 100, 289–98.

Redden, R., van Leur, J., Zong, X., Bao, S., Yujioa, L., Paull, J. and Leonforte, T. (2008), Evaluation and utilisation of pea and faba bean germplasm from China. In Global issues, paddock action. Unkovich, M. (Eds), *Global Issues Paddock Action*. Proceedings of the 14th Australian Agronomy Conference. September 2008, Adelaide South Australia. Australian Society of Agronomy. www.regional.org.au/au//pdf/asa/2008/5804_redden.pdf. (Accessed 10 July 2016).

Rispail, N., Kalo, P., Kiss, G. B., Ellis, T. H. N., Gallardo, K., Thompson, R. D., Prats, E., Larrainzar, E., Ladrera, R., Gonzalez, E. M., Arrese-Igor, C., Ferguson, B. J. and Gresshoff, P. M. (2010), Model legumes to contribute to faba bean breeding. *Field Crop Res.* 115, 253–69.

Robertson, L. D. (1984), A note on the ILB source of *Botrytis fabae* resistance. In Chapman, G. P. and Tarawali, S. A. (Eds), *Systems for Cytogenetic Analysis in Vicia Faba L.* Martinus Nijhoff Dr W. Junk Publishers, p. 80.

Rubiales, D., Avila, C. M., Sillero, J. C., Hybl, M., Narits, L., Sass, O. and Flores, F. (2012), Identification and multi-environment validation of resistance to *Ascochyta fabae* in faba bean *Vicia faba*. *Field Crop Res.* 126, 165–70.

Rubiales, D., Pérez-de-Luque, A., Fernández-Aparico, M., Sillero, J. C., Román, B., Kharrat, M., Khalil, S., Joel, D. M. and Riches, C. (2006), Screening techniques and sources of resistance against parasitic weeds ingrain legumes. *Euphytica* 147, 178–99. https://doi.org/10.1007/s10681-006-7399-1

Rubiales, D., Verkleij, J., Vurro, M., Murdoch, A. J. and Joel, D. M. (2009), Parasitic plant management in sustainable agriculture. *Weed Res.* 49, 1–5.

Satovic, Z., Avila, C. M., Cruz-Izquierdo, S., Díaz-Ruíz, R., García-Ruíz, G. M., Palomino C., Gutiérrez, N., Vitale, S., Ocaña-Moral, S., Victoria Gutiérrez, M., Cubero, J. I. and Torres, A. M. (2013), A reference consensus genetic map for molecular markers and economically important traits in faba bean *Vicia faba* L. *BMC Genomics* 14, 932.

Siddique, K. H. M., Regan, K. L., Tennant, D. and Thomson, B. D. (2001), Water use and water use efficiency of cool season grain legumes in low rainfall Mediterranean-type environments. *Eur. J. Agron.* 15, 267–80. doi:10.1016/S1161-0301(01)00106-X.

Sillero, J. C., Avila C. M., Moreno, M. T. and Rubiales, D. (2001), Identification of resistance to *Ascochyta fabae* in *Viciai fabai* germplasm. *Plant Breed.* 120, 529–31.

Sillero, J. C., Villegas-Fernandez Angel, M., Thomas, J., Rojas-Molina, M. M., Emeran, A. A., Fernandez-Aparicio, M. and Rubiales, D. (2010), Faba bean breeding for disease resistance. *Field Crop Res.* 115, 297–307.

Solh, M. B. and Pala, M. (1990), Weed control in chickpea. *Options Mediterr. Ser. Semin.* 9, 93–9.

Stoddard, F. L., Balko, C., Erskine, W., Khan, H. R., Link, W. and Sarker, A. (2006), Screening techniques and sources of resistance to abiotic stresses in cool-season food legumes. *Euphytica* 147, 167–86. doi:10.1007/s10681-006-4723-8.

Subbarao, G. V., Johansen, C., Slinkard, A. E., Nageswara, R. C., Saxena, N. P. and Chauhan, Y. S. (1995), Strategies for improving drought resistance in grain legumes. *Crit. Rev. Plant Sci.* 14, 469–523. doi:10.1080/07352689509701933.

Temesgen, T., Keneni, G., Sefera, T. and Jarso, M. (2015), Yield stability and relationships among stability parameters in faba bean (*Vicia faba* L.) genotypes. *Crop J.* 3, 258–69.

Tivoli, B., Baranger, A., Avila, C. M., Banniza, S., Barbetti, M. C. W., Davidson, J., Lindeck, K., Kharrat, M., Rubiales, D., Sadiki, M., Sillero, J. C., Sweetingham, M. and Muehlbauer, F. J. (2006), Screening techniques and sources of resistance to foliar diseases caused by major necrotrophic fungi in grain legumes. *Euphytica* 147, 223–53.

Torres, A. M., Avila, C. M., Gutierrez, N., Palomino, C., Moreno, M. T. and Cubero, J. I. (2010), Marker-assisted selection in faba bean (*Vicia faba* L.). *Field Crop Res.*, 115, 243–52.

Villegas-Fernández, A. M., Sillero, J. C., Emeran, A. A., Winkler, J., Raffiot, B. and Tay, J. (2009) Identification and multi-environment validation of resistance to Botrytis fabae in *Vicia faba*. *Field Crop Res.* 114, 84–90.

Villegas-Fernández, A. M., Sillero, J. C. and Rubiales, D. (2011), Screening faba bean for chocolate spot resistance: evaluation methods and effects of age of host tissue and temperature. *Eur. J. Plant Pathol.* 132, 443–53.

Wang, J., Gan, Y. T., Clarke, F. and McDonald, C. L. (2006), Response of chickpea yield to high temperature stress during reproductive development. *Crop Sci.* 46, 2171–8.

Webb, A., Cottage, A., Wood, T., Khamassi, K., Hobbs, D., Gostkiewicz, K., White, M., Khazae, H., Ali, M., Street, D., Duc, G., Stoddard, F. L., Maalouf, F., Ogbonnaya, F., Link, W., Thomas, J. and O'Sullivan, D. M. (2015), A SNP-based consensus genetic map for synteny – based trait targeting in faba bean *Vicia faba* L. *Plant Biotechnol. J.* 14(1), 177–85. doi:10.1111/pbi.12371.

Wery, J., Silim, S. N., Knights, E. J., Malhotra, R. S. and Cousin, S. (1994), Screening techniques and sources and tolerance to extremes of moisture and air temperature in cool season food legumes. *Euphytica* 73, 73–83. doi:10.1007/BF00027184.

Zahran, H. H. (1999), Rhizobium-legume symbiosis and nitrogen fixation under severe conditions and in arid climate. *Microbiol. Mol. Biol. Rev.* 63: 968–89.

Chapter 13

Efficient and sustainable production of faba bean

R. Redden, RJR Agricultural Consultants, Australia; X. Zong, Chinese Academy of Agricultural Sciences (CAAS), China; R. M. Norton, International Plant Nutrition Institute and The University of Melbourne, Australia; F. L. Stoddard, University of Helsinki, Finland; F. Maalouf, International Centre for Agricultural Research in the Dry Areas (ICARDA), Lebanon; S. Ahmed and M. El Bouhssini, International Centre for Agricultural Research in the Dry Areas (ICARDA), Morocco; Y. Tao and L. Rong, Chinese Academy of Agricultural Sciences (CAAS), China; and L. Ling, Liaoning Academy of Agricultural Science, China

1 Introduction
2 Faba bean production in China
3 Faba bean production in Central and West Asia, North Africa regions (CWANA)
4 Faba bean breeding
5 Faba bean genetic and genomic resources
6 Faba bean management with limited water resources
7 Faba bean BNF
8 Faba bean mineral nutrient requirements
9 Conclusion
10 Where to look for further information
11 References

1 Introduction

Faba bean (*Vicia faba* L.) is grown in many different cropping systems as a dry grain (pulse), green grains and pods, animal feed and a green manure worldwide. Faba bean contributes to the sustainability of cropping systems through its ability to fix nitrogen that can be used by succeeding non-legume crops. Cropping systems range from intercropped with maize to cultivation as a broad acre field crop as part of a rotation with other arable crops.

Dry faba bean is cultivated on 2.39 Mha worldwide with a production of 4.34 Mt of dry grains and China is the largest producer (FAOSTAT, 2016), with yields ranging from 0.76 t/ha (Morocco) to 8.0 t/ha (Argentina). Faba bean is produced as vegetables (green pods and green grain) notably in China (0.43 Mha, MOA China, 2015), India, Morocco, Algeria,

http://dx.doi.org/10.19103/AS.2017.0023.32

Tunisia, Syria, Turkey, Iraq and Iran. Production is predominantly under rain-fed growing conditions, while irrigated and supplementary watered production is important in Egypt, Sudan the Middle East and Central Asian river basins. Faba bean is preferentially irrigated in China, both where local water supply is limited as well as in river basins and canal networks, the latter include growing faba bean on irrigation bunds around rice paddies and harvested as green vegetable pods (Li et al., 2017).

The main focus of this chapter is faba bean cultivation under rain-fed growing conditions, with sections on regional production of faba bean; diseases, weeds and pests; faba bean breeding: genetic resources and markers for breeding; water deficit management; tillage systems; biological nitrogen fixation (BNF); and mineral nutrient requirements.

2 Faba bean production in China

Chinese food legume crops can be divided into three groups according to their cropping seasons: cool-, temperate- and warm-season food legumes (Zheng et al., 1997; Zong, 2002). Cultivated in China for over 2000 years, faba bean is autumn-sown in south China and early spring-sown in north China as a cool-season food legume (Zong et al., 2006). Faba bean has played an important role in Chinese cropping systems for traditional and sustainable agriculture since ancient times (Zheng et al., 1997; Zong, 2002), and is also one of the important nutritional components for dietary protein in China, ensuring the basic health of the Chinese people (Jin et al., 2000; Zong, 2002). Although 20 species of cultivated pulse crops can be found in China, faba bean is one of the most commonly preferred in cropping systems (Long et al., 1989; Zheng et al., 1997; Zong, 2002). The major varieties in the production are Yundou 324, Yundou 147, Fengdou 6, Fengdou 10, Chenghu 14, Chenghu 15, Chenghu 16, Tongcan 5, Tongcan (xian) 6, Chonglicandou, Qinghai 9, Qinghai 10, Qinghai 12, Qinghai 13, Qinghai 14, Lincan 4, Lincan 5, Tongcanxan 6 and Tongcanxuan 7 (Li et al., 2006a).

Yundou 147 was released by crossing the ICARDA line ILB8047 with ko285 from China and now occupies 50% of the faba cultivation in Yunnan province, the major area for faba bean production in China (Cheng and Wang, 2009). Chenghu 19 introduced from ICARDA as 41207/92 has been recently released in Chengdu province. It is characterized by high (31%) content protein and high yield (4.5 t/ha) (Cheng and Wang, 2009).

2.1 Area and production of faba bean in China

Faba bean is cultivated in Southwest, East, Southeast and North China, with 85% sown in the winter cropping areas of the semi-tropical climate zone (Liu, 1984; Ye, 2003; Hu et al., 1994; National Center for Extension of Agronomic Techniques, Ministry of Agriculture of China, 2015) in the provinces of Yunnan, Sichuan, Chongqing, Guizhou, Jiangsu, Zhejiang, Anhui and Hubei (Long et al., 1989; Zheng et al., 1997), while faba bean is spring sown in Gansu, Qinghai, Hebei, Ningxia, Inner Mongolia and Xinjiang provinces of northern China (Long et al., 1989; Zheng et al., 1997). Faba bean is a favoured daily food and has been important historically in China (Zong, 2002). Dry faba bean cultivation rapidly decreased from above 4 Mha in 1962 to around 1 Mha from 1991 to the present (FAOSTAT, 2016) (Fig. 1).

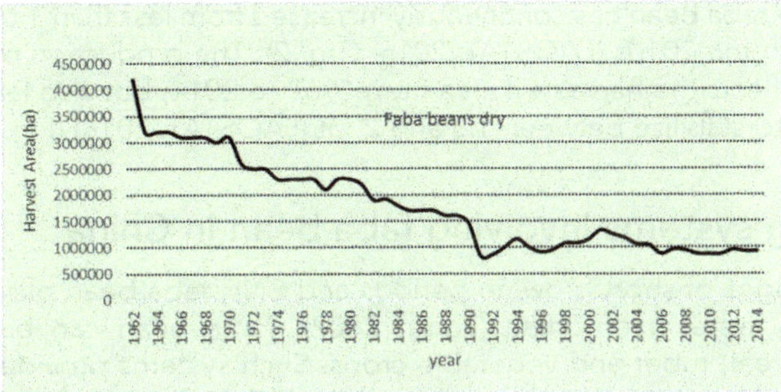

Figure 1 Harvest area of dry faba bean in China since 1962 (Bahmani et al., 2012).

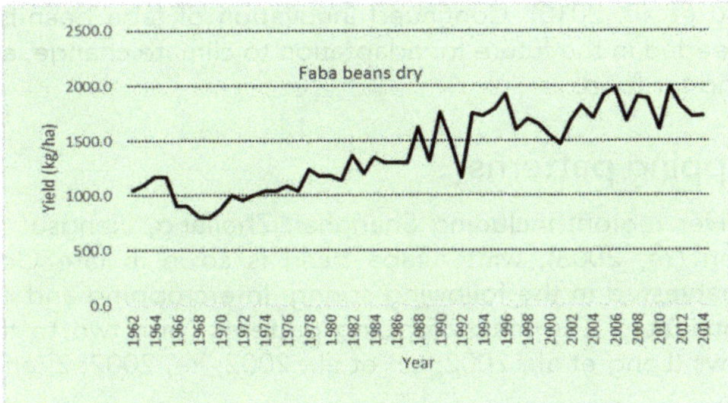

Figure 2 Yield of dry faba bean in China since 1962 (Bahmani et al., 2012).

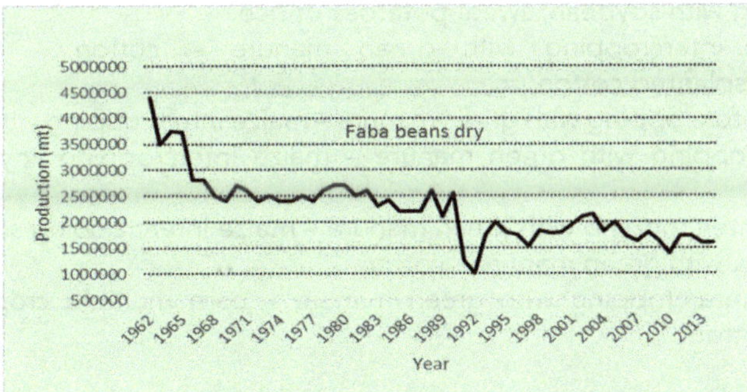

Figure 3 Production of dry faba bean in China since 1962 (FAOSTAT 2016).

Cultivation of vegetable faba bean has increased rapidly and reached 0.428 Mha in 2014 (National Center for Extension of Agronomic Techniques, Ministry of Agriculture of China, 2015), representing 31.6% of the total faba bean harvest area of 1.35 Mha in China.

The yield of dry faba bean has continuously increased from less than 1 t/ha in the 1960s to almost 2 t/ha in the 2010s (FAOSTAT, 2016) (Fig. 2). The production of dry faba bean in China partly reflects the harvested area from 1962 to 2014, but also large increases in yield from 1988, to stabilize between 1.5 and 2 Mt (FAOSTAT, 2016) (Fig. 3).

2.2 Cropping systems involving faba bean in China

With the advantages of short growing periods and BNF, faba bean plays an important role in cropping systems of China (Zong, 2002). Faba bean can be integrated in rotations with cereal, tuber and vegetable crops. Such systems provide a quality food harvest after non-leguminous crops and improve soil fertility on farms (Zong, 2002). To obtain full farm productivity in favourable environments, intercropping or relay intercropping of faba bean with other crops is another good choice (Zong, 2002). As an ideal intercrop, faba bean can increase crop yields and economic benefits in these systems (Mucheru et al., 2010). Continued innovation of faba bean-based cropping systems will be needed in the future for adaptation to climate change, and to meet the increasing demand for food.

Southern cropping patterns

In the Yangtze river region, including Shanghai, Zhejiang, Jiangsu, Jiangxi, Anhui, Hubei and Hunan (Ye, 2003), winter faba bean is sown in late October to early November and harvested in the following spring. Intercropping and rotation of faba bean now predominates. The main cropping patterns (with two to three crops per year) are as follows (Long et al., 2002; Lin et al., 2002; Ye, 2002; Zheng, 2009; Zong, 2010):

1 Faba bean – rice – barley/faba bean intercropping with green manure – cotton
2 Faba bean – transplanted cotton – faba bean intercropping with green manure – maize intercropping with soybean, sweet potatoes or rice
3 Faba bean intercropping with green manure – cotton – faba bean or barley – transplanted cotton
4 Faba bean intercropping with green manure – maize intercropping with cotton – Faba bean intercropping with green manure – maize intercropping soybean or sweet potatoes
5 Faba bean intercropping with green manure – maize intercropping soybean – barley intercropping with green manure – cotton
6 Faba bean intercropping with green manure – corn multiple cropping with rice – barley – cotton

South-western cropping patterns

In the south-western region, including Yunnan, Sichuan, Chongqing and Guizhou (Bolland et al., 1999), winter faba bean is sown from the beginning of October to early November and harvested in the next April and May. The main biennial cropping patterns, generally

Table 1 Characteristics of soils from the main faba bean growing regions of China

Measurement*	Mean	Min	Max	SD	CV (%)
pH	7.3	4.9	8.3	0.8	11
SOM (%)	2.3	0.6	6.1	1.3	56.5
AN (mg/kg)	99.3	28	331	56.7	57.1
AP (mg/kg)	51	3.3	113.5	38.6	75.7
AK (mg/kg)	154.7	38.5	301.1	57.3	37

* pH, soil pH; SOM, soil organic matter; AN, available nitrogen; AP, available phosphorus; AK, available potassium.

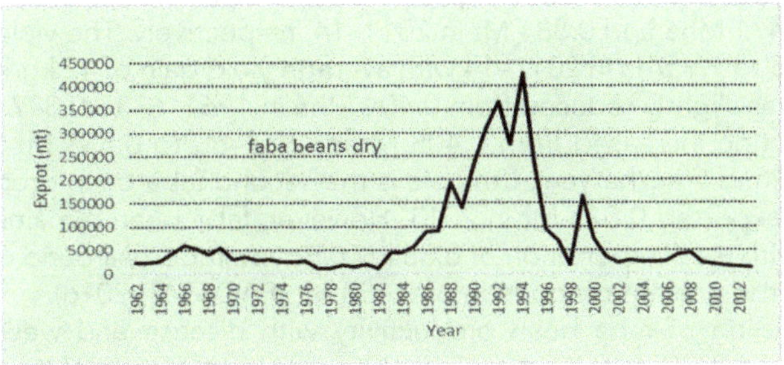

Figure 4 Export of dry faba bean from China since 1962 (Faostat 2016).

with two crops a year, are as follows (Long et al., 2002; Lin et al., 2002; Ye, 2002; Zheng, 2009; Zong, 2010):

1 Faba bean – corn relay intercropping with sweet potatoes – wheat (canola) – peanut
2 Wheat – cotton – faba bean – corn relay intercropping with sweet potatoes
3 Wheat (barley) – corn relay intercropping with sweet potatoes – faba bean – peanut
4 Faba bean – early rice – late rice – canola – semi-late rice
5 Wheat – semi-late rice – faba bean – semi-late rice – barley

2.3 Soil fertilities in faba bean production areas of China

In 2014, soil samples from faba bean production areas in China were collected. Soil pH, soil organic matter (SOM), available nitrogen (AN), available phosphorus (AP) and available potassium (AK) of the samples were analysed (Li et al., 2017). Values were 4.9–8.3 for soil pH, 0.6–6.1% for SOM, 28–331 mg/kg for AN, 3.3–113.5 mg/kg for AP and 38.5–301.1 mg/kg for AK in faba bean production areas (Table 1). The minimum CV was 11% for pH in surface soil and the maximum CV was 75.7% for soil AP content, much higher than that for other elements. The CVs of soil pH and available nutrients were in the order AP > AN > SOM > AK > soil pH.

2.4 Export and import of faba bean in China

Faba bean exports were very limited, at less than 0.050 Mt per year from 1961 to 1982 (Fig. 4). Since 1982, exports of dry faba bean to Europe and Japan were sharply increased until 1994, after which they reduced to around 0.050 Mt again, and now have stabilized to around 0.025 Mt. There has been almost no importation of faba bean into China.

3 Faba bean production in Central and West Asia, North Africa regions (CWANA)

3.1 Status of faba bean

In East Africa, the dry faba bean area and production increased from 0.348 Mha and 0.378 Mt in 1993–95 to 0.471 Mha and 0.884 Mt in 2011–14, respectively. The yield increased from 1 t/ha in 1993–95 to 1.9 t/ha in 2011–14 with average yield gain of 42 kg/ha/year. In North Africa the area was slightly reduced from 0.404 Mha in 1961–63 to 0.387 Mha in 2010–14 while the production increased from 0.405 Mt in 1961–65 to 0.553 Mt in 2011–14 with average yield gain of 7 kg/ha/year. Ethiopia is the second faba bean producer worldwide after China and exported 0.025 Mt in 2013. However, faba bean remains in short supply in North Africa with total exportation of 0.354 Mt. Egypt imports around 64% of its annual needs to add to the present production of 0.158 Mt (FAOSTAT, 2016).

The near doubling of faba bean productivity with disease and weed management, despite shifts to marginal growing area have been important parts of the development of this crop in CWANA. In addition there are new initiatives in breeding for heat tolerance to assist adaptation to climate change.

The regionally traditional diversified agriculture has been converted into heavy reliance on a few cereal and cash crops. While the area under cereals has doubled to 21 Mha during the last five decades in North Africa, the area under legumes remains at 0.90 Mha with about 0.53 Mha for dry faba bean. This has led to depletion of soil nutrients and underground water; a rise in the contribution of agriculture to greenhouse gases; the emergence of new pests, diseases, and a decline in response to farm inputs. However, faba bean and other grain legumes which used to be the mainstay of cropping systems in the region in early 1960s are now limited to those areas where cereals could not be grown successfully.

3.2 Principal diseases and weeds

Under these conditions, productivity of faba bean is mainly affected by fungal diseases, annual weeds, parasitic weeds, viruses and insect pests (Stoddard et al., 2010).

The dominant foliar diseases affecting faba bean are chocolate spot (*Botrytis fabae*), rust (*Uromyces fabae*), Ascochyta blight (*Ascochyta fabae*) and powdery mildew (*Erysiphe pisi*). Chocolate spot can cause major yield reductions in Egypt, Maghreb countries and Ethiopia (Tivoli et al., 2006; El-Sayed et al., 2011). Ascochyta blight can cause up to 90% yield losses as well as quality losses on susceptible faba bean cultivars (Hanounik, 1980). Faba bean rust can cause moderate to high yield losses in North Africa and Ethiopia (Emeran et al., 2011; Shifa et al., 2011). The emerging new faba bean gall (*Olpidium viciae*) disease is devastating faba bean crops in Ethiopia, causing up to 100% yield losses in the highlands of Ethiopia (Hailu et al., 2014).

Faba bean is also affected by soil-borne pathogens (*Fusarium solani*, *F. oxysporum*, *F. avenaceum*, *Rhizoctonia solani*) in East African highlands, Egypt and Sudan where the losses could reach 40% or more (Bogale et al., 2009; Belete et al., 2015). A number of insect pests such as *Sitona* weevil (*Sitona lineatus* L.), cowpea aphid (*Aphis craccivora* Koch) and black bean aphid (*Aphis fabae* Scopoli) cause damage by direct feeding as well as by transmission of viruses (Mwanauta et al., 2015). The broad bean beetle, *Bruchus rufimanus*, starts its infestation of grain in the field and continues in post-harvest storage, thus causing losses of seed weight, and a reduction in seed germination. *B. rufimanus* larvae can destroy between 20 and 50% of bean seeds harvested in the Middle East (Huignard et al., 2011). Faba bean necrotic yellow virus (FBNYV) is considered to be the most important virus disease on faba bean causing up to 90% yield losses in Egypt (Kumari and Makkouk, 2007).

Parasitic weeds (*Orobanche* and *Phelipanche* spp.) are key biological constraints that drastically reduced production and area of faba bean in West Asia and North Africa (Gressel et al., 2004; Mesa-Garcia and Garcia-Torres, 1991; Khalil et al., 2004b; Abang et al., 2007; Maalouf et al., 2011). Parasitic weeds are expanding their ecological range to Ethiopia and Sudan where farmers have thus abandoned faba bean production in many areas (Abebe et al., 2013). In addition to *Orobanche*, dodder (Cuscuta spp) started to be a big problem for faba bean and other legumes in CWANA. In addition to parasitic weeds, annual weeds are a major constraint for legume production in CWANA. Faba bean is a slow growing crop in the early stages and prone to weed competition that may reduce grain legume yields by 25–40% (Pandey et al., 1998).

3.3 Management of faba bean diseases and insect pests

In order to minimize the yield gaps caused by foliar diseases and insect pests, different management strategies are being followed by ICARDA and national research programmes in the regions. The major emphasis is on development of resistant faba bean cultivars but the level of resistance is not enough to protect losses during favourable weather conditions for disease and insect pest development. Uses of pesticides, agronomic practices, biological control methods and an integration of two or more methods are being used to manage key pests of faba bean in the regions. To manage foliar fungal diseases like chocolate spot, Ascochyta blight, faba bean gall and rust, fungicides are used. Many fungicidal compounds are reported to be used to control chocolate spot, Ascochyta blight, faba bean gall and rust (Stoddard et al., 2010; Emeran et al., 2011; El-Sayed et al., 2011; Ahmed et al., 2015). Since most foliar and soil-borne diseases are occurring at the same time, fungicides targeting more than one disease are recommended (Gossen et al., 2016). Seed- and soil-borne pathogens can be managed using seed treatment fungicides and biological control agents (Davidson et al., 2007; Chang et al., 2014; Belete et al., 2015; Gossen et al., 2016).

Cultural practices like adjusting planting date, proper drainage to remove excess water, crop mixtures, proper irrigation, crop rotation with non-host crops and use of pathogen-free seeds are recommended to manage faba bean diseases (Stoddard et al., 2010). Control of diseases such as chocolate spot (*Botrytis fabae*) and ascochyta (*Ascochyta fabae*) requires rotation to avoid repeat faba bean crops on the same field and spatial separation of current from previous faba bean fields, and use of fungicides including for rust (*Uromyces viciae-fabae*) (Mathews and Marcellos, 2003). Another strategy is mixed cropping of faba bean with cereals, which has been found to be effective in reducing epidemics of chocolate spot (Fernández-Aparicio et al., 2011). The most effective methods of managing FBNYV were through an integrated approach adjusting planting dates, application of insecticides (1–2 times) to control aphids and seed treatment with imidacloprid and rogueing infected plants early in the crop season (Makkouk and Kumari, 2001). Crop rotation is recommended to manage stubble-borne fungal pathogens and parasitic nematodes.

Integrated pest management is the approach adopted to limit damage caused by insect pests. Delaying planting date helps reduce *Sitona* infestation. Seed dressing insecticides using carbofuran and imidacloprid have been shown to reduce *Sitona* damage (Steene et al., 1999). A number of predators and parasitoids play an important role in regulating populations of faba bean pests in the production system and limiting the use of pesticides. More efforts should be given to alternatives to pesticides in managing faba bean pests, such as host plant resistance, botanical pesticides and biological control (entomopathogenic fungi, endophytes and beneficial insects).

3.4 Weed management

For parasitic weeds, integrated management using partially resistant faba bean cultivars with 1–2 sub-lethal glyphosate applications at flowering stage is the most practical method to reduce negative impacts on crop yield and reduce the weed seed bank over time (Abang et al., 2007; Rubilaes and Fernández-Aparicio, 2012).

Weeds should be controlled within 50–70 days after sowing to minimize grain yield loss (Díaz and Peñaloza, 1995) in food legumes. Chemicals have come to play a key role in weed management and without them it is unlikely that faba bean would have been able to reach its current production levels in extensive agriculture. Pre-emergence herbicides are efficient, and post-emergence herbicides are important in North Africa, but weed control is now becoming more difficult due to widespread evolution of herbicide-resistant weed biotypes. Therefore, the scarcity of effective herbicides is one of the most serious constraints to legume production. ICARDA recently identified faba bean sources with resistance to combined metribuzin and imazethpyr that can be used to develop faba bean lines with resistance to multiple herbicides (Maalouf et al., 2016c).

4 Faba bean breeding

4.1 Breeding for resistance to foliar and soil-borne diseases

Major diseases worldwide are Ascochyta blight, chocolate spot and rust, which are major breeding priorities (Redden et al., 2014). Major and minor genes control resistance to Ascochyta blight, while resistances to chocolate spot and to rust are quantitative (Duc et al., 2015).

The first effective Ascochyta blight and chocolate spot resistance sources were identified in the ICARDA faba bean breeding programme (Hanounik and Robertson, 1988, 1989). These lines were used by ICARDA and National Agricultural Research Systems (NARS) to help develop lines with good levels of disease resistance and high yield potential. As a result, cultivars were released in Australia, Canada, China, Egypt, Ethiopia and Spain (Sillero et al., 2010; Trivoli et al., 2006; Villegas-Fernández et al., 2009; Rubiales et al.,

2012; Temesgen et al., 2015). Although most breeding programmes are focusing on developing resistant genotypes for a single disease of economic importance, faba bean lines with multiple disease resistance have been identified (Villegas-Fernández et al., 2011). Breeding programmes in Australia have released *Ascochyta*-resistant cultivars Farah, Nura, PBA Kareema and PBA Rana, and rust-resistant cultivars Doza and PBA Warda (Paull et al., 2013). The parents of Nura had chocolate spot and Ascochyta blight resistances (Paull et al., 2006).

Currently, ICARDA has developed new sources for multiple disease resistance for Ascochyta blight, chocolate spot and rust (Maalouf et al., 2016a) for further use in the breeding programme. In addition, more than 180 new sources for resistance to chocolate spot, Ascochyta blight and rust were identified at ICARDA in 'sick plots' of heavy soils infested by a mixture of the most virulent pathogens collected in Syria. Varietal genetic resistance to rust has also been developed by Mathews and Marcellos (2003). Lines with combined resistance have been developed at ICARDA and sent to different NARS to observe the response of the resistant lines to different races in varying environments. Several cultivars released in Ethiopia with a high-level resistance to chocolate spot, derived directly or indirectly from the ICARDA breeding programme, were released by the Ethiopian Institute of Agricultural Research (EIAR) (details described in chapter 31 of this book).

Through extensive collaborative research, EIAR researchers have released several high-yielding faba bean varieties through direct selection from germplasm supplied by ICARDA, and by transferring good levels of resistance from ICARDA germplasm into locally adapted varieties. Among the faba bean varieties released with good levels of chocolate spot partial resistance are 'Moti' (ILB 4432 X Kuse-2-27-33), 'Gebelcho' (ILB 4726 X 'Tesfa'), Obsie (ILB 4427 X CS20DK) and 'Walki' (ILB 4615 X Bulga 70). The variety 'Walki' was developed for waterlogged areas and is gaining popularity in the central highlands of Ethiopia, and more recently the variety Gora with large seed type.

Cercospora leaf spot (*Cercospora zonata*) is an important disease of faba bean in southern Australia, and major gene resistance has been identified which will be incorporated into new cultivars (Kimber and Paull, 2011). For faba bean production in northern Australia a number of viruses have been detected, with BLRV resistance from Yunnan used in the breeding programme, and sources of BYMV resistance are being jointly investigated with ICARDA (van Leur et al., 2006).

4.2 Breeding for *Orobanche* resistance

Efforts to breed faba bean resistant to *Orobanche* have resulted in the release of cultivars with useful levels of incomplete resistance combined with a degree of tolerance (Cubero et al., 1994; Khalil and Erskine, 1999; Maalouf et al., 2011). The resulting resistance, which might be based on a combination of resistance mechanisms, is more likely to last longer than resistance that is based on a single gene (Rubiales et al., 2006; Perez-de-Luque et al., 2010). The first significant finding of resistance was in the identification of family 402 derived from a three-year cycle of individual plant selection in an F_7 from the cross (Rebaya 40 X F216) made at ICARDA (Cubero et al., 1994). Different cultivars have been developed in collaboration with NARS partners in different ecological systems, for example, Giza 402, Cairo 843, Misr1 and Misr3 in Egypt, Najah and Chourouk in Tunisia, and Ashengie in Ethiopia. New improved lines for resistance to *Orobanche* have been also identified in Morocco and are now being validated.

4.3 Breeding for heat, drought and waterlogging stresses

Heat, drought, waterlogging and frost are the major abiotic stresses affecting the faba bean productivity in CWANA, terminal drought, frost and heat. Drought, an interval of water deficiency leading to a significant reduction in yield, is widely considered to be the most important environmental constraint to crop productivity (Fischer and Turner, 1978; Borlaug and Dowswell, 2005). In Ethiopia, waterlogging is a major problem in areas dominated by vertisols (Keneni et al., 2001). In many production regions in the Mediterranean basin, the crop is seldom if ever irrigated and generally relies on stored soil moisture and current rainfall for its growth and development (Sau and Mınguez, 2000). Variation in the amount and distribution of rainfall is generally considered the major reason for variability in the grain yield of faba bean (Bond et al., 1994; Siddique et al., 2001).

Extreme heat is the major threat to faba bean production in south Egypt, Sudan and the Ethiopian lowlands. This adverse effect could be attributed mainly to the high temperature during the vegetation period which checked growth and led to the development of a small, short-stemmed crop with few branches and pods. Abdelmula and Abuanja (2007) concluded that the genotype C.52/1/1/1 could be used to improve heat tolerance in faba bean and make it possible to extend production of faba bean in the non-traditional areas of Sudan. Moreover, the traits that exhibited strong and positive association with yield could be used as selection criteria for improving faba bean under heat stress conditions. Breeding efforts at ICARDA resulted in the identification of sources of resistance to heat by growing faba bean under high temperatures in a spring sowing at Tel Hadya (Syria) and a summer sowing in Terbol (Lebanon), where temperatures reached as high as 40°C during the flowering and pod filling stages. Among accessions evaluated for heat pollen germination under heat stress, 135 were screened for heat tolerance and seven were selected (Maalouf et al., 2016b). Artificially induced terminal heat stress can significantly reduce yield and the yield components of faba bean genotypes (Abdelmula and Abuanja, 2007).

Extreme crop stresses for drought, waterlogging and heat are likely to be more frequent with climate change. Improved genetic resilience will assist in the management of these stresses.

5 Faba bean genetic and genomic resources

5.1 Major faba bean gene banks

ICARDA safeguards the largest collection of faba bean worldwide (32% of the total world collection). This global collection conserves materials from 71 countries with a high percentage of unique accessions. A total of 8628 of these accessions comprise the international collection held in trust for the global community. The collection held at ICARDA also conserves over 6000 accessions of other *Vicia* species, including about 3000 accessions of wild species of *Vicia*. The accession type and source data in Table 2 provide an indication of the uniqueness of the collections. Collections with a high percentage of wild relatives, landraces and materials originally collected by ICARDA are most likely to encompass unique accessions prioritized in a rational global system (Maalouf et al., 2013). The ICARDA collection was established in Aleppo, Syria, in 1977. The genetic resources unit moved to Terbol, Lebanon, after war broke out in Syria in 2012.

Table 2 Gene banks with more than 500 faba bean accessions

Country/city	Organization	No. of accessions
Lebanon/Terbol	ICARDA	10,045
China/Beijing	CAAS	5,200
Australia/Victoria	AGG	2,445
Germany/Gatersleben	Genebank IPK	1,920
France/Dijon	INRA	1,900
Russia/St. Petersburg	VIR	1,881
Italy/Bari	Genebank	1,876
Morocco/Rabat	INRA	1,715
Spain/Madrid	CNR	1,622
Poland/Poznam	IOPG-PAS	1,258
Ethiopia/Addis Ababa	PGRC	1,118
Spain/Córdoba	IFAPA	1,091
Poland/Radzików	PBAI	856
Portugal/Oeiras	INRB-IP	788
USA/Pullman	USDA	750
Netherlands/Wageningen	DLO	726
Bulgaria/Sadovo	IIPGR	692

Genetic resources provide the diversity required for addressing the ever-changing challenges of resistances to diseases and pests and the new abiotic stress challenges with climate change. The primary wild relative gene pool(s) of *V. faba* have not been identified (Duc et al., 2015), hence the most accessible genetic resources are in the domestic gene pools with a wide range of seed sizes.

5.2 Genome mapping and markers

A composite map of the *V. faba* genome based on morphological markers, isozymes, seed protein genes and microsatellites was constructed by Román et al. (2004). The map incorporates data from 11 F2 families for a total of 654 individuals all sharing the common female parent *Vf* 6. The integrated map is arranged in 14 major linkage groups (LGs; 5 of which were located in specific chromosomes). These LGs included 192 loci and cover 1559 cM with an overall average marker interval of 8 cM. By joining data of a new F2 population segregating for resistance to *Ascochyta* and broomrape, other traits of agronomic interest were revealed. Ellwood et al. (2008) used 151 intron-targeted amplified polymorphic markers to construct a comparative genetic map of faba bean (length 1685.8 cM). Linkage analysis revealed seven major and five small LGs, one pair and 12 unlinked markers, with 3–30 markers per LG. Faba bean possesses one of the largest genomes among cultivated legumes (~13 000 Mbp), which may explain why the number of LGs exceeds that for chromosomes.

The combination of trisomic segregation, linkage analysis among loci from different families with a recurrent parent and the analysis of new physically located markers has allowed the establishment of a *V. faba* map with wide coverage. These maps provide an efficient tool in breeding applications, such as disease-resistance mapping, quantitative trait loci analyses and marker-assisted selection (MAS). Comparative genomics and synteny analysis with closely related legumes will reveal new candidate genes and selectable markers for use in MAS.

One hundred and four of the 127 mapped markers in the 12 LGs, which were previously assigned to *Medicago truncatula* genetic and physical maps, were found in regions syntenic between the faba bean and *M. truncatula* genomes. However, chromosomal rearrangements were observed that could explain the difference in chromosome numbers between faba bean, lentil and *M. truncatula*. Multiple polymerase chain reaction amplicons and comparative mapping were suggestive of small-scale duplication events in faba bean. They provided a preliminary indication of finer scale macro-synteny between *M. truncatula*, lentil and faba bean. Markers originally designed from genes on the same *M. truncatula* bacterial artificial chromosomes were found to be grouped together in corresponding syntenic areas in lentil and faba bean (Ellwood et al., 2008), which may facilitate a more efficient selection of new cultivars free of anti-nutritional compounds. Synteny of the *V. faba* genome with other legumes and with the *M. truncatula* reference genome along with establishment of DNA libraries has led to important biotechnology advances in development of markers for resistances to diseases and weeds, while those for frost and for drought tolerances are being validated across environments and different genetic backgrounds (Duc et al., 2015). These advances are now being implemented in breeding programmes.

6 Faba bean management with limited water resources

6.1 The management of faba bean response to water deficit

Faba bean is considered to be less tolerant to water deficit than the other cool season food legumes, lentil, chickpea and pea, but more tolerant of waterlogging (Stoddard et al., 2006). In the worst case, stomatal closure occurs too late and the plants lose water, the root system does not explore the soil sufficiently deeply or widely so the plants do not take up water that would be available to deeper-rooting crops and there is no evidence that osmotic adjustment occurs, whereby solutes such as proline and glycinebetaine protect the cells against oxidative damage (Khan et al., 2010). There is substantial genetic variation in stomatal activity (Khazaei et al., 2013) and in root volume (work in progress), so improving resilience to water deficit is a realistic objective.

This susceptibility to water deficit has probably contributed to the general view that faba bean grows best on heavy, water-retentive soils, and it is likely that the breeding focus on moist environments has exacerbated the intolerance to water deficit. A further confounding factor is the association of pH with soil texture, clay-rich soils being more alkaline and some sandy soils more acid, and faba bean is stressed when soil pH is below five (French and White, 2005).

The timing of water deficit is important. Early deficit is detrimental to seedling emergence and establishment, as in all crops. Moderate deficit during the juvenile phase, between establishment and flowering, slightly delays the onset of flowering (Lizarazo et al., 2017).

The species is considered to be most susceptible to water deficit during flowering, when it inhibits pod setting and leads to premature termination of flowering (Khan et al., 2010). Mild deficit late in the flowering phase can, however, be beneficial to crop management, as it terminates flowering so the plant can mature rather than continuing to grow in its indeterminate fashion. Terminal drought is common in Mediterranean climates, and premature deficit during seed filling reduces final seed size (Adisarwanto and Knight, 1997).

Maximizing productivity of faba bean thus requires minimizing water deficit through most of the growing season, by choosing sowing time appropriately, matching the maturity time of the crop to the expected duration of the growing season and by alleviating transient drought by irrigation. Unfortunately, where the crop is primarily grown for feed use, its market value is generally too low to repay investment in irrigation, but where it is a food crop, irrigation is worthwhile.

Irrigation water use efficiency for faba bean sole crop and wheat/faba intercropping may be increased by alternate watering on both sides of crop rows, though these small-scale results still need to be tested at the farm scale (Yang et al., 2015).

6.2 Soil moisture conservation with reduced tillage

Availability of soil moisture throughout the growing season can be a limitation to crop growth with rain-fed agriculture. In-crop drought severely reduces BNF (Elias and Herridge, 2014). No-till crop production (no tillage before sowing) retains crop residue *in situ*, thereby increases SOM and water infiltration, and reduces evaporation and rainfall runoff (Huggins and Reganold, 2008). No-till improves effectiveness of rainfall especially when this is low, thereby improving nodule longevity, BNF and grain yields of legumes. Although less energy is expended with no-till, the increased cropping intensity means more investment in control of weeds with chemicals, and specialized machinery is required (McRobert and Rickards, 2005). Uptake of no-till technology in Australia varies regionally, affected by direct benefits of reduced soil erosion and greater flexibility in operations, and by motivation. Uptake was associated with engagement of farm advisors and becoming dependent on them, whereas non-adoption reflected concerns with costs of transition from conventional tillage and of herbicides, little opportunity for farmers to conduct their own comparison of tillage methods and often a reduction in livestock production with less on-farm diversification (McRobert and Rickards, 2005).

In the winter-dominant rainfall conditions of southern Australia, early sowing of faba bean is recommended since this results in higher grain yields, and use of no-till enables this to be achieved (Mathews and Marcellos, 2003; Elias and Herridge, 2014). However, the advantages of no-till may be less apparent on heavier soils (McRobert and Rickards, 2005). Broadly no-till has advantages for faba bean cultivation in Australia with a high level of mechanization and large-scale farming, but evaluation is needed to test its applicability on small farms with <20 ha of cropping, and sometimes <1 ha, in Africa and Asia.

In southern Africa, reduced tillage systems provided additional soil moisture to maize crops, in comparison with conventional tillage, in low and uncertain rainfall environments (Mupangwa et al., 2016).

Speratti et al. (2015) reviewed uptake of conservation agriculture based on no-tillage cropping/rotation systems in Latin America, from the 1970s. Adoption has been most rapid in Brazil and Argentina, with 50% and 90% of cropping, respectively, on over 25.5 Mha in each country, and variously 600 000–2.4 Mha in Paraguay, Bolivia, Uruguay, Chile, Venezuela and Mexico. This includes 80% adoption in southern Brazil including 90% of smallholders

(Gianessi, 2014). Drivers of adoption included improved soil moisture availability, SOM and soil structure, reduced soil erosion, improved yields of rotation cereal crops, reduced disease incidence and reduced liming on acid soils. Constraints were the costs of transitioning to no-till planters, lack of credit and increased costs of herbicides. Local manufacturing has provided a range of no-till planters from single furrow horse drawn, to cropping on raised beds with irrigation (irrigation furrows are reshaped between seasons) in Mexico. Farmer associations, government extension services, agri-industry commerce and information sharing networks have all contributed to adoption of no-till (Speratti et al., 2015; Gianessi, 2014). Also important was farmer participation in trials and development of farming systems to suit local socio-economic conditions and crop choices.

In the tropics maize rotations including the legume '*Mucuna*' have been adopted by smallholders in central America to give sustainable cropping systems. Key features have been no-till/legume rotation benefits as a source of nitrogen, with 30% increased net returns for cereal and legume crops over time after the initial costs of converting to conservation agriculture, reduced labour and reduced soil erosion. Bottom-up and top-down approaches were integrated to realize successful outcomes (Speratti et al., 2015; Gianessi, 2014).

Faba bean has been one of the crops to benefit from these recent improvements in farm management in Latin America. Giambalvo et al. (2012) reviewed a long-term tillage trial with faba bean over 18 years, finding a 31% yield benefit of no-till over conventional tillage (23% over reduced tillage), especially in low rainfall seasons. Broomrape weed infestation was reduced; however, total weed biomass increased, emphasizing the need for effective weed control with herbicides.

7 Faba bean BNF

Irrespective of the many different agronomic systems for faba bean production, there are some common principles concerning managing the mineral nutrition of this important crop. As a legume, faba bean can access dinitrogen (N_2) from the atmosphere and convert it to organic nitrogen through the activity of the soil bacteria *Rhizobium leguminosarum* bv. *viciae* that infects the roots of this species (Jensen et al., 2010). This symbiosis results in enhanced growth and yield of the crop because of an improved supply of N in legumes in general (Ludwig and de Vries, 1986) and specifically in faba bean (Carter et al., 1994).

Traditional agricultural production of legumes for subsistence and limited regional trading of produce has generally been secondary to cereals as the staple foods. Thus, faba bean as a minor crop was mainly grown in rotation with cereals. Now with a greater emphasis on sustainable agricultural systems, more attention is being given to the incorporation of faba bean and other legumes into long term rotations. There are economic benefits from nitrogen fixation for reduction of synthetic nitrogen fertilizer application to following crops and from the disease break effect for cereals (Herridge, 2013). There are many examples of both faba/cereal rotations and complex faba/cereal/canola intercropping and relay-cropping long-term systems that have been developed to suit respective regional cropping environments such as in China (Li et al., 2017) and in cotton rotations in eastern Australia (Rochester et al., 1998).

Even though there have been studies on the response of faba bean to applied N fertilizers (e.g. Richards and Soper, 1982), in comparison to other legumes, faba bean has a large capacity to fix N, so it has a significant role in nitrogen cycling in farming systems.

In general, legume N fixation is around 20–25 kg N per tonne of shoot biomass, so a high biomass crop like faba bean will fix relatively more N than a slower growing crop such as lentil (Peoples et al., 2009). Jensen et al. (2010) reported that mean faba bean fixation was 154 kg N/ha compared to *Pisum sativum* (86 kg N/ha), *Cicer arietinum* (51 kg N/ha) and *Lens culinaris* (71 kg N/ha). As well, faba bean showed the highest percentage of N derived from the atmosphere (Ndfa) at 75%. The %Ndfa is lower where there is a high concentration of mineral N. A summary of 18 studies of N fixation by faba bean showed that the proportion of Ndfa varied from 34 to 99% depending mainly on the amount of mineral N available (Peoples et al., 2009; Jensen et al., 2010). Where N fertilizer is applied, the proportion of Ndfa declines (Richards and Soper, 1982).

The net N benefit to a cropping system will depend then on the amount of N removed, and losses of N as denitrification under waterlogging or leaching on light textured soils. In comparing removal and fixation, Peoples et al. (2001) indicated that the faba bean legume phase contributed up to 270 kg N/ha to soil N after harvest (Rochester et al., 1998). Enhanced N fixation ability has been reported where faba bean and maize are intercropped, with exudates from the maize roots reported to increase N_2 fixation in the intercropped faba bean (Li et al., 2015).

Most soils contain indigenous rhizobia, although these strains may not be as effective in fixing N_2 (Dean et al., 1979; Armarger, 1986; Carter et al., 1994) and there also appear to be differences in compatibility between faba bean genotypes and different strains of rhizobia (Mytton et al., 1977; Dean et al., 1979). Rhizobia are able to persist between host crops on most soils, so that inoculation is generally only required if the land has not been sown with faba bean previously, although the persistence of inoculated strains may be poor on acid soils (Unkovich et al., 1997). Even so, on neutral or alkaline soils, competition among indigenous and supplied rhizobia may not always result in infection with the improved and desired strain (Carter et al., 1995). Selecting rhizobia inoculants appropriate for the soil type and crop genotype is an important aspect of improving the potential to enhance the percentage of Ndfa. The slurry-on-seed method of *Rhizobium* inoculation is regarded as the best for nodulation, nitrogen fixation and grain yield (Elias and Herridge, 2014).

Especially with climate change likely to result in higher C:N ratios in plant biomass, thus reducing the protein content in cereal grain and subsequent bread quality (Fernando et al., 2014), it is increasingly important to manage soil fertility and structure for a more sustainable agriculture. Legume crops, notably faba bean and lupins, have high rates of BNF and low sensitivity to levels of soil nitrogen (Hardarson and Atkins, 2003; Herridge, 2009), so can make an important contribution to sustainable farming as agriculture moves into a new era of permanent climate change. In a review of 900 trials, Angus et al. (2015) found that wheat yields showed the greatest benefit of 1 t/ha (range 0.7–1.6 t/ha) following a legume crop, as well as for oats (0.5 t/ha) and canola (0.8 t/ha), compared with wheat after wheat, and even extending to a second wheat crop a benefit of 0.2–0.3 t/ha. With strong BNF capacity faba bean can help reduce the fertilizer cost and improve the profitability of wheat in rotation with faba bean.

8 Faba bean mineral nutrient requirements

Seed mineral nutrient compositions taken from a range of faba bean studies is presented in Table 3, with comparisons to canola, wheat and field pea. Despite the range of values from the literature, there is reasonable consistency in a value of 50 kg N/t and 4–6 kg P/t for faba bean. Both concentrations are higher in faba bean than wheat, canola and pea. The variation in faba bean seed P is not unexpected as there are differences in P uptake due to cultivar and season (Daoui et al., 2011). Potassium values vary widely probably in response to the variable supply of K in different soils, and plants generally accumulate K to luxury levels when it is available. There is little information in the literature on survey values for the S concentration in faba bean despite its importance in supplying the sulphur-containing essential amino acids cysti(e)ne and methionine which cannot be synthesized by higher animals (Brosnan and Brosnan, 2006). In terms of management of S, because of its mobility, especially on light soils, splitting applications will generally provide yield benefits.

There is also little data on seed B concentrations, but the literature suggests that faba bean is the most tolerant of the pulse crops to boron toxicity (Anon, 1998) and where low soil B occurs, usually on acid, sandy, leached soils, responses to rates of foliar B to 100 g/ha can be beneficial (Shaaban et al., 2006). In terms of other nutrients, deficiencies of calcium and magnesium are rare (Mahler et al., 1988). Co-authors RMN and FLS have also seen iron deficiency in faba bean when grown on highly calcareous soils.

The symbiosis between rhizobia and faba bean is also determined by adequate supply of phosphorus and the micronutrient molybdenum, with nodules being strong sinks for these nutrients (Brodrick and Giller, 1991). Molybdenum is an essential component of the Mo-Fe sub-unit of nitrogenase, which is essential in N fixation but in acid soils (pH <5.0) can be strongly bound to Fe-Al complexes and so become unavailable. Liming acid soils will raise pH and increase Mo availability so that growth and yield are enhanced. For example, Quertatani et al. (2011) reported better nodulation on faba bean roots with liming, which in turn induced higher N, P and K contents in plant tissues and improved growth and yield in Tunisia on acid soils. Van Zweiten et al. (2015) reported that 10 t/ha dry weight of biochar derived from the pyrolysis of paper mill by-product were able to provide pH amendment, enhanced P status and reduced Al toxicity similar to 3 t/ha of lime. In that study there were no significant differences between those two treatments for %Ndfa and the total amount of N fixed.

High levels of Mn and Al can be harmful to legume growth as well as inhibiting the development of rhizobia (Unkovich et al., 1997). Faba bean is more affected by acidity than other pulse crops, canola and wheat, with the largest difference where soil pH is less than 5.2 (Lake, 2000). Manganese is a particular problem where soils are waterlogged and the redox state induces high levels of soluble and bioavailable Mn^{2+}. Legumes such as faba bean are more at risk of Mn toxicity than cereals because of the rhizosphere acidification due to root exudates (Pearse et al., 2006). Faba bean roots exude predominantly malate and citrate (Pearse et al., 2006) and while this assists in acquiring P from soil (Nuruzzaman et al., 2005), it may also have an adverse effect on Mn^{2+} efflux into the plant.

Cobalt is an essential nutrient for rhizobia as it is a component of leghaemoglobin. This protein functions to provide oxygen to respiring rhizobia while keeping oxygen nodule levels low enough for nitrogenase enzymes to function. Cobalt deficiency will appear as green non-functioning root nodules. Deficiencies have been reported in pulse crops (Tisdale et al., 1985) and treatment is typically as a foliar application of cobalt sulphate at rates of 250 g/ha or as a seed dressing (Mahler et al., 1988).

Correcting for soil pH using lime is an appropriate strategy to address the low pH, and the use of calcium derived from gypsum (calcium sulphate dihydrate) can provide improved soil structure and therefore drainage, especially on sodic soils. Improved drainage may also assist with salinity management, as faba bean is generally considered intolerant of soil salinity, although there are differences among accessions (Raheshwaran et al., 2016). Differences among faba bean genotypes have been demonstrated for root exudation (Rose et al., 2010), production of antioxidants (Arya and Roy, 2011) and for other coping mechanisms in heavy metal environments such as the upregulation of metallothioneins that can strongly bind to transition metals such as Mn (Foley and Singh, 1994). Given the diversity of adaptations and their variability within faba bean germplasm, there are opportunities to improve acid soil adaptation of this species.

Due attention should be given to both potassium and phosphorus nutrition, as limited supply of either can restrict growth and yield as well as N fixation (Sangakara et al., 1996). Potassium in particular can improve tolerance to drought and other abiotic stresses including salinity (Ma et al., 2015). On neutral or alkaline soils, P is probably the most important nutrient to supply to faba bean (Lewis and Hawthorn, 1996; Bolland et al., 1999; Bolland et al., 2000) and adequate P nutrition is critical for both crop yield and ensuring effective N fixation. Research has shown that P is most efficiently supplied when banded at sowing at 5 cm or banded 2.5 cm to the side and below the seed rather than broadcast and incorporated by cultivation (Henry et al., 1995). Phosphorus demand will depend on soil P levels and also on crop demand, and responses can be poor if there is low soil pH or deficiencies or toxicities of other nutrients. Faba bean was reported by Bolland et al. (1999) to be the more responsive to P than chickpea, lentil or lupin in Western Australia. In Canada, Henry et al. (1995) also found faba bean to be more responsive than field pea or lentil. Both studies suggested rates of P for faba bean to be 10–12 kg P/ha depending on soil P status and P binding ability. This rate is close to the P removal in seed of a 2.5–3.0 t/ha faba bean crop (Table 3). When harvested for fresh vegetables, removal rates would be higher as the pods also contain significant amounts of P (Jensen et al., 2010).

During early growth, faba bean shows little response to added P which has been attributed to the large seed reserves (Bolland et al., 1999), but P demand and uptake can continue during flowering (Daur et al., 2011). Phosphorus acquisition from the soil by the roots of faba bean is enhanced by the root exudates that solubilize generally less available P pools (Gardener et al., 1983). This process not only aids the legume but also acts to carry P over to subsequent crops as P decomposes from legume roots or microorganisms (Nuruzzaman et al., 2005), although the release may be too slow to completely supply the needs of the crop grown with a low P background (Nachimuthu et al., 2009).

There are reports of differences among faba bean genotypes in phosphorus use efficiency (Rose et al., 2010; Daoui et al., 2011). These differences may be a consequence either in acquisition (root architecture and/or ability of plant to extract different pools of nutrients) and/or utilization – the conversion of P accessed to seed yield. Different mechanisms may be more important in high or low P soils and in different soil pH ranges (Rose et al., 2010). There are opportunities to develop plants with higher P efficiency, so reducing the need for P fertilizer, although to maintain a soil P resource at least P replacement would be needed. However, given the acidifying root exudates of citrate, malate and protons from faba bean, there may be the opportunity to solubilize lower solubility P sources such as rock phosphate rather than single or triple superphosphate or the ammoniated phosphates.

Faba bean, like 80% of plant species, develops arbuscular mycorrhizas on its roots, which are an association between a fungus and the roots of a host plant (AMF). Several different fungi cause the development of these structures and the fungi occur naturally in most

Table 3 Mineral composition of four common crops expressed at 0% moisture content

N kg/t	P kg/t	K kg/t	S kg/t	B mg/kg	Cu mg/kg	Mn mg/kg	Zn mg/kg	Reference
Faba bean seed								
48	4.6	2.8	–	–	15	11	53	Singh et al., 2014
41	4.0	10.0	1.5		10	28	33	Anon., 1998
50	3.6	–	–	–	–	–	>25	Lewis and Hawthorn, 1996
42	4.0	10.8	1.8	–	–	–	–	DJ Reuter, pers. comm.
56	7.3	23.5			12	22	53	Westermann et al., 2011
46	5.8	12.4			–	–	–	Jensen et al., 2010
Faba bean empty pods								
13	0.7	32.8	–	–	–	–	–	Jensen et al., 2010
Wheat grain								
22	3.3	4.5	1.7	2.2	4.8	44	23	Norton, 2012
Canola seed								
43	5.7	6.9	4.1	12	2.9	28	38	Norton, 2014
Field pea seed								
38	3.4	9	1.8	–	5.0	30	28	Anon., 1998

soils. The most significant effect of AMF is to increase the host plants' ability to access immobile nutrients such as P and Zn (Abbott and Robson, 1991). For example, faba bean inoculated with *Glomus mosseae* showed increased shoot growth and P uptake compared to uninoculated plants, but as additional P is available from the soil, the impact of the symbiosis declines (Kucey and Paul, 1983; Almethyeb et al., 2013).

In addition to AMF, there has been interest in free-living microbial inoculants such as *Penicillium bilaii* to solubilize P from less available soil pools (Asea et al., 1988) or the plant-growth-promoting *P. radicum* (Kucey, 1987). Organic acids (Veneklaas et al., 2003) and phosphatases (Dinkelaker and Marschener, 1992) from soil microbes are reported to assist with mineral nutrient uptake. Similarly, free-living N fixers (Rodelas et al., 1999) and various plant-growth-promoting microbes (El-Azeem et al., 2007) are under investigation, but there is only limited information on the field efficiency or the ecological competence of these bio-stimulants in a soil environment (Richardson et al., 2011).

Zinc is also a critical nutrient that is often in short supply in agricultural systems, although faba bean has a low requirement compared to other arable crops (Brennan et al., 2001). Balancing P and Zn supply is important for nodulation, growth and yield in environments with calcic soils such as southern Australia (Lewis and Hawthorn, 1996) and northern

Ethiopia (Weldua and Habtegriel, 2013). Where soil available P is high, crops commonly suffer zinc deficiency (Snehi Dwivedi et al., 1975). This syndrome is a consequence of immobilization of Zn in the roots and nodes of the plants. Zinc nutrition also affects the uptake of the heavy metal cadmium (Lombi et al., 2000) which causes oxidative stress in plants and has a high level of toxicity for plants, animals and humans (Bahmani et al., 2012). Added zinc can reduce the uptake of cadmium in faba bean through competition for uptake (Bystricka et al., 2013).

Balanced nutrition is fundamental to the successful cultivation of faba bean, and this will include due attention to moderating soil pH to promote conditions for rhizobial activity to ensure adequate N is supplied to the faba bean and which also flows through to subsequent crops. Adequate P and Zn are critical to ensuring rhizosphere health and due consideration to other nutrients such as K and S will be important where soil supply is limited.

Other aspects of improving faba bean production include improved pollination with honey bees, the selection of deep well-drained soils with neutral to alkaline pH though faba bean can manage wet conditions, establishing plant populations of 35–40 plants/m^2, but as low as 20 plants/m^2 in wide 50 cm rows which are suited to no-till with no yield penalty, and seed inoculation with rhizobia (Mathews and Marcellos, 2003).

9 Conclusion

Faba bean is a very productive legume for dry grain and for vegetable, especially green seed in unripe pods. China, which already has the world's largest production, is likely to increase both area sown and productivity through more intensive inter-crop farming systems as the length of growing seasons increase in northern China due to climate change. With more informed plant nutrition and use of rhizobial inoculum, biological nitrogen fixation (BNF) can be optimized for each region, thereby substituting for applied nitrogen fertilizer and also providing a nitrogen benefit to succeeding cereal crops. The end result is better economic efficiency, reduced downstream pollution and increased use of faba bean in holistic farming systems.

Important advances have been made in breeding for multiple resistances to Ascochyta blight, and chocolate spot and rust diseases, as well as to *Cercospora* and to the parasitic weed *Orobanche*. Sources of heat tolerance are being identified in specific nurseries in Sudan, for incorporation into climate change–resilient varieties. These genetic gains will be aided in future with the development of genetic markers and the identification of linkage groups for key traits, and synteny of the faba bean genome with other legumes will advance the application of genetic tools. Eventually these could lead to identification of the wild crop progenitors of faba and provide access to a larger genepool. Thus increased opportunities for genetic improvement are becoming available.

These will be complemented by agronomic strategies such as reduced tillage to conserve soil moisture, strategic irrigation pre-flowering – grain filling, and weed control. Intercropping presents both challenges and opportunities for mechanized agriculture, notably with precision agriculture. This may also deliver more widespread benefits to companion/relay crops. Faba bean has an important role in holistic farming systems as climate changes.

10 Where to look for further information

Additional information on faba bean cropping systems and ecoonomics can be found in

Murphy-Bokern, D., Stoddard, F. L. and Watson, C. A., eds (2017), *Legumes in Cropping Systems*. CAB International, Oxford, UK. ISBN 978-1-78064-498-1.

Reckling, M., Bergkvist, G., Watson, C. A., Stoddard, F. L., Zander, P. M., Walker, R., Pristeri, A., Toncea, I. and Bachinger, J. (2016), Trade-offs between economic and environmental impacts of introducing legumes into cropping systems. *Frontiers in Plant Science* 7: 669. doi: 10.3389/fpls.2016.00669.

Watson, C. A., Reckling, M., Preissel, S., Bachinger, J., Bergkvist, G., Kuhlman, T., Lindström, K., Nemecek, T., Topp, C. F. E., Vanhatalo, A., Zander, P., Murphy-Bokern, D. and Stoddard, F. L. (2017), Grain legume production and use in European agricultural systems. *Advances in Agronomy* 144: 235-303. https://doi.org/10.1016/bs.agron.2017.03.003.

For further information on breeding see:

R. Redden, Paull, J., Zong, X., Sass, O., Yang, T. and Li, L. (2014), Ch 7. Faba bean. In 'Broadening the genetic base of grain legumes', In Singh Mohar, Bisht Ishwari Singh, Dutta Manoranjan (Eds.), *Broadening the Genetic Base of Grain Legumes*, Publication by Springer.' This is complementary to the reference of Duc et al. (2015), in this chapter. The latter also provides information on the complex search to identify the wild relatives of faba.

The genetic diversity of faba bean in China in relation to the world is covered by Wang, H., Zong, X., Guan, J., Yang, T., Sun, X., Ma, Y. and Redden, R. (2011), Genetic diversity and relationship of global faba bean (Vicia faba L.) germplasm revealed by ISSR markers *Theoretical and Applied Genetics*. DOI 10.1007/s00122-011-1750-1.

Further information on BNF is found in the references for Herridge (2009), Li et al. (2017) and Peoples et al. (2009), in this chapter.

Climate change may affect bee pollination of faba:

Stoddard, F. L. (2017), Climate change can affect crop pollination in unexpected ways. *Journal of Experimental Botany* 68: 1819–21. doi:10.1093/jxb/erx075.

11 References

Abang, M. M., Bayaaa, B., Abu-Irmaileh, B. and Yahyaoui, A. (2007), A participatory farming system approach for sustainable broomrape (*Orobanche* spp.) management in the Near East and North Africa, *Crop Protection*, 26, 1723–32. http://www.sciencedirect.com/science/journal/02612194/26/12.

Abdelmula, A. A. and Abuanja, I. K. (2007), Genotypic Responses, Yield Stability, and Association between Characters among some of Sudanese Faba bean (*Vicia faba* L.) Genotypes under Heat stress. Conference on International Agricultural Tropentag Research for Development University of Kassel-Witzenhausen and University of Göttingen, pp. 1–7.

Abebe, T., Beyene, H. and Nega, Y. (2013), Distribution and economic importance of broomrape (*Orobanche crenata*) in food legumes production of south Tigray, Ethiopia, *ESci Journal of Crop Production*, 2(3), 101–6.

Adisarwanto, T. and Knight, R. (1997), Effect of sowing date and plant density on yield and yield components in the faba bean, *Australian Journal of Agricultural Research*, 48, 1161–8.

Ahmed, S., Abang, M. M. and Maalouf, F. (2015), Integrated management of Ascochyta blight (*Didymella fabae*) on faba bean under Mediterranean conditions, *Crop Protection*, 81, 65–9.

Almethyeb, M., Ruppel, S., Paulsen, H.-M., Vassilev, N. and Eichler-Löbermann, B. (2013), Single and combined applications of arbuscular mycorrhizal fungi and Enterobaceria radicincitans after nutrient uptake of faba bean and soil biological characteristics, *Applied Agriculture and Forestry Research*, 3, 229–34.

Angus, J. F., Kirkegaard, J. A., Hunt, J. R., Ryan, M. H., Ohlander, L. and Peoples, M. B. (2015), Break crops and rotations for wheat, *Crop and Pasture Science*, 66(6), 523–52.

Anon. (1998), *The Grain Legume Handbook*. Grains Research and Development Corporation, Canberra, p. 68.

Armarger, N. (1986), Nodulation competition among Rhizobium leguminosarum strains, *Vorträge für Pflanzenzüchtung*, 11, 186–94.

Arya, S. K. and Roy, B. K. (2011), Manganese induced changes in growth, chlorophyll content and anti-oxidant activity of seedling broad bean (Vicia faba L), *Journal of Environmental Biology*, 32, 707–11.

Asea, P. E. A., Kucey, R. M. N and Stewart, J. W. B. (1988), Inorganic phosphate solubilisation by two Penicillium species in solution culture and soil, *Soil Biology and Biochemistry*, 20, 459–64.

Bahmani, R., Bihamta, M. R., Habibi, D., Forozesh, P. and Ahmadvand S. (2012), Effect of cadmium chloride on growth parameters of different bean genotypes (Phaseolus vulgaris L.), *Journal of Agricultural and Biological Science*, 7(1), 35–41.

Belete, E., Ayalew, A. and Ahmed, S. (2015), Evaluation of Local Isolates of *Trichoderma* Spp. against Black Root Rot (*Fusarium solani*) on Faba Bean, *Journal of Plant Patholology and Microbiology*, 6, 279. doi:10.4172/2157-7471.1000279.

Belete, E, Ayalew, A. and Ahmed, S. (2013), Associations of biophysical factors with faba bean root rot (*Fusarium solani*) epidemics in the northeastern highlands of Ethiopia, *Crop Protection*, 52, 39–46. http://www.sciencedirect.com/science/article/pii/S0261219413001154.

Bogale, M., Steenkamp, E. T., Wingfield, M. J. and Wingfield, B. D. (2009), Diverse *Fusarium solani* isolates colonise agricultural environments in Ethiopia. *European Journal of Plant Pathology*, 124, 369–78.

Bolland, M. D. A., Siddique, K. H. M. and Brennan, R. F. (2000), Grain yield responses of faba bean (Vicia faba L.) to applications of fertiliser phosphorus and zinc, *Australian Journal of Experimental Agriculture*, 40, 849–57.

Bolland, M. D. A., Siddique, K. H. M., Loss, S. P. and Baker, M. J. (1999), Comparing responses of grain legumes, wheat and canola to applications of superphosphate, *Nutrient Cycling in Agroecosystems*, 53, 157–75.

Bond, D. A., Jellis, G. J., Rowland, G. G., Le Guen, J., Robertson, L. D., Khalil, S. A. and Li-Juan, L., (1994), Present status and future strategy in breeding faba beans (*Vicia faba* L.) for resistance to biotic and abiotic stresses, *Euphytica*, 73, 151–66.

Borlaug, N. and Dowswell, C. R. (2005), Feeding a world of ten billion people: A 21st century challenge. In Tuberosa, R., Phillips, R. L. and Gale, M. D. (Eds), *In the Wake of the Double Helix: From the Green Revolution to the Gene Revolution*, Avenue Media, Bologna, Italy, 27–31.

Bray, T. (2006), Benefits of pulses in irrigation rotations, *GRDC Grains Research Update for Irrigation Croppers*, pp. 1–4.

Brennan, R. F., Bolland, M. D. A. and Siddique, K. H. M. (2001), Responses of coo-season grain legumes and wheat to soil-applied zinc, *Journal of Plant Nutrition*, 24, 727–41.

Brodrick, S. J. and Giller, K. E. (1991), Root nodules of Phaseolus: Efficient scavengers of molybdenum for N2-fixation, *Journal of Experimental Botony*, 42, 679–86.

Brosnan, J. T. and Brosnan, M. E. (2006), The sulfur containing amino acids: An overview, *The Journal of Nutrition*, 136, 1636S–40S.

Bystricka, J., Tomas, J., Toth, T., Trebichalsky, P. and Kavalcoa, P. (2013). Potential antagonistic Zn effect on faba bean (Faba vulgaris m.) contaminated by Pb and Cd, *Environmental Protection and Natural Resources*, 24, 21–4.

Carter, J. M., Gardner, W. K. and Gibson, A. H. (1994), Improved growth and yield of faba bean (Vicia faba cv. Fiord) by inoculation with strains of Rhizobium leguminosarum bv. viciae in acid soils in South-west Victoria, *Australian Journal of Agricultural Research*, 45, 613–23.

Carter, J. M., Tieman, J. S. and Gibson A. H. (1995), Competitiveness and persistence of strains of Rhizobia for faba bean on acid and alkaline soils, *Soil Biology and Biochemistry*, 27, 617–23.

Cazzato, E., Tufarelli, V., Ceci, E., Stellacci, A. M. and Laudadio, V. (2012), Quality, yield and nitrogen fixation of faba bean seeds as affected by sulphur fertilization, *Acta Agriculturae Scanindavica Section B, Soil and Plant Science*, 62(8), 732–8.

Chang, K. F., Conner, R. L., Hwang, S. F., Ahmed, H. U., McLaren, D. L., Gossen, B. D. and Turnbull, G. D. (2014), Effects of seed treatments and inoculum density of *Fusarium avenaceum* and *Rhizoctonia solani* on seedling blight and root rot of faba bean, *Canadian Journal of Plant Science*, 94, 693–700.

Cheng, X. Z. and Wang, S. M. (2009), *Records of Chinese Food Legumes Cultivars*. China Agricultural Science and Technology Press, Beijing, pp. 1–16 (in Chinese).

Cornish, P. S. (2009), Research directions: Improving plant uptake of soil phosphorus and reducing dependency on input of phosphorus fertilizer, *Crop and Pasture Science*, 60, 190–6.

Cubero, J. I., Pieterse H., Khalil S. A. and Sauerborn J. (1994), Screening techniques and sources of resistance to parasitic angiosperms, *Euphytica*, 74, 51–8.

Daoui, K., Karrou, M., Mrabet, R., Fatemi, Z., Draye, X. and Ledent, J. F. (2012), Genotypic variation of phosphorus use efficiency among Moroccan faba bean varities (Vicia faba) under rainfed conditions, *Journal of Plant Nutrition*, 35, 34–48.

Daur, I., Septoglu, H. and Sindel, B. (2011), Dynamics of faba bean growth and nutrient uptake and their correlation with grain yield nutrient uptake patterns of beans, *Journal of Plant Nutrition*, 34, 1360–71.

Davidson, J. A. and Kimber R. B. E. (2007), Integrated disease management of ascochyta blight in pulse crops, *European Journal of Plant Pathology*, 119, 99–110.

Dean, J. R., Toomsan, B. and Clark, K. W. (1979), Rhizobium strain selection for faba bean, *Canadian Journal of Plant Science*, 60, 385–97.

Díaz, S. J. and Peñaloza H. E. (1995). Critical period of weed interference in dry peas (Pisum sativum L) c.v. Progreta and lentil (Lens culinaris M) cv Araucana-INIA, *Agricultura Técnica Santiago*, 55, 176–82.

Dinkelaker, B. and Marschner, H. (1992), In vivo demonstration of acid phosphatase activity in the rhizosphere of soil-grown plants, *Plant Soil*, 144, 199–205.

Duc, G., Aleksić, J. M., Marget, P., Mikić, A, Paull, J, Redden, R. J., Sass, O., Stoddard, F. L., Vandenberg, A., Vishnyakova, M. and Torres, A. M. (2015), Faba bean, Chapter 5. In DeRon, A. (Ed.), *Handbook for Grain Legumes*. Springer, New York, Heidelberg, Dordrecht, London, pp. 144–78.

El-Azeem, S. A., Mehana, T. A. and Shabayek, A. A. (2007), Response of Faba Bean (Vicia faba L.) to Inoculation with Plant Growth-Promoting Rhizobacteria, *CATRINA*, 2(1), 67–75.

Elias, N. V. and Herridge, D. (2014), Crop – available water and agronomic management, rather than nitrogen supply, primarily determine grain yield of commercial chickpea in northern New South Wales, *Crop and Pasture Science*, 65, 442–52.

El-Sayed, A. S., El-Shennawy, R. Z. and Ismail, A. I. (2011), Fungicidal management of chocolate spot of faba bean and assessment of yield losses due to the disease, *Annals of Agricultural Science*, 56, 27–35.

Ellwood, S. R., Phan, H. T., Jordan, M., Hane, J., Torres, A. M., Avila, C. M. and et al. (2008), Construction of a comparative genetic map in faba bean (Vicia faba L.); conservation of genome structure with Lens culinaris, *BMC Genomics*, 9, 380.

Emeran, A. A. and Sillero, J. C., Fernández-Aparicio, M. and Rubiales, D. (2011). Chemical control of faba bean rust (*Uromyces viciae-fabae*), *Crop Protection*, 30, 907–12.

FAOSTAT. (2016), http://faostat3.fao.org/download/Q/QC/E.

Fernández-Aparicio, M., Shtaya, M. J. Y., Emeran, A. A., Allagui, M. B., Kharrat, M. and Rubiales, D. (2011), Effects of crop mixtures on chocolate spot development on faba bean grown in Mediterranean climates, *Crop Protection*, 30, 1015–23. https://doi.org/10.1016/j.cropro.2011.03.016

Fernando, N., Panozzo, J., Tausz, M., Norton, R. M., Neumann, N., Fitzgerald, G. J. and Seneweera, S. (2014), Elevated CO2 alters grain quality of two bread wheat cultivars grown under different environmental conditions, *Agriculture, Ecosystems and Environment*, 185, 24–33.

Fischer, R. A. and Turner N. C., (1978), Plant productivity in the arid and semiarid zone, *Annual Review of Plant Physiology*, 29, 277–317.

Foley, R. C. and Singh, K. B. (1994), Isolation of a Vicia faba metallothionein-like gene: Expression in foliar trichomes, *Plant Molecular Biology*, 26, 435–44.

Franz, A., Makkouk, K. M. and Vetten, H. J. (1998), Acquisition, retention and transmission of Faba bean necrotic yellows virus by two of its aphid vectors, *Aphis craccivora* (Koch) and *Acyrthosiphon pisum* (Harris), *Journal of Phytopathology*, 146, 347–55.

French, B. and White, P. (2005), Soil and environmental factors affecting pulse adaptation in Western Australia, *Australian Journal of Agricultural Research*, 50, 375–87.

Gardner, W. K., Barber, D. A. and Pabery, K. G. (1983), The acquisition of phosphorus by Lupinus albus L III. The probable mechanism by which phosphorus movement in the soil/root interface is enhanced,. *Plant and Soil*, 70, 107–24. https://doi.org/10.1007/bf02374754

Giambalvo, D., Ruisi, P., Saia, S., Di Miceli, G., Frenda, A. S. and Amato, G. (2012), Faba bean grain yield, N2 fixation, and weed infestation in a long-term tillage experiment under rainfed Mediterranean conditions, *Plant and Soil*, 360, 215–27. doi:10.1007/s11104-012-1224-5.

Gianessi, L. P. (2014), International pesticide benefits case study no. 111, *Crop Life*.

Gossen, B. D., Conner, R. L., Chang, K.-F., Pasche, J. S., McLaren, D. L. Henriquez, M. A., Chatterton, S. and Hwang, S.-F. (2016), Identifying and Managing Root Rot of Pulses on the Northern Great Plains. *Plant and Disease*, 100(10), 1965–78. http://dx.doi.org/10.1094/PDIS-02-16 0184-FE.

Gressel, J., Hanafi, A., Head, G., Marasas, W., Obilana, B., Ochanda, J., Souissi, T. and Tzotzos, G. (2004), Major heretofore intractable biotic constraints to African food security that may be amenable to novel biotechnological solutions, *Crop Protection*, 23, 661–89.

Hadarson, G. and Atkins, C. (2003), Optimising biological N2 fixation by legumes in farming systems, *Plant and Soil*, 252, 41–54.

Hailu, E., Getaneh, G., Sefera, T., Tadesse, N., Bitew, B., Boydom, A., Kassa, D. and Temesgen, T. (2016), Faba bean gall: A new threat for faba bean (Vicia faba) production in Ethiopia, *Advances in Crop Science and Technology*, 2, 144. doi:10.4172/2329-8863.1000144. ISSN: 2329–8863. ACST, an open access journal.

Hanounik, S. B. (1980), Effect of chemical treatments and host genotypes on disease severity/yield relationships of Ascochyta blight in faba beans, FABIS Newsletter 2: 50 (Hanounik, S. B., Roberston, L. D.), Resistance of faba bean germplasm to blight caused by *Ascochya fabae, Plant Disease*, 73, 202–5.

Hanounik, S. B. and Robertson, L. D. (1988), New sources of resistance in *Vicia faba* to Chocolate Spot Caused by *Botrytis fabae, Plant Disease*, 72, 696–8.

Hanounik, S. B. and Robertson, L. D. (1989). Resistance in *Vicia faba* germplasm to blight caused by *Ascochyta fabae, Plant Disease*, 73, 202–5.

Henry, J. L., Slinkard, A. E. and Hogge, T. J. (1995), The effect of phosphorus fertilizer on establishment, yield and quality of pea, lentil and faba bean, *Canadian Journal of Plant Science*, 75, 395–8.

Herridge, D. (2009), Nitrogen fixation benefits of pulsse crops. *Grains Research & Development Corporation, Fact Sheet*, pp. 1–7.

Hu, X. and Guo, G. Q. (1994), *High-yield Cultivation of Faba Bean and Pea*. Jindun Publishing House, Beijing, pp. 1–7, 75–9 (in Chinese).

Huignard, J, Glitho, I. A., Monge, J.-P. and Regnault, R. C. (2011), Insectes ravageurs des graines de légumineuses. Biologie des Bruchinae et lutte raisonnée en Afrique. Editions Quae, p. 146. ISBN: 978-2-7592-1659-8, référence 02274. Collection Update Sciences & Technologies.

Jensen, E. S., Peoples M. B. and Haugaard-Nielsen H. (2010), Faba bean in cropping systems, *Field Crops Research*, 115, 203–16.

Jensen, E. S. (1987), Inoculation of pea by application of Rhizobium in the planting furrow, *Plant and Soil*, 97, 63–70.

Jin, W. L. and Zong, X. X. (2000), *High-Yield and Super-Quality Cultivation Techniques for Food Legumes*. China Braille Publishing House, Beijing, pp. 1–18 (in Chinese).

Keneni, G., Asmamaw, B. and Jarso, M. (2001), Efficiency of drained selection environments for improving grain yield in faba bean under undrained target environments on vertisol, *Euphytica*, 122, 279–85.

Khalil, S. and Erskine, W. (1999), Breeding for Orobanche resistance in faba bean and lentil. In Cubero, J. I., Moreno, M. T., Rubiales, D. and Sillero, J. C. (Eds), *Resistance to Broomrape, the State of the Art*. Junta de Andalucıa, Spain, pp. 63–76.

Khalil, S., Kharrat, M., Malhotra, R., Saxena, M. and Erskine, W. (2004), Breeding faba bean for Orobanche resistance. In Dahan, R. and El-Mourid, M. (Eds), *Integrated Management of Orobanche in Food Legumes in the Near East and North Africa*. Proceedings of the Expert Consultation on IPM for Orobanche in Food Legume Systems in the Near East and North Africa, Rabat, pp. 1–18.

Khan, H. R., Paull, J. G., Siddique, K. H. M. and Stoddard, F. L. (2010), Faba bean breeding for drought-affected environments: A physiological and agronomic perspective, *Field Crops Research*, 115, 279–86. doi:10.1016/j.fcr.2009.09.003.

Kharrat, M., Abbes, Z. and Amri, M. (2010), A new faba bean small seeded variety najeh tolerant to orobanche registered in the Tunisian catalogue, *Tunisian Journal of Plant Protection*, 5, 125–30.

Khazaei, H., Street, K., Santanen, A., Bari, A. and Stoddard F. L. (2013), Do faba bean (Vicia faba L.) accessions from environments with contrasting seasonal moisture availabilities differ in stomatal characteristics and related traits? *Genetic Resources and Crop Evolution*, 60, 2343–57. doi:10.1007/s10722-013-0002-4.

Kimber, R. B. E. and Paull J. G. (2011). Identification and genetics of resistance to cercospora leaf spot (*Cercospora zonata*) in faba bean (*Vicia faba), Euphytica*, 177, 419–29. https://doi.org/10.1007/s10681-010-0288-7

Kucey, R. M. N. and Paul, E. A. (1983), Vesicular arbuscular mycorrhizal spore populations in various Saskatoon soils and the effect of inoculation with Glomus mosseae on faba bean growth in greenhouse and field trials, *Canadian Journal of Soil Science*, 63, 87–95.

Kucey, R. M. N. (1987), Increased phosphorus uptake by wheat and field beans inoculated with a phosphorus solubilizing Penicillium bilaji strain and with vesicular-arbuscular mycorrhizal fungi, *Applied and Environmental Microbiology*, 53, 2699–703.

Kumari, S. G. and Makkouk, K. H. M. (2007), Virus diseases of faba bean (*Vicia faba* L.) in Asia and Africa, *Plant Viruses*, 1, 93–105.

Lake, B. (2000), Understanding soil pH, *Leaflet No 2. New South Wales Agriculture, Yanco*, p. 4.

Li, B., Li, Y., Wu, H., Zhang, F., Li, C., Li, X., Lambers, H. and Li, L. (2015), Root exudates drive interspecific facilitation by enhancing nodulation and N2 fixation, *Proceedings of the National Academy of Science*, 113, 6496–501.

Li, L., Yang, T., Liu, R., Redden, B., Maalouf, F. and Zong, X. X. (2017), Food legume production in China, *The Crop Journal*, 5(2), 115–26. doi:10.1016/j.cj.2016.06.001.

Lin, R. F., Chai, Y., Liao, Q. and Sun, S. X. (2002), *Minor Grain Crops in China*. China Agricultural Science and Technology Press, Beijing, pp. 192–333 (in Chinese).

Liu, Q. F. (1984), *Faba Bean Cultivation*. Yunnan People's Publishing House, Kunming, pp. 1–3 (in Chinese).

Lizarazo, C. I., Isotalo, J., Lindfors, A. V. and Stoddard, F. L. (2017), Progress towards flowering of faba bean (Vicia faba L.) is more than photothermal. *Journal of Agronomy and Crop Science* 203: in press. doi:10.1111/jac.12200.

Lombi, E., Zhao, F. J. and Dunham, S. J. (2000), Cadmium accumulation in populations of Thlaspi caerulescen and Thlapsi geosingense, *New Phytologist*, 1, 11–20.

Long, J. Y., Lin, L. F., Hou, X. S., Duan, X. N. and Duan, H. Y. (1989), *Food Legume Crops*. Science Press, Beijing, pp. 1–15, 83–315 (in Chinese).

Long, J. Y. and Wang, Z. Q. (2002), *Cultivation Techniques for Food Legumes*. Jindun Publishing House, Beijing, (in Chinese).
Ludwig, R. A. and de Vries, G. E. (1986), Biochemical physiology of Rhizobium dinitrogen fixation. In Broughton, W. J. and Pühler, S. (Eds), *Nitrogen Fixation, Vol. 4: Molecular Biology*. Clarendon University Press, Oxford, UK, pp. 50–69.
Ma, Q., Bell, R., Scanlan, C., Sarre, G. and Brennan, R. (2015), Growth and yield responses in wheat and barley to potassium supply under drought or moderately saline conditions in south-western Australia, *Crop and Pasture Science*, 66, 135–44.
Maalouf, F., Khalil, S., Ahmed, S., Akintunde, N. A., Kharrat, M., Hajjar, S. and El Shama'a, K. (2011), Yield stability of faba bean lines under diverse broomrape prone production environments, *Field Crop Research*, 124, 288–94.
Maalouf, F., Nawar, M., Hamwieh, A., Amri, A., Zong, X., Bao, S. and Yang, T. (2013), Faba bean. In Singh, M., Upadhyaya, H. D. and Bisht, I. S. (Eds), *Genetic and Genomic Resources of Grain Legume Improvement*. Elsevier Insight, Waltham, 113–36. ISBN: 978-0-12-397.
Maalouf, F., Ahmed, S., Shaaban, K., Bassam, B., Nawar, F., Singh, M. and Amri, A. (2016a), New faba bean germplasm with multiple resistances to Ascochyta blight, Chocolate spot and rust diseases, *Euphytica*, 211(2), 157–67. online 10.1007/s10681-016-1726-y.
Maalouf, F., Chalak, L. and Hammoud, M. 2016b. Evaluation of faba bean accession under heat prone conditions. International Conference on Pulses for health nutrition and sustainable agriculture in Drylands. Marrakesh, Morocco, 18–20 April, p. 84.
Maalouf, F., Somanagouda, P., Karthika, R., Aladdin, H., Aakash, G. and Kumar, S. (2016c), Breeding for post-emergence herbicide tolerance in cool-season food legumes. International Conference on Pulses for health nutrition and sustainable agriculture in Drylands. Marrakesh, Morocco, 18–20 April, p. 62.
Mahler, R. L., Saxena, M. C. and Aeschlimann, J. (1988), Soil fertility requirements of pea, lentil, chickpea and faba bean. In Summerfield, R. (Ed.), *World Crops, Cool Season Food Legumes*. Kluwer Academic Publishers, Dordrecht, the Netherlands, pp. 279–87.
Makkouk, K. M. and Kumari, S. G. (1995), Transmission of broad bean stain comovirus and broad bean mottle bromovirus by weevils in Syria. *Journal of Plant Diseases and Protection*, 102, 136–9.
Makkouk, K. M. and Kumari, S. G. (2001), Reduction of spread of three persistently aphid transmitted viruses affecting legume crops by seed-treatment with Imidacloprid (Gaucho), *Crop Protection*, 20, 433–7.
Makkouk, K., Pappu, H. and Kumari, S. G. (2012), Virus diseases of peas, beans, and faba bean in the Mediterranean region, *Advances in Virus Research*, 84, 367–402. https://doi.org/10.1016/b978-0-12-394314-9.00011-7
Mathews, P. and Marcellos, H. (2003), Faba bean. Agfact P4.2.7, 2nd edition 2003,. Agdex 164, NSW Agriculture, pp. 1–12.
McRobert, J. and Rickards, L. (2005), Social research: Insights into farmers' conversion to no-till farming systems, *Extension Farming Systems Journal*, 6(1), 43–52.
Mesa-García, J. and Gacría-Torres, L. (1991), Status of *Orobanche crenata* in faba bean in the Mediterranean region and its control. *Options Méditerranéennes – Série Séminaires* – N 10 - 1991, 75–8.
MOA (2015), *Statistics Book of Minister of Agriculture China*.
Mucheru, M. M., Pypers, P., Muendi, D., Mugwe, J., Merckx, R. and Vanlauwe, B. (2010), A staggered maize-legume intercrop arrangement robustly increases crop yields and economic returns in the highlands of central Kenya, *Field Crops Research*, 115, 132–9.
Mwanauta, R. W., Mtei, K. M. and Ndakidemi, P. A. (2015), Potential of controlling common bean insect pests (Bean Stem Maggot (*Ophiomyia phaseoli*), ootheca (*Ootheca bennigseni*) and aphids (*Aphis fabae*)) using agronomic, biological and botanical practices in field, *Agricultural Sciences*, 6, 489–97. http://dx.doi.org/10.4236/as.2015.65048.
Mytton, L. R., El-Sherbeeny, M. H. and Lawes, D. A. (1977), Symbiotic variability in Vicia faba. III. Genetic effects of host plant rhizobium strain and of host x strain interaction, *Euphytica* 26, 785–791.

Nachimuthu, G., Guppy, C., Kristiansen, P. and Lockwood, P. (2015), Isotopic tracing of phosphorus uptake from corn from 33P labelled legume residues and 32P labelled fertilizers applied to a sandy loam soil, *Plant and Soil*, 314, 303–10.

National Center for Extension of Agronomic Techniques. (2008–14), National-wide information on extension of agronomic techniques of grain and oil crops. Ministry of Agriculture of China, Beijing, 2015, pp. 108, 132, 160 (in Chinese).

Norton, R. M. (2012), Wheat grain nutrient concentrations for south-eastern Australia. 'Capturing Opportunities and Overcoming Obstacles in Australian Agronomy'. Edited by I. Yunusa, *Proceedings of 16th Australian Agronomy Conference*, Armidale, NSW. http://www.regional.org.au/au/asa/2012/nutrition/7984_nortonrm.htm.

Norton, R. M. (2014), Canola seed nutrient concentrations for southern Australia. In Ware, A. H. and Potter, T.D. (Eds), 18th Australian Research Assembly on Brassicas (ARAB 18). Tanunda, 2014. Proceedings of Australian Oilseed Federation, pp. 1–6.

Nuruzzaman, M., Lambers, H., Bolland, M. D. A. and Veneklaas, E. J. (2005), Phosphorus uptake by grain legumes and subsequently grown wheat at different levels of residual phosphorus fertiliser, *Australian Journal of Agricultural Research*, 56, 1041–7.

Pandey, A. K., Prasad, K., Singh, P. and Singh, R. D. (1998), Comparative yield loss assessment and crop – weed association in major winter crops of mid hills of N-W Himalayas, *Indian Journal of Weed Science*, 30, 54–7.

Paull, J. G., Rose, I., van Leur, J. A. G., Kimber R. B. E. and Seymour M. (2006), Breeding faba beans for the Australian environment, International workshop on faba bean breeding and agronomy, Cordoba, Spain, pp. 67–9.

Paull, J., Kimber, R., van Leur, J., Ware, A., Armstrong, E. and Adhikari, K. (2013). PBA faba bean breeding: Aims and progress, PBA Inaugural Pulse Conference, Adelaide, Australia.

Pearse, S. J., Veneklaas, E. J., Cawthray, G. R., Bolland, M. D. A. and Lambers, H. (2006), Carboxylate release of wheat, canola and 11 grain legumes as affected by phosphorus status, *Plant and Soil*, 288, 127–39.

Peoples, M. B., Bowman, A. M., Gault, R. R., Herridge, D. F., McCallum, M. H., McCormick, K. M., Norton, R. M., Rochester, J. J., Scammell, G. and Schwenke, G. D. (2001), Factors regulating the contributions of fixed nitrogen by pastures and crop legumes in different farming systems of eastern Australia, *Plant and Soil*, 228(1), 29–41.

Peoples, M. B., Brockwell, J., Herridge, D. F., Rochester, I. J., Alves, B. J. R., Urquiaga, S., Boddey, R. M., Dakora, F. D., Bhattari, S., Maskey, S. L., Sampet, C., Rerkesam, B., Khan, D. F., Hauggaard-Nielsen, H. and Jensen, E. S. (2009), The contributions of nitrogen-fixing crop legumes to the productivity of agricultural systems, *Symbiosis*, 48, 1–17.

Perez-de-Luque, A., Eizenberg, H., Grenz, J. H., Sillero, J. C., Avila, C. M., Sauerborn, J. and Rubiales, D. (2010), Broomrape management in faba bean, *Field Crops Research*, 115, 319–28.

Quertatani, S., Regaya, K., Ryan, J. and Gharbi, A. (2011), Soil liming and mineral fertilization for root nodulation and growth of faba bean in an acid soil in Tunisia, *Journal of Plant Nutrition*, 34, 850–60.

Rashid, K. Y., Bernier, C. C. (1991), The effect of rust on yield of faba bean cultivars and slow rusting populations, *Canadian Journal of Plant Science*, 71, 967–72.

Redden, R., van Leur, J., Zong, X., Bao S., Yujioa, L., Paull, J. and Leonforte, T. (2008), Evaluation and utilisation of pea and faba bean germplasm from China. In Unkovich M. (Ed.), Global issues, paddock action, *Proceedings of the 14th Australian Agronomy Conference*, Adelaide, South Australia, Australian Society of Agronomy. Available at: www.regional.org.au/au//pdf/asa/2008/5804_redden.pdf

Rhameshwaran, P., Qadir, M., Ragab, R., Arslann, A., Majid, G. A. and Abdallah, K. (2016), Tolerance of faba bean, chickpea and lentil to salinity, accessions salinity response functions, *Irrigation and Drainage*, 65, 49–60.

Richards, J. E. and Soper R. J. (1982), N fertilization of field grown faba beans in Manitoba, *Canadian Journal of Plant Science*, 62, 21–30. https://doi.org/10.4141/cjss82-003

Richardson, A. E., Lynch, J. P., Ryan, P. R., Delhaize, E., Smith, F. A., Smith, S. E., Harvey, P. E., Ryan, M. H., Veneklaas, E. J., Lambers, H., Oberson, A., Culvenor, R. A. and Simpson, R. J. (2011), Plant and microbial strategies to improve the phosphorus efficiency of agriculture, *Plant and Soil*, 349, 121–56.

Rochester, I. J., Peoples M. B., Constable G. A. and Gault R. R. (1998), Faba beans and other legumes add nitrogen to irrigated cotton cropping systems, *Australian Journal of Experimental Agriculture*, 38, 253–60.

Rodelas, B., Gonzales-Lopez, J., Martinez-Toledo, M. V., Pozo, C. and Salmeron, V. (1999), Influence of Rhizobium/Aztobacter and Rhizobium/Azospirilum combined inoculation on mineral composition of faba bean (Vicia faba L), *Biology and Fertility of Soils*, 29, 165–9.

Román, B., Satovic, Z., Pozarkova, D., Macas, J., Dolezel, J., Cubero, J. I. and et al. (2004). Development of a composite map in Vicia faba, breeding applications and future prospects, *Theoretical and Applied Genetics*, 108(6), 1079–88. https://doi.org/10.1007/s00122-003-1515-6

Rose, T. J., Damon, P. and Rengel Z. (2010), Phosphorus-efficient faba bean (Vicia faba L.) genotypes enhance subsequent wheat crop growth in an acid and an alkaline soil, *Crop and Pasture Science*, 1, 1009–16.

Rubiales, D., Perez-de-Luque, A., Fernandez-Aparicio, M., Sillero, J. C., Roman, B., Kharrat, M., Khalil, S., Joel, D. M. and Riches, Ch. (2006), Screening techniques and sources of resistance against parasitic weeds in grain legumes, *Euphytica*, 147, 187–99.

Rubiales, D., Avila, C. M., Sillero, J. C. and et al. (2012). Identification and multi-environment validation ofresistance to Ascochyta fabae in faba bean (Vicia faba). *Field Crops Research*, 126, 165–70.

Rubiales, D. and Fernández-Aparicio, M. (2012b), Innovations in parasitic weeds management in legume crops, A review, *Agronomy and Sustainable Development*, 32, 433–49.

Sangakkara, R., Hartwig, U. A. and Nosberger J. (1996), Growth and symbiotic nitrogen fixation by Vicia faba and Phaseolus vulgaris as affected by fertilizer potassium and temperature, *Journal of Food Science and Agriculture*, 70, 315–20.

Sau, F., Mínguez, M. I. (2000), Adaptation of indeterminate faba beans to weather and management under a Mediterranean climate, *Field Crops Research*, 66, 81–99.

Shaaban, M., El-Sayed, F., El-Nour, A., El-Zanaty, A. M. A. and El-Saady, A. K. M. (2006), Boron/Nitrogen interaction effects on growth and yield of faba bean plants grown under sand soil conditions, *International Journal of Agricultural Research*, 1, 322–30.

Shifa, H., Hussien, T. and Sakhuja, P. K. (2011), Association of faba bean rust (*Uromyces viciae fabae*) with environmental factors and cultural practices in the Hararghe Highlands, Eastern Ethiopia, *East African Journal of Sciences*, 5, 58–65.

Siddique, K. H. M., Regan, K. L., Tennant, D. and Tomson, B. D. (2001), Water use and water use efficiency of cool season grain legumes in low rainfall Mediterranean-type environments, *European Journal of Agronomy*, 15, 267–80.

Sillero, J. C., Villegas-Fernandez Angel, M., Thomas, J., Rojas-Molina, M. M., Emeran, A. A., Fernandez-Aparicio, M. and Rubiales, D. (2010), Faba bean breeding for disease resistance, *Field Crop Research*, 115, 297–307.

Singh, A. K., Bharadwa, J. and Is Sing, H. (2014), Assessment of nutritional quality of developed faba bean (Vicia faba L) lines, *Journal of Agrisearch*, 1, 96–101.

Snehi Dwived, R., Randhawa, N. S. and Bansal, R. L. (1975), Phosphorus-zinc interaction I. Sites of immobilization of zinc in maize at a high level of phosphorus, *Plant and Soil*, 43, 639–48.

Speratti, A., Turmel, M. S., Calegari, A., Araujo-Junior, C. F., Violic, A., Wall, P. and Govaerts, B. (2015), Conservation agriculture in Latin America, Chapter 16. In Farooq, M. and Siddique, K. H. M. (Eds), *Conservation Agriculture.* Springer International Publishing, Switzerland, pp. 391–415.

Steene van de, F., Vulseteke, G., de Proft, M. and Callewaert, D. (1999), Seed coating to control the pea leaf weevil, *Sitona lineatus* (L.) in pea crops, *Zeitschrift für Pflanzenkrankheiten und Pflanzenschutz*, 106, 633–7.

Stoddard, F. L., Balko, C., Erskine, W., Khan, H. R., Link, W. and Sarker, A. (2006), Screening techniques and sources of resistance to abiotic stresses in cool season food legumes, *Euphytica*, 147, 167–86. doi: 10.1007/s10681-006-4723-8.

Stoddard, F. L., Nicholas, A. H., Rubiales, D., Thomas, J. and Villegas-Fernandez, A. M. (2010), Integrated pest management in faba bean, *Field Crops Research*, 115, 308–18.

Temesgen, T., Keneni, G., Sefera, T. and Jarso, M. (2015), Yield stability and relationships among stability parameters in faba bean (*Vicia faba* L.) genotypes, *The Crop Journal*, 3, 258–69. https://doi.org/10.1016/j.cj.2015.03.004

Tisdale, S. L., Nelson, W. L. and Beaton J. D. (1985), *Soil Fertility and Fertilizers*. MacMillan, New York, pp. 358–60.

Tivoli, B, Baranger, A., Avila, C. M., Banniza, S., Barbetti, M., Chen, W. D., Davidson, J., Lindeck, K., Kharrat, M., Rubiales, D., Sadiki, M., Sillero, J. C., Sweetingham, M. and Muehlbauer, F. J. (2006), Screening techniques and sources of resistance to foliar diseases caused by major necrotrophic fungi in grain legumes, *Euphytica*, 147, 223–53.

Unkovich, M. J., Pate, J. S. and Sanford, P. (1997), Nitrogen fixation by annual legumes in Australian Mediterranean agriculture, *Australian Journal of Agricultural Research*, 48, 267–93.

Van Leur, J. A. G., Kumari, S. G., Makkouk, K. M. and Rose, I. (2006). Viruses on faba bean in north-east Australia and strategies for virus control, International workshop on faba bean breeding and agronomy, Cordoba, Spain, pp. 129–31.

Van Zwieten, L., Rose, T., Herridge, D., Kimber, S., Rust, J., Cowie, A. and Morris, S. (2015), Enhanced biological N2 fixation and yield of faba bean (VIcia faba L) in an acid soil following biochar addition: Dissection of causal mechanisms, *Plant and Soil*, 395, 7–20.

Villegas-Fernández, A. M., Sillero, J. C. and Rubiales, D. (2011), Screening faba bean for chocolate spot resistance: Evaluation methods and effects of age of host tissue and temperature, *European Journal of Plant Pathology*, 132, 443–53. https://doi.org/10.1007/s10658-011-9889-9

Villegas-Fernández, A. M., Sillero, J. C., Emeran, A. A., Winkler, J., Raffiot, B. and Tay, J. (2009), Identification and multi-environment validation of resistance to Botrytis fabae in *Vicia faba, Field Crop Research*, 114, 84–90.

Webb A, Cottage A, Wood T, Khamassi K, Hobbs D, Gostkiewicz K, White M, Khazaei H, Ali M, Street D, Duc G, Stoddard FL, Maalouf F, Ogbonnaya FC, Link W, Thomas J, O'Sullivan DM. 2016. A SNP-based consensus genetic map for synteny-based trait targeting in faba bean (Vicia faba L.). *Plant Biotechnol J*.14(1):177–85. doi: 10.1111/pbi.12371. Epub 2015 Apr 10.

Weldu, Y. and Habtegriel, K. (2013), Effect of zinc and phosphorus fertilizers application on nodulation and nutrient concentration of faba bean (Vicia faba L) grown in calcaric cambisol of semi-arid Northern Ethiopia, *Journal of Agricultural Research*, 1, 220–6.

Westermann, D. T., Teran, H., Munoz-Perea, C. G. and Shree, P. (2011), Plant and seed nutrient uptake in common bean in organic and conventional production systems, *Canadian Journal of Plant Science*, 91, 1089–99.

Yang, C., Chai, Q., Li, G., Feng, F. and Wang, L. (2015), Water use efficiency of controlled alternate irrigation on wheat/faba bean intercropping, *African Journal of Agricultural Research*, 10(48), 4348–55.

Ye, Y. (2003), *Monograph on Faba Bean in China*. China Agriculture Press, Beijing, pp. 5–6, 301–21 (in Chinese).

Zheng, Z. J., Wang, S. M. and Zong X. X. (1997), *Monograph on Food Legumes in China*. China Agriculture Press, Beijing, pp. 3–24, 53–365 (in Chinese).

Zheng, D. S. and Fang J. H. (2009), *High-quality Varieties and Their Cultivation of Minor Legumes and Cereals*. China Agriculture Press, Beijing, pp. 169–315 (in Chinese).

Zong, X. X. (2002), *High-yield Cultivation and Processing of Food Legumes*. China Agricultural Science and Technology Press, Beijing, pp. 2–10 (in Chinese).

Zong, X. X., Bao S. Y. and Guan J. P. (2006), *Descriptors and Data Standard for Faba Bean (Vicia faba L.)*. China Agriculture Press, Beijing, pp. 1–3 (in Chinese).

Zong, X. X. (2010), *Integrated Cultivation of Food Legumes*. China Agriculture Press, Beijing, pp. 1–153 (in Chinese).

Chapter 14

Developing improved varieties of pigeonpea

K. B. Saxena, United Arab Emirates; Y. S. Chauhan, Department of Agriculture and Fisheries, Australia; C. V. S. Kumar, A. J. Hingane, R. V. Kumar, R. K. Saxena and G. V. R. Rao, International Crops Research Institute for the Semi-Arid Tropics (ICRISAT), India

1 Introduction
2 Pigeonpea for nutritional security
3 Factors affecting stability of pigeonpea production
4 Genetic factors influencing sustainability of pigeonpea production
5 Enhancing pigeonpea sustainability through crop modelling
6 Enhancing sustainability through an efficient seed system
7 Enhancing sustainability through plant breeding
8 Pigeonpea hybrids for greater productivity and sustainability
9 Future trends and conclusion
10 Where to look for further information
11 References

1 Introduction

By 2050 the world population is likely to reach nine billion, with most of the increase in the semi-arid tropics where many of the world's poor reside. It has been estimated that this will require a 70% increase in food production (Alexandratos and Bruinsma, 2012). Increasing food production to this extent in the face of looming climate changes, decreasing water resources, escalating production cost and limitation of arable land will be a Herculean task. In this context, a food legume, popularly called as pigeonpea (*Cajanus cajan* (L.) Millsp.), red gram (*tuar*) or yellow lentil, could play a significant role in providing food and nutritional security. This is possible because this crop has potential to grow well in warmer conditions with limited water and inputs.

Pigeonpea commands respect among farming communities due to its special traits and role in sustainable agriculture. The plants fix their own nitrogen, enhance the release of soil-bound phosphorus, require less fertilizers, withstand intermittent droughts and recover more quickly from the damage caused by various biotic and abiotic stresses (Saxena, 2008). It is, therefore, a favourite crop in low-input, rain-fed farming and, for these reasons,

http://dx.doi.org/10.19103/AS.2017.0023.33

is an appropriate choice to meet the challenge of providing food security for the growing global population. This pulse not only suits the taste buds but also provides quality protein for building body tissues. It is one of the most popular pulses used in Indian and African cuisines. It is often said that no Indian vegetarian meal is complete without a bowl of pigeonpea *dal* (thick spicy soup).

According to the FAO (2013), the estimated global area for pigeonpea is over six million ha with a total production of 4.74 million tons and average yield of 762 kg/ha. The crop is well adapted to diverse soils and climatic conditions of semi-arid tropical regions of India (4.04 m ha), Myanmar, and eastern and southern Africa (with about 0.5 m ha each). India is the largest (3.02 m tons) producer of pigeonpea (DES, 2015), but to meet the domestic requirements, another 500,000 tons of pigeonpea is imported annually from Myanmar and Africa. In rural areas, pigeonpea is considered a multipurpose crop because it is used as food (fresh as vegetable and decorticated dry splits as *dal*), fodder, feed, fuel wood and even as construction material (Saxena, 2008). In this chapter, besides listing the factors influencing sustainability of production, the achievements of crop improvement efforts geared towards attaining the overall goal of sustainable legume production are reviewed.

2 Pigeonpea for nutritional security

2.1 Human nutrition

A sufficient quantity of protein in the diet is essential for normal growth and development. A lack of this vital growth component among people living under subsistence level is common, and it leads to severe malnutrition particularly among children and women. Animal protein is becoming less accessible day-by-day. The availability of home-grown vegetable protein is insufficient to meet the domestic needs of many farmers due to their limited smallholdings, low productivity and high input costs. This problem is now growing beyond dangerous proportions, especially in the underdeveloped and developing countries. For example, the per capita protein availability in India in the decade ending 2009 has seen a significant reduction from the recommended 46 g/head/day to <25 g (NIN, 2010). To overcome this shortage, there is a need to produce more protein per unit of land area. This is possible either by enhancing yield or by breeding high-protein cultivars. In this respect, pigeonpea stands ahead of most pulses due to its tolerance to various stresses and production of protein-rich (20–22%) grains. At ICRISAT, high-protein (28–30%) pigeonpea lines were bred with no loss of productivity and these lines yielded additional protein at 100,000 g/ha (Saxena and Sawargaonkar, 2015). Immature pigeonpea seeds also form a nutritive vegetable. Such types, mostly cultivated around cities and towns, have attractive pods with large sweet seeds and better nutrition than dry seeds (Saxena et al., 2010).

2.2 Livestock nutrition

Domestic animals are an integral part of sustainable agriculture. Farmers depend on draft animals for field operations and rural transportation, and mulch animals for generating additional income. Hence, animal health issues are also important for any sustainable agricultural system. Pigeonpea by-products such as fresh tops, pod shells and broken seeds offer a good fodder/feed supplement (Table 1) for domestic animals. During the

Table 1 Nutritional quality of pigeonpea fodder and feed

Nutrient (%)	Fresh fodder	Dried fodder	Silage	Seed	Seed coat
Moisture	70	11	67	10	–
Crude protein	7	15	15	18	5
Crude fibre	11	29	67	5	32
N-free extract	8	40	33	–	–
Fat	2	2	–	–	1
Ash	2	4	–	–	4

Source: various ICRISAT reports.

dry summer season, when fodder is scarce, farmers release their animals to graze in the harvested pigeonpea fields. The regenerated foliage on the pigeonpea stumps provides valuable fodder during the harsh summer. For stall-feeding, pigeonpea biomass also makes quality leguminous fodder. The nitrogen-deficient grasses are mixed with chopped pigeonpea fodder in the ratio of 3:1 for dairy animals. Whiteman and Norton (1981) recorded edible forage yields of 20–25 t/ha in Australia, while Yang et al. (2001) reported 30 t/ha fresh fodder yield in the Guangxi province of China. According to Embong and Ravoof (1978), pigeonpea leaves can provide a good substitute for alfalfa in animal feed formulations, particularly in areas that are not suitable for growing alfalfa. An important point in the forage usage of pigeonpea is that it is available at times of the year when there are shortages of energy and protein for the animals. Alongside this, the pigeonpea pod shells, left after threshing, also make good animal feed. The processes of de-hulling and splitting generate about 20–30% of by-products. These by-products, including husk, powder, and broken or under-sized seed, then form protein-rich concentrate for feeding both ruminant and non-ruminant livestock. These special components of pigeonpea make it a crucial component of sustainable agriculture in the tropics and subtropics.

3 Factors affecting stability of pigeonpea production

3.1 Physiological factors

Physiology is one of the least researched aspects of the pigeonpea crop (Sheldrake, 1984; Lawn et al., 1990). As large sections of the poor living in semi-arid tropics are increasingly dependent upon this crop for livelihoods and nutrition, it is imperative to reassess the physiology of pigeonpea to ensure tangible gains in yield, which currently are dismally low. Since the yield of this crop has remained low for decades, it appears that the key yield-forming factors have not been addressed adequately in the improvement programmes of the past.

3.1.1 Biomass

Grain yield in any crop is dependent on how efficiently it produces dry matter and converts it into yield. Unfortunately, neither dry matter production efficiency nor partitioning efficiency was appropriately assessed in pigeonpea to identify superior genotypes for these key physiological traits. The variety development, as in other legumes (Hay et al., 1995), has largely resulted in an indirect selection of genotypes that are superior only in biomass production (Chauhan et al., 1995). In the case of pigeonpea, breeders have unconsciously selected for higher biomass production capacity, the main determinant of yield. Most of this enhanced biomass production capacity among improved varieties results from longer growing duration or heterosis, and very little of its superiority arises from its better partitioning into yield.

Although genotypic differences in radiation use efficiency have been documented (Nam et al., 1998), there is little evidence of any increase in biomass having occurred due to improvement in radiation use efficiency *per se.* Often vigorous individual plants, arising from natural out-crossing in the preceding generation in farmer's fields, have attracted breeder's attention for exercising selection. However, such plants tended to lose this advantage due to inbreeding depression and competition for light and water when grown as a population. However, a number of pigeonpea varieties have been released from such individual plant selections (Singh et al., 2016). The biomass production among pigeonpea varieties varies considerably due to a number of agronomic and production factors. These include climatic conditions, cropping systems, duration and plant population, and this has resulted in a wide variety of yields being observed.

The enhanced capacity of biomass production in improved pigeonpea varieties and hybrids (Chauhan et al., 1995) ensured that such genotypes produced good yields when growing conditions were favourable, and ensured some yield even when growing conditions became slightly less favourable. However, in the event of being grown on marginal soils with low water-holding capacity, the extra rapid water utilization required to support greater growth tended to negate this advantage. If improved biomass production can be combined with better partitioning, it will lead to a greater potential and homeostasis in yield.

3.1.2 Harvest index

The dry matter production potential of pigeonpea is about 10 t/ha, and even a conservative harvest index of 20–30% suggests that up to 2–3 t/ha of yield should be harvested compared to the world average of only 746 kg/ha of this crop. The ideal route to realize higher yield in pigeonpea is to convert a greater amount of accumulated dry matter into yield. This approach to increasing yield has fascinated pigeonpea scientists, but little progress has been made in improving the harvest index on a consistent basis.

Low harvest index in pigeonpea is attributed to extensive flower drop (Sheldrake et al., 1979). Some of this flower drop could be due to indeterminate nature of flowering, where the earlier formed pods out-compete the later formed flowers for attracting assimilates. The focus on developing determinate cultivars in this crop has not paid desired dividends as these genotypes also behaved like indeterminate plants in terms of reproductive physiology (Sheldrake and Narayanan, 1979).

It appears that the key to increased partitioning could be to reduce indeterminateness through increased synchrony of flowering. Similar observations were also made in soybean, where the earlier formed flowers have greater chance of developing into pods, and more pods can be formed when flower synchrony is increased (Egli and Bruening, 2002). Hence, there is a need to develop varieties with greater synchrony in flower production in pigeonpea because it tends to equalize the competition among developing pods. Alternatively, cultivars in which seed development initially is very slow could defer serious intra-plant competition for assimilates to a later stage when most pod set has already occurred. The latter could also reduce insect pest pressure on the crop, as they will not be able to obtain much energy from slower growing grains until the pod walls become strong enough to resist their invasion. All those legumes, where pod wall constrictions (e.g. garden peas) are absent, have this trait. A wild relative of pigeonpea *Rhynchosia bracteata* also has this trait, which needs consideration in future breeding programmes.

3.1.3 Photoperiod sensitivity and flowering

Pigeonpea is a quantitative short-day plant and its flowering is delayed as days become longer than 13 h (Silim et al., 2007). The photoperiod sensitivity is also a source of indeterminateness and non-synchronous flowering in this crop. While warm weather combined with long photoperiods promotes longer vegetative phase, the short photoperiod and milder mean ambient temperatures (>18 and <25°C) promote flowering. Below the critical photoperiod, the temperature hastens or delays flowering. There are genetic differences in sensitivity to photoperiod which have been exploited to develop varieties which take less than 85 days to over 300 days. The traditional long-duration varieties are more photoperiod sensitive as compared to more recent extra-early varieties. Phenological plasticity achieved through genetic manipulation of photoperiod sensitivity has provided valuable means of developing varieties for various niches and has helped expanding its cultivation in a wider range of cropping systems and latitudes. There are also varietal differences on phenological effects of temperature. The optimum temperature for rapid flowering was 24.7°C for extra-early, 23.1°C for early, 22.2°C for medium and 18.3°C for long-duration varieties (Silim et al., 2007). A few studies also highlighted the carry-over effects of temperature and photoperiod on the period between flowering and maturity. Generally, optimum temperatures (~24°C) that hasten flowering also reduce the period between flowering and maturity. The cooler suboptimum period (<18°C) particularly tends to prolong this period.

Flowering in pigeonpea is more synchronous under inductive short days and optimum temperatures. In the post-rainy season, pigeonpea flowering is more synchronous. Harvest index of pigeonpea in this season is far greater than the crop planted in the regular season. This results in yield often being similar to that obtained in the main season even with half the dry matter production (Naryanan and Sheldrake, 1979). However, the strong negative relationship between dry matter production and yield in pigeonpea makes it difficult to increase the harvest index component of yield without reducing biomass production and vice versa (Chauhan et al., 1995). It is not clear if the negative relationship between dry matter and harvest index could be weakened through an appropriate selection strategy as has been achieved in wheat (Ding et al., 2016). To achieve this in pigeonpea, genotypes will need to be developed that are able to produce more biomass before flowering and partition more to grain after flowering, exhibiting a higher degree of determinateness.

Another approach to increase floral synchrony is to reduce photoperiod sensitivity. With the purpose of introducing greater photoperiod insensitivity in pigeonpea, the development of short-duration pigeonpea cultivars was taken up in the 1980s in Australia and India. Short-duration varieties not only take less time to flower and mature, but also have slightly higher harvest index. They are indeed less photoperiod sensitive too (Wallis et al., 1981; McPherson et al., 1985) and could be grown at higher plant population as a monocrop and fit into rotation with wheat (Dahiya et al., 2002). Breeding of such genotypes was undertaken in northern India as these genotypes were considered to be better adapted to wheat growing environments in terms of rainfall pattern and cropping windows. The evaluations done in tropical environments suggested that they could also grow well in central and southern India (Chauhan et al., 1987; Chauhan, 1990). Chauhan et al. (1987) Chauhan et al. (1987) compared early pigeonpea varieties grown at Hisar (29°N) in northern India, with those grown in tropical environments at Patancheru (17°N) near Hyderabad. Here, day length before flowering during the growing season was shorter. Varieties grown at Patancheru exhibited up to 42% harvest index and gave up to 2.7 t/ha in < 100 days. The yield obtained was also higher compared to those at Hisar where the yield was 2.35 t/ha and the biomass was 42% more due to almost one extra month of growing period.

3.2 Agronomic factors

3.2.1 Maturity and cropping systems

Major agronomic factors influencing yield of pigeonpea are appropriate maturity, sowing time, seed rates, row arrangements and nutrition. The short-duration types (<140 days) are cultivated as a sole crop and harvested before the sowing of post-rainy-season crops. Both the medium (170–180 days) and late-maturing (>250 days) types are sown after the longest day, and are harvested a few months after the shortest day. Because of their protracted growing periods, these types are cultivated in inter-crops with fast-growing cereals such as millets, sorghum and maize. Over the long growing period, they develop a deep root system (Chauhan, 1993) which helps the crop to exploit residual moisture left after harvesting of the companion crop. In pigeonpea-based inter-cropping systems, very little agronomic or genetic innovation, apart from hybrids, has been introduced in this century. Since the pigeonpea yields in such systems depend upon moisture that is present at the harvest of the companion crop, the 'decision support tools' based on crop simulation models could assist in maximizing and stabilizing the yield (Smith et al., 2016). These tools include vital parameters such as soil moisture holding capacity, utilization of light by cereal crops, and selection of appropriate pigeonpea genotypes. In pigeonpea, there are diverse plant types within each maturity type and therefore it would be ideal to further understand pigeonpea responses to plant population and row arrangements. This could help to enhance crop productivity and sustainability.

Rao and Wiley (1983) demonstrated the significant contribution of planting configuration and plant populations in the inter-crops. The plant population effect was, however, negative on companion crops leading to the recommendation of a conservative population of 50,000 plants/ha. However, in the shorter-duration genotypes the importance was higher and 300,000 plants/ha was recommended (Chauhan, 1990). This recommendation is based on the facts that such genotypes had less time to grow and were sown as a sole crop. The other benefits arising from higher plant population were greater production of stem biomass per unit area and leaf fall. Both these factors positively contributed to the sustainability of following crops (Chauhan et al., 2004).

3.2.2 Adaptation

Differences in sensitivity to day length experienced by short-duration genotypes in a given environment seems to play a crucial role in determining yield and to influences adaptation to different environments. Chauhan et al. (2002) evaluated pigeonpea genotypes that were selected under long day conditions of 45°N latitude (Minnesota, USA) and those selected under shorter day conditions in India at Patancheru (17°N). They observed that the genotypes selected in Minnesota were relatively less photoperiod-sensitive, and their harvest index was less affected by extended photoperiods. This study showed that changes in photoperiod influenced the dry matter partitioning in pigeonpea. They concluded that the selection environments representing different photoperiods had a profound influence on the level of partitioning in pigeonpea.

In this study, flowering and maturity durations were positively related to yield under natural days, but both were negatively related to yield under extended day length.These findings suggest a profound influence of photoperiod on pigeonpea physiology (Chauhan et al., 2002). The effect of photoperiod on harvest index does not appear unique to pigeonpea but seems a general response of other legumes as well (Wallace et al., 1993).

Studies by Chauhan et al. (2002) suggested that in pigeonpea photoperiod, controlls partitioning through the control of indeterminateness. In pigeonpea it has always been tempting to breed determinate cultivars, as these were expected to have higher partitioning in addition to facilitating mechanized management in terms of sprays and harvesting. However, Sheldrake and Naraynan (1979) showed that both cultivar types with respect to partitioning and the differences in flower and pod formation were morphological rather than physiological. In a later study, Chauhan et al. (1998) reported that the indeterminate pigeonpea types had a greater advantage over determinate varieties in the environments where growing period was longer, and less so in shorter growing period conditions.

Future physiological research could therefore focus more on the partitioning physiology of pigeonpea crops. An important area of research is the identification of genes related to photoperiod sensitivity, and the clarification of conditions affecting the expression of these genes. Conducting such experiments on medium- and long-duration pigeonpea will be a complicated task that must consider the confounding effect of different management factors. Hence, identification of the relevant genes/markers and their use in the selection of new genotypes will be a positive step forward.

4 Genetic factors influencing sustainability of pigeonpea production

4.1 Susceptibility to biotic and abiotic stresses

Under natural conditions, a number of biotic and abiotic factors adversely affect pigeonpea growth and development. The losses, however, vary from one environment to another. Among biotic constraints, insects (mainly pod borers and pod fly) and diseases (mainly wilt and sterility mosaic) are the common yield reducers in most pigeonpea areas. In general, breeding-tolerant cultivars tend to lack any effective genetic resistance against insects. However, the situation with respect to disease-resistance is more promising, with the availability of both cultivars and hybrids showing high levels of stable resistances to wilt and sterility mosaic.

4.2 Understanding the genetics of key traits

For successful plant breeding, it is important that genetic information regarding key traits is available to breeders. In pigeonpea, such information on yield and related traits is limited and inconclusive (Saxena and Sharma, 1990; Sawargaonkar, 2011). Lack of information on the aspects such as number of genes, their mode of action, heritability and genotype–environment interaction slows the process of genetic gain. Therefore, it is important to strengthen efforts in this direction using both traditional as well as genomic-based approaches.

4.3 Limited genetic variability

Genus *Cajanus* contains a tremendous amount of phenotypic variability within the primary gene pool for both quantitative as well as qualitative traits (Table 2). However, the studies conducted at molecular level to quantify the diversity in this group of materials showed contrasting results of limited genetic diversity (Yang et al., 2006; Odeny et al., 2007; Yadav et al., 2014). This limitation appears to be the main reason for the restricted genetic enhancement of yield and its stability in pigeonpea. Shiv Kumar et al. (2003) studied the diversity among the parents of released pigeonpea cultivars in India. They found that of the 86 cultivars released, 50% of them had ten parents in common. Further, Singh et al. (2016) reported that within the released cultivars, more than one-third originated directly from landraces. These include the present-day popular cultivars (Table 3) such as Maruti, BDN 2, LRG 30 and LRG 36 in the medium-duration group, and Gwalior-3, Bahar, MAL 13 and NA1 in the long-duration group. All these studies showed that the genetic diversity within the primary gene pool of pigeonpea is not large enough for the genetic enhancement of yield. The genetic diversity among crossable wild species, however, is relatively high (Yang et al., 2006; Yadav et al., 2014), and it could be used to diversify the genetic variation within the primary gene pool.

4.4 Limited use of genomics in breeding

In modern plant breeding, the genomics tools are helpful in achieving genetic gains with greater accuracy at a faster pace. In pigeonpea, research initiatives in this area gained momentum only after its draft genome was sequenced (Varshney et al., 2012). This provided necessary genomics resources in terms of molecular markers and information on genes controlling important traits in pigeonpea. Furthermore, this research also facilitated genetic mapping for the traits using advanced genotyping approaches such as whole-genome re-sequencing, genotyping by sequencing, high-density genotyping arrays and so on.

Table 2 Variability observed in *Cajanus cajan* germplasm

Trait	Variability	Trait	Variability
Days to flower	High	*Maruca* tolerance	Very low
Days to maturity	High	Wilt resistance	High
Plant type	Moderate	Sterility mosaic resistance	High
Plant height	High	*Alternaria* blight resistance	Very low
Seeds/pod	High	*Phytophthora* resistance	Very low
Seed colour	High	*Helicoverpa* resistance	Very low
Protein	Moderate	Pod fly resistance	Very low
Cooking time	Very low	Waterlogging resistance	Moderate

Table 3 Some popular pigeonpea cultivars developed by different breeding methods in three maturity groups in India

Maturity	Total releases	Introduction	Hybridization and selection	Mutation	Popular cvs	Year of release
Early	33	6	24	3	UPAS 120	1976
					AL 15	1981
					Manak	1983
					Pusa 992	2002
					ICPL 88039	2011
Medium	37	17	19	1	BDN 2	1979
					LRG 30	1980
					Maruti	1985
					Asha	1993
					BSMR 853	2001
Late	16	9	7	0	Gwalior 3	1980
					Bahar	1980
					NA 1	1997
					MAL 13	2004
					IPA 203	2012
Total	86	32	50	4	–	–

Source: Singh et al. (2016).

This knowledge can facilitate breeding of the traits of interest including (i) incorporation of resistance to various biotic and biotic stresses, (ii) molecular tagging of male fertility restoring (*Rf*) genes, (iii) transfer of targeted genes from wild species and (iv) genomics-based seed quality control. Tagging of the targeted genes can help breeders in identifying genotypes carrying the particular gene of interest within segregating populations or new germplasm. With this technology, the selection of individuals carrying the desired genes is possible without testing their progeny. The processes are accurate, economic and rapid. This technology is also useful for pigeonpea breeders to determine the seed quality of hybrids and their parents, to help maintain the valuable genetic stocks. Since the processes of genomics research in pigeonpea started recently, this technology in the past could not be exploited to breed cultivars with greater yield and stability.

4.5 Genetic contamination of seed

To reap the benefits of released cultivars, it is essential that their genetic purity is maintained year after year. Overall, it is not a serious issue in pulses, but in pigeonpea, where a considerable extent of natural out-crossing occurs, the maintenance of genetic purity is a significant issue. In a recent review, Saxena et al. (2016) reported that in pigeonpea the out-crossing occurs in most places, and it ranged from 0 to 48% in India, 0 to 60% in China, 14 to 19.64% in Sri Lanka, 13 to 70% in Kenya, 8 to 22% in Uganda, 15 to 39% in

Cameroon and 2 to 40% in Australia. In cases where the breeders/seed producers failed to maintain the purity of released cultivars, their performance with respect to disease resistance and productivity deteriorated rapidly. This factor has a negative impact on the productivity and stability of the cultivars.

5 Enhancing pigeonpea sustainability through crop modelling

Traditional pigeonpeas are cultivated as intercrop with short-aged cereals. Their long growing duration complements the resource use with companion crops, both in terms of time and space (Natarajan and Wiley, 1980). The sole crop cultivation in pigeonpea–wheat rotation is also popular with substantial residual benefit to wheat (Chauhan et al., 2004). A key question arises as to whether the advantages these systems could be further increased. In this context, conducting field experiments at the same scale to answer this question may not be feasible. In fact, a plethora of cropping systems and combinations exists which involves pigeonpea for rotations. This makes it difficult, if not impossible, to practically investigate them individually. Further, it would also be important to quantify the influence of other crops used in the systems. Moreover, such experiments, even on a limited scale, could only be done over a limited number of locations and seasons and may not generate useful information.

Given that all factors that contribute to sustainability of cropping systems are highly variable, results from such experiments will need to be examined for their applicability to other situations and seasons. These assessments need to be beyond the consideration of trade-offs in economic terms, as has been done in the past. In this context, accurate crop simulation models (Moller et al., 2007) which have already been demonstrated with wheat, could prove useful (Moller et al., 2014). The development of a pigeonpea model (Robertson et al., 2001a) has created the possibility of investigating the productivity/sustainability parameters in a more objective way.

5.1 The pigeonpea model

Robertson et al. (2001a) for the first time described a pigeonpea model that was capable of simulating its development, growth, nitrogen accumulation and yield in response to weather, agronomy and soil conditions. This model can simulate a range of maturity types and is based on extensive summarization of data collected at ICRISAT (Robertson et al., 2001b; Carberry et al., 2001). The model, when applied to a number of experiments covering a range of factors including sowing times, population densities, soil types and seasons effects on pigeonpea, could simulate yield of extra-short to medium-duration pigeonpea with a notable accuracy of 92–96%. However, for long-duration pigeonpea, simulation for smallholding farming systems was not possible due to limited data availability (Smith et al., 2016).

5.2 Crop modelling for developing sustainable cropping systems

One of the most common applications of models is to add value to experimentation and demonstration. Models, once properly parameterized and tested, will allow extrapolation of crop performance to other seasons (Whitbread et al., 2010). This could be helpful in explaining why farmers persist with certain practices. For example, farmers might be more inclined to grow long-duration pigeonpea intercropped with sorghum or maize, rather than pigeonpea followed by wheat or a single cereal crop. Such assessments based on the results of one or two seasons might help to understand the factors behind one crop being more favorable. However, under long-term rainfall conditions, a different cropping system might appear more stable. The relative profitability of different farming systems such as those which are contingent on prevailing climate conditions must be considered. Also, the availability of resources supported by modeling outcomes, and those outcomes which are more rigidly based on farmer's long-standing experience could be compared (Rodriguez et al., 2011).

The other application of pigeonpea models could be in assisting breeding efforts. The modelling thus far has been used in benchmarking yield, and assessing risk to production of a particular practice. Its use in facilitating breeding is just beginning to occur. Currently, the pigeonpea adaptation zones are largely based on agro-ecological regions that were defined by experts (Sameer Kumar et al., 2014). The frequencies with which droughts or thermal regimes are experienced in these target environments are not well defined.

Models could also be used to characterize pigeonpea growing environments which breeders can take into consideration to develop different genotypes. For example, the ideal production niches for the newly developed super-early pigeonpea types could be defined. The model could also assist in defining agro-ecological regions that are characterized by different frequencies of drought and temperature regimes as has been done in mung bean in Australia (Chauhan and Rachaputi, 2014). This could be followed by developing landscapes to identify superior genotype × environment × management interactions as done in sorghum by Hammer et al. (2014). It could be further used to identify appropriate maturity types for each environment. Following this, best local management practices with a broadly adapted genotype, or genotypes with specific adaptation to local agronomy could be determined.

6 Enhancing sustainability through an efficient seed system

Seed is a basic unit in agriculture because it contains all vital factors (genes) necessary to achieve productivity and sustainability during farming. Hence, it becomes imperative to produce and distribute quality seeds year after year. To achieve this goal on a sustainable basis, an efficient seed chain for different grades of seed with strict quality control needs to be established (Saxena, 2006). Unlike most pulses, the maintenance of seed quality in pigeonpea is rather difficult and resource-intensive. This is primarily due to a considerable extent (20–50%) of natural out-crossing. The maintenance of cultivars for the key traits that are under the control of recessive gene(s) is very critical, because the natural out-crossing can destroy them rapidly. There are examples where, in the absence of a good seed system, some previously disease-resistant pigeonpea cultivars have become highly susceptible over a period of a few years. At the time of its release, a late-maturing variety 'Bahar' was high yielding, widely adapted and highly resistant to sterility mosaic virus. After a few years, it became a mixture of resistant and susceptible plants and consequently, lost yield potentional and wide adaptation abilities. Similarly, there are other causalities (e.g.

cv. 'C11' and 'BDN 1'), which became highly susceptible under poor seed management. Therefore, for sustaining the high productivity of pigeonpea cultivars across locations and years, it is necessary to invest sufficient resources into the maintenance of quality seeds.

7 Enhancing sustainability through plant breeding

7.1 Breeding early maturing cultivars for high yield and adaptation

For expanding pigeonpea cultivation and ensuring sustainability, the diversification of cropping systems involving pigeonpea is essential. The traditional pigeonpea cultivars and landraces do not fit the bill due to their strong photosensitive nature. These types have a strict short-day requirement for the induction of flowering (also see Section 3), and it has restricted the adoption of traditional pigeonpea cultivars between 30°N and 30°S latitudes. This adaptation scenario started changing with the development of early maturing cultivars such as Prabhat, Pusa Ageti, UPAS 120 and a series of pigeonpea genotypes developed in India at Punjab, Haryana, Delhi, Pantnagar, Kanpur and ICRISAT. Wallis et al. (1981) suggested that earliness of pigeonpea genotypes was closely related to its photo-insensitivity reaction. After conducting a series of experiments at ICRISAT and in Australia, a number of relatively photo-insensitive cultivars were identified. To test their adaptation, these were tested at nine locations representing a wide range (7–46°N) of latitudes in an international nursery. Some of the genotypes such as ICPL 83015, 85030 and 85010 produced over 2 t/ha of grain even at 46°N (Table 4). Vales et al. (2012) reported breeding of some super-early pigeonpea genotypes, which mature in < 90 days. They also demonstrated that these types were adapted to higher altitudes and latitudes. In cases where grain yield and crop duration are considered together (yield per unit of time consumed), the super-early pigeonpea could be an attractive option. Besides yield, these types also have synchronous flowering and maturity to facilitate better insect management

Table 4 Seed yield (t ha^{-1}) of extra-short-duration pigeonpea lines at different latitudes, 1988–89

	Latitude (°N)								
ICPL	7	9	17	23	29	31	32	34	46
83015	2.32	1.48	2.35	1.75	1.06	1.74	3.73	1.86	2.06
83019	2.21	1.39	1.46	1.43	1.00	1.36	3.58	1.67	1.76
84023	2.34	1.14	1.42	1.83	1.37	1.87	2.99	2.49	1.59
85010	2.79	1.55	1.59	1.88	1.17	1.25	3.16	2.33	2.15
85030	1.52	1.11	1.21	1.29	0.98	1.16	2.52	0.83	2.46
Mean	2.17	1.27	1.65	1.59	1.16	1.67	3.19	1.70	1.77
CV%	22.8	23.3	14.5	29.2	12.9	4.70	17.1	NA	NA

Source: Saxena (2002).

Table 5 Summary of generation turnover in four early maturing pigeonpea genotypes

		Date of			Days			
ICPL No.	Generation number	Sowing	Flowering	Harvesting	Flowering	Pod age	Cumulative	Germ. (%)
85024	1	15 June	4 Aug	1 Sept	50	28	78	83.3
	2	9 Sept	24 Oct	17 Nov	43	25	146	81.5
	3	17 Nov	15 Jan	10 Feb	59	26	231	86.7
	4	16 Feb	14 Apr	9 May	57	25	313	72.7
	Mean				**52.3**	**26.0**	–	**81.05**
87093	1	15 June	4 Aug	1 Sept	55	28	78	100.0
	2	9 Sept	1 Nov	26 Nov	53	25	161	95.0
	3	26 Nov	29 Jan	24 Feb	64	26	251	80.0
	4	26 Feb	27 Apr	24 May	60	27	338	100.0
	Mean				**56.8**	**26.5**	–	**93.7**
00004	1	15 June	15 Aug	10 Sep	61	26	87	90.0
	2	16 Sept	11 Nov	11 Dec	56	30	173	96.0
	3	15 Dec	17 Feb	15 Mar	64	28	265	80.0
	4	20 Mar	15 May	11 June	57	27	349	100.0
	Mean				**59.5**	**27.8**	–	**91.5**
00151	1	15 June	12 Aug	16 Sep	58	35	93	96.0
	2	16 Sept	15 Nov	14 Dec	60	29	182	85.0
	3	15 Dec	20 Feb	20 Mar	67	28	277	86.2
	4	20 Mar	20 May	17 June	52	28	367	100.0
	Mean				**61.8**	**30.0**	–	**98.80**

Source: Saxena et al. (unpublished).

and mechanized culture. This material would allow farmers to introduce pigeonpea in new niches for enhancing sustainability and nutritional security.

The early maturing inbred cultivars can be bred quickly using a rapid generation turnover schedule as proposed by Saxena et al. (in press). This approach is an extension of the popular 'single seed descent' method. For generation advancement, instead of single seeds this approach uses single pods from each plant, and germination of immature seeds. This scheme ensures conservation of genetic variability for desirable traits in subsequent generations. Adoption of this technology will considerably reduce the breeding time, as pigeonpea breeders can easily produce four generations within a year (Table 5).

7.2 Breeding cultivars for fragile ecosystems

7.2.1 Wheat-based cropping systems

A key development in pigeonpea agronomy occurred when it was realized that pigeonpea could be grown in rotation with wheat in northern India. This rotation became even more popular with the development of extra-short-duration varieties such as ICPL 88039, which took less time to harvest than the ruling varieties UPAS 120 and Manak. This provided greater turnaround time for wheat sowings (Dahiya et al., 2002). This concept was further extended to diversify the cropping system in Indo-Gangetic plains, where rice–wheat crop rotation is extensively practised. The heavy use of chemical fertilizers and liberal use of irrigation have made this traditional rotation unsustainable, primarily due to increases in soil salinity and poor response to added fertilizers (Dahiya et al., 2002). ICPL 88039, the extra early pigeonpea variety, has allowed farmers in Indo-Gangetic plains, Punjab and Haryana, to replace the water-demanding rice crops. Unlike rice, pigeonpea only required two irrigations for crop establishment in these regions. Alongside this, it was also observed that in the pigeonpea–wheat cropping sequence yields of wheat were increased by a margin of about 1,000 kg/ha. These increases were due to the increased residual benefits of returning organic materials rich in nitrogen and other nutrients to the soil, as well as permitting the timely sowing of wheat crops.

7.2.2 Rain-fed low and mid-hills

In general, the slopping agriculture lands in hillside regions suffer from repeated heavy erosion of top soils. This makes the soils deficit in organic matter and important macro- and micro-nutrients. Agriculture in such areas entirely depends on rainwater, and it often suffers from post-rainy season drought. Hence, the productivity of traditionally grown cereals and pulses is very low (300–400 kg/ha). This has contributed to overall poverty and protein malnutrition among those living below the poverty line. To overcome this problem pigeonpea cultivar ICPL 88039 was introduced into the low to mid-hills of Uttarakhand. The on-farm trials conducted in the state demonstrated that pigeonpea could be grown successfully under rain-fed conditions up to elevations of 1,580 m (Table 6). This variety was found useful for both grain production (up to 1,878 kg/ha) as well as soil conservation (Saxena et al., 2011). The cultivation of pigeonpea has now been extended to all the districts of low and mid-lands. Variety ICPL 88039 is also growing successfully even in the rocky terrains, where no crop can produce economic yield. The adoption of this technology is easy and profitable, and it could pave the way for overall sustainability of agriculture and

Table 6 Yield recorded in different districts of Uttarakhand

District	Altitude (m)	Yield (kg/ha)
Champawat	1580	1400
Almora	1480	1275
Uttarkashi	1310	1266
Rudraprayag	1280	1250
Chamoli	1270	1878
Tehri Garhwal	1200	1266
Range	1200–1580	1250–1878

prosperity of hilly areas. Pigeonpea is a good source of home-grown high-protein food and the direct beneficiaries from this initiative are smallholder Himalayan farmers.

7.2.3 Arid environments

Frequent droughts and land degradation are common features of arid areas in Rajasthan, India. Here, a lack of soil-enhancing legume crops has led to further soil degradation. However, the introduction of the pigeonpea variety ICPL 88039 has paved the way for long-term sustainability in this region. At present, over 12,000 farmers are cultivating pigeonpea on about 10,000 ha of land, with an average production of 719 kg/ha over 120–130 days. The ICPL 88039 variety yielded a net average income of Rs. 60,000/ha under rain-fed conditions. Farmers recognized that this variety is ideal for their light soils, in a region where the annual rainfall only reaches about 300 mm. ICPL 88309 has become well-established among farmers in the region who value the variety's role in developing long-term sustainable agricultural solutions. Its appeal is not only due to its adaptation and yield, but also its provision of protein and other important nutrients. The dry stems can also be used as a fuel source, with 10 t/ha of stems providing 3,000 calories/kg in energy. The cultivation of ICPL 88039 has now restricted deforestation for domestic fuel purposes, and thus directly contributed towards environmental conservation, as well as the benefits already outlined above.

7.2.4 Wastelands

Successful attempts were made to introduce pigeonpea cultivars for rehabilitating the hilly wastelands of southern China by exploiting the cultivar's soil amelioration and perennial properties (Saxena, 2008). The long-duration pigeonpea varieties can survive under rain-fed, low-fertility conditions, whilst producing the significant amount of biomass needed to protect the soil. In these sloping terrains, the recovery of ecology is not easy due to heavy-intensity rains, long-term deforestation, and the fragile nature of soils. Traditionally, in the absence of any fast-growing forest tree species, shrubs such as *Phyllanthus emblica* and *Dodonaea viscosa* were transplanted to reduce soil erosion. However these have no economic value, and so two long-duration pigeonpea cultivars ICPL 87119 and ICP 7035 were introduced. These cultivars adapted well to the harsh conditions, produced a significant amount of biomass and grains, and covered the barren land within a few months. Due to these qualities, pigeonpea has been identified as a key species to help restore areas damaged by deforestation.

7.3 Breeding pigeonpea for resistance

7.3.1 Disease resistance

Since in medium- and late-maturing pigeonpea both *Fusarium wilt* and sterility mosaic diseases are two of the challenges faced by both medium- and late-maturing pigeonpea crops. ICRISAT's strategy aimed to breed both pure-line and hybrid cultivars with genetic resistance to both these diseases. Therefore, to breed such varieties and to attain the target sustainability, ICRISAT launched a strong breeding programme in which several cultivars and hybrid parents were bred. This was possible because of an effective field screening technology invented at ICRISAT by Nene et al. (1981). To achieve the target of resistance breeding, pedigree selection within the landraces has been very effective, both in India and Africa. In India for example, variety Maruti, a selection from ICP 8863 collected from Maharashtra, has been proven to be of benefit to the farmers of wilt-prone areas. In some districts, its adoption is as high as 60% (Bantilan and Joshi, 1996). Similarly, in eastern and southern Africa, the most popular pigeonpea variety is Nandolo wa nswana, which is also a selected from a Tanzanian landrace (ICP 9145). In Malawi, it occupies a considerable area (Simtowe et al., 2016). These genotypes have been used extensively as donor parents in resistance breeding programmes.

In 1992, ICRISAT used a bulk-pedigree method to develop a widely adapted, medium-maturing pigeonpea variety called ICPL 87119 (Asha). This variety, besides recording 20% high yield, had a high level of resistance to wilt and sterility mosaic diseases. The adoption of this variety is very popular in peninsular and central India. Similarly, both the released and unreleased hybrids have high levels of resistance to the two diseases (Table 7). Such cultivars will help to reduce yield losses and contribute significantly towards the sustainability of pigeonpea production in rain-fed agriculture.

7.3.2 Insect resistance

The most serious pigeonpea pests are the pod borers (*Helicoverpa armigera* and *Maruca vitrata*) and pod fly (*Melanagromyza obtusa*). Among these, *H. armigera* is prevalent throughout the tropics and subtropics, costing an estimated pigeonpeas loss of over $310 million annually (Ranga Rao et al., 2013). In most countries, the present-day plant protection systems are embedded with chemicals. In the past five decades, the use of chemical pesticides has increased by 170-fold, from 2.2 g/ha of active ingredient in 1950 (Vasantharaj,

Table 7 Some promising medium-duration pigeonpea hybrids

Hybrid (ICPH)	Yield (kg/ha)	Standard heterosis (%)	100-seed weight (g)	Wilt (%)	Sterility mosaic (%)
3371	3013	62**	11.50	0	0
2740	2900	57**	12.30	0	0
3762	3000	62**	11.90	0	0
Check	1864	–	11.10	0	0
CV (%)	11.9	–	3.98	–	–

Disease data recorded in sick nursery. Source: ICRISAT.
** Significant at 1% probability.

1995) to 381 g/ha in 2007 (Anonymous, 2009). To control this pest, pigeonpea farmers usually apply 5–8 chemical sprays. The repeated use of some insecticides may lead to the development of inherent resistance in the target insects. Alongside this, potential dangers of excessive dependence on pesticides can include outbreaks of secondary pests, chemical residues contaminating the food chain, and problems relating to loss of biodiversity.

The insect-control programmes should include both the recently developed high-yielding pure-line/hybrid cultivars (which offer a certain level of genetic tolerance), as well as an effective IPM package. Based on the mechanisms involved, Painter (1958) classified plant resistance to insects into three categories. These include (i) non-preference for oviposition, food or shelter, (ii) antibiosis – adverse effect of plants on the biology of insects, and (iii) tolerance – repair, recovery or ability to withstand infestation. In pigeonpea, so far there are no reliable genetic solutions for pod borers, and in spite of diverting huge resources to develop an effective host plant resistance, success up to this point has proved elusive. Field screening of over 10,000 germplasm for tolerance to pod borers at ICRISAT did not yield any genotype with a significant level of genetic resistance. A few accessions, however, with relatively less pod damage were identified, and one such selection (ICPL 332), was released as variety 'Abhaya' in India. This variety, in spite of good yield and tolerance to pod borers (Table 8), could not make any impact due to its high susceptibility to *Fusarium* wilt disease. Recently, this weakness of ICPL 332 was eliminated and it was renamed as ICPL 332WR (Sharma, 2016). This improved variety is likely to contribute significantly towards the sustainable production of pigeonpea.

M. vitrata (Geyer) is another serious insect pest of tropical legumes. It causes substantial losses to early maturing genotypes under humid conditions. Some pigeonpea genotypes with moderate levels of resistance (Table 9) were identified in Sri Lanka (Saxena et al., 2002), but no targeted resistance breeding programme followed. Long-duration genotypes PDA 88-2E, PDA 89-2E, PDA 92-2E and PDA 93-2E have been identified as resistant to pod fly, a serious insect pest of northern India (Lal et al., 1999).

7.3.3 Resistance to waterlogging

In soils characterized by high water-holding capacity, temporary waterlogging poses a worrying threat to pigeonpea productivity. According to Choudhary et al. (2011), about 1.1 m ha of land is waterlogged annually, inflicting losses of about 25–30% onto the productivity.

Table 8 Pod damage and yield of *Helicoverpa*-tolerant cv. Abhaya and control

	Selection		Control	
Year	Damage (%)	Yield (t/ha)	Damage (%)	Yield (t/ha)
1984	49.0	2.27	76.0	1.83
1985	11.6	1.84	33.4	1.44
1986	22.5	1.05	71.4	0.58
1987	70.6	2.73	94.2	1.54
1988	19.0	1.48	48.1	0.89
Mean	34.54	1.87	64.62	1.25

Source: Compiled from various ICRISAT reports.

Table 9 Yield (t/ha) of three lines selected for resistance to *Maruca vitrata* in Sri Lanka

Selection	Sprayed	Unsprayed	Loss (%)
MPG 664-M1-2-M2	2.41	1.99	17.4
MPG 664-M1-2-M13	2.64	2.19	17.1
MPG 664-M1-2-M27	2.22	1.92	13.5
UPAS 120 (control)	2.32	0.67	68.9
SE (spray)	2.41		

Source: Saxena (2002).

Under waterlogged conditions, the useful aerobic bacteria become inactive while their anaerobic counterparts (facultative/obligate bacteria) are able to thrive. This results in a shortage of oxygen in the soil which has a negative effect on general plant health. Chauhan et al. (1997) developed a laboratory technique to screen pigeonpea genotypes for their waterlogging tolerance. Recently, Sultana et al. (2013) identified a number of tolerant genotypes. As well as waterlogged soils, drought, soil salinity, aluminium toxicity and low temperature are other abiotic stresses affecting pigeopea, but very little information is available to understand various relevant aspects of these stresses.

7.4 Breeding cultivars with high-protein content

The protein malnutrition among poverty-stricken populations is usually high, especially in underdeveloped and developing countries. Pigeonpea seeds, besides providing valuable protein, also contain certain amounts of anti-nutritional compounds such as oligosaccharides, enzyme inhibitors, phenols and tannins. Pigeonpea seeds, whilst providing valuable protein, also contain compounds such as oligosaccharides, enzyme inhibitors, phenols and tannins which do not offer nutritional benefits. Where arable lands are limited, the only option available is to overcome this problem is to produce more protein per unit of land area. This means breeding new high-protein varieties for cultivation. Pal (1939) reported that in comparison to other pulses, pigeonpea has the best combination of nutrition-related traits with high biological value.

A considerable variety of for protein content among pigeonpea genotypes has been reported by various researchers, but it was not large enough for direct use in breeding programmes (see review by Saxena and Sawargaonkar, 2015). Therefore, wild relatives of pigeonpea were selected as alternative donors for breeding high-protein pigeonpea genotypes. These included *C. albicans* (30.5%), *C. sericeous* (29.4%) and *C. scarabaeoides* (28.4%). Dahiya et al. (1977) reported that 3-4 genes controlled the protein content in pigeonpea. Durga (1989) and Saxena and Sharma (1990) observed that protein content was under additive and complementary gene effects, and that low-protein genes were dominant or partially dominant over high protein genes.

The breeding for high-protein pigeonpea cultivars was undertaken at ICRISAT using a pedigree breeding method. A large population of approximately 3,000 plants was raised to form the F_2 generation, and as expected, the populations segregated for various other morphological traits. The primary selection criteria was for protein content. Therefore, seed samples from each F_2 plant were analysed to determine their protein content, with the desirable plants remaining in the field. The high-protein (>25%) segregants were selected for further breeding. Within each progeny a mild selection for plant type was exercised and the individuals carrying wild species traits such as creeping and abnormal growth were rouged. Interestingly, in the F_5 generation, some segregants with 28% protein were recovered. In the next four generations (F_6 to F_9) a few selections with protein ranging from 28 to 32% were also recovered. During this period, selection for seed shape and size was exercised. In the F_{10} generation, the yield trials of high-protein (> 28%) lines were conducted and the results were very encouraging (Table 10). Among the non-determinate selections, yield of the top two lines (HPL 40-5 and HPL 40-17) was similar to that of the control BDN 1 (2.02 t/ha), but the selections were significantly higher than the control (23.2%) in their protein content. The advantage of the high-protein lines was reflected in the 'total protein' harvest from unit land. Similar results were recorded from the evaluation of determinate high-protein lines. These results demonstrated that in pigeonpea, seed yield and protein content can be enhanced simultaneously.

Studies conducted to understand the effect of diverse environmental factors on the protein content of high-protein lines yielded very useful results. The evaluation of the lines across wide Indian locations in the south (Andhra Pradesh, Karnataka); centre (Gujarat, Maharashtra, Madhya Pradesh); and north (Haryana), showed that although minor (2–3 protein units) differences were observed, the differences between the high and low protein cultivars were maintained (Table 11). HPL 24 appeared to be the best with >30% protein recorded at each place. Similarly, evaluation of high-protein lines over a six year period showed that the protein content of each high-protein selection was higher in each year, and the values were much higher than the control. These observations indicated that high-protein traits derived from the wild relatives of pigeonpea were stable, and will not pose any difficulty in breeding high-yielding, high-protein pigeonpea cultivars in future.

Biological assessment of the protein-rich selections was conducted to determine if the additional protein can be utilized in growth and development of the individuals. The data (Table 12) showed that the high-protein lines were significantly superior to the control cultivar in terms of available protein. It was also reported that the high-protein lines were nutritionally superior to normal cultivars because of their greater amount of sulphur-rich amino acids.

Table 10 Seed yield and protein harvest from high-protein F_{10} lines

Genotype	Maturity (days)	100-seed wt (g)	Yield (t/ha)	Protein (%)	Protein yield (kg/ha)
HPL 40-5	169	9.6	2.10	26.9	452
HPL 40-17	169	8.5	2.07	26.5	440
BDN 1 (c)	168	9.6	2.02	23.2	373
SE	±0.9	±0.18	±0.16	±0.46	±37.3
CV (%)	0.9	3.4	17.3	3.0	17.0
HPL 8-10	163	10.5	1.66	26.5	353
HPL 8-16	162	10.5	1.57	27.4	344
ICPL 211(C)	162	14.3	1.46	21.6	251
SE	±1.1	±0.15	±0.19	±0.21	±38.5
CV (%)	13	2.5	27.0	1.7	25.8

Source: Singh et al. (1990).

Table 11 Stability of four high-protein selections at six diverse locations in India

Location	State	HPL 24	HPL 25	HPL 26	HPL 28	Control	SE
Patancheru	Andhra Pradesh	31.3	28.6	29.7	27.8	23.3	±0.26
Jalna	Maharashtra	32.2	28.9	29.7	30.4	23.1	±0.69
SK Nagar	Gujarat	30.9	28.4	29.0	27.3	21.4	±0.36
Gulbarga	Karnataka	32.1	29.9	–	27.6	23.0	±0.49
Gwalior	Madhya Pradesh	32.3	30.4	28.2	27.3	22.0	±0.71
Hisar	Haryana	31.1	29.6	31.7	29.2	24.5	±0.51

Saxena et al. (2002).

Table 12 Comparison of high-protein pigeonpea line and control for protein and its biological parameters

Item	High-protein line HPL 8	High-protein line HPL 40	Control line (ICPL 211)	SE
Constituents				
Starch (%)	54.3	55.6	59.3	±0.30
Protein (%)	28.7	31.1	23.1	±0.09
Albumin (%)	9.1	8.0	8.6	±0.34
Globulin (%)	63.5	66.2	60.3	±1.08
Glutelin (%)	20.2	19.7	22.8	±0.75
Prolamin (%)	2.9	3.2	2.1	±0.06
Cysteine	0.8	0.8.	0.7	±0.01
Biological parameters				
Total protein digestibility	83.7	82.9	85.7	±2.14
Biological value	67.0	65.3	62.9	±1.68
Net protein utilization	56.1	54.1	53.9	±1.06
Utilization protein	15.5	16.7	12.3	±0.25

Source: Singh et al. (1990).

8 Pigeonpea hybrids for greater productivity and sustainability

8.1 Evolution of hybrid technology

The discovery of cytoplasmic nuclear male sterility (Saxena et al., 2005) encouraged pigeonpea breeders to explore the possibility of developing commercial hybrids by using pigeonpea's inherent partial out-crossing and break the decades-old productivity barrier. Soon a multi-institutional research and development programme was launched

Table 13 Performance (yield kg/ha) of hybrid ICPH 2671 in the on-farm trials

State	Farmers (no.)	Hybrid yield	Control yield	Standard heterosis (%)
Maharashtra	782	969	717	35
Andhra Pradesh	399	1411	907	55
Jharkhand	288	1460	864	69
Madhya Pradesh	360	1940	1326	46
Total/mean	1829	1445	954	51

Source: Saxena et al. (2013).

by ICRISAT and ICAR. A breakthrough in this endeavour was achieved in 2010, when the world's first commercial pigeonpea hybrid ICPH 2671 was released in Madhya Pradesh (Saxena et al., 2013). Prior to its release, the hybrid was extensively evaluated in farmer's field. A total of 1829 on-farm trials were conducted (Table 13) in the states of Maharashtra (782 trials), Andhra Pradesh (399 trials), Madhya Pradesh (360 trials) and Jharkhand (288 trials). In this extended exercise, the hybrid ICPH 2671 recorded 30–60% superiority over the best local cultivar. Overall, the superiority of the hybrid ICPH 2671 over the control was 51%.

The release of hybrid ICPH 2671 was followed by the release of ICPH 2740 and ICPH 3762. A yet-to-be-released white-seeded hybrid named ICPH 4104 has also been identified, (and this will be released by a private seed company in Gujarat). These hybrids recorded 30–50% more yields in farmer's fields. To cater to the needs of different agronomic niches and cropping systems, a number of new hybrids in early and medium groups are in the advance stages of testing. These include ICPH 2433, ICPH 2438 and ICPH 2363 in the early-group category, which showed a yield advantage of 54%, 42% and 36%, respectively. Similarly, in the most popular medium maturity category, hybrids ICPH 3371 (3013 kg/ha) and ICPH 3491 (2919 kg/ha) were found to be highly promising with, respectively, 62% and 57% superiority over the national control variety Asha. All the medium-duration hybrids are free from both wilt and sterility mosaic diseases. This performance data showed that by cultivating these pigeonpea hybrids, a significantly higher productivity levels can be achieved by farmers.

Saxena (2015) have extensively reviewed this subject. They concluded that besides high yields, the pigeonpea hybrids also have the following advantages over the inbred cultivars, and can play a great role in enhancing both the productivity and sustainability of the crop.

- Hybrid seeds germinate faster and have rapid seedling growth
- Hybrid plants have >30% greater biomass both above and below ground
- Hybrids need 30% less seeding rate than varieties
- Hybrids maintain superiority under both high- and low-input situations
- Hybrids have extra resilience to encounter various stresses
- Hybrids compete well with inbred cultivars for various nutritional and marketing traits

8.2 Adoption of hybrids

For the adoption of hybrids, there are three primary prerequisites. These include (i) high and stable yields, (ii) economically viable seed technology, and (iii) a good promotion and marketing network. A brief overview follows.

8.2.1 High yields

The high yield potential of hybrids has been well documented across a range of field environments, including under rain-fed and high-input conditions (Saxena, 2015). In rain-fed situations, hybrid yields are around 1,500–2,500 kg/ha. Interestingly, under high-input conditions, the hybrid's productivity levels approached 4,000–5,000 kg/ha (Table 14). Such commercial productivity levels are encouraging, and more farmers are adopting this technology to reap high profits from this crop.

8.2.2 Seed production

The production of hybrid seed is the most important component of the breeding programme. The hybrid seed technology in pigeonpea has now been perfected (Saxena, 2015) and its on-farm validation has produced encouraging results. This programme was organized at 94 locations and, on average, 1,019 kg/ha hybrid seeds were harvested (Table 15). This productivity level is attractive and economical (MK Saxena et al., 2011). Furthermore, a number of key areas in Telangana and Madhya Pradesh, India, have been identified with hybrid yields of over 1,500 kg/ha. These results demonstrated that if the site selection is appropriate, and the crop management is effective, then reasonably high hybrid yields are possible. Also, a seed-to-seed ratio of 1:200 to 1:300 makes the seed production and marketing appealing to both the growers and seed companies. This

Table 14 Demonstration of exceptionally high yields (kg/ha) of ICPH 2740 by some farmers in Maharashtra

Location	Area (m^2)	Hybrid yield	Control yield	Standard heterosis (%)
Salod	450	3956	2044	94
Nimgaon	1012	3951	2469	60
Kothoda	450	4667	3556	31
Tamoli	450	3889	2278	71
Mean	–	4116	2587	59

Table 15 Hybrid seed production (kg/ha) recorded in six states

State	Locations	Mean yield	Highest yield
Andhra Pradesh	34 (6)	998	1750
Madhya Pradesh	9 (3)	1674	3040
Gujarat	4 (2)	1179	1669
Maharashtra	5 (2)	603	1017
Odisha	40 (1)	523	1040
Karnataka	2 (2)	1138	1900
Total/Mean	94	1019	3040

( )number of years. Source: ICRISAT.

information confirmed that (i) a perfect hybrid seed technology is now available and (ii) it is ready for large-scale adoption.

8.2.3 Promotion and marketing

Farmers in Maharashtra and Telangana, India, have realized the importance of hybrid technology in enhancing pigeonpea yields, and that its demand is on increase. Therefore, the attention has now been shifted to expand the area under hybrid cultivation. In this endeavour, the release of hybrid ICPH 2740 in Telangana has been an important development. This technology has also made a strong impact in the neighbouring state of Maharashtra. The hybrid has met the farmer's requirements in both the peninsular and central areas of India. By offering a yield advantage of approximately 700–1,000 kg/ha, with a market value of Rs. 50/kg, one hectare of hybrid cultivation can generate an additional Rs. 40,000–50,000 for farmers. In 2016, ICRISAT and the Agriculture Research Station (Palem, Telangana,) estimated that over 700,000 kg of hybrid pigeonpea seed was available. This quantity could replace more than 140,000 ha of the traditional pigeonpea cultivars. This will benefit farmers by providing an additional 70,000 tons of grain. This level of profitability from the pigeonpea hybrids is comparable with many high value field crops. Its commercial cultivation will be fundamental in improving pigeonpea sustainability.

8.3 Two-parent hybrid breeding technology

In pigeonpea, both temperature and photoperiod were understood to influence the initiation and appearance of floral buds. However, their role in determining male fertility/sterility still needed to be established. In studies conducted by Saxena (2014), temperature

Table 16 Number of male fertile and sterile plants recorded under July-sown field conditions in September, November and February (representing different temperature regimes) in four temperature-sensitive selections

Year	Selection/ type of plants	September		November		February	
		Sterile	Fertile	Sterile plants	Fertile plants	Sterile	Fertile
2008	Envs S-1	22	0	1	21	22	0
	Envs S-2	8	0	1	7	8	0
	Envs S-3	10	0	0	10	7	0
	Envs S-5	18	0	3	15	16	0
	Total	58	0 (0.0 %)	5	53 (91.4%)	53	0 (0.0 %)
2009	Envs S-1	37	0	0	37	37	0
	Envs S-2	32	0	0	32	32	0
	Envs S-3	27	0	0	27	25	0
	Envs S-5	23	0	0	22	21	0
	Total	119	0 (0.0 %)	0	118 (100.0%)	115	0 (0.0 %)

( ) per cent fertile plants.
Source: Saxena (2014).

was found to influence the fertility status of pigeonpea plants. Under the temperature regime of ≥25°C the plants were completely male sterile. In contrast, when daily mean temperatures dropped down to <24°C, the male-sterile plants turned fully fertile and produced self-pollinated pods (Table 16). In early generations of breeding this material, Saxena et al. (2004) observed that some male-sterile pigeonpea plants converted to male fertility much earlier than the rest, and these male-sterile plants were classified as 'early' and 'late' converters. This suggested the presence of more than one gene with different temperature thresholds to produce fertile plants. All the 'converted male fertile' plants reverted back to male sterility when these plants encountered high temperatures (Table 16).

9 Future trends and conclusion

According to the FAO (1983), the definition of 'food security' is that everyone has both physical and economic access to the basic food that they need. By this definition, most of the developing countries are 'nutritionally insecure'. The scenario of nutritional security in India is quite interesting, as it is self-sufficient in calorie production but is lacking in protein and other vital nutrients. The per capita protein availability in India has witnessed a sharp decline from 27.3 kg/year in 1950 to 10 kg/year in 2000 (www.commodityonline.com, 2009). The escalating prices of pulses and other nutrient-filled vegetables, fruits and dairy products have further added to this difficult situation. A large proportion of rural and urban populations, in particular women and children, are suffering from malnutrition and nutrient deficiencies. Therefore, a dramatic improvement in breeding techniques and sustainability of nutrient-rich crops is urgently needed. In this context, pigeonpea can play an important role as it can grow easily with minimal inputs, rejuvenate soils and produce high-protein grains. Therefore, its production needs to be increased 2-3-fold, and this can be achieved by increasing its area and/or productivity per unit of land.

In this context, the promotion of an early maturing cultivar ICPL 88039 (= VLA 1) in the non-traditional niches is an interesting development. Firstly, to attain sustainability and self-sufficiency, concerted efforts will be needed in the near future. Secondly, the recent breakthrough achieved in hybrid pigeonpea technology has shown a great promise with 30–50% yield increase experienced by farmers. The extensive on-farm testing of the hybrids in six states, has generated awareness among farmers as to the benefits provided by pigeonpea hybrids which include increased yields and improved profits. The seed technology relating to pigeonpea hybrids is now well developed, and with an attractive seed-to-seed ration of between 1:200 to 1:300, confidence in the technology continues to grow.

With a yield advantage of 1,000–1,500 kg/ha and a value of Rs. 50/kg, pigeonpea farmers can fetch additional profit of about Rs. 50,000 from one hectare of hybrid crop. The magnitude of realized heterosis for yield and its economics in pigeonpea is similar to those of other crops (Singhal, 2013) and therefore, pigeonpea hybrids are attracting scores of farmers in Maharashtra and Telangana states. The cultivation of hybrid ICPH 2740 is now receiving greater attention by the agriculture departments of these states. According to the estimates of ICRISAT, National Seeds Corporation, State Seed Development Corporations, State Agricultural Universities and private seed companies, in 2016 at least 700,000 kg of hybrid seed was available on the market to replace over 140,000 ha of traditional pigeonpea cultivars.

Development of a sustainable quality seed chain is the key for success, particularly in crops like pigeonpea which is prone to genetic contamination through a variety of nectarivore insects. The efforts of both public and private seed sectors along with appropriate agronomy are necessary to achieve the target of a healthy crop production.

The promotion of early maturing pigeonpea in new niches and hybrid technology in the traditional areas will not only raise the national pigeonpea production but also provide easily digestible quality protein. Pigeonpea is capable of fulfilling various social, nutritional, economic and environmental needs of the smallholder farming communities, and therefore, will pave the way for food security and long-term agricultural sustainability.

10 Where to look for further information

The first research paper on pigeonpea breeding was published by Howard et al. (1919) from Indian Agricultural Research Institute, Pusa (Bihar), in Botanical Series. The authors wrote that genetic improvement of yield in pigeonpea inbred cultivars will be a Herculean task. Unfortunately, the predictions of the authors proved right as, in spite of extensive national breeding programmes, there has been very little geniune yield enhancement in this crop in the last 100 years. These observations were based on the atypical breeding behaviour and long generation turnover time of pigeonpea germplasm. Besides these, this paper also highlighted other key breeding constraints which still persist even after a century. I believe that the present generation of pigeonpea breeders may not be aware of this quality research, and therefore, the authors would like to recommend a thorough reading of this valuable piece of work.

The celebrations of International Year of Pulses in 2016 provided an excellent opportunity to summarise the breeding accomplishments, challenges and opportunities of different aspects of genetic enhancement of the crop (Saxena et al., 2016). As discussed in the present paper, a breakthrough in pigeonpea productivity was achieved by developing a hybrid breeding technology, the first in any pulse crop. The entire scenario of the evolution of hybrid pigeonpea technology including its adoption and impact has recently been reviewed by Saxena et al. (2018), and soon it will appear in volume 41 of Plant Breeding Reviews.

The question of narrow genetic diversity in pigeonpea has often been related to the issues of yield plateauing. To overcome this constraint, utilization of the wild relatives of the crop has been recommended and a considerable progress has been made in this direction. For those researchers who are interested in this area, the review published by Mallikarjuna et al. (2011) will be an ideal publication.

The genomics scientists have also made significant progress in recent times and the technologies are being used by breeders to enhance the precision and pace of breeding new cultivars and hybrids. Varshney et al. (2017) have elegantly summarized these aspects and achievements, and the present-day pigeonpea scientists are advised to read this paper to help them in planning their high-profile research programmes.

11 References

Alexandratos, N. and Bruinsma, J. (2012), 'World agriculture towards 2030/2050: The 2012 revision'. *ESA Working Paper*, 12–03. FAO, Rome.

Anonymous. (2009), *'Indian Pesticides Industry – Vital for Ensuring Food Security'*. Bharat Book Bureau, 30A, Vashi, Navi Mumbai – 400703, INDIA. www.bharatbook .com.

Bantilan, M. C. S. and Joshi, P. K. (1996), 'Returns to research and diffusion investments on wilt resistance in pigeonpea'. *Impact Series no. 1*. International Crops Research Institute for the Semi – Arid Tropics (ICRISAT), Patancheru 502 324, Andhra Pradesh, India, p. 34.

Carberry, P. S., Ranganathan, R., Reddy, L. J., Chauhan, Y. S. and Robertson, M. J. (2001), 'Predicting growth and development of pigeonpea: Flowering response to photoperiod'. *Field Crops Research*, 69: 151–62.

Choudhary, A. K., Sultana, R., Pratap, A., Nadarajan, D. and Jha, U. C. (2011), 'Breeding for abiotic stresses in pigeonpea'. *Journal Food Legumes*, 24: 165–74.

Chauhan, Y. S. (1990), 'Optimum agronomic management', pp. 257–78. In *The Pigeonpea* (Nene, Y. L., Hall, S. D. and Sheila, V. K. (Eds)). CAB International, Wallingford, UK.

Chauhan, Y. S. (1993), 'Pigeonpea', pp. 79–91. In *Rooting Pattern of Tropical Crops* (Salam, A. and Wahid, W. A. (Eds)). Tata McGraw Hill, India.

Chauhan, Y. S., Apphun, A., Singh, V. K. and Dwivedi, B. S. (2004), 'Foliar sprays of concentrated urea at maturity of pigeonpea to induce defoliation and increase its residual benefit to wheat'. *Field Crops Research*, 89: 17–25.

Chauhan, Y. S., Johansen, C. and Saxena, K. B. (1995), 'Physiological basis of yield variation in short-duration pigeonpea grown in different environments of the semi-arid tropics1'. *Journal of Agronomy and Crop Science*, 174: 163–71.

Chauhan, Y. S., Johansen, C., Moon, J., Lee, L., Lee, Y. and Lee, S. (2002), 'Photoperiod responses of extra-short-duration pigeonpea lines developed at different latitudes'. *Crop Science*, 42: 1139–46.

Chauhan, Y. S. and Rachaputi, R. C. N. (2014), 'Defining agro-ecological regions for field crops for variable target production environments: A case study of mung bean in the northern grains region of Australia'. *Agricultural and Forest Meteorology*, 194: 207–17.

Chauhan, Y. S, Silim, S. N, Kumar Rao, J. V. D. K. and Johansen, C. (1997), 'A pot technique to screen pigeonpea cultivars for resistance to water logging'. *Journal of Agronomy and Crop Science*, 178: 179–83.

Chauhan, Y. S., Venkataratnam, N. and Sheldrake, A. R. (1987), 'Factors affecting growth and yield of short duration pigeonpea and its potential for multiple harvests'. *Journal of Agricultural Science, Cambridge*, 109: 519–29.

Chauhan, Y. S., Wallace, D. H., Johansen, C. and Singh, L. (1998), 'Genotype-by-environment interaction effect on yield and its physiological bases in short-duration pigeonpea'. *Field Crops Research*, 59: 141–50.

Dahiya, S. S., Chauhan, Y. S., Johansen, C., Waldia, R. S., Sekhon, H. S. and Nandal, J. K. (2002), 'Extra short duration pigeonpea for diversifying wheat based cropping systems in sub-tropics'. *Experimental Agriculture*, 38: 1–11.

DES. (2015), '*Directorate of Economics and Statistics*'. Government of India, New Delhi, India. http://eands.dacnet.nic.in/.

Ding, J., Zi, Y., Li, C., Peng, Y., Zhu, X. and Guo, W. (2016), 'Dry matter accumulation, partitioning, and remobilization in high-yielding wheat under rice – wheat rotation in China'. *Agronomy Journal*, 108: 604–14.

Egli, D. B. and Bruening, W. P. (2002), 'Synchronous flowering and fruit set at phloem-isolated nodes in soybean', *Crop Science*, 42: 1535–40.

Embong, W. M. W. and Ravoof, A. A. (1978), 'Investigation on pigeonpea (*Cajanus cajan*) as a legume forage', p. 78. In *Proceedings of Symp. Feeding Stuffs for Livestock in South East Asia* (Devendra, C. and Hetagalung, R. I. (Eds)). National University of Malaysia, Kuala Lumpur, Malaysia.

FAO. 1983. World Food Security: a Reappraisal of the concepts and Approaches. Director General's Report, Rome.

FAO. (2013), http://faostat3 .fao.org.

Hammer, G. L., McLean, G., Chapman, S., Zheng, B., Doherty, A., Harrison, M. T., van Oosterom, E. and Jordan, D. (2014), 'Crop design for specific adaptation in variable dryland production environments'. *Crop and Pasture Science*, 65: 614–26.

Hay, R. K. M. (1995), 'Harvest index: A review of its use in plant breeding and crop physiology'. *Annals of Applied Biology*, 126: 197–216.

Howard, A., Howard, G. L. C. and Khan, A. R. (1919), Studying the pollination of Indian crops I. *Memoirs. Dept. Agr.* (Bot. Series) 10: 195–200.

Lal, S. S., Singh, I. P. and Jeet, V. (1999), 'Heterosis and combining ability analysis for resistance to podfly and lepidopteron borers in pigeonpea [*Cajanus cajan* (L.)Millsp.]'. *Indian Journal of Agriculture Science*, 69: 786–8.

Lawn, R. J. and Troedson, R. J. (1990), 'Pigeonpea: Physiology of yield formation', pp. 179–208. In *The Pigeonpea* (Nene, Y. L., Hall, S. D. and Sheila, V. K. (Eds)). CAB International, Wallingford, UK.

Mallikarjuna, N., Saxena, K. B. and Jadhav, D. R. (2011), Cajanus. In: *Wild Crop Relatives: Genomics and Breeding Resources, Legume Crops and Forages*. Springer-Verlog, Berlin, pp. 21–33.

McPherson, H. G., Warrington, I. J. and Turnbull, H. L. (1985), 'The effects of temperature and daylength on the rate of development of pigeonpea'. *Annals of Botany*, 56: 597–611.

Moeller, C., Sauerborn, J., de Voil, P., Manschadi, A. M., Pala, M. and Meinke, H. (2014), 'Assessing the sustainability of wheat-based cropping systems using simulation modelling', *Sustainability Science*, 9: 1–16.

Moeller, C., Pala, M., Manschadi, A. M., Meinke, H. and Sauerborn, J. (2007), 'Assessing the sustainability of wheat-based cropping systems using APSIM: Model parameterisation and evaluation'. *Crop and Pasture Science*, 58: 75–86.

Nam, N. H., Subbarao, G. V., Chauhan, Y. S. and Johansen, C. (1998), 'Importance of canopy attributes in determining dry matter accumulation of pigeonpea under contrasting moisture regimes'. *Crop Science*, 38: 955–61.

Narayanan, A. and Sheldrake, A. R. (1979), 'Pigeonpea (*Cajanus cajan*) as a winter crop in peninsular India'. *Experimental Agriculture*, 15: 91–5.

Natarajan, M. and Willey, R. W. (1980), 'Sorghum-pigeonpea intercropping and the effects of plant population density'. *The Journal of Agricultural Science*, 95: 51–8.

Nene, Y. L., Kannaiyan J. and Reddy, M. V. (1981), 'Pigeonpea disease resistance screening technique'. *ICRISAT Information Bulletin*. 4. International Crops Research Institute for the Semi – Arid Tropics (ICRISAT), Patancheru, 502 324, Andhra Pradesh, India, p. 14.

NIN. (2010), '*Dietary Guidelines for Indians – A Manual*'. National Institute of Nutrition, Hyderabad, Andhra Pradesh. India.

Odeny, D. A., Jayashree, B., Ferguson, M., Hoishington, D., Crouch, J. and Gebhardt, D. (2007). 'Development, characterization, and utilization of micro-sattelite markers in pgeonpea'. *Plant Breeding*, 125:130–6.

Painter, R. H. (1958), 'Resistance of plants to insects'. *Annual Review of Entomology*, 3: 267–90.

Pal, R. K. (1939), 'A review of the literature on the nutritive value of pulses'. *Indian Journal of Agriculture Science*, 9: 133–7.

Ranga Rao, G. V., Rao, R. and Ghaffar, M. A. (2013). 'Hand book on Chickpea and Pigeonpea: Insect Pests Identification and Management'. *Information Bulletin:* 57. International Crops Research Institute for the Semi – Arid Tropics (ICRISAT), Patancheru 502 324, Andhra Pradesh, India, p. 90.

Rao, M. R. and Willey, R. W. (1983), 'Effects of pigeonpea plant population and row arrangement in sorghum/pigeonpea intercropping'. *Field Crops Research*, 7: 203–12.

Robertson, M. J., Carberry, P. S., Chauhan, Y. S., Ranganathan, R. and O'leary, G. J. (2001a), 'Predicting growth and development of pigeonpea: A simulation model'. *Field Crops Research*, 71: 195–210.

Robertson, M. J., Silim, S., Chauhan, Y. S. and Ranganathan, R. (2001b), 'Predicting growth and development of pigeonpea: Biomass accumulation and partitioning'. *Field Crops Research*, 70: 89–100.

Rodriguez, D., Power, B., Cox, H., Crimp, S. and Meinke, H. (2011), 'The intrinsic plasticity of farm businesses and their resilience to change. An Australian example'. *Field Crops Research*, 124: 157–70.

Sameer Kumat, C. V., Mula, M. G., Singh, P., Mula, R. P., Saxena, R. K., Ganga Rao, N. V. P. R. and Varshnry, R. K., (2014), 'Pigeonpea perspectives in India'. Paper presented at *1st Philippine Pigeonpea Congress*, December 16–18, 2014. Mariano Marcos State University, Batak, Ilacos Norte, Philippines.

Sawargaonkar, S. L. (2011), Study of heterosis, combining ability, stability and quality parameters in CGMS based pigeonpea [*Cajanus cajan* (L.)Millsp.] hybrids. *Ph. D. Thesis*, Marathwada Agricultural University, Parbhani, Maharashtra, India, p. 151.

Saxena, K. B. (2006), 'Seed production systems in pigeonpea'. *Technical Bulletin*. International Crops Research Institute for the Semi – Arid Tropics, Patancheru, 502 324, p. 76.

Saxena, K. B. (2008), 'Genetic improvement of pigeonpea – a review'. *Tropical Plant Biology*, 1: 159–78.

Saxena, K. B. (2015). 'From concept to field: Evolution of hybrid pigeonpea technology in India'. *Indian Journal of Genetics and Plant Breeding*, 75: 279–93.

Saxena, K. B., Chandrasena, G. D. S. N., Hettiarachi, K., Iqbal, Y. B., Fonseka, H. H. D. and Jayasekera, S. J. B. A. (2002), 'Evaluation of pigeonpea accession of selected lines for reaction to *Maruca*', *Crop Science*, 42: 615–18.

Saxena, K. B, Kumar, R. V., Srivastava, N. and Shiying, B. (2005), 'A cytoplasmic-nuclear male-sterility system derived from a cross between *Cajanus cajanifolius* and *Cajanus cajan*'. *Euphytica*, 145: 289–94.

Saxena, K. B., Kumar, R. V. and Sultana, R. (2010), 'Quality nutrition through pigeonpea – a review'. *Health*, 2: 1335–44.

Saxena, K. B., Kumar, R. V., Tikle, A. N., Saxena, M. K., Gautam, V. S., Rao, S. K., Khare. D., Chauhan, Y. S., Saxena, R. K., Varshney, R. K., Reddy, B. V. S., Sharma, D., Reddy, L. J., Green, J. M., Faris, D. G., Mula, M., Sultana, R., Srivastava, R. K., Gowda, C. L. L. and Sawargaonkar, S. L. (2013), 'ICPH 2671 – The world's first commercial food legume hybrid', *Plant Breeding*, 132:479–85.

Saxena, K. B. and Sawargaonkar, S. L. (2015), 'Genetic enhancement of seed proteins in pigeonpea – methodologies, accomplishments, and opportunities'. *International Journal of Scientific Research*, 4: 3–7. ISSN No.: 2277–8179.

Saxena, K. B., Saxena, R. K. and Varshney, R. K. (2017), 'Germination of immature seeds for rapid generation advances to accelerate genetic gains in pigeonpea'. *Plant Breed.* doi:10.1111/pbr.12538

Saxena, K. B., Sultana, R., Bhatnager-Mathur, P., Saxena, R. K., Chauhan, Y. S., Kumar, R. K., Singh, I. P., Raje, R. S. and Tikle, A. N. (2016), Accomplishments and challenges of pigeonpea breeding in India. *Ind J. Genet.*, 76 (4): 467–82.

Saxena, K. B., Sharma, D. and Vales, M. I. (2018), 'Development and commercialization of CMS pigeonpea hybrid'. *Plant Breed. Rev.* 41: (in press).

Saxena, K. B. and Sharma, D. (1990), 'Pigeonpea genetics', pp. 137–58. In *The Pigeonpea* (Nene, Y. L., Hall, S. D. and Sheila, V. K. (Eds)). CAB International, Wallingford, UK.

Saxena, K. B., Singh, G., Gupta, H. S., Mahajan, V., Kumar, R. V., Singh, B., Vales, M. I. and Sultana, R. (2011b), 'Enhancing the livelihoods of Uttarakhand farmers by introducing pigeonpea cultivation in hilly areas'. *Journal Food Legumes*, 24: 128–32.

Saxena, K. B., Tikle, A. N., Kumar, R. V., and Choudhary, A. K. and Bahadur, B. (2016), 'Nectarivore-aided cross-pollination and its exploitation in genetic enhancement of productivity in pigeonpea'. *International Journal of Scientific Research*, 6: 321–8.

Saxena, M. K., Saxena, U., Saxena, K. B., Khandalkar, V. S. and Sultana, R. (2011a), 'Profitability and production cost of hybrid pigeonpea seed'. *Electronic Journal of Plant Breeding*, 2: 409–12.

Sharma, H. C. (2016), 'Host plant resistance to insect pests in pigeonpea: Potential and limitations'. *Legume Perspectives*, 11: 26–31.

Sheldrake, A. R., (1984), 'Pigeonpea', pp. 385–417. In *The Physiology of Tropical Field Crops* (Goldsworthy, P. R. and Fisher, N. M. (Eds)). Academic Press, Chichester, UK.

Sheldrake, A. R. and Narayanan, A. (1979), 'Comparisons of earlier-and later-formed pods of pigeonpea (*Cajanus cajan* (L.) Millsp.)'. *Annals of Botany*, 43: 459–66.

Sheldrake, A. R., Narayanan, A. and Venkataratnam, N. (1979), 'The effects of flower removal on the seed yield of pigeonpea (*Cajanus cajan*)'. *Annals of Applied Biology*, 91: 383–90.

Shiv Kumar, Gupta, S., Chandra, S. and Singh, B. B., (2003), 'How wide is the genetic base of pulse crops', pp. 211–21. In *Pulses in New Perspective* (Ali, M. (Ed.)). Indian Institute of Pulses Research, Kanpur, 208 204, India.

Silim, S. N., Gwata, E. T., Coeb, R. and Omanga, P. A. (2007), 'Response of pigeonpea genotypes of differrent maturity duration to temperature and photoperiod in Kenya'. *African Crop Science Journal*, 15: 73–81.

Simtowe, F., Asfew, S. and Abete, T. (2016), 'Determinants of agricultural technology adoption under partial population awareness: The case of pigeonpea in Malawi'. *Agricultural and Food Economics*, 4: 7. doi:10.1186/s40100-016-0051-z.

Singh, I. P., Bohra, A. and Singh, F. (2016), 'An overview of varietal development program of pigeonpea in India'. *Legume Perspectives*, 11: 39–42.

Singhal, N. C. (2013), '*Hybrid Seed Production in Field Crops – Principle and Practices*'. Oxford IBH Publication, New Delhi, India.

Smith, A., Snapp, S., Dimes, J., Gwenambira, C. and Chikowo, R., (2016), 'Doubled-up legume rotations improve soil fertility and maintain productivity under variable conditions in maize-based cropping systems in Malawi'. *Agricultural Systems*, 145: 139–49.

Sultana, R., Vales, M. I., Saxena, K. B., Rathore, A., Rao, S., Rao, S. K. Mula, M. and Kumar R. V. (2013), 'Water-logging tolerance in pigeonpea (*Cajanus cajan* (L.) Millsp.): Genotypic variability and identification of tolerant genotypes'. *Journal Agriculture Science*, 151: 659–71.

Vales, M. I., Srivastava, R. K., Sultana, R., Singh, S., Singh, I., Singh, G., Patil, S. B. and Saxena, K. B. (2012), 'Breeding for earliness in pigeonpea: Development of new determinate and non-determinate lines'. *Crop Science*, 52: 2507–16.

Varshney, R. K., Chen, W., Li, Y., Bharti, A. K., Saxena, R. K., Schlueter, J. A., Donoghue, M. A. T. and et al. (2012), 'Draft genome sequence of pigeonpea (*Cajanus cajan*. L), an orphan legume crop of resource-poor farmers'. *Nature Biotechnology*, 30: 83–9.

Varshney, R. K., Saxena, R. K., Upadhyaya, H. D., Khan, A. W., Yu, Y., Kim, C., Rathore, A., Kim, D., Kim, J., An, S., Kumar, V., Anuradha, G., Yamini, K. N., Zhang, W., Muniswamy, S., Kim, J. S., Penmetsa, R. V., von Wettberg, E. and Datta, S. K. (2017), Whole-genome re-sequencing of 292 pigeonpea accessions identifies genomic regions associated with domestication and agronomic traits. *Nat. Genet.* doi:10.1038/ng.3872.

Vasantharaj, D. B. (1995), 'The Pesticides Industry'. *Kothari's Desk Book series*, HC Kothari Group, Publications Division, Kothari Buildings, Nungambakkam, Madras - 600034, India. p. 464.

Wallace, D. H., Yourstone, K. S., Masaya, P. N. and Zobel, R. W. (1993), 'Photoperiod gene control over partitioning between reproductive and vegetative growth'. *Theoretical and Applied Genetics*, 86: 6–16.

Wallis, E. S., Byth, D. E. and Saxena K. B. (1981), 'Flowering responses of thirty-seven early maturing lines of pigeonpea', pp. 143–59. In *Proceedings International Workshop on Pigeonpeas. Volume 2, International Crops Research Institute for the Semi-arid Tropics*. ICRISAT, Patancheru, 502 324.

Whitbread, A. M., Robertson, M. J., Carberry, P. S. and Dimes, J. P. (2010), 'How farming systems simulation can aid the development of more sustainable smallholder farming systems in southern Africa'. *European Journal of Agronomy*, 32: 51–8.

Whiteman, P. C. and Norton, B. W. (1981), 'Alternative uses of pigeonpea', pp. 365–77. In *Proceedings International Workshop on Pigeonpeas. Volume 1. International Crops Research Institute for the Semi-arid Tropics*. ICRISAT, Patancheru, 502 324.

Yadava, K., Yadava, S. K., Yadava, A, Pandey, V. P. and Dwivedi, U. N. (2014), 'Comparative analysis of genetic diversity among cultivated (*Cajanus cajan*) and its wild relatives (*C. albicans* and *C. lineatus*) using randomly amplified polymorphic DNA (RAPD) and inter single sequence repeats (ISSR) finger printing'. *American Journal of Plant Science*, 5: 1668–78.

Yang, S., Wen, P., Zong, X., Li. Z., Chaohong, Z., Saxena, K. B. and Hanchao, L. (2001). 'Pigeonpea: A potential fodder crop for Guangxi province of China', *International Chickpea and Pigeonpea Newsletter*, 8: 54–5.

Yang, S., Pang, W., Ash, G., Harper, J., Carling. J., Wenzl, P., Huttner, E., Zong, X. and Kilian, A. (2006), 'Low level of genetic diversity in cultivated pigeonpea compared to its wild relatives is revealed by diversity arrays technology'. *Theoretical and Applied Genetics*, 113: 585–95.

Chapter 15

Improving the cultivation of pigeonpea

K. R. Latha and L. Vimalendran, Tamil Nadu Agricultural University, India

1 Introduction
2 Seeds and sowing
3 Intercropping/mixed cropping
4 Nutrient management
5 Water management
6 Efficient use of fertilizer and water
7 Weed management
8 Agronomic management of pigeonpea hybrids
9 Early-maturing pigeonpea varieties
10 Rabi and summer pigeonpea cultivation
11 Harvesting, threshing and storage
12 Conclusion
13 Guidelines for future research
14 Where to look for further information
15 References

1 Introduction

Pigeonpea (*Cajanus cajan* L. Millspaugh) is grown by millions of resource-poor farmers on marginal land across the semi-arid regions of Asia, Africa, Latin America, and the Caribbean region. The production of pigeonpea is constrained by poor agronomic management viz. improper method of sowing, incorrect sowing time, inadequate seed rate, insufficient weeding, imbalance in fertilization, inadequate intercultural operations, insufficient irrigation and inadequate use of fertilizers. In Africa, poor production practices such as low plant densities, low soil fertility, insufficient weeding, and insufficient/inappropriate use of fungicides and herbicides are the key constraints (Odeny, 2007). In recent years, a large number of improved varieties of pigeonpea have been developed and are cultivated in different conditions. To realize the yield potential of these varieties under varying situations, optimum agronomic management is essential. This chapter analyses integrated crop

http://dx.doi.org/10.19103/AS.2017.0023.34

management techniques with particular emphasis on efficient fertilizer and water use to maximize sustainable yields.

Pigeonpea [*Cajanus cajan* (L.) Millsp.] is a short-lived perennial shrub that is traditionally cultivated as an annual crop in developing countries. Pigeonpea is valued by farmers and consumers as a food and forage crop. It is also a major contributor to food security in areas facing the early effects of global climate change. In India pigeonpea is grown in an area of 5.06 m.ha, with a production of 3.29 m.t and an average productivity of 650 kg ha^{-1} (Food and Agriculture Organization statistics (FAOSTAT), 2014), which is 11% lower than average global productivity. India contributes 68% of the global pigeonpea production. India has the largest area under pigeonpea, followed by Myanmar (611 600 ha), Tanzania (276 400 ha), Kenya (276 124 ha), Malawi (229 790 ha), Haiti (111 950 ha), Uganda (33 000 ha), Dominican Republic (23 088 ha) and Nepal (17 006 ha) (FAOSTAT, 2014).

India produces 3 290 000 tonnes of pigeonpea followed by Myanmar (575 100 tonnes), Malawi (301 010 tonnes), Kenya (274 523 tonnes), Tanzania (9 249 250 tonnes) and Haiti (90 480 tonnes) (FAOSTAT, 2014). In Myanmar, pigeonpea is grown mainly for export to India. As a result, production trends in Myanmar have a direct effect on domestic pigeon pea prices in India. In southern China, pigeonpea is primarily used for soil conservation, food and fodder. The International Crops Research Institute for the Semi-Arid Tropics (ICRISAT) variety ICP 7035 has shown high adaptation in different provinces.

Pigeonpea can be grown between 14°N and 28°N latitude, with a temperature ranging from 26 to 30°C in the rainy season (June–October) and from 17 to 22°C in the post-rainy (November–March) season (Reddy and Virmani, 1981). The amount of daily global solar radiation varies from 400 to 430 cal cm^{-2} day^{-1} in the rainy season and from 380 to 430 cal cm^{-2} day^{-1} in the post-rainy season. Mean annual rainfall ranges from 600 to 1400 mm, of which 80–90% is received in the rainy season.

In India, pigeonpea is cultivated on Entisols, Alfisols, Inceptisols and Vertisols. The Entisols found in the alluvial-soil belt of the Indo-Gangetic region are deep loams, slightly alkaline (pH 7.5–8.5), with about 150–200 mm available water storage capacity in 2 m of soil. Pigeonpea can be grown on a range of soil and climatic conditions and play an important role in crop rotation, mixed cropping and inter-cropping by maintaining soil fertility through N fixation and release of soil-bound phosphorus, and thus contribute significantly to sustainability of the farming systems (Gowda et al., 2013). A number of management practices can increase pigeonpea production and productivity.

Pigeonpea does not require special land preparation but its production varies greatly depending on the depth and moisture-holding capacity of the soil. In the traditional system, fields are prepared with a *desi* plough. Deep ploughing to a depth of 15 cm is sufficient to obtain a good crop (Lugo-mercado et al., 1987). However, where a hard pan exists in the rooting zone, sub-soiling has been found beneficial in improving root development and infiltration rate (Reddy et al., 1978). Maintenance of proper soil tilth is essential to ensure sufficient air and water in the soil. In a properly aerated soil with sufficient moisture, development of the root system is generally good and crop yields are higher. Organic manure may be applied 2–4 weeks before sowing. In acidic soils, 2000–4000 kg ha^{-1} of lime is incorporated 3–4 weeks before sowing to neutralize the acidity (ICRISAT, 1992).

At the time of sowing, the field should be well levelled, free from weeds and clods, and should have sufficient moisture. Pigeonpea is very sensitive to waterlogging particularly during the seedling stage. Contour beds or a ridge-and-furrow system are useful in preventing waterlogging by draining excess surface water, and in preventing soil erosion. A crop sown on ridges gives 26–31% higher yield than one sown on flat beds (Tayo, 1985b; Reddy et al., 2009).

2 Seeds and sowing

2.1 Selection of variety

One of the reasons often cited for the low yields of pigeonpea is the lack of improved genotypes available to farmers. The choice of variety depends upon the growing season and the pattern of rainfall distribution. In North India, tall bushy varieties are grown as a rainfed crop, sown in June-July and harvested in March-April. Maturity duration (at 17°N latitude) varies from about 90 days for extra-early varieties to more than 260 days for long-duration (LD) varieties. Attributes such as short stature and early maturity (120–130 days) make cultivars like ICPL 87 suitable for sole cropping and multiple harvesting. Extra-short-duration (ESD) and SD pigeonpea varieties fit well in various cropping systems, providing greater crop diversity. Pigeonpea varieties suitable for cultivation in different pulse-growing regions of India are shown in Table 1.

Selection of drought-tolerant varieties can provide cost-effective long-term solutions against the adverse effect of drought. A wider dissemination of drought-tolerant material would provide sustenance to the livelihoods of farmers who are more vulnerable to shocks due to crop failure (Gowda et al., 2013).

Five pigeonpea varieties, based on ICRISAT-bred material have been released in Myanmar. In South Africa, based on the studies carried out by the Lowveld Research Unit-Department of Agriculture and Land Administration (LRU–DALA) in Mpumalanga during 1998–2002 and across three other provinces (Eastern Cape, Northern Cape and North West) during the 2000–2001 seasons, varieties viz. ICPL 85010 (extra-short, 110–120 days), ICPL 87 (short, 120–130 days), ICPL 87051 (medium, 170–180 days), ICEAP 00040 (long spreading, 230–240 days), and ICEAP 00053 (long compact, 230–240 days) were recommended for wider adoption in South Africa (Mathews, 2003). Preliminary studies also showed the potential of ICPL 87 as a vegetable type. ICRISAT pigeonpea varieties

Table 1 Pigeonpea varieties suitable for different pulse-growing zones of India

Pigeonpea duration	Varieties
North East Plain zone:	
Kharif long duration	Bahar, Amar, Narendra Arhar 1, Pusa 9, MAL 13, NDA 2, MA 6
North West Plain zone:	
Kharif short duration	Paras, AL 201, UPAS 120, Pusa 992, TJT 501
Central zone:	
Kharif	
Medium duration	MA 3 JKM 7, BSMR 175, JA 4, ICPL 87119, BSMR 736
Short duration	TJT 501, TT 401, GT100, AKT 8811
South zone:	
Kharif	
Medium duration	VBN 2, CO 6, ICPL 87119 (Asha), ICPL 85063 (Laxmi), LRG 41
Short duration	VBN 3, CO (Rg) 9701, ICPL 84031 (Durga)

Source: (Latha, 2012).

resist wilt, have high yields and large seeds, and are widely grown in eastern and southern Africa (ESA), increasing farmers' incomes by up to 80% (ICRISAT, 2012). In Eastern Kenya, about 20% of the farmers have adopted new varieties. Farmers have also started adopting the medium-duration (MD) pigeonpea varieties (ICEAP 00554 and 00557) both as a grain and a vegetable.

In some districts of Tanzania (such as Babati), around 50% of farmers have adopted new varieties, a practice that has spread to neighbouring districts such as Karatu and Mbulu. In some areas, farmers are adopting the LD compact growth habit genotype (ICEAP 00053) in maize intercropping systems. The use of the LD fusarium wilt-resistant variety (ICEAP 00040) in northern and central Tanzania, Kenya and Malawi resulted in increased grain yields and lowered production costs in comparison to local genotypes. NARS partners have released pigeonpea varieties in Malawi (4), Kenya (3), Tanzania (3), Uganda (2) and Mozambique (1).

To break the yield barrier in pigeonpea, ICRISAT and partners have developed medium-maturity hybrids ICPH 2671 and ICPH 2740, which have produced 30–40% greater grain yields than the popular varieties across farmers' fields in India. However, there is not enough seed production and distribution of these hybrids due to high cost of production of seeds and low ability of farmers to purchase at a higher price. Hence, the adoption rate of these hybrids has not picked up (Reddy et al., 2013).

2.2 Seed treatment

Seed treatment includes use of rhizobium, phosphorus solubilizing bacteria (PSB) or plant growth promoting rhizobacteria (PGPR), or treating the seed with chemicals for better nodulations, crop establishment, protection against soil-borne pathogens, and yield. Seeds are treated with carbendazim or thiram (at a rate of 2 g/kg of seed) 24 hours before sowing (or) with talc formulation of *Trichoderma viride* 4 g/kg of seed (or) *Pseudomonas fluorescens* at 10 g/kg seed. Biocontrol agents are compatible with biofertilizers. Initially the seeds are treated with biocontrol agents and then with rhizobium. Fungicides and biocontrol agents are incompatible (CPG, 2012). Rhizobial strains viz. CC 1, CRR 6 and CPR 9 are effective on pigeonpea (Gopalaswamy, 2012). Phosphorus solubilizing microorganisms solubilize inorganic insoluble phosphorus by excreting organic acids that directly dissolve phosphatic materials and/or they chelate cationic partners of phosphorus ion.

In pigeonpea, seed treatment with micronutrients (Zn, Mo and Co: 4 g, 1 g and 0.5 g/kg seed) and combined inoculation of *Rhizobium*, *Phosphobacteria* and *Pseudomonas fluorescens* (PGPR) significantly increased the growth and yield over control. In another experiment, combined inoculation of *Rhizobium*, PSB and PGPR registered 24.1% increase in yield over control (Gopalaswamy, 2012).

2.3 Time of sowing

In the rainfed and dry areas, pigeonpea is sown with the onset of the monsoon. Earlier sowing gives higher yields in India. In Tamil Nadu, farmers raise MD varieties viz. CO 6, VBN 2 and LRG 41 during June-July in the districts which receive bimodal rainfall. In districts which receive mono-modal rainfall, sowing can be done with the little rainfall or with supplemental irrigation for initial establishment of the crop during the rainy season (kharif). Thereafter the crop is raised in rainfed conditions. The optimum time of sowing is between July and August (late kharif). However, the SD cultivars viz. CO(Rg) 9701 and VBN 3 that are photo-insensitive can be sown between January and May as summer-irrigated crop and the yield level ranges from 1000 to 2500 kg ha^{-1}. If the sowing gets delayed beyond May, harvesting coincides with the North East Monsoon resulting in yield loss.

When sowing extra-early and early-maturing varieties in the first fortnight of June, the field is available for post-rainy season crops by the end of November. Therefore, sowing should not be delayed beyond June.

The date of sowing is one of the most important agronomic practices in determining yields in pigeonpea (Pavan et al., 2009). Growth and yield of pigeonpea varies from location to location due to variability in agro-climatic and soil-water-related parameters. Even in the same location, variation in growth takes place due to different sowing dates, planting density, irrigation methods and frequency, nutrient and weed management, and other cultural and management practices (Ahlawat and Rana, 2005).

Comprehensive studies on the date of sowing of pigeonpea started with the initiation of the All India Co ordinated Pulses Improvement Project (AICPIP) in 1967. Numerous studies conducted on the date of sowing of pigeonpea revealed that mid-May to mid-June is the best time of sowing for SD pigeonpea under irrigated conditions (Sandhu et al., 1981).

Studies at Chandra Shekhar Azad University of Agriculture and Technology (CSUAT), Kanpur (Uttar Pradesh) showed that a higher productivity from early pigeonpea could be obtained by advancing its planting to mid-April at wider row spacing (75–100 cm) and intercropping with two rows of mungbean (Panwar and Yadav, 1980). Experiments conducted at the Punjab Agricultural University (PAU), Ludhiana, showed that the second fortnight of May was the best time of sowing for MD cultivars like T 21 and the first fortnight of June for SD cultivars like AL 15 and AL 201. For hybrid PPH 4, the last week of May to the first week of June is the suitable time of sowing for harvesting higher yield as well as for timely sowing of the succeeding wheat (PAU, 1995). The SD cultivars (AL 15 and AL 201) when sown at the end of April or the first week of May bear flowers by the end of June. During the South West monsoon period, these flowers shed and the crop reverts to vegetative growth, which ultimately leads to poor yield. The crop sown in May experiences less competition from weeds than the crop sown in June (Sekhon et al., 1993). The crop sown late in July not only provides low grain yield but also matures late due to declining temperature and solar radiation.

For pre-post rainy season (*rabi)* pigeonpea in Bihar state of India, the optimum time of sowing is 1st September for cultivar IE, 8th September for Bahar, and 15th September for Basant at Sabour. Sowing after these dates causes sharp decline in grain yield. Similarly, at Dholi, there was a decline in yield with delay in sowing beyond 10th September with cultivar Bahar. At Kanke (Ranchi), which is situated in the dryland shallow-soil zone at Chhotanagpur, cultivar Bahar sown on 5th September gave the highest yield. However, yield dropped markedly with the crop sown on 15th October (Sekhon and Singh, 2005). Desai et al. (2000) have reviewed the cultivation of post-rainy season pigeonpea in rotation with rice in Gujarat. The crop sown in the third week of October was significantly superior to the crop sown in the first and third weeks of November. In Bihar, due to introduction of *Alternaria*-resistant varieties viz. Sharad and Pusa 9, the maize-pigeonpea crop rotation became the choice of the area (Sharma et al., 1980). Its optimum time of sowing was found to be the first fortnight of September with a plant density of 300 000–350 000/ha, 40 kg N/ha and two irrigations. For achieving higher productivity, crop should be sown in September at a row distance of 30 cm and plant spacing of 15 cm accommodating about 200 000 plants per hectare. The fields vacated by early rice varieties, maize, minor millets, rice seedlings etc. may be utilized for the cultivation of pre-rabi pigeonpea (Ali et al., 2012).

In South Africa, it was observed that late planting of the LD and MD varieties after December resulted in smaller canopy and lower yields. The SD types planted later in the season (after January) led to poor establishment, early flowering, smaller canopy and lower yields, probably due to below-optimum temperature (Mathews and Saxena, 2005).

2.4 Seed rate, crop geometry and method of sowing

Optimum plant population is essential for achieving higher yield. The seeding rate of pigeonpea depends on the desired plant density for a genotype (early, medium or late), cropping system (pure crop, mixed crop or intercrop), germination rate of seed and mass of seeds. The plant density recommended for extra-early sole crop pigeonpea is 330 000 plants ha^{-1}, whereas for early- and medium-maturity varieties it ranges from 80 000 to 100 000 plants ha^{-1}. The plant density for tall and highly branched late varieties could be 50 000–60 000 ha^{-1} (ICRISAT, 1992). The seed rate recommended for early, medium and late varieties are 12–15, 10–12 and 8–10 kg ha^{-1}, respectively.

The optimum plant density can be maintained by using adequate seed rate in a proper planting geometry. Some factors like seeding depth, seed dressing, seed grading, variety, seeding rate, moisture and temperature may also influence emergence and establishment of pigeonpea (Sekhon and Singh, 2005). For hybrid PPH 4, a spacing of 67.5 × 25 cm (60 000 plants ha^{-1}) was found to be the best (Sekhon et al., 2000). Saxena et al. (1992) also reported an increase in grain yield of the hybrid ICPH 8 at wider spacing as compared to its parental lines.

In 2011, Mula et al. reported that a spacing of 75 × 30 cm in 3:1 row ratio with irrigation at 14-day interval during reproductive period produced increased seed yield of pigeonpea. As the plant density increases, there was a progressive reduction on yield and yield traits of individual plants (Mula et al., 2010, 2011). Tuppad et al. (2012) found that 90 × 20 cm spacing was optimum for getting significantly higher pigeonpea grain yield and net return. The agronomic requirements of SD varieties were worked out at Indian Agricultural Research Institute, New Delhi; Punjab Agricultural University, Ludhiana; Govind Ballabh Pant University of Agricultural Technology, Pantnagar; and Hissar Agricultural University, Hissar under AICPIP as well as in the institutional programmes of the Indian Institute of Pulses Research (IIPR). A plant population of 100 000–150 000 ha^{-1} was found optimum. The crop was amenable for intercropping with SD pulses like mungbean, urdbean and cowpea (Ali et al., 2012).

The concept of broad bed furrows (BBF) was introduced by ICRISAT during early 1980s (ICRISAT, 1981, 1982) for pigeonpea, but did not become popular due to high costs and difficulties in management. Later during 1994–95, ridge planting of pigeonpea was conceptualized and evaluated under AICPIP on pigeonpea, which showed very encouraging results in terms of maintenance of optimal plant populations and consequently higher productivity due to proper drainage. Ridges are made at a distance of 60–75 cm, leaving 30-cm-wide furrows for drainage of rainwater. Two to three rows of SD legumes such as mungbean/urdbean can be successfully planted on ridges. This system helps in reducing the quantity of irrigation water (IW) and also minimizes the incidence of *Phytophthora* blight in pigeonpea. The rainfed bed system of planting has a similar beneficial effect (Ali et al., 1998).

On heavy soils, sowing on raised beds seems to be quite beneficial as it helps avoid waterlogging. This practice also saves IW. Ali et al. (1998) reported significant increase in the grain yield of pigeonpea sown on raised beds over the flat beds at Hissar and Pantnagar in India. At Ludhiana (Punjab state, India), flat sowing with making bunds 50 days after sowing (DAS) produced higher yield than the flat bed sowing. Similarly, in South Gujarat (India), Desai et al. (2000) observed that bed sowing (1 furrow after 4 rows of pigeonpea; 2.5 m top and 50 cm furrow) was significantly superior to flat bed under high rainfall area and on soils having high clay content.

2.5 Depth of sowing

Depth of sowing is also an important factor influencing the emergence and establishment of crops grown especially under dryland conditions. Various factors influencing seeding depth include soil texture, date of sowing, climatic conditions and seed size. In pigeonpea, the optimum depth of seeding seemed to be 5–6 cm. Tayo (1983) found that emergence of pigeonpea decreased with increase in sowing depth to 8 cm because the food reserves in cotyledons are exhausted. Shallow depths also showed poor emergence as seedlings are exposed to water deficits if dry and hot temperature prevails soon after sowing. A uniform seeding depth can easily be achieved with the use of mechanical planters or seed drills.

2.6 Method of establishment

Normally pigeonpea is sown either by broadcasting ot by drilling/dibbling. However, in dry farming areas such as Karnataka State, rainfall is both limited and erratic. Farmers use transplanting as an alternative. This technique involves raising of seedlings in polythene bags in the nursery for a period of one month and then transplanting these seedlings in the main field, immediately after soil wetting rains (Pavan et al., 2011).

Pigeonpea seedlings at the age of 28 days transplanted in the inter-row spacing of 120 cm had recorded maximum gross return, net returns and benefit ratio, which is comparable to 35-day-old seedlings transplanted in the inter-row spacing of 120 cm (Jamadar, 2009). Transplanting of 5-week-old pigeonpea seedlings as sole crop as well as intercropped with finger millet gave higher net returns and B:C ratio than direct-sown sole pigeonpea and finger millet + direct-sown pigeonpea (Poornima et al., 2010). Studies at Raichur showed that transplanted pigeonpea had significantly higher grain yield and net income than conventional sowing after a period of good rainfall. However, the B:C ratio was higher when pigeonpea was sown after the receipt of good rainfall (Anilkumar et al., 2011).

Dibbled pigeonpea with a spacing of 90 × 20 cm proved better than other treatment combinations of transplanted pigeonpea, evaluated at Raichur in the North Eastern Dry Zone of Karnataka during kharif season (Pavan et al., 2011). The study conducted at Annigeri (Dharwad district) indicated that hybrid pigeonpea has better response to transplanting with wider planting geometry. The hybrid pigeonpea performed better than check Maruti under both planting methods (Mallikarjun, 2012). Transplanting of varying aged pigeonpea seedling under sole cropping recorded significantly higher pigeonpea yield over direct-sown crop. However, higher net return and B:C ratio were recorded with direct-sown sole pigeonpea (Goud and Andhalkar, 2012).

In the state of Tamil Nadu, India, this technology is also being followed in districts which receive bimodal rainfall. In districts which receive monomodal rainfall (northeast monsoon, NEM), supplemental irrigation is a must. Under drip system, transplanting CO 6 pigeonpea at 150 × 60 cm spacing recorded higher mean grain yield of 1799 kg ha^{-1}, while direct sowing registered 1615 kg ha^{-1} at the same spacing level. Though transplanted pigeonpea (CO 6) gives higher mean grain yield (1799 kg ha^{-1}) than direct-sown crop (1615 kg ha^{-1}), it is not economical due to increased cost of cultivation (515.19$ ha^{-1}) when compared to the cost of cultivation of direct-sown pigeonpea (367.26$ ha^{-1}). The mean net income and B:C ratio were the highest (342.81$ ha^{-1}and 1.94) when pigeonpea was dibbled at 150 × 60 cm spacing. The water use efficiency (WUE) was 3.7 kg ha^{-1} mm. Under drip-irrigated situation, direct sowing of pigeonpea (CO 6) at 150 × 60 cm spacing was found to be economical (Latha et al., 2012).

3 Intercropping/mixed cropping

Legumes have beneficial effects over non-leguminous crops when grown as intercrop. Intercropping facilitates use of higher moisture and WUE leading to higher production than pure crop. Differential durations and growth habit of pulse crops make it possible to grow them as intercrop in cereals and other crops. Willey et al. (1981) grouped pigeonpea intercrops into three broad categories:

1. With cereals (sorghum, maize, pearl millet, setaria, finger millet and rainfed rice)
2. With legumes (groundnut, cowpea, mungbean, urdbean and soybean)
3. With LD annuals (castor, cotton, sugarcane and cassava)

Intercropping has been recognized as beneficial cropping system under rainfed condition. Pigeonpea with its LD and wider row spacing forms an ideal intercrop with SD component crops like mungbean (Dudhade et al., 2009) with early-maturing and shallow-rooted finger millet (Padhi et al., 2010).

The pigeonpea is most common intercrop grown with cotton in Vidarbha region in a combination of cotton + pigeonpea in 6:1 and 4:1 ratios. Paired row planting of sorghum or pearl millet alternated with single row of mid-late-maturing pigeonpea Vipula is beneficial. Similarly, soybean + pigeonpea (3:1), mungbean + pigeonpea (3:1) and urdbean + pigeonpea (3:1) are also most common and beneficial intercropping systems. They not only cover the risk of failure but also give higher total production of 60–70% over a sole crop of cereal. It has been proved that at lower levels of nitrogen, association of legumes as an intercrop in non-legumes increases the production of base crop. A large area of dryland farming is still under sole crop. Efforts should be made to bring all such area under intercropping of pulses in cereals and other rainfed crops.

Trials conducted under the All India Co ordinated Research Project (AICRP) on pigeonpea showed that, in the North West Plain Zone (NWPZ), intercropping with mungbean produced higher pigeonpea equivalent yield than the sole pigeonpea cropping; in Central Zone (CZ), intercropping with groundnut or soybean was better; and in South Zone (SZ), mungbean or urdbean intercrops showed higher pigeonpea equivalent yield (Table 2) (IIPR, 2003–04).

Studies conducted at ICRISAT showed that sorghum intercrop reduced the incidence of pigeonpea wilt. In addition, a fast-growing intercrop not only gives additional yield benefits but also reduces the need for weeding (Shetty and Rao, 1981). Intercropping is more appropriate under poor growing conditions and it has very little scope where high-cost inputs are used. Intercropping may also be disadvantageous in the form of a yield decrease because of adverse competitive effects. Allelopathic effects may also occur. A major disadvantage of this method lies in the practical management of intercropping, especially where there is a high degree of mechanization or where the component crops have different requirements for fertilizers, irrigation, herbicides, etc.

Studies on spatial arrangement in pigeonpea + sorghum intercropping in 2:1 row ratio showed that ridge planting system recorded higher pigeonpea equivalent yield and B:C ratio as compared to 1:1 and mixed planting system. However, the spatial arrangement varied with the variation in plant types of the base crop (pigeonpea). In early pigeonpea + groundnut intercropping, 2:2 row ratio with North-South orientation was found to be most productive (1238 kg PEY ha^{-1}) with 90% higher land use efficiency. In LD pigeonpea + groundnut intercropping system, semi-spreading variety Pusa 9 was more productive under 5:2 row ratio but compact type variety Bahar recorded 67% yield advantage and 46% higher land use efficiency under 5:1 row ratio. As regards spatial arrangements, 2:1 row ratio was found better than 2:2 for pigeonpea + urdbean intercropping (IIPR, 2009).

Intercropping of early pigeonpea with groundnut (var. Chitra) provided additional yield of groundnut, which led to significantly higher pigeonpea equivalent yield and land

Table 2 Intercropping studies in pigeonpea

Zone/location	Treatment	Grain yield (tonnes ha^{-1}) Pigeonpea	Grain yield (tonnes ha^{-1}) Intercrop	Pigeonpea equivalent yield (tonnes ha^{-1})
NWPZ				
Ludhiana	Pigeonpea sole	2.09	–	2.09
	Pigeonpea + mungbean	2.04	0.45	2.60
CZ				
Junagadh	Pigeonpea sole	2.30	–	2.30
	Pigeonpea + groundnut	2.31	0.61	3.08
Sardarkrushinagar	Pigeonpea sole	1.45	–	1.45
	Pigeonpea + groundnut	0.85	1.11	2.01
Badnapur	Pigeonpea sole	1.37	–	1.37
	Pigeonpea + soybean	1.10	0.66	1.94
SZ				
Bangalore	Pigeonpea sole	2.15	–	2.15
	Pigeonpea + mungbean	2.03	0.20	2.28
Warangal	Pigeonpea sole	2.02	–	2.02
	Pigeonpea + mungbean	2.32	0.37	2.64
	Pigeonpea + urdbean	2.26	0.31	2.53

Source: (IIPR, 2003–04).

equivalent ratio (LER). Intercropping under replacement system drastically reduced the yield of pigeonpea and groundnut. Out of two early genotypes, UPAS 120 was found more compatible than ICPL 151. Intercropping of pigeonpea cv. Pusa 9 (semi-spreading type) and groundnut cv. Chitra in 5:2 row ratio was found to be more productive (4422 kg PEY ha^{-1}) followed by Bahar (compact) + Chitra in 5:1 row ratio. These intercropping systems were significantly superior to sole crop of pigeonpea (IIPR, 2009).

4 Nutrient management

Nitrogen, phosphorus and potassium are the three major and primary nutrients required for crop growth and productivity. A small amount of nitrogen is very important during early phase for seedling vigour. Pigeonpea being a rich source of protein helps to maintain soil fertility through biological nitrogen fixation (BNF) in soil and thus plays a major role in furthering sustainable agriculture. It enriches soil fertility by fixing atmospheric nitrogen in the root nodules if proper inoculants and sufficient amounts of P, K, sulphur and micronutrients are present.

4.1 Nitrogen

Pigeonpea nitrogen requirements are largely met through BNF. With the inception of AICPIP in 1966, a systematic and comprehensive research programme was launched to work out nutrient requirements of pulse crops. Kulkarni and Panwar (1981) reviewed the studies in India on the pigeonpea response to N. However, in some situations a starter dose of 20–25 kg N ha^{-1} was beneficial. Apart from N, adequate phosphorus is essential particularly on soils poor in nitrogen and organic matter for obtaining optimum yield and protein content of pigeonpea. Since pigeonpea is cultivated largely as a rainfed crop, the response to fertilizer is limited by the rainfall pattern and available moisture in the soil profile. Balanced and efficient use of nutrients are essential to obtain higher grain yield of an improved variety.

Maximum N uptake occurs during branching and early bud formation (Malhi et al., 2007). Therefore, during the early growing period, availability of P and K is critical not only for their direct incorporation into plant biomass but also to ensure early N fixation. Total nutrient uptake by pigeonpea depends on yield obtained, which will vary with season, variety, soil, nutrient status and cultural practices. Pigeonpea takes up relatively small amounts of nutrients early in the season, but as it grows and develops the daily rate of nutrient uptake increases and it requires an adequate supply of nutrients at each developmental stage for optimum growth. High-yielding LD pigeonpea removes substantial amount of nutrients from the soil, and this should be taken into account in an overall nutrient management plan.

Application of N as urea source at 20 and 40 kg ha^{-1} significantly increased the shoot, root and total plant dry matter production (Durante and Carpne, 1981). Singh and Bahadur (1990) opined that an addition of 45 kg N ha^{-1} promoted maximum vegetative growth and produced the highest yield of green pods. Balanced supply of nitrogenous fertilizer improves not only crop yield and nutrient uptake but also nodulation in pigeonpea (Thakuria and Loikham, 1991). Significant improvement in plant height was observed with increasing levels of N at 60 DAS (Chittapur et al., 1994). Kale et al. (1997) found that 75 kg N ha^{-1} was the optimum level for getting the highest yield of quality pods.

Application of 30 kg N ha^{-1} produced significantly higher yield through higher number of pods in pigeonpea (Meyyazhagan et al., 1999).

Being a smallholder's crop, pigeonpea does not receive significant fertilizer inputs in Africa. However, it has the ability to fix up to 235 kg Nitrogen (N) ha^{-1} (Peoples et al., 1995) and produces more N per unit area from plant biomass than many other legumes. Nitrogen is one of the most abundant elements on earth (Vance, 2001) and yet the most limiting nutrient for increasing crop productivity (Wani et al., 1995; Graham and Vance, 2003). The N-fixing ability of pigeonpea is desirable for environmentally sustainable agricultural production (Peoples et al., 1995). A recent study (Myaka et al., 2006) in Tanzania demonstrated that the yield of an unfertilized inter-cropped maize (*Zea mays*)–pigeonpea crop generally equalled the yield of a moderately fertilized sole maize yield.

4.2 Phosphorus

Phosphorus is an essential element for grain legumes as it helps in plant growth, nodulation, biological nitrogen fixation (BNF) and better grain development. It is necessary for energy storage and transfer and is essential for vigorous growth and seed development. Legumes need P in adequate amounts, but their response to fertilizer application is variable and often poor. The P requirement is in the range of 0.3–0.5% of dry weight during the vegetative stage of the crop growth. Plants obtain their P in soluble ionic forms (HPO_4^- and $H_2PO_4^-$). Response to phosphorus applications has been positive and in some cases highly significant in pigeonpea (Pathak, 1970). Kulkarni and Panwar (1981) summarized the phosphorus response studies in India. They concluded that applications of 17–26 kg P ha^{-1} increased seed yield by 300–600 kg ha^{-1}.

Phosphorus is highly immobile in soils and can become a limiting factor in pigeonpea production. A common difficulty in recovering P from the soil is that it is not readily available to plants because P reacts with aluminium, iron and calcium in the soil to form complexes. These nutrients are essentially insoluble resulting in very little movement of P in the soil solution, and none of the complexes can be taken up directly by roots (Sinclair and Vadez, 2002). The use of PSB (strains from the genera of *Pseudomonas*, *Bacillus* and *Rhizobium* are among the most powerful P solubilizers) as inoculants simultaneously increases P uptake by the plant and thus crop yields (Khan et al., 2009).

Phosphorus-deficient plants suffer from reduced leaf expansion, reduced surface size and reduced number of leaves. Respiration and photosynthesis are both reduced in P-deficient plants. Stress from P deficiency at early growth stage has considerable negative influence on crop production (Grant et al., 2001). Crop response to P fertilizer depends on genetic and physiological characteristics of the plant that helps in efficient P uptake and utilization, or it will be facilitated by the development of more P-efficient crop cultivars that will produce more yield per unit of phosphorus applied. Pigeonpea is a highly exhaustive pulse crop and requires higher amount of P nutrition for optimum production (Kumar and Kushwaha, 2006).

Pigeonpea has a special mechanism to release soil-bound phosphorus to meet its own needs as well as those of subsequent crops. However, a dose of 40–50 kg P_2O_5 ha^{-1} is considered optimum for pigeonpea. However, its requirement varies with the availability of P in the soil.

Pigeonpea has been shown to be more efficient at utilizing the iron-bound P in Alfisols. This ability is attributed to root exudates in particular piscidic acid and its p-O-methyl derivative which release P from iron-bound P by chelating Fe[34]. This may enhance P utilization and improve P-use efficiency. Cultivation of pigeonpea increases total P availability in cropping systems with low available P (Ae et al., 1990). Adu-Gyamfi (1999) found that deep banding was the best method for P application in pigeonpea as it helps to develop deep-root system. Responses of pigeonpea to P fertilization would not be expected in soils with more than 5 mg kg^{-1} extractable P by Olsen's bicarbonate extraction; however lower values would not necessarily predict a response (Johansen, 1990). The growth and yield parameters significantly improved by application of FYM, PSB and phosphorus fertilization (Mahetele and Kushwaha, 2011; IIPR, 2009).

Unlike in other legumes where growth has been reported to be limited by P (Ae et al., 1990), pigeonpea is cited as one of the few crop species that can utilize iron (Fe)-bound P efficiently (Ae et al., 1990; Subbarao et al., 1997). The LD genotypes (Vesterager et al., 2006) typically grown in Africa (Mergeai et al., 2001) utilize various sources of P more efficiently (Ishikawa et al., 2002) than their SD counterparts. The critical requirement of P concentration for dry matter production in pigeonpea is low compared to other major protein crops like soybean (Adu-Gyamfi et al., 1989) because it efficiently incorporates external orthophosphates (Pi) into residue P (Adu-Gyamfi et al., 1990). Compared with other crops, pigeonpea is also more efficient in P uptake when grown on low-P soils containing Aluminium (Al) (Ae et al., 1990).

4.3 Potassium

Potassium maintains water balance and therefore stalk strength. In addition, K promotes energy generation which is required for translocation of nutrients in the plant, N uptake and protein synthesis. Crops that are K deficient result in low total N uptake, yield and protein.

Potassium is known to play a vital role in osmoregulation and in the activation of several enzymes. Besides this, it helps the plant to adapt under terminal moisture stress/abiotic stresses mostly experienced during the reproductive stage. Potassium is involved in many physiological processes which affect crop quality. It activates more than 60 enzyme systems, aids photosynthesis, favours high energy status, regulates opening of leaf stomata, maintains cell turgor, promotes water uptake, regulates nutrient translocation, favours carbohydrate transport, and hence enhances protein and starch synthesis (Brar and Imas, 2010). Pigeonpea does not respond to potassium chloride applications unless it is grown on low available potassium soils (Kulkarni and Panwar, 1981). In a field experiment on an alluvial soil at Aligarh (in Uttar Pradesh state of India), the response of pigeonpea (cv Type 21) to potassium was studied at different irrigation levels. With two or three irrigations, optimum yield was obtained at 50 kg ha^{-1} K under rainfed conditions and 75 kg ha^{-1} K proved to be the best under rainfed conditions (Afridi et al., 1990). Response to applied K was either absent or very low. However, over the years with depletion of reserve soil K, response to 20–40 kg K_2O ha^{-1} has been observed, especially in light-textured soils.

Pigeonpea cultivated in the arid and the semi-arid tropics is a high-risk crop. Kulkarni and Panwar (1981) concluded that an application of 20 kg N, 17 kg P and 17 kg K ha^{-1} could give a satisfactory yield on medium black, red, and mixed black and red soils. This gave a cost-benefit ratio of 0.02–0.06 \$ \$$^{-1}$ cost of fertilizer. However, in red and laterite soils, 20-17-17 kg NPK ha^{-1} gave a cost-benefit ratio of 0.10–0.14 \$ \$$^{-1}$ cost of fertilizer (ICAR, 1979).

4.4 Sulphur

Sulphur is also an essential nutrient, which plays an important role in improving the quality and yield of pulses. Being a part of proteins (as –SH groups in enzyme) it acts as a co-factor in oxidation of pyruvic acid, in chlorophyll synthesis and also in nodule development. There is considerable evidence of S-deficiency in soils, particularly due to intensive cropping systems and low use of S-containing fertilizers (Tandon, 1995). Sekhon et al. (1997) reported improvement in nodulation, growth and grain yield of pigeonpea with 40 kg S ha^{-1}. In Tamil Nadu 10 and 20 kg S ha^{-1} is recommended for rainfed and irrigated pigeonpea, respectively (CPG, 2012). The single superphosphate has the advantage of containing sulphur.

4.5 Micronutrients

A recent study by ICRISAT indicated that soils in many states are deficient in micronutrients such as boron, sulphur, zinc and magnesium (Wani et al., 2012). Application of small quantities (0.5–2.0 kg) of micronutrients ha^{-1} resulted in 40–120% increase in grain yield. Hence, making these micronutrient fertilizers easily available to smallholder farmers in remote areas will go a long way in enhancing the productivity and yield of pigeonpea. Under a mission to increase productivity of rainfed agriculture through science-led interventions in the Karnataka state of India (called the Bhoochetana project), the improved management practices (including application of micronutrients) have increased the yield of pigeonpea by 26–38% during 2010–11 (Gowda et al., 2013).

High pH reduces the availability of zinc, iron, manganese, copper and boron. On the contrary, the availability of molybdenum increases with increase in pH. The Alfisols and Vertisols are generally low in organic matter and invariably have lime in their profiles. Therefore, this combination is known to cause a greater loss of all these micronutrients except for molybdenum (Katyal, 1980).

Experiments conducted at Bangalore, Badnapur and Gulbarga (India) showed significantly higher yields of pigeonpea with 15 kg $ZnSO_4$ ha^{-1} while borax, sodium molybdate and iron application failed to further enhance the productivity (Table 3) (IIPR, 2002–2003).

Most of the pigeonpea cultivars are susceptible to zinc deficiency. Therefore, foliar applications of 2–4 ppm zinc as 0.5% zinc sulphate with 0.25% lime have been effective to overcome zinc deficiencies (Saxena and Singh, 1970). The nutrients required for selected levels of grain yield in pigeonpea are given in Table 4.

There is a need to clarify the effect of integrated nutrient management in pigeonpea, as chemical fertilizers/composts can influence productivity and soil fertility status. Organic manures play a direct role in supplying macro- and micronutrients and an indirect role by improving physical, chemical and biological properties of the soil (Acharya and Biswas, 2002).

4.6 Nutrient management in cropping systems

The difficulty of nutrient management arises due to the differences in nutrient requirement of intercrops, and also due to biological nitrogen contribution from pigeonpea. Variations in plant populations also add to the problem (Yadav et al., 1997). In the past, efforts were

Table 3 Grain yield of pigeonpea as influenced by micronutrients at different locations

Treatment	Grain yield (tonnes ha^{-1})		
	Badnapur	Bangalore	Gulbarga
Absolute control	0.84	0.83	0.71
NPKS (18 + 46+20 + 20 kg ha^{-1})	1.34	1.11	1.05
PKS + $ZnSO_4$ (15 kg ha^{-1})	1.66	1.32	1.25
NPKS + $ZnSO_4$ (25 kg ha^{-1})	1.64	1.40	1.26
NPKS + Borax (5 kg ha^{-1})	1.45	1.37	1.24
NPKS + Sodium molybdate (1 kg ha^{-1})	1.37	1.45	1.22
C.D. 5%	0.25	0.31	0.20

Source: (IIPR, 2002–2003).

Table 4 Estimated nutrients required (kg ha^{-1}) to produce selected levels of grain yield of pigeonpea

Nutrients	Grain yield (kg ha^{-1})							
	500	1000	1500	2000	2500	3000	3500	4000
N	132	148	164	181	197	213	230	246
P	10	11	13	15	17	18	20	22
K	31	36	41	47	52	58	63	69
Ca	38	39	40	41	42	43	44	45
Mg	27	28	29	29	30	31	32	33
Mn	0.10	0.11	0.11	0.12	0.12	0.13	0.13	0.13
Zn	0.16	0.19	0.22	0.25	0.28	0.31	0.35	0.38
Cu	0.07	0.07	0.07	0.08	0.09	0.09	0.10	0.11

(Source: Calculation based on Dalai, 1980).

made to better understand nutrient requirements but these studies were mainly confined to growing crop in a single crop system. As 90% area of pigeonpea is under intercropping, massive efforts are needed to clarify nutrient requirements in a cropping system mode. The nutrient demand of pigeonpea is determined by its biomass production potential in the system, which in turn is determined by the various biotic and abiotic stresses affecting the crop. Nutrient supply to pigeonpea can be reduced by competition for uptake by companion crops or weeds (Johansen, 1990). In intercropping system, due to high plant population, nutrient requirements are generally more. Ahlawat et al. (1986) suggested that in cereal/pigeonpea intercropping systems, nitrogen to cereal should be applied in two splits, that is half as a basal dose and the remaining half as a top dressing after one month. The basal dose of nitrogen meets the initial demand of the cereal and legume. Afterwards, biologically fixed N is available to both the pigeonpea and intercrop components. When legumes are associated with cereal crop in an intercropping system, a portion of nitrogen requirement of cereal is supplemented by the legume. The amount may be as small as a few kilograms to 20 kg ha^{-1}. Cereal + legume intercropping is therefore mainly advantageous under low fertilizer application. Considering all the factors, it is suggested that the nitrogen dose recommended for base crop as pure crop is sufficient for intercropping system with cereals + legume or legume + legume intercropping system with cereals + legume, or legume + legume. With regard to phosphorus and potassium, one-eighth to one-fourth of the recommended dose of intercrop is also added in addition to the recommended dose of base crop to meet the extra demand.

Ahlawat and Shivakumar (2002) reported that application of the recommended dose of nutrients to all the crops in a system under irrigated conditions with narrow row proportion (1:1) for sorghum and pigeonpea recorded the highest productivity at Indore. Further, in a pigeonpea + legume intercropping system, a higher dose of phosphorus has been found to increase productivity. Soil application of the recommended dose of fertilizer, along with foliar application of 2% urea twice at 15 and 30 days after intercrop harvest, recorded a higher productivity and profitability from pigeonpea (Sharma et al., 2010). Pigeonpea + mungbean (1:2) intercropping systems under set furrow with vermicompost 2500 kg ha^{-1} showed significantly higher WUE (2.99 kg ha^{-1} mm) and pigeonpea equivalent yield of 2500 kg ha^{-1} (Sharma and Guled, 2012).

5 Water management

Water is essential to achieve the desired growth and yield of any crop. Pulses are generally grown under rainfed conditions. They are capable of extracting moisture from deeper layers of soil, however under prolonged moisture stress pulses respond to irrigation. Chauhan (1990) reported that in northern and central India, soil moisture levels are usually sufficient to meet pigeonpea requirements. There is often a need to remove excess water rather than to irrigate. Moisture availability in peninsular India is relatively poor as the length of crop growing season in many areas is restricted to 90–120 days. Consequently, the crop suffers mainly from terminal drought stress.

5.1 Moisture stress and its effect on pigeonpea

In semi-arid regions, pigeonpea is often subjected to terminal water deficit during flowering and pod-filling phases. Water stress is responsible for rapid leaf shedding, and reduction in grain size, weight and grain yield of crops (Becker et al., 1986). Water deficit may also affect the canopy development through its effect on total leaf number (Arkin et al., 1983). The leaf area decreases with the proportional increase in water stress. The levels of total chlorophyll has been found to be affected by a reduced supply of water, irrespective of the plant parts, specifically leaves, flowers and pods. Chlorophyll 'a' has been affected to a greater extent than chlorophyll 'b' levels.

Mukherjee and Kumar (2005) analyzed a pot culture study involving pigeonpea exposed to three moisture regimes. They found a sharp reduction in growth parameters and maximum leaf shedding under the severe water stress conditions. Monty and James (2011) observed that plant morphology, biomass production and root diameters were significantly affected by moisture regimes. They also reported that plants which were frequently supplemented with water were found to grow taller with a greater number of leaves, accumulate maximum dry weight, and have larger vessel diameters within pigeonpea root tissue.

5.2 Response of pigeonpea to irrigation

Pigeonpea responds well to limited irrigation under moisture stress conditions. Various approaches such as crop growth stage (Saxena and Yadav, 1975; Panwar, 1979; Dastane et al., 1971; Singh et al., 1979), IW/cumulative pan evaporation (CPE) ratio (Yadav, 1975) and cumulative evaporation (Panwar, 1979) have been used in irrigation schedules. Gajera and Ahlawat (2006) concluded that for the maximum production of pigeonpea, the irrigation could be scheduled based on a climatological approach by adopting an IW/CPE ratio of 0.30. Pramod (2007) reported that seed yield of pigeonpea (1651 kg ha^{-1}) with two protective irrigations was significantly higher than with one protective irrigation (1433 kg ha^{-1}) and control (1096 kg ha^{-1}) under Raichur conditions. Duraisamy and Manickasundaram (2008) reported that the perennial pigeonpea (BSR 1) with 0.45 IW/CPE irrigation schedule showed the highest WUE of 1.05 kg ha^{-1} mm^{-1} when compared to other treatments.

The consumptive use of kharif season pigeonpea was estimated to be 406 mm at Hissar, 416 mm at Sri Ganganagar and 491 mm at Pantnagar in India (Agarwal et al., 1997). Irrigation based on a meteorological approach showed that an IW/CPE ratio of 0.7 at Morena was optimum, however at Navsari there was no difference in seed yield between an IW/CPE ratio of 0.25 and 0.75 (Patel and Patel, 1995). Tomar et al. (2001) observed that scheduling irrigation at an IW/CPE ratio of 0.5 during branching, flowering and pod development stages, resulted in a significantly higher seed yield of pigeonpea when two irrigations at branching and pod development stages were used.

Pigeonpea productivity is enhanced through irrigation (Chauhan et al., 1987; Mula et al., 2010). The crop's irrigation requirement depends upon many factors including time of sowing, soil type, weather conditions and duration of variety. One or two protective irrigations should be applied when the soil moisture reaches the lowest availability. Furthermore, irrigation at critical stages such as seedling, flowering and pod formation should not be avoided. Moisture can be conserved during crop growth by bunding, basin listing, cris-cross ridging and chisel trenching. During an acute drought, irrigating the crop at flower initiation and pod filling stages increased the yield up to 303% (Bhan, 1976). Irrigation at flower initiation and pod filling stages, increased pigeonpea yield more significantly when compared to irrigations limited to either flower initiation, peak flowering or pod filling stages (Bhan and Khan, 1979). However, the response of irrigation varied with the seasonal rainfall and soil type.

Irrigation every 14 days in a row ratio of 4:1 resulted in a higher seed yield (Mula et al., 2010). In many areas, pigeonpea is grown under moisture stress conditions. Yields are therefore limited by the amount of water available to support growth. Supplemental irrigation with a limited amount of water, if applied to rainfed crops during critical stages, can result in substantial improvement in yield and water productivity (Gowda et al., 2013). A recent review by Sinclair and Vadez (2012) has quantified this relationship. Results have shown that doubling the available water from 150 to 300 mm will double the yield to 3520 kg ha^{-1}. This shows the great potential for enhancing the legume crop yields by providing irrigation.

Studies conducted at Ludhiana on a loamy sand soil, revealed that for crops sown in May, the first irrigation applied two weeks after sowing resulted in a significantly higher yield than when it was applied four weeks after sowing (Rana and Malhotra, 1992).

Studies were conducted at Tirupati (Andhra Pradesh) during the post-rainy season, on a sandy loam soil with SD cultivar ICPL 288. They showed that no irrigation under water stress conditions at 50% flowering (80 DAS), resulted in a significantly reduced flower formation and an increase in flower drop, when compared to non-stressed conditions. A decrease in pods, 100-seed weight, harvest index and seed yield was also observed under water stress conditions without irrigation (Rao and Narayanan, 1997). Application of three irrigations doubled pigeonpea's seed yields in Alfisols (Chauhan, 1990). Application of one or two protective irrigations during the critical stages for pigeonpea (the flowering and pod formation stages), produced higher yields (Rao et al., 1983; Saritha et al., 2012).

Pigeonpea + mungbean (1:2) intercropping system under set furrow with vermicompost 2.5 t ha^{-1} recorded significantly higher WUE (2.99 kg ha^{-1} mm) and pigeonpea equivalent yield of 2.5 t ha^{-1} (Sharma and Guled, 2012). Mula et al. (2010) observed that irrigation every 14 days from flower initiation to pod development stage of pigeonpea, was optimum for producing ample quantity seeds of ICPA 2043 in Alfisols at ICRISAT, Patancheru, India. Thanki and Solanki (2010) recorded the highest seed yield (1429 kg ha^{-1}) and stalk yield (3921 kg ha^{-1}) in pigeonpea by scheduling irrigation at an IW/CPE ratio of 0.9.

6 Efficient use of fertilizer and water

Water is becoming scarce due to erratic rainfall. There is a need for water conservation and improvement in WUE to achieve 'more crop per drop of water'. To increase the productivity of pulses, key inputs namely water, seeds and nutrients need to be applied in the right quantity to enhance the production. Since 85% of pulses are rainfed, it is necessary to develop technology to save water and nutrients, and to provide irrigation at the most critical stages of crop growth. Drip irrigation can be considered an efficient irrigation system as it only causes wetting of the soil, and helps to maintain an optimum moisture content in the root zone.

6.1 Micro irrigation in pigeonpea

Water is essential for achieving the desired growth and yield of any crop. A crop's water requirement is defined as the quantity of water needed for normal growth, development, and yield, and this may be supplied through precipitation, irrigation or via both methods. On average, pigeonpea uses about 200–250 mm water to produce approximately one tonne grain ha^{-1} (Saxena and Yadav, 1975).

Micro irrigation is just one of the many irrigation and water management technological tools, but it is a tool that has a number of advantages. It can reduce the water wastage to a negligible amount. Irrigation events can be fine-tuned to spoon-feed water and nutrients just in time to avoid plant stress. Micro irrigation can optimize crop production (maximize economic yield for each unit of water,) and in many cases increase the quality of agricultural products. It's benefits could include increased beneficial use of water, enhanced plant growth and yield, improved efficiency of fertilizer and other chemicals, reduced salinity hazard, limited weed growth, decreased energy requirements, and improved cultural practices.

6.2 Drip fertigation

Fertigation is a technology where water-soluble fertilizer or chemical can be applied in precise amounts in synchrony with the plant needs, directly into the root zone of the crop. This not only economizes the water use, but also improves nutrient use efficiency as the applied fertilizer remains confined to the root zone of the crop. Drip fertigation also enables accurate adjustment of water. The Indian government is giving subsidies of 75–100%, depending upon the social and economic status of farmers, to enhance pigeonpea production. Therefore, farmers are beginning to use this method more frequently.

Fertigation technology is also becoming more popular because of the advantages it has over other methods. It increases the fertilizer use efficiency by reducing nutrient loss due to volatilization, leaching and run-off. It reduces fertilizer requirement by increasing nutrient availability, and is seen to be a convenient and labour-saving method.

In recent years, drip irrigation has been routinely followed for pigeonpea crops. At IIPR, Kanpur, drip fertigation, at branching, or branching and pod development stages, (with NK 10:10 kg ha^{-1} in 5–10 kg splits) was found to be the most efficient (Praharaj and Kumar, 2011). Mahalakshmi et al. (2011) reported that pigeonpea under drip irrigation with 0.8 Epan throughout the crop period showed higher plant height (61.6 cm), leaf area index (LAI) (2.04) and grain yield (3731 kg ha^{-1}). The research work on drip fertigation carried out in Tamil Nadu, revealed that a lateral spacing of 150 × 60 cm was optimum for MD transplanted pigeonpea. A fertilizer dose of 25 kg N as urea, 25 kg K as muriate of potash ha^{-1} through fertigation, and 50 kg P as superphosphate through soil application, produced a higher grain yield (Latha et al., 2012).

6.3 Irrigation scheduling

For the drip fertigation trials conducted at Tamil Nadu Agricultural University (TNAU) Coimbatore on pigeonpea, first irrigation was given immediately after sowing and life irrigation was given on 3 DAS. Subsequent irrigations were scheduled once in seven days based on the following formula and applied each time as per the schedule. For drip fertigation system, the operational pressure was maintained at 1.0 kg cm^{-2}. The computed water requirement of crop (WRc) was calculated based on the formula

$$WRc = CPE \times Kp \times Kc \times Wp \times A$$

where WRc – Computed water requirement (lit $plant^{-1}$), CPE – Cumulative pan evaporation for seven days (mm), Kp – Pan factor, Kc – Crop factor, Wp – Wetting percentage (0.8) and A – Area per plant (150 × 60 cm).

The pan co efficient and crop co efficient are given in Table 5.

Time of operation of drip system to deliver the required volume of water per plot was computed based on the formula

$$\text{Time of operation} = \frac{\text{Volume of water required}}{\text{Emitter discharge} \times \text{number of emitters}}$$

Considering the nutrient uptake pattern of pigeonpea at phenological growth phases, the fertigation schedule with water-soluble fertilizers and conventional fertilizers was worked out, and is presented in Tables 6 and 7.

Table 5 Pan coefficient and crop coefficient for pigeonpea

Period	Value of Kp	Crop growth stage	Crop factor (Kc)
November to February	0.6	Initial stage	0.40
March to April	0.7	Mid stage	1.15
September to October	0.7	Final stage	0.35
May to August	0.8	Singandhupe et al. (2005)	
Ravikumar et al.(2011)			

Table 6 Fertigation schedule for pigeonpea at 100% RDF through water-soluble fertilizers

			Quantity of nutrients – fertilizer grade (kg ha^{-1})		
Stage	Duration (days)	Application time (DAS)	Urea	MAP	SOP
Seedling	1–30	10, 17 and 24	2.32	32.78	0
Vegetative	31–90	31, 38, 45, 52, 59, 66, 73, 80 and 87	9.87	24.58	12.50
Flowering	91–120	94, 101, 108 and 115	9.87	24.58	20.00
Pod development	121–140	122, 129 and 136	10.85	0	17.50
		Total	32.91	81.94	50.00

Source: (Vimalendran, 2013).

Table 7 Fertigation schedule for pigeonpea at 100% RDF through conventional fertilizers

			Quantity of nutrients – fertilizer grade (kg ha^{-1})	
Stage	Duration (days)	Application time (DAS)	Urea	MOP
Seedling	1–30	10, 17 and 24	10.85	0
Vegetative	31–90	31, 38, 45, 52, 59, 66, 73, 80 and 87	16.28	10.38
Flowering	91–120	94, 101, 108 and 115	16.28	16.60
Pod development	121–140	122, 129 and 136	10.85	14.53
		Total	54.30	41.50

Source: (Vimalendran, 2013).

6.4 Efficiency of fertilizers

Nitrogen applied through drip fertigation appeared to favour fertilizer efficiency when compared to solid surface application. Uptake of fertilizer nitrogen was improved by supplying it frequently through the drip irrigation system. Although nitrogen fertilizer use efficiency was nearly doubled by fertigating frequently once in a week, the improvement in fertilizer use efficiency was, however, not accompanied by a correspondingly significant increase in yield. The agronomic efficiency, partial factor productivity, water used and WUE are given in Table 8.

6.5 Economics of drip fertigation

One of the main objectives of fertigation is to reduce the cost of cultivation and to increase the economic product as high as possible. Therefore getting maximum benefits from each unit of water and nutrients applied to crops is important. A technically feasible level of fertigation with conventional and water-soluble fertilizer through a drip method would be economically viable for its successful adoption. Trials were conducted at TNAU, Coimbatore, Tamil Nadu analyzing drip irrigation at 100% crop water requirement, with fertigation at 125% recommended dose of fertilizer, (25:50:25 kg NPK ha^{-1}) through water-soluble fertilizer once every seven days. This resulted in a significantly higher grain yield and net return from pigeonpea crops (see Table 9). Though water-soluble fertilizers provided higher net income, the return per rupee invested was only 2.83. However, drip fertigation with conventional fertilizers once every seven days showed a slightly lesser net return and the return per rupee invested was 3.01 (Vimalendran, 2013).

Table 8 Effect of drip fertigation on agronomic efficiency, partial factor productivity, water used and water use efficiency

Treatments	Agronomic efficiency (kg kg^{-1} nutrient applied)	Partial factor productivity (kg kg^{-1} nutrient applied)	Water used (mm)	WUE (kg ha^{-1} mm^{-1})
Drip fertigation with water soluble fertilizers	15.9	25.7	427	6.05
Drip fertigation with conventional fertilizers	10.8	20.7	427	4.87
Surface irrigation with conventional fertilizers applied as basal	8.7	18.5	523	3.54

Source: (Vimalendran, 2013).

Table 9 Comparison of irrigation methods and source of fertilizers in terms of yield

Treatments	Grain yield (kg ha^{-1})	% yield increase over surface irrigation	Net income (Rs. ha^{-1})	B:C ratio
Drip fertigation with water-soluble fertilizers	2571	39	77839	2.83
Drip fertigation with conventional fertilizers	2069	12	64898	3.01
Surface irrigation with conventional fertilizers applied as basal	1851	–	48981	2.90

Source: (Vimalendran, 2013).

7 Weed management

Pigeonpea, on account of its low initial vigour and slow growth rate, often suffers heavily from weed infestation. Yield losses due to weeds can be as high as 90% (Saxena and Yadav, 1975). Manual weeding is common, but due to the amount of labour required and difficulties in weeding during the rainy season, herbicides are also used.

Among the pulses, pigeonpea is the only crop that has a characteristically slow initial growth rate. This makes pigeonpea less competitive with weeds. Infestation of weeds is one of the major constraints in achieving higher yield as they compete with crop plants for moisture, light, nutrients and space, and can thus considerably reduce the yields by 40–100%. Weeds caused 29% losses in grain yield of pigeonpea (Sekhon et al., 1993). The extent of yield lost depends upon a number of factors including the variety of weed species, quantum of seed flora, and duration of crop-weed competition. Ali (1992) reported yield losses of pigeonpea to the range of 40–64% in different zones of India due to weeds.

Traditionally, hoeing or interculture is taken up for controlling weeds. However, cultural methods of weed control are sometimes not feasible due to rains and unworkable soil conditions. Before 1970, one hoeing four weeks after sowing of the crop and another hoeing four weeks thereafter, or interculture with country *(desi)* plough, particularly in line-sown crop, was the common practice to control weeds. In another experiment, Patro and Tosh (1973) tried varitox (Na salt of trichloroacetic acid), propotox (Na salt of 4-(2-methyl-4-chlorophenoxy) butyric acid), and enbutox (Na salt of 2,4-dichlorophenoxy butyric acid) at a dose of 2.24 kg ai ha^{-1} as pre-emergence and post-emergence sprays. Varitox both as pre- and post-emergence was found to be an effective herbicide and recorded maximum grain yields. Following this, various herbicides were used in pigeonpea to control weeds (Table 10).

Sekhon et al. (1993) observed that intercropping of SD mungbean in pigeonpea suppressed weed growth by 23% over sole pigeonpea (un-weeded check,) and produced a higher grain yield. The herbicides used to control weeds in a crop may adversely affect germination and growth of the succeeding pigeonpea crop, as reported in the case of atrazine (Patil et al., 1981). So there is a need to select not only safe herbicides for use in pigeonpea crops, but also care should be taken to avoid any adverse effect of herbicides applied on a previous crop in the rotation.

At ICRISAT, excellent control of weeds in pigeonpea experimental plots is achieved by using a mixture of prometryn, fluchloralin and paraquat at 1.0 kg ai each in 350 L water ha^{-1} on Vertisols, and with 0.75 kg ai ha^{-1} each on Alfisols (Chauhan, 1990). Prometryn was quite effective in controlling broad-leaved weeds and fluchloralin in controlling grass weeds. Both these herbicides when used in combination showed significantly higher pigeonpea yield in treated plots than that from an un-weeded control, and from plots where either herbicide was used alone. A hand weeding at 30–40 DAS was still required, even after the use of herbicides to keep fields entirely weed free. The use of paraquat on all fields was found necessary to control weeds that had already started germinating at the time of sowing pigeonpea, even on freshly prepared land. There was no toxicity apparent in the crop even when paraquat was applied at 5 DAS. Irrigation or rain after sowing enhanced the effect of these herbicides under the cropping systems used at ICRISAT. Moreover, these chemicals did not affect subsequent crops since prometryn loses its effectiveness after about ten weeks, and fluchloralin after about 16 weeks.

Table 10 Herbicides used in sole and intercropped pigeonpea

Herbicide	Rate (kg ha^{-1})	Sole/intercrop	Reference
Alachlor**	2	Pigeonpea + urdbean	Ali (1992)
Dinitramine**	1 and 1.5	Pigeonpea + soybean	
Nitrofen**	1.25 and 1.75	Sole pigeonpea	Kolar et al. (1985)
Oxadiazon**	0.5 and 0.75		
Terbutryn**	0.5 and 1.00		
Alachlor**	I and 3	Sole pigeonpea	Bhalla et al. (1988)
Nitrofen**	I and 2		
Fluchloralin*	I and 2		
Prometryne** + Paraquat***	0.60 + 0.50		
Oxyfluorfen	0.144	Sole pigeonpea	Sinha et al. (1989)
Fluchloralin*	1.5	Pigeonpea + soybean Pigeonpea + groundnut	Tiwari et al. (1992)
Alachlor**	1.0	Sole pigeonpea	Maheshwarappa and Nanjappa (1994)
Metolachlor**	1.0		
Fluchloralin**	1.0		
Oxyfluorfen**	0.075		
Pendimethalin**	0.5		
Fluchloralin*	1.0	Pigeonpea + mungbean Pigeonpea + cowpea	Rana and Paul (1994)
Imazethapyr***	0.60	Sole pigeonpea Pigeonpea + urdbean/mungbean/groundnut	CPG (2012)
Quizalofop-ethyl***	0.50	Sole pigeonpea Pigeonpea + urdbean/mungbean/groundnut	

*Pre-sowing; **Pre-emergence; ***Post-emergence.

Use of early post-emergence herbicides has been tried in AICRP trials. The work on post-emergence herbicides initiated during 2008 at IIPR, Kanpur under AICRP showed that some of the post-emergence herbicides like quizalofop-ethyl and imazethapyr have great potential for tackling of weeds in pulse crops (Kumar, 2010).

To reduce labour requirements, weeds were shown to be effectively controlled by the following herbicide schedule: Pendimethalin 0.75 kg ai ha^{-1} on 3 DAS, followed by an early post-emergence application of imazethapyr 60 g ai ha^{-1} on 15 days after emergence (DAE) of weeds, (2–3 leaves stage of weeds), and quizalofop-ethyl 50 g ai ha^{-1} on 20 DAE of weeds (2–3 leaves stage of weeds). Imazethapyr was effective in controlling broad-leaved weeds, and quizalofop-ethyl in controlling grass weeds. If both the weeds are present, tank mix application of imazethapyr (60 g ai ha^{-1}) and quizalofop-ethyl (50 g ai ha^{-1}) at 15–20 days after emergence of weeds (2–3 leaves stage of weeds) can control weeds effectively. Irrigation or rainfall after application of herbicides increases its efficacy (CPG, 2012). Pre-emergence herbicide pendimethalin (0.75 kg ai ha^{-1}), + early post emergence application of imazethapyr (100 g ai ha^{-1}) on 10–15 DAS (2–3 leaves stage of weed), + one-hand weeding at 50 DAS had relatively more weed control and increased the pigeonpea yield.

An integrated weed-management approach produced better results than a solitary herbicide application. Most of the herbicides used in pigeonpea are applied at pre-sowing or pre-emergence, and thus their effect on controlling weeds does not last up to crop maturity. These herbicides can keep weeds under check for the early part of the crop season (up to 30–40 DAS). However, new weed species may emerge at later stage. Several studies showed that two-hand weedings gave higher yield than herbicides alone. Some herbicides are very costly and have an adverse on crops at a higher dose. So herbicide use at lower rates of application, along with other mechanical and cultural methods of weed control, is perhaps the best way for reducing weeds and thereby achieving higher crop yields. Supplementing herbicides with hand weeding/interculture gave a higher pigeonpea yield when compared to only applying herbicides (Tiwari et al., 1992; Maheshwarappa and Nanjappa, 1994). Recent studies conducted at Ludhiana showed that the pre-emergence application of 0.45 kg ha^{-1} pendimethalin and ridging 50 DAS, gave significantly higher yields over the sole application of pendimethalin (PAU, 2002).

8 Agronomic management of pigeonpea hybrids

The hybrids in pigeonpea, (the first leguminous hybrids in the world,) are the result of the crop improvement efforts by ICRISAT based in Hyderabad. The first pigeonpea hybrid ICPH 8 was released during 1991 by ICRISAT, in collaboration with AICPIP of ICAR, and is recommended for cultivation in the central zone comprised of Maharashtra, Madhya Pradesh and Gujarat (India). The hybrid matures in 115–135 days giving a yield advantage of 41–52% over other varieties in cultivation (Srivastava et al., 1994). The other genetic male sterility-based hybrids developed are PPH 4 (PAU, Ludhiana), COPH1 (TNAU, Coimbatore), AKPH 4101 and AKPH 2012 (PDKV, Akola) (DPD, 2016).

The cytoplasmic genetic male sterility-based hybrid GTH-1 developed by Sardarkrushinagar Dantiwada Agricultural University, Sardarkrushinagar, Gujarat (India), has recently been released for cultivation. The CMS-based pigeonpea hybrids namely ICPH 2671 and ICPH 2740 have been developed from ICRISAT, Hyderabad. It is believed that hybrids have more canopy than traditional varieties, and consequently less per-hectare seed rate, compensating for the higher cost of hybrid seed. The hybrids have tremendous scope for popularization in northern India, and in other areas where wheat crop can be taken after harvest of hybrid pigeonpea. Similarly, in central and southern India, early and MD pigeonpea hybrids could play an important role in populating the additional area by way of replacement of traditional poor performers. With the adoption of hybrids, the cost of cultivation would also be affordable for many farmers. In over 2000 on-farm trials conducted in five Indian states, the hybrids ICPH 2671 and ICPH 2740, exhibited a 47% and a 42% yield advantage respectively, compared with the best local variety. Seed production of hybrids, mediated by honeybees, is straightforward.

Under congenial growing conditions, 700–1200 kg ha^{-1}of hybrid seed can be produced (DPD, 2016). However, efforts are required for the development of commercially viable pigeonpea hybrids for their promotion in India (Chaturvedi and Sandhu, 2016).

A field experiment was conducted during kharif season in 2009, at the Main Agricultural Research Station, University of Agricultural Sciences, Dharwad, Karnataka state (India). The study analysed the response of hybrid pigeonpea to planting geometry and fertility levels. Nine treatment combinations were tested in hybrid pigeonpea ICPH-2671 comprising three planting geometry (60 × 30 cm, 90 × 30 cm and 90 × 45 cm), and three fertility levels (25:50 N:P kg ha^{-1}, 37.5:75 N:P kg ha^{-1} and 50:100 N:P kg ha^{-1}), along with pigeonpea variety Maruti (ICP-8863) as a control with the recommended package of practices. The results revealed that the hybrid pigeonpea ICPH-2671 showed a significantly higher grain yield of 2404 kg ha^{-1}, and the magnitude of yield increase was 41.7% higher than that of variety Maruti (1679 kg ha^{-1}). The growth, yield and physiological parameters were higher with hybrid pigeonpea ICPH-2671 compared to variety Maruti. Significant difference among the plant geometry and fertility-level treatment combination with respect to growth and yield parameters was observed. Plant spacing of 90 × 45 cm along with fertilizer dose of 50:100 N:P kg ha^{-1} provided significantly higher grain yield (2565 kg ha^{-1}), which was 42.8% higher than that of the control (1679 kg ha^{-1}) with higher growth and yield parameters. The study indicated that hybrid pigeonpea ICPH-2671 has a better response to wider geometry and higher dosage of fertilizer in terms of improving its yield (Meena, 2010).

Over 70 smallholder farmers and seed producers from Andhra Pradesh, Odisha, Maharashtra, Madhya Pradesh, Rajasthan and Gujarat (India), collectively endorsed and recognized the contribution of hybrid pigeonpea in achieving food and nutrition security, resilience, and improved livelihood. Despite being faced with challenges such as resource scarcity, climate change and degradation of soil fertility, farmers from the above states cultivating the ICRISAT hybrids reported significant increase in pigeonpea yields. An MD ICRISAT hybrid, ICPH 2740, demonstrated 38% higher yields than other local varieties. Bringing new hope to farmers, the scientists at Acharya N. G. Ranga Agricultural University's (ANGRAU) Agricultural Research Station (ARS) in Andhra Pradesh (India) reported a pigeonpea yield of nearly 3000 kg ha^{-1} in their on-station fields using the ICPH 2740 hybrid (Sushil Rao, 2014).

Fifteen pigeonpea hybrids were introduced from ICRISAT, India, to Myanmar and were evaluated at three locations. Hybrids ICPH 2671, ICPH 2673, ICPH 2740 and ICPH 3497 were found to be stable over the three environments and produced 1846–1967 kg ha^{-1} yield with 30.4–41.7% standard heterosis. ICPH 3461 (2217 kg ha^{-1}) was found to be suitable for Zaloke with 42.0% standard heterosis. In 36 on-farm trials, hybrid ICPH 2671 was 11.9–53.1% superior in yield over the control. The other promising hybrid ICPH 2740 (2247 kg ha^{-1}) also exhibited 70.0% standard heterosis in an on-farm trial conducted in Monywa township. Based on farmers' preference and yield, ICPH 2671 and ICPH 2740 were selected for further promotion. The large-scale hybrid seed production programme was also successful with 1596–2931 kg ha^{-1} hybrid seed yield. The results of current investigations showed that hybrids have potential for enhancing production and productivity of pigeonpea in Myanmar (Kyu et al., 2011).

In Myanmar, during 2013, ICPH 2433 (2507 kg ha^{-1}) showed 62.41% yield superiority over the control, Yezin-3. In 2014, ICPA 2039 × Line-25 (2239 kg ha^{-1}) exhibited a 57% higher yield than ICPL 88039. These findings showed the potential of hybrids in enhancing the productivity of pigeonpea in Myanmar (Kyu et al., 2016).

9 Early-maturing pigeonpea varieties

In the last decade, pigeonpea cultivation has grown significantly in the states of Andhra Pradesh, Gujarat, Haryana, Karnataka and Maharashtra (India). This is mainly due to the introduction of SD genotypes in new cropping systems. Cultivation of pigeonpea in the NWPZ (Haryana, Punjab, Delhi and Uttar Pradesh in India) began only with the development of SD cultivars. At present, all genotypes grown in this region are SD. ESD genotypes of pigeonpea, maturing in 110–130 days, have a yield potential of 1500–2000 kg ha^{-1} and possess resistance to the major diseases affecting this crop. This may generate an estimated 0.8 million extra hectares of pigeonpea in the irrigated wheat belt of the NWPZ, and another 0.2 million hectares in the rainfed coastal peninsular region with bimodal rainfall.

SD pigeonpea varieties have helped establish a new pigeonpea-wheat cropping system in the northwestern plains of India. However, because wheat sowing is often delayed in this rotation, ESD pigeonpea (ESDP) that will mature 10–15 days earlier than SD genotypes are being developed. Some of the important SD and ESD varieties of pigeonpea released in India are discussed below.

9.1 Short-duration varieties

Pusa Ageti (150–160 days, determinate), T 21(150–170 days, indeterminate), HY 2 (140–150 days), Pusa 84 (140–150 days), AL 15 (135 days), AL 201(140 days), CO 1 (135–140 days, photo-insensitive) and Paras (H 82–1) (140–145 days). Currently, ICRISAT SD varieties namely ICPL 88039, ICPL 87, UPAS 120, ICPL 98010, ICPL 85010 (145 days, determinate), ICPL 88039 (150 days, indeterminate), ICPL 84031 (determinate), and ICPL 98015 have been replaced in the drought-prone area of Magway and Mandalay regions of Myanmar (Kyu et al., 2016).

9.2 Extra-short-duration varieties

UPAS-120 (120–140 days, indeterminate), Pant A3 (120–130 days, determinate), Prabhat (110–120 days, determinate), ICPL 87 or Pragati (120–130 days, determinate), Manak (122 days), ICPL 151 or Jagriti (120–130 days, determinate), APK 1 (95–100 days, determinate), CORG 9701 (120–130 days), VBN 3 (120–135 days), Pusa Arhar 16 (120 days, determinate, synchronous maturity also suitable for combine harvesting) and Durga (115–120 days, determinate) are some of the ESDP varieties.

These super-early lines are expected to be important for cropping systems in Asia, Africa and the Americas. This new maturity class will open additional niches for pigeonpea such as wheat-based cropping systems, rice (*Oryza sativa*) fallows, high hills and short growing environments affected by terminal drought, or frost. This may allow for sustainable intensification of various farming systems worldwide.

9.3 Climate and soil

ESDP grow well in warm tropical and subtropical climates. They prefer warm weather (maximum 30–40°C) during germination and pre-flowering growth, and a lower temperature (maximum 25–30°C) during the flowering and pod filling stages. Waterlogging is very harmful which means the fields should be well drained to allow better aeration. ESDP thrive best on well-drained loamy soils. Saline/alkaline, highly acidic (pH<5) and waterlogged soils are unfit for their cultivation (Johansen, 1990).

ESDP require field preparation similar to SD and LD pigeonpea.

9.4 Seed rate and seed treatment

The seed rate for early pigeonpea is 12–15 kg ha^{-1}. Seed treatment with thiram (3 g kg^{-1} of seed) is helpful to protect against dry root rot and collar rot (Reddy et al., 1990). Seed treatment with *Rhizobium* culture is beneficial for nodulation and N fixation.

9.5 Time of sowing

Sowing dates were found to be significant for obtaining good grain yield of pigeonpea. This is the most important non-monetary agronomic input for realizing high yield from an improved cultivar. For SD pigeonpea genotypes maturing in 140–160 days, sowing at the onset of rainy season, which is in the first or second week of July in northwest India, is not the best way to obtain high yields (Ahlawat et al., 1975; Kaul et al., 1980; Rathi and Tripathi, 1978; Singh and Kalra, 1980; Sandhu et al., 1981). Sowing in the first fortnight of June with irrigation or pre-monsoon rains generally results in higher yields, while delayed sowings cause gradual reductions in grain yield. The reason for this is the decline in biomass production due to early flowering and slow growth (Dahiya et al., 1974). The optimum time of sowing ESDP in the western Gangetic Plain (India), is from the last week of May to mid-June. T 21 was the best variety for early sowing (March-May). UPAS 120 and Manak were found to be suitable for sowing from the end of May to June. These sowings produced significantly higher grain yields than crops sown in the first week of July. Mid-July sowing further reduced yields, which were 40–50 kg day^{-1} ha^{-1} lower than the yields of a crop sown on 15 June (Tomer, and Ram Dhari, 1996). Sowing of extra-early and early-maturing varieties in the first fortnight of June made the field available for post-rainy season crops by the end of November. Therefore, sowing should not be delayed beyond June. In Tamil Nadu (India), since the duration of CORG 9701 and VBN 3 is 120–130 days, they have to be sown in such a way that flowering/harvesting does not coincide with monsoon rains.

9.6 Method of sowing

ESDP seed should be sown using a plough or a seed drill. Line sowing or a drilling-based method with tractor-drawn ferti-cum-seed drill or by bullock-drawn ferti-cum-seed drill is recommended. Seeds can be sown via a dibbling method using a hand dibbler to maintain optimum plant stand and higher productivity. Ferti-cum-seed drills should be used for the application of fertilizers where seed sowing occurs at the same time. Seeds should not be sown at a depth of more than 5–6 cm. Sowing through broadcasting, which is quite popular as it saves labour and time, leads to uneven plant stand (Dahiya et al., 2001).

9.7 Plant population density

Plant population density is an important factor which can affect the yield of SD pigeonpea. This variety is more responsive to close sowing as the small plants have a shorter growth duration. Furthermore, SD genotypes grow under relatively assured soil moisture conditions. Plant population requirements for SD genotypes vary with location, time of sowing, and genotype. Requirements depend mainly on the extent of vegetative growth. The interactions between plant population and sowing date, and plant population and genotype, are often significant (Ahlawat et al., 1985). In Australian conditions, a population density of 40–50 plants m^{-2} is required for higher yields from photo-insensitive cultivars (Wallis et al., 1981). In peninsular India, the optimum plant population for SD genotypes is about 30 plants m^{-2} and this can be obtained by maintaining a row-to-row spacing of 30 cm and plant-to-plant spacing of 10 cm (Chauhan et al., 1987b).

9.8 Nutrient management

Farmers are advised to add fertilizer on the basis of a soil test value. In the absence of this information, a fertilizer dose of 15–20 kg N and 40–60 kg P ha^{-1} ensures a good crop. Conveniently, 100 kg DAP is applied. Use of PSB culture improves the available P status in the soil. In addition, application of 20 kg S ha^{-1}, 15 kg $ZnSO_4$ ha^{-1}, 10 kg borax ha^{-1}, 1 kg sodium molybdate ha^{-1} and 1 kg Fe ha^{-1} is also recommended. In rainfed and dry areas, pigeonpea crops are sown at the onset of the monsoon. Contact between fertilizer and seeds should be avoided, as germinating pigeonpea seeds are sensitive to salt toxicity (Dahiya et al., 2001).

9.9 Intercropping and rotations

During the kharif (rainy) season, SD varieties of cowpea, mungbean or urdbean, and maize can be successfully grown as an intercrop in normal stands of SD and ESD pigeonpea varieties. This intercropping system can occur without reducing the pigeonpea's grain yield. For this purpose, one row of an intercrop between two rows of pigeonpea 50 cm apart can be accommodated. In spring/summer (March to mid-April), two rows of cowpea, mungbean or urdbean can be raised between two rows of pigeonpea sown 75 cm apart. This system yielded 700–1000 kg ha^{-1} grain of mungbean or urdbean, or 2000–2500 kg ha^{-1} of green fodder for cattle during the dry season when fodder was scarce. In this system, intercropped pigeonpea was found to be more remunerative than sole-cropped pigeonpea. However, the whole package of practices for intercrops should be followed for this system to be successful (Tomer and Ram Dhari, 1996).

Field observations have revealed that intercropping of maize with pigeonpea (genotype ICPL 88039) has shown promise in Sonepat (Haryana, India). For this purpose, one row of maize can be accommodated between the two rows of pigeonpea spaced 60 cm apart. The seed rate of component crops for intercropping should remain the same as for the pure crop. ESDP fits well for rotation with wheat, late-sown potato and chickpea (Dahiya et al., 2001).

9.10 Weed management

Short-season crops offer the advantage of high plant density even when grown alone. It has been found that the ability of pigeonpea to suppress weeds improved when the plant population was increased from 1 to 10 plants m^{-2} (Rao and Shetty, 1976). The timing of weed control also plays an important role. For SD pigeonpea, the first 30 days appear critical, although this period may vary with genotype and time of sowing (Diaz-Rivera et al., 1985). In the early stages of crop growth, weeds cause the maximum yield loss.

Hand weeding once or twice was found to be superior to chemical weed control. Of the herbicides tested, pendimethalin (1.00 kg ai/ha) (pre-emergence) controlled weeds most effectively, resulting in higher pigeonpea grain yields than other herbicides (Tomer and Ram Dhari, 1996).

9.11 Water management

Pigeonpea is a rainfed crop grown in areas of reliable rainfall. This means it doesn't usually require any irrigation. If there is water stress, protective irrigation may be given in alternate rows during budding, flowering, and pod formation stages. Pigeonpea requires 35–40 cm water, during its entire growth period. Optimum moisture is necessary during these three stages. Intensively managed pigeonpea systems that involve SD pigeonpea have a higher water requirement because they are grown at higher densities (Singh et al., 1983). The crop sown in May needs a pre-sowing irrigation. There is no need for further irrigation after the start of monsoon rains, but irrigation may be quite beneficial in cases of drought during the reproductive stage. However, if rainfall is well distributed, a mid-June-sown crop may not require any irrigation (Dahiya et al., 2001).

9.12 Harvesting

Early varieties are harvested in the middle of November. The plants are cut close to the ground and bundled. Bundles are then taken to a threshing floor and staked upright. Following a few days of drying, they are shaken vigorously to separate the pods. The resulting materials, (pods and dried leaves), are beaten with sticks or trampled by bullocks.

Short-statured ESD pigeonpea lines such as ICPL 83006, ICPL 83015, ICPL 84023 and ICPL 85010 have created the opportunity to make pigeonpea amenable to commercial, mechanized field production. Because the newly developed ESD lines are similar in height to wheat, there is no problem in adopting mechanical interculture and insecticide spraying operations using equipment designed for cereal crops. However, because of their thick and woody stem, the feasibility of mechanical harvesting and threshing of pigeonpea by a combine harvester was considered doubtful especially at higher latitudes, where the growth is usually more than that at lower latitudes. Despite this concern, Wallis et al. (1981) have successfully demonstrated the mechanized harvesting of SD pigeonpeas in Australia. They used leaf defoliants before harvesting using a combine harvester. A study conducted at Hissar in northern India (29°N latitude) indicated the feasibility of using wheat combine harvesting for three ESD pigeonpea lines at higher latitudes (Gupta, 1990). This offers the wheat-growing farmers the option of using their wheat combines for harvesting pigeonpea. This practice can save both time and labour, which in October/November prior to sowing wheat are at premium demand. Consequently, farmers usually harvest the pigeonpea plants and store them for threshing later when time permits. The prolonged storage of this harvested material in the open usually results in the loss of grain quantity and quality because of high humidity.

9.13 Yield

The average yield of early pigeonpea is 1200–1500 kg ha^{-1}. The highest recorded yield from ESD pigeonpea (5000 kg ha^{-1}) was from the experimental plots in Tifton, Georgia, USA, at 34°N (Gupta et al., 1991b). In India, the maximum recorded yields in some small-plot experiments are 4000 kg ha^{-1} at Hissar (29°N) (ICRISAT, 1989) and 2900 kg ha^{-1} at Patancheru (17°N) (ICRISAT, 1990). The greater yield potential in northern India is largely related to the profuse growth of ESD genotypes. The generally realized yield of the ESD genotypes in experimental plots in both northern and peninsular India is 1000–2000 kg ha^{-1}. At appropriate plant populations, the yield potential and the generally realized yield of ESD pigeonpea compare well with those of SD cultivar UPAS 120 in these environments (Chauhan et al., 1996).

10 Rabi and summer pigeonpea cultivation

10.1 Rabi pigeonpea

Cultivation of pigeonpea during rabi season is practised in flood-prone areas where fields get flooded or waterlogged during the rainy season. The states of Uttar Pradesh (eastern parts), Bihar, West Bengal, Odisha, Gujarat and Madhya Pradesh (India) may exploit this potential with the following practices/recommendations for successful cultivation of rabi pigeonpea:

- Sowing must be carried out in the second or third week of September. The crop can also be taken after harvest of early maize or paddy.
- Sowing at closer spacing (30 × 20 cm), having up to 200 000 plants ha^{-1}.
- High seed rate of 40–50 kg ha^{-1} should be followed.
- Sowing depth should not exceed 5 cm. The seeds should be treated with bacterial culture.
- Apply 30 kg N and 50 kg P (20 kg N and 50 kg P as basal and 10 kg N as top dressing at 30 DAS). Also apply 20 kg $ZnSO_4$ and 10 kg S if previous crop is not supplied with Zn and S. Fertilizer application is based on soil testing.
- The crop should be irrigated three times, once each at branching (30 DAS), pre-flowering (70 DAS) and pod filling (110 DAS) stages.
- Heptachlor 6% (25 kg ha^{-1}) should be mixed with soil at the time of the last tillage operation before sowing.
- Lasso/Tok E-25 should be applied at the rate of 1 kg ha^{-1} soon after sowing to ward off weeds.
- Spraying of malathion 0.05% or carbaryl 0.1% at the pod formation stage controls pod borers (DPD, 2016).

10.2 Summer pigeonpea

Summer-sowing of pigeonpea, alongside summer mungbean and carefully scheduled wheat crops, can provide an alternative way of increasing cropping intensity. This method was studied in northern India over an area of approximately 200 000 hectares. Under this situation, advanced sowing of pigeonpea may be done during mid-April, keeping row-to-row spacing of 90 cm, inter-cropping with three rows of green gram at 20 cm row spacing. Mungbean becomes ready for harvest by the end of June after two pickings. Immediately, in the space vacated by mungbean, interplanting of urdbean can be done between pigeonpea rows. While mungbean will be ready for harvest by the end of September, pigeonpea reaches maturity to be harvested by mid-November. Early harvesting of pigeonpea therefore facilitates wheat sowing at optimum times to harness the best yield.

Consequently, summer-sown pigeonpea may be harvested along with other kharif crops in November, and short-statured crops of mungbean and urdbean will be an additional source of income (DPD, 2016).

11 Harvesting, threshing and storage

11.1 Harvesting

Pigeonpea plants are ready for harvest when 80% of the pods have turned brown. Delayed harvesting, during bad weather, may increase the risk of damage to mature seeds (ICRISAT, 1992). For single harvest, cutting plants at ground level is a common practice, but where ratoon harvests are intended, the method of harvesting greatly influences subsequent flush yields. After harvesting, the plants are bundled and transported to the threshing floor. These are staked in upright bundles and allowed to dry in the sun. Another way to harvest pigeonpea is by hand picking the mature pods. This allows the crop to flower and pod for a second harvest. Hand picking may not be economical beyond a second flush. When hand picking of pods is not feasible, the upper branches with mature pods are cut (a good method for determinate types) (ICRISAT, 1992). This method delays the second harvest and usually results in a lower yield than hand picking. Traditional pigeonpeas are tall and bushy, which means their mechanized harvest is difficult or impossible.

11.2 Threshing

After complete sun drying, the crop is threshed by beating with wooden sticks or with threshers. The seeds are cleaned and dried in the sun to 6–8% moisture content before recording the yield.

11.3 Storage

The seeds are stored in gunny or cloth bags for short-term storage (8–9 months) with a seed moisture content of 8–9%. The seeds are stored in polylined gunny bag for medium-term storage (12–15 months) with a seed moisture content of 8–9%. The seeds in 700-gauge polythene bag for long-term storage (more than 15 months) have a seed moisture content of less than 8%.

12 Conclusion

The productivity of pigeonpea in India is low, so to make pigeonpea production internationally competitive, the average yield levels need to be increased to at least 1000 kg ha^{-1}. Because of the high level of fluctuation in pigeonpea production, farmers are not interested in taking up pigeonpea cultivation, despite high wholesale pigeonpea prices in recent years. To meet the demands of its growing population, India is importing pulses in the range of 3.0–4.0 million tonnes every year. This strong upward trend in the import of pulses is a cause for concern since an increase in demand from India has shown to have a cascading effect on international prices, thus draining the precious foreign exchange. To feed the estimated population, India needs to produce the required quantity of pigeonpea, but also remain competitive to protect indigenous pigeonpea production. Improved production technologies including proper land management, quality seeds with seed treatment, timely sowing, intercropping, proper nutrient management, water management and weed management have the potential to double the yield of pigeonpea. Bio-physical constraints, lincluding a lack of water and unfavourable temperatures, are the main factors to be alleviated to significantly increase the yield in low productivity areas. This may be achievable through better land management, or by matching technologies to withstand the constraints without much reduction in yield. Even a small jump of 100 kg ha^{-1} from one million hectares of low-productivity area of pigeonpea would have a tremendous effect on pulse supply in the country. However, a concerted effort by farmers, researchers, development agencies and the government are needed to ensure self-sufficiency in pigeonpea in the next decade.

13 Guidelines for future research

Listed below are various suggestions for future areas of research:

- Identification of high-yielding drought-tolerant varieties suitable for different agro climatic conditions under sole and intercropped situations to enhance production and productivity. In addition to developing varieties that are suitable for rainfed cropping, there is a need for developing varieties responsive to high input conditions. This will require substantial restructuring of plant types.
- Even though several genotypes have been developed, bringing more area under pigeonpea by introduction of extra short duration, photo-insensitive, high-yielding, and excess moisture-tolerant genotypes in rice fallows is urgently needed.
- There is a need to develop genotypes which can grow well at mean temperatures below 20°C, because growth of post-rainy season pigeonpea is limited by low temperatures. There is a need to develop varieties amenable to mechanical harvesting.
- Developing genotypes tolerant to herbicides, drought and heat stresses.
- Development of designer seeds which include coating of seeds with seed hardening chemicals, pesticides, bio-control agents and bio-fertilizers and making them available in the market will benefit farmers.
- Optimum time of sowing for SD, ESD and MD varieties in the changing climate needs to be identified for different pigeonpea growing areas to get higher productivity.
- Crop-modelling studies need to be strengthened to predict yield under varying environmental conditions which will in turn facilitate the determination of optimum time of sowing and population to enhance yields.
- Altering crop geometry to suit mechanization.
- Though transplanting pigeonpea is adopted in some areas as a risk management technology, its suitability under different agro climatic conditions and its cost effectiveness have to be thoroughly studied.
- In intercropped situations, the row ratios have to be optimised in different zones in order to achieve higher productivity and land use efficiency.
- Introduction of long duration perennial pigeonpea in the agroforestry/alley cropping system if suitable genotypes and agro techniques are developed.
- Quantity of N fixed and made available by varying maturity groups of pigeonpea *viz.*, extra short, short, medium and long duration in various systems of cultivation such

as sole, inter, and mixed cropping, need to be assessed as it is one of the important characteristics of legumes.

- The fertilizer recommendations for different duration groups (ESD, SD and MD varieties) of pigeonpea under different fertility levels and cropping system have to be studied in order to higher productivity.
- P efficient pigeonpea genotypes that will produce more yield per unit of P applied should be used. Soil test based fertilizer recommendation has to be encouraged in order to have balanced nutrition.
- Possibilities have to be explored for providing supplemental irrigation during critical stages for enhancing productivity.
- Weeds are one of the major yield limiting factors. Newer molecules of pre- and early post-emergence herbicides have to be developed in order to have broad spectrum weed control.
- Agronomic management of pigeonpea hybrids, ESD, SD, rabi and summer pigeonpea need to be standardized in order to achieve enhanced yields.
- Mechanizing pigeonpea cultivation will be a boon to farmers in the context of labour shortage. Since intercultural operations, harvesting and threshing are laborious, suitable machineries have to be designed to overcome the drudgery.
- Information on ratooning of pigeonpea is limited. Wide-ranging studies are needed to explore the feasibility and economic viability of this system in different pigeonpea growing regions.
- Reducing storage losses and improving market information and infrastructures.

14 Where to look for further information

The important organizations connected with this topic are given below:

- International Crops Research Institute for the Semi-Arid Tropics (ICRISAT) is an international organization headquartered in Patancheru (Hyderabad, Telangana, India) with several regional centres (Bamako (Mali), Nairobi (Kenya)) and research stations (Niamey (Niger), Kano (Nigeria), Lilongwe (Malawi), Addis Ababa (Ethiopia), Bulawayo (Zimbabwe)). Pigeonpea is one of the mandate crops. For more information, see: (www.icrisat.org).
- CGIAR (Consultative Group for International Agricultural Research) is a global partnership that unites organizations engaged in research for a food-secure future. CGIAR research is carried out by 15 centers, that are members of the CGIAR Consortium, in close collaboration with hundreds of partners, including national and regional research institutes, civil society organizations, academia, development organizations and the private sector.
- CGIAR Research Program on Grain Legumes: This program was led by ICRISAT in partnership with Centro Internacional de Agricultura Tropical (CIAT), International Center for Agricultural Research in Dry Areas (ICARDA), International Institute of Tropical Agriculture (IITA) and public and private institutes and organizations, governments and farmers worldwide. The eight major food legumes for focus in this research program are chickpea, common bean, cowpea, faba bean, groundnut, lentil, pigeonpea and soybean. For more information, see website: Grain Legumes Flyer:PDF

- Indian Institute of Pulses Research (IIPR) is a national centre for basic and applied research on pulse crops. The mandate crops are chickpea, pigeonpea, urdbean, mungbean, lentil, fieldpea, rajmash and lathyrus. The All India Research Project on Pigeonpea (AICRPP) with its headquarters at Indian Institute of Pulses Research, Kanpur was created by ICAR through trifurcation of All India Co-ordinated Pulses Improvement Project in 1993. The project has a network of nine main centres and thirteen sub-centres and ten voluntary centres on pigeonpea. For more information, see: www.iipr.res.in.

15 References

Acharya, C. L. and Biswas, A. K. (2002), Integrated nutrient management in pulse based cropping systems: Present status and future prospects, pp. 1–12. In *Proceedings of National Symposium on Pulses for Sustainable Agriculture and Nutritional Security* (M. Ali, S. K. Chaturvedi and S. N. Gurha (Eds)), 17–19 April 2001. Indian Agricultural Research Institute, New Delhi, Indian Society of Pulses Research and Development, Kanpur, India.

Adu-Gyamfi, J. J. (1999), 15 years of Progress, pp. 5–6. In *Government of Japan Project at ICRISAT 1984–99*. International Crops Research Institute for the Semi-Arid Tropics (1CRISAT), Patancheru, India.

Adu-Gyamfi, J. J., Fujita, K. and Ogata, S. (1989), Phosphorus absorption and utilization efficiency of pigeonpea (*Cajanus cajan* (L.) Millsp.) in relation to dry matter production and dinitrogen fixation. *Plant Soil*, 119: 315–24.

Adu-Gyamfi, J. J., Fujita, K. and Ogata, S. (1990), Phosphorus fractions in relation to growth in pigeonpea (*Cajanus cajan* (L.) Millsp.) at various levels of P supply. *Soil Sci. Plant Nutr.*, 36: 531–43.

Ae, N., Arihara, J., Okada, K., Yoshihara, T. and Johansen, C. (1990), Phosphorus uptake by pigeonpea and its role in cropping systems of the Indian subcontinent. *Science (Washington)*, 4954(248): 477–80.

Afridi, M. M. R. K., Inam, A., Samiullah. and Yahiya, M. (1990), Morphological studies on mustard and pigeonpea with special reference to K nutrition under irrigated and un-irrigated conditions. *Third Progress Report*, Potash Research Institute of India, Department of Botany, Aligarh Muslim University, Aligarh.

Agarwal, M. C., Dhindwal, A. S., Jaiswal, A. S. C. S., Prabhakar, A. and Aujla, M. S. (1997), *Status of Research on Agricultural Water Management of Northeren Region*. AICRP on Water Management, Directorate of Water Management Research (ICAR), Patna, p. 138.

Ahlawat, I. P.S., Ali, M. and Singh, A. (1986), Research needs and directions in pigeonpea based cropping system, pp. 183–209. In *Proceedings of the National Symposium on Cropping Systems*, 3–5 April 1985. Central Soil Salinity Research Institute, Karnal, Indian Society of Agronomy, New Delhi, India.

Ahlawat, I. P. S. and Rana, D. S. (2005), Concept of efficient water use in pulses. In *Pulses* (G. Singh, H. S. Sekhon and J. S. Kolar (Eds)). Agrotech Publishing Academy, Udaipur, India, p. 320.

Ahlawat, I. P. S., Saraf, C. S. and Singh, A. (1975), Studies on the performance of pigeonpea varieties to dates of planting and row spacing. *Indian J. Agron.*, 20: 245–7.

Ahlawat, I. P. S., Saraf, C. S. and Singh, A. (1985), Production potential of summer and rainy season pigeonpea intercropped with cowpea and greengram. *Indian J. Agric. Sci.*, 55, 565–9.

Ahlawat, I. P. S. and Shivakumar, B. G. (2002), Cropping systems research in pigeonpea, pp. 23–32. In *Proceedings of National Symposium on Pulses for Sustainable Agriculture and Nutritional Security* (M. Ali, S. K. Chaturvedi and S. N. Gurha (Eds)), 17–19 April 2001. Indian Agricultural Research Institute, New Delhi, Indian Society of Pulses Research and Development, Kanpur, India.

Ali, M., Kumar, N. and Ghosh, P. K. (2012), Milestones on agronomic research in pulses in India, *Indian J. Agron.*, 57(3rd IAC Special issue): 52–7.

Ali, M., Mishra, J. P., Ahlawat, I. P. S., Kumar, R. and Chauhan, Y. S. (1998), Effective management of legumes for maximizing biological nitrogen fixation and other benefits, pp. 107–28. In *Residue*

Effects of Legumes in Rice and Wheat Cropping Systems in the Indo-Gangetic Plain (J. V. D. K. Kumar Rao, C. Johansen and T.J. Rego (Eds)). International Crops Research Institute for the Semi-Arid Tropics (ICRISAT), Patencheru, India.

Ali, M. (1992), Weeds are a great threat to *kharif* pulses, *Indian Farming*, 42(5): 27, 29–30.

Anilkumar, S. N., Pujari, B. T., Vishwanatha, S. and Kopalkar, B. G. (2011), Effect of different methods of establishment on the yield, yield components and economics of Pigeonpea (*Cajanus cajan* L. Millsp.), *Res. J. Agric. Sci.*, 2(2): 325–7.

Arkin, G. F., Rosenthal, W. D. and Jordan, W. R. (1983), A sorghum leaf area model, ASAE Technical Paper 83-2098, *American Society of Agric. Engineers*, St. Joseph, MI.

Bains, S. S. (1970), Multiple cropping sequences, *World Science News*, 7–17.

Becker, T. W., Moria, H. and Fock, H. P. (1986), Evidence for the participation of dissimilatory process in maintaining high carbon fluxes through the photosynthetic carbon reduction and oxidation cycles in water stress *Phaseolus* leaves, *Photosynthetica*, 20: 153–7.

Bhalla, C. S., Jain, H. C., Ubadhyay, V. B. and Aggarwal, S. B. (1988), Controlling weeds in pigeonpea with herbicides, *Indian J. Weed Sci.*, 20(8): 78–80.

Bhan, S. (1976), Water management problems in Uttar Pradesh, *Indian Farming*, 26(2): 11–13, 20.

Bhan, S. and Khan, S. A. (1979), Response of *kharif* crops to irrigation in light textured alluvium of Uttar Pradesh, *Indian J. Agron.*, 24(4): 410–16.

Brar, M. S. and Imas, P. (2010), *Potassium and Crop Quality*. International potash Institute, India, pp. 1–8.

Chaturvedi, S. K. and Sandhu, J. S. (2016), *Strategies to Increase Productivity of Pulses in India*, Pulses Hand Book 2016. Commodity India.com, India, pp. 84–92.

Chauhan, Y. S, Venkataratnam, N. and Sheldrake, A. R. (1987b), Factors affecting growth and yield of short duration pigeonpea and its potential for multiple harvests, *J. Agric. Sci., Cambridge*, 109: 519–29.

Chauhan, Y. S. (1990), Pigeonpea: Optimum agronomic management, pp. 257–78. In *The Pigeonpea* (Y. L. Nene, S. D. Hall and V. K. Sheila (Eds)). CAB International, Wallingford, UK.

Chauhan, Y. S., Johansen, C. and Singh, L. (1996), Enhancing adaptation of extra-short-duration pigeonpea by improving plant type, pp. 62–8. In *Prospects for Growing Extra-short-duration Pigeonpea in Rotation with Winter Crops Proceedings of the IARI/ICRISAT Workshop and Monitoring Tour*, 16–18 October 1995, New Delhi, India (Singh, L., Chauhan, Y. S., Johansen, C. and Singh, S. P. (Eds)). Indian Agricultural Research Institute and International Crops Research Institute for the Semi-Arid Tropics, New Delhi 110 012 and Patancheru 502 324, Andhra Pradesh, India.

Chittapur, B. M., Kulkarni, B. S., Hiremath, S. M. and Hosmani, M. M. (1994), Influence of nitrogen and phosphorus on growth and yield of short duration pigeonpea (*Cajanus cajan* (L.) Millsp.), *Indian J. Agron.*, 39(4): 657–9.

Crop Production Guide. (2012), Published by Directorate of Agriculture, Chennai and Tamil Nadu Agricultural University, Coimbatore, India.

Dahiya, B. S., Brar, J. S. and Kaul, J. N. (1974), Changes in growth habit of pigeonpeas (*Cajanus cajan* (L) Millsp.) due to late sowing. *J. Agric. Sci., Cambridge*, 83: 379–80.

Dahiya, S. S., Chauhan, Y. S., Srivastava, S. K., Sekhon, H. S., Waldia, R. S., Gowda, C. L. L. and Johansen, C. (2001), Production practices for ESDP. In *Growing Extra Short Duration Pigeonpea in Rotation with Wheat in the Indo-Gangetic Plain*, Natural Resource Management Program, Report No. 1. ICRISAT, Patencheru 502 324, Andhra Pradesh, India, pp. 1–40.

Dalai, R. C. (1980), Nutrient element concentration and dry matter production by field grown pigeonpea (*Cajanus cajan* (L.) Millsp.), *Trop. Agric. (Trinidad)*, 57(2): 107–12.

Dastane, N. G., Yusuf, M. and Singh, N. P. (1971), Performance of different *rabi* crops under varying frequencies and timing of irrigation, *Indian J. Agron.*, 16: 483–6.

Desai, N. C., Ardeshna, R. B. and Intwala, C. G. (2000), Increasing pigeonpea productivity by providing land configuration in Vertisols, *International Chickpea and Pigeonpea Newsletter*, 7: 66–7.

Diaz-Rivera, M., Hepperly, P. R., Riveros, G. and Almedovar-Vega, L. (1985), Weed – crop competition in pigeonpeas in Puerto Rico, *J. Agr. U. Puerto Rico*, 69: 201–13.

DPD/Pub.1/Vol. 2/2016, Pulses in India: Retrospect and Prospects. Compiled & edited by Dr. A. K. Tiwari, & Dr. A. K. Shivhare, Government of India, Ministry of Agriculture & Farmers Welfare (Department of Agriculture, Co-operation and Farmers Welfare) Directorate of Pulses Development, Vindhyachal Bhavan, Bhopal (M. P.) 462004. India. http://dpd.dacnet.nic.in. p.1–317.

Dudhade, D. D., Deshmukh, G. P., Harer, P. N. and Patil, J. V. (2009), Studies on intercropping of pulse crops with pigeonpea under rainfed condition, *Legume Res.*, 32(3): 215–17.

Duraisamy, V. K. and Manickasundaram, P. (2008), Agronomic management for perennial pigeonpea through irrigation and mulching, *Madras Agric. J.*, 95(1–6): 205–7.

Durante, A. and Carpne, F. (1981), Effect of phosphorus levels on nitrogen and phosphorus uptake by *kharif* pulses and available nitrogen and phosphorus content of the soil after harvest, *Agric. Sci. Dig.*, 6: 153–5.

FAOSTAT (2014), www. faostat.fao.org.

Gajera, M. S. and Ahlawat, R. P. S. (2006), Optimization of irrigation and evaluation of consumptive water use efficiency for *rabi* pigeonpea (*Cajanus cajan* (L) Millsp), *Legume Res.*, 29(2): 140–2.

Gopalaswamy, G. (2012), Effective use of biofertilizers for sustained pulse production. In *Pulses Production Strategies* (P. Murugesa Boopathy and M. Paramathma (Eds)). Printed by Sri Sakthi Promotional Litho Press, Edayarpalayam, Coimbatore 25. ISBN: No. 81-9233041-3.

Goud, V. V. and Andhalkar, A. S. (2012), Feasibility studies in transplanted pigeonpea + soybean intercropping system, *J. Food Legumes*, 25(2): 128–30. Online published on 18 June 2012.

Gowda, L. C. L., Srinivasan, S., Gaur, P. M. and Saxena, K. B. (2013), Enhancing the productivity and production of pulses in India. In *Climate Change and Sustainable Food Security* (P. K. Shetty, S. Ayyappan and M. S. Swaminathan (Eds)). National Institute of Advanced Studies, Bangalore and Indian Council of Agricultural Research, New Delhi, ISBN: 978-81-87663-76-8.

Graham, P. H. and Vance, C. P. (2003), Legumes: Importance and constraints to greater use. *Plant Physiol.*, 131: 872–7.

Grant, C. A., Flaten, D. N., Tamasiewicz, D. J. and Sheppard, S. C. (2001), The importance of early season phosphorous nutrition, *Can. J. Plant Sci.*, 81: 211–24.

Gupta, S. C. (1990), *Mechanical Harvesting of Extra-shortduration Pigeonpea*. IPN 12, August 1990 (ICRISAT Center, India), pp. 14–15.

Gupta, S. C., Singh, L. and Ariyanayagam, R. P. (1991b), Adaptation of short duration pigeonepa to higher latitudes and altitudes. *Int. Pigeonpea Newslett.*, 13: 10–11.

ICRISAT. (1981), *Annual Report 1980*. ICRISAT, Patencheru, A. P., India.

ICRISAT. (1982), *Annual Report 1981*. ICRISAT, Patencheru, A. P., India.

ICRISAT. (1992), *Pigeonpea Botany and Production Practices*, Skill Development Series No. 9, (F. Singh and D. L. Oswalt (Eds)). ICRISAT, Human Resource Development Program, International Crops Research Institute for the Semi Arid Tropics, Patencheru 502 324, India.

ICRISAT. (2012), *Science With a Human Face*, August, 2012. ICRISAT, Patencheru, A. P., India.

IIPR. (2009), *25 Years of Pulses Research at IIPR* (S. Kumar and M. Singh (Eds)). Published by Director, IIPR, Kanpur, pp. 1–211.

IIPR. (2002–2003), All India Coordinated Research Project on Pigeonpea, *Annual Report 2002*.

IIPR. (2003–2004), All India Coordinated Research Project on Pigeonpea, *Annual Report* 2003.

Ishikawa, S., Adu-Gyamfi, J. J., Nakamura, T., Yoshihara, T., Wantanabe, T. and Watgatsuma, T. (2002), Genotypic variability in phosphorus solubilizing activity of root exudates by pigeonpea grown in lownutrient environments, *Plant Soil*, 245: 71–81.

Jamadar. (2009), Effect of nursery techniques, seedling age and spacing on seed yield and quality in transplanted red gram. M. Sc. (Ag) Thesis submitted to the Department of Seed Science and Technology, College of Agriculture, University of Agricultural Sciences, Dharwad – 580 005.

Johansen. (1990), Pigeonpea-Mineral Nutrition. In *The Pigeonpea* (Y. L. Nene, S. G. Hall and V. K. Sheila (Eds)). CAB International, International Crops Research Institute for Semi-Arid Tropics, India, pp. 209–31.

Kale, R. D., Sreenivasa, M. N. and Bagyaraj, D. J. (1997), Growth, development and yield of pigeonpea in the low land tropics, *J. Agr. Sci.*, 35: 110–15.

Katyal, J. C. (1980), Micronutrient research in pigeonpea, pp. 221–6. In *Proceedings of the International Workshop on Pigeonpea*, Volume 1, 15–19 December 1980. International Crops Research Institute for the Semi-Arid Tropics (ICRISAT), Patancheru, India.

Kaul, J. N., Sekhon, H. S., Dhingra, K. K and Tripathi, H. P. (1980), Date of planting and row-spacing studies with early maturing pigeon pea (Arhar) varieties under irrigated conditions for double cropping. *Trop. Grain legume Bull.*, 19: 20–3.

Khan, A. A., Jilani, G., Akhtar, M. S., Naqvi, S. M. S. and Rasheed, M. (2009), Phosphorus solubilizing bacteria: Occurrence, mechanism and their role in crop production, *J. Agric. Biol. Sci.*, 1, 48–58.

Kolar, J. S., Anand, S. and Singh, G. (1985), Comparative efficacy of chemical and mechanical methods of weed control in Pigeonpea, *Indian J. Weed Sci.*, 17(3): 40–5.

Kulkarni, K. R. and Panwar, K. S. (1981), Response of pigeonpea to fertilizers in India: a critical review. In *Proceedings of the International Workshop on Pigeonpeas*, Vol. 1, 15–19 December 1980. ICRISAT Center, India, Patencheru, A. P., India, pp. 212–20.

Kumar, A. and Kushwaha, H. S. (2006), Response of pigeonpea (*Cajanus cajan*) to sources and levels of phosphorus under rainfed condition, *Indian J. Agron.*, 51(1): 60–2.

Kumar, N. (2010). Imazethapyr: A potential post-emergence herbicide for *kharif* pulses, *Pluses Newslett.*, 21(3): 5.

Kyu, K. L., Saxena, K. B., Kumar, R. V. and Rathore, A. (2011), Prospects of hybrids in enhancing production and productivity of pigeonpea in Myanmar, *J. Food Legumes*, 24(1): 1–7. ISSN: 0976-2434.

Kyu, K. L., Shwe, T. and Sameer Kumar, C. V. (2016), Overview of pigeonpea research in Myanmar, *Legume Perspectives*, 11 January 2016, pp. 41–3.

Latha, K. R., Vimalendran, L. and Muthiah, A. R. (2012), Feasibility studies on transplanted pigeonpea under drip fertigation in Western Zone of Tamil Nadu, Extended Summaries, Vol. 3, *Third Intl. Agron. Congress- Agriculture Diversification, Climate Change management and Livelihoods* held during November 26–30, pp. 1384–5.

Latha, K. R. (2012), Production technologies of pulse crops. In *Pulses Production Strategies* (Dr. P. Murugesa Boopathi and M. Paramathma (Eds)). TNAU, Coimbatore, pp. 401–23.

Lugo-Mercado, H. M., Badillo-Feliciano, J. and Ortiz-Alvarado, F. H. (1987), Yield response of watermelon, tomato and pigeonpea to land preparation techniques in Southern Puerto Rico, *J. Agr. U. Puerto Rico*, 71, 203–8.

Mahalakshmi, K., Kumar, K. A., Reddy, M. D. and Devi, M. U. (2011), Response of *rabi* pigeonpea [*Cajanus cajan*(L.)] to different levels of drip irrigation, *J. Res. ANGRAU.*, 39(4): 101–3.

Maheshwarappa, H. P. and Nanjappa, H. V. (1994), Relative efficacy of herbicides in controlling the weeds infesting pigeonpea*(Cajanus cajan), J. Agron.*, 39(4): 662–4.

Mahetele, D. and Kushwaha, H. S. (2011), Productivity and profitability of pigeonpea as influenced by FYM, PSB and phosphorus fertilization under rainfed condition, *J. Food Legumes*, 24(1): 72–4.

Malhi, S. S., Johnston, A. M., Schoenau, J. J., Wang, Z. H. and Vera, C. L. (2007), Seasonal biomass accumulation and nutrient uptake of P and Lentil on a black chernozem soil in Saskatchewan, *J. Plant Nutr.*, 30: 721–37.

Mallikarjun, C. (2012), Performance of hybrid pigeonpea [*Cajanus cajan* (L.) Millsp.] (cv. ICPH-2671) under different planting methods and geometry in Northern dry zone of Karnataka. M. Sc. (Ag) Thesis submitted to Dept. of Agronomy, University of Agricultural Sciences, Dharwad – 580005, Karnataka State, India.

Mathews, C. (2003), Pigeonpea variety evaluation in Mpumalanga 1998–2000. Summary Proceedings of the 2nd pigeonpea workshop , 1–11 April 2003, Nelspruit, South Africa. Department of Agriculture and Land administration, Mpumalanga, South Africa (Limited publication).

Mathews, C. and Saxena, K. B. (2005), Prospects for pigeonpea cultivation in drought-prone areas of South Africa. In *Proceedings of the First International Edible Legume Conference in Conjunction with the IVth World Cowpea Congress*, 17–21 April 2005, Durban, South Africa.

Meena, B. K. (2010), Response of hybrid pigeonpea [*Cajanus cajan* (L.) *Millsp.] to planting geometry and fertility levels. M.Sc. (Ag) thesis submitted to the University of Agricultural Sciences, Dharwad, Department of Agronomy, College of Agriculture, Dharwad University of Agricultural Sciences, Dharwad* – 580 005, pp. 1–76.

Mergeai, G., Kimani, P., Mwangombe, A., Olubayo, F., Smith, C., Audi, P., Baudoin, J. P. and Le Roi, A. (2001), Survey of pigeonpea production systems, utilization and marketing in semi-arid lands of Kenya, *Biotechnol. Agron. Soc. Environ.*, 5(3): 145–53.

Meyyazhagan, N., Velayutham, A. Arunachalam, N. and Krishnarajan, J. (1999), Effect of intercrops and fertility levels on the yield and crop equivalent yield of Pigeonpea, *Madras Agric. J.*, 86(1–3): 146–7.

Monty, P. A. and James, B. E. (2011), Morphology, biomass and vessel diameter of pigeonpea subjected to water stress, *Comm. Soil Sci. Plant Anal.*, 42: 2334–43.

Mukherjee, D. and Kumar, R. (2005), Water stress induced physiological and biochemical changes in different plant parts of pigeonpea, *Bull. Natl. Inst. Ecol.*, 15: 191–9.

Mula, M. G., Saxena, K. B., Rathore, A. and Kumar, R. V. (2010), Response to spacing and irrigation in medium-duration CMS line of Pigeonpea, *J. Food Legumes*, 23(3&4): 186–90.

Mula, M. G., Saxena, K. B., Rathore, A. and Kumar, R. V. (2011), Influence of spacing and irrigation on seed production of medium duration pigeonpea hybrid, *Green Farming*, 2(1): 24–6.

Myaka, F. M., Sakala, W. D., Adu-Gyamfi, J. J., Kamalongo, D., Ngwira, N., Odgaard, R., Nielsen, N. E. and Høgh-Jensen, H. (2006), Yields and accumulation of N and P in farmer-managed intercrops of maize, pigeonpea in semi-arid Africa, *Plant Soil*, 285: 207–20.

Odeny, D. A. (2007), *The Potential of Pigeonpea (Cajanus cajan (L.) Millsp.) in Africa*, Natural Resources Forum 31: 297–305. Blackwell Publishing Ltd, 9600 Garsington Road, Oxford, OX4 2DQ, UK and 350 Main Street, Malden MA 02148, USA.

Padhi, A. K., Panigrahi, R. K. and Jena, B. K. (2010), Effect of planting geometry and duration of intercrops on performance of pigeonpea–fingermillet intercropping systems, *Indian J. Agric. Res.*, 44(1): 43–7.

Panwar, K. S. (1979), *Summer Institute on Pulse production Technology*, 4 June– 3 July, HAU, Hisar.

Panwar, K. S. and Yadav, H. L. (1980), Response of short duration pigeonpea to early planting and phosphorus levels in different cropping systems, pp. 37–44. In *Proceedings of the International Workshop on Pigeonpea*, Volume 1, 15–19 December 1980. International Crops Research Institute for the Semi-Arid Tropics (ICRISAT), Patancheru, India.

Patel, J. R. and Patel, Z. G. (1995), Effect of post monsoon irrigation, rhizobium inoculation and nitrogen on yield attributes of Pigeonpea, *Indian J. Agron.*, 40: 220–2.

Pathak, G. N. (1970), Pigeonpea, pp. 14–53. In *Pulse Crops* of India. Indian Council of Agricultural Research (ICAR), New Delhi, India.

Patil, B. N., Panchal, Y. C., Patil, V. S. and Krishnamurthy, K. A. (1981), Residual toxicity of triazine, phenoxy and amide herbicides on succeeding crops, *Indian J. Weed Sci.*, 13(2): 103–10.

Patro, G. K. and Tosh, G. C. (1973), Relative efficiency of herbicides on control of weeds in red gram *(Cajanus cajan*(L) Millsp.), *Andhra Agric. J.*, 20(3–4): 65–70.

PAU (1995), *Package of Practices for Crops of Punjab, Kharif. (1995).* Punjab Agricultural University, Ludhiana, India, 45p.

PAU (2002), Annual Progress Report, *Research on Kharif Pulses*, Punjab Agricultural University, Ludhiana, India, 23p.

Pavan, A. S., Nagalikar, V. P., Halepyati, A. S. and Pujari, B. T. (2009), Effect of planting on the yield, yield components and economics of transplanted Pigeonpea, *Karnataka J. Agric. Sci.*, 22(2): 433–4.

Pavan, A. S., Nagalikar, V. P., Pujari, B. T. and Halepyati, A. S. (2011), Influence of planting geometry on the growth characters, seed yield and economics of transplanted pigeonpea. *Karnataka J. Agric. Sci.*, 24(3): 390–2.

Peoples, M. B., Herridge, D. F. and Ladha, J. K. (1995), Biological nitrogen fixation: An efficient source of Nitrogen for sustainable agricultural production? *Plant Soil*, 174: 3–28.

Poornima, D. S., Shankaralingappa, B. C., Murthy, K. K. N. and Savitha, H. R. (2010), Economics of transplanted pigeonpea in sole cropping and finger millet based intercropping system, *Intl. J. Agric. Sci.*, 6(2): 501–3.

Praharaj, C. S. and Kumar, N. (2011), Drip fertigation in pigeonpea, *Pulses Newslett.*, 22(2): 5.

Pramod, G. (2007), Response of pigeonpea (*Cajanus cajan* (L.) Millsp.) genotypes to planting geometry under different protective irrigation, *Karnataka J. Agric. Sci.*, 2(4): 2007.

Rana, K. S. and Malhotra, O. P. (1992), Effect of water supply and date of sowing on the yield and water use efficiency of pigeonpea *(Cajanus cajan), Indian J. Agron.*, 37(1): 194–5.

Rana, K. S. and Paul, M. (1994), Productivity and water use in pigeonpea *(Cajanus* ca;Vm)-based intercropping systems as affected by weed control in rainfed conditions. *Indian J. Agron.*, 39(1): 27–30.

Rao, I. M., Venkataratnam, N., Faris, D. G. and Sheldrake, A. R. (1983), Response to irrigation in post rainy-season Pigeonpea, *Int. Pigeonpea Newslett.* 2: 35–6.

Rao, J. S. P. and Narayanan, A. (1997), Effect of water, light stress and foliar application of auxins on flower abscission and seed yield of Pigeonpea, *Madras Agr. J.*, 84(3): 127–9.

Rao, M. R. and Shetty, S. V. R. (1976), Some biological aspects of intercropping systems on crop- weed balance, *Indian J. Weed Sci.*, 8: 32–43.

Rathi, K. S. and Tripathi, H. N. (1978), Studies on time of sowing, row spacing and phosphate fertilization on early maturing arhar variety T-21 in central Uttar Pradesh, *Indian J. Agron.*, 23: 233–7.

Ravikumar, V., Ranghaswami, M. V., Appavu, K. and Chellamuthu, S. (2011), *Micro Irrigation and Irrigation Pumps*. Kalyani publishers, India, p. 15.

Reddy, A. A., Bantilan, M. C. S. and Mohan, G. (2013), Pulses production scenario: Policy and technological options, *Policy Brief*, 26, 1–7.

Reddy, B. S., Maruthi, V., Adake, R. V. and Madal, U. K. (2009), Effect of different land configuration practices on productivity of sorghum-pigeonpea intercropping system in shallow alfisols, *Indian J. Dryland Agric. Res. Devt.*, 24(1): 57–62.

Reddy, D. S., Reddy, D. R. and Acharya, G. V. N. (1978), Studies on water intake and hydraulic conductivity of shallow and deep ploughing operating under rainfed agriculture at Anantapur, *Ann. Arid Zone*, 17: 291–7.

Reddy, M. V., Sharma, S. B. and Nene, Y. L. (1990), Pigeonpea: Disease management, pp. 303–48. In *The Pigeonpea* (Nene, Y. L., Hall, S. D. and Sheila, V. K. (Eds)). CAB International, patencheru, Wallingford, UK; ICRISAT, Andhra Pradesh, India.

Reddy, S. J. and Virmani, S. M. (1981), Pigeonpea and its climatic requirement, In *Proceedings of the International Workshop on Pigeonpeas*, Vol. 1, 15–19 December 1980. ICRISAT Center, India, Patencheru, A.P., India: ICRISAT, pp. 259–70.

Sandhu, H. S., Brar, S. S., Dhillon, M. S., Gill, G. S. and Singh, G. (1981), Effect of sowing time, row spacing and plant population on the performance of *arhar* and succeeding wheat in arhar-wheat rotation. *Indian J. Agron.*, 26: 154–7.

Saritha, K. S., Pujari, B. T., Basavarajappa, R., Naik, M. K., Babu, R. and Desai, B. K. (2012), Growth of pigeonpea (*Cajanus cajan*(L.)Millsp.) and nutrient status of soil after the harvest of crop as influenced by plant densities, different irrigation and nutrient levels, *Karnataka J. Agric. Sci.*, 25(1): 134–6.

Saxena, K. B., Johansen, Y. S. and Singh, L. (1992), Recent development in hybrid pigeonpea research, pp. 58–9. In *Proceedings of the National Symposium on New Frontiers in Pulses Research and Development.* Indian Institute of Pulses Research (IIPR), Kanpur, India.

Saxena, M. C. and Singh, Y. (1970), Relative susceptibility of important varieties of some pulses and soybean to zinc deficiency, Presented at the *Third Annual Workshop of Co-ordinated scheme on Micro-nutrients of soils*, 5–7 October 1970, Ludhiana, Punjab, India.

Saxena, M. C. and Yadav, D. S. (1975), Some agronomic considerations of pigeonpea and chickpea, pp. 32–62. In *Proceedings of International Workshop on Grain Legumes*, 13–16 January 1975. International Crops Research Institute for the Semi-Arid Tropics (1CR1SAT), Patancheru, India.

Saxena, M. C. and Yadav, D. S. (1975), Multiple cropping with short duration pulses, *Indian J. Genet. Plant Breed.*, 35: 194–208.

Saxena, M. C. and Yadav, I. S. (1975), Some agronomic consideration of pigeonpea and chickpea. In *Proceedings of International Workshop on Grain Legumes*, 13–16 January. ICRISAT, Hyderabad, pp. 31–61.

Sekhon, H. S., Guriqbal, S. and Amandeep. (2000), Response of pigeonpea hybrid PPH 4 to varying plant densities, *Int. Chickpea Pigeonpea Newslett.*, 7: 43–5.

Sekhon, H. S., Guriqbal, S. and Brar, J. S. (1993), Effect of chemical, mechanical and cultural manipulations on weed growth and grain yield of various pulse crops, pp. 141–6. In *Proceedings of the International Symposium on Integrated Weed Management for Sustainable Agriculture*, Volume 111, 18–20 November 1993. CCS HAU, Hisar, India.

Sekhon, H. S., Guriqbal, S. and Poonam, S. (1997), Effect of sulphur nutrition in different pulses, p. 23. In *Proceedings of Third Agricultural Science Congress*, Volume 2, 12–15 March 1997. Punjab Agricultural University, Ludhiana, India.

Sekhon, H. S. and Singh, G. (2005), Agronomic research and management in pigeonpea. In *Advances in Pigeonpea Research* (M. Ali and S. Kumar(Eds)). Indian Institute of Pulses Research, Kanpur, India, pp. 144–63.

Sharma, A., Rathod, P. S. and Basavaraj, K. (2010), Agronomic management of pigeonpea (*Cajanus cajan*) based intercropping systems for improving productivity under rainfed conditions, *Karnataka J. Agric. Sci.*, 23(4): 570–4.

Sharma, A. and Guled, M. B. (2012), Effect of set furrow method of cultivation in pigeonpea + greengram intercropping system in medium deep black soil under rainfed conditions, *Karnataka J. Agric. Sci.*, 25(1): 18–24.

Sharma, R. R. P., Thakur, H. C. and Sharma, H. M. (1980), Pigeonpea as a *rabi* crop in India, pp. 26–36. In *Proceedings of the International Workshop on Pigeonpea*, Vol. 1, 15–19 December 1980. International Crops Research Institute for Semi Arid Tropics (ICRISAT), Patencheru, India.

Shetty, S. V. R. and Rao, A. N. (1981), Weed management studies in sorghum/pigeonpea and pearl millet/groundnut intercrop systems: Some observations, pp. 238–48. In *International Workshop on Intercropping*, 10–13 January 1979. International Crops Research Institute for the Semi-Arid Tropics (ICRISAT), Patancheru, India.

Sinclair, T. R. and Vadez, V. (2002), Physiological traits for crop yield improvement in low N and P environments, *Plant Soil*, 245: 1–15.

Sinclair, T. R. and Vadez, V. (2012), The future of grain legumes in cropping systems. *Crop Pasture Sci.*, 63(6): 501–12.

Singh, A., Prasad, R. and Saraf, C. S. (1983), Effect of variety, row spacing and phosphate fertilization on yield and moisture extraction pattern in pigeonpea. *Indian J. Agron.*, 28, 33–6.

Singh, R. C., Faroda, A. S. and Sharma, H. C. (1979), Effect of irrigation and fertility levels on the yield and water requirement of lentil, *Indian J. Agron.*, 24: 351–3.

Singh, S. and Bahadur. (1990), Planting pattern, population density and fertilizer effect in pigeonpea intercropping system, *Indian J. Dryland Agric. Res. Develop.*, 1: 39–49.

Sinha, A. C., Mandel, B. B. and Jana, P. K. (1989), Effect of time of sowing, row spacing and weed control treatment on weed and grain yield of pigeonpea *(Cajanus cajan)*, *Indian J. Agric. Sci.*, 59(6): 353–8.

Srivastava, D. P. and et al. (1994), In *Int. Symp. Pulses Res.*, ISPRD, IIPR and IARI, New Delhi, India, pp. 63–4.

Subbarao, G. V., Ae, N. and Otani, T. (1997), Genetic variation in acquisition, and utilization of phosphorus from iron-bound phosphorus in pigeonpea. *Soil Sci. Plant Nutr.*, 43: 511–19.

Sushil Rao. (2014), Hybrid pigeonpea technology achieves record yield. *Times of India*, 29 January 2014. http://timesofindia.indiatimes.com/city/hyderabad/Hybrid-pigeonpea-technology-achieves-record-yield/articleshow/29558633.cms.

Tandon, H. L. S. (1995), Sulphur deficiencies in soils and crops: Their significance and management, pp. 1–23. In *Sulphur Fertilisers for* Indian *Agriculture* (H. L. S. Tandon (Ed.)). Fertiliser Development and Consultation Organisation, New Delhi, India.

Tayo, T. O. (1983), Growth, development and yield of pigeonpea *(Cajanus cajan* (L) Millsp.) in the lowland tropics. 4, Effect of sowing depth, *J. Agric. Sci., Cambridge*, 101: 435–40.

Tayo, T. O. (1985b), The effect of ridging on the performance of pigeonpea*(Cajanus cajan*(L) Millsp.) in the lowland tropics, *J. Agric. Sci.*, Cambridge 104, 243–5.

Thakuria, K. and Loikham, E. (1991), Effect of nitrogen and phosphorus on growth and yield of pigeonpea (*Cajanus cajan*, L.), *Indian J. Agron.*, 36: 602–4.

Thanki, J. D. and Solanki, R. K. (2010), Response of *rabi* pigeonpea (*Cajanus cajan* (L.) Millsp) to moisture regimes and fertilizer management with and without FYM under south Gujarat condition, *Crop Res.*, 39(1, 2&3): 39–42.

Tiwari, R. B., Pariharand, R. S. and Tripathi. (1992), Weed management in pigeonpea (*Cajanus cajan*) based intercropping system. *Indian J. Agron.*, 37(1): 145–7.

Tomar, R. S., Kanzaria, M. V. and Jain, V. K. (2001), Response of chickpea to potassium in a calcareous soil, *J. Potassium Res.*, 17: 98–100.

Tomer, Y. S. and Ram Dhari. (1996), Extra-short-duration pigeonpea in Haryana, pp. 11–16. In *Prospects for Growing Extra-short-duration Pigeonpea in Rotation with Winter Crops: Proceedings of the IARI/ICRI -SAT Workshop and Monitoring Tour*, 16–18 Oct 1995, New Delhi, India (L. Singh, Y. S. Chauhan, C. Johansen and S. P. Singh (Eds)). Indian Agricultural Research Institute and International Crops Research Institute for the Semi-Arid Tropics, New Delhi 110 012 and Patancheru 502 324, Andhra Pradesh, India.

Tuppad, G. B., Koppalkar, B. G., Halepyati, A. S. and Desai, B. K. (2012), Yield and economics of pigeonpea genotypes as influenced by planting geometry under rainfed condition, *Karnataka J. Agric. Sci.*, 25(2): 179–82.

Vance, C. P. (2001), Symbiotic nitrogen fixation and phosphorus acquisition: Plant nutrition in a world of declining renewable resources, *Plant Physiol.*, 127: 390–7.

Vesterager, J. M., Nielsen, N. E. and Hogh-Jensen, H. (2006), Variation in phosphorus uptake and use efficiencies between pigeonpea genotypes and cowpea, *J. Plant Nutr.*, 29: 1869–88.

Vimalendran, L. (2013), Optimizing drip irrigation and fertigation scheduling for long duration pigeonpea [*Cajanus cajan* (l.) millsp.], Ph.D. Thesis, Tamil Nadu Agricultural University, Coimbatore.

Wallis, E. S., Byth, D. E. and Whiteman, P. C. (1981), Mechanized dry seed production of pigeonpea. In *Proceedings of the International Workshop on Pigeonpeas*, 1, 15–19 December 1980, ICRISAT centre, India, Patancheru, Andhra Pradesh, India, pp. 51–60.

Wani, S. P., Rupela, O. P. and Lee, K. K. (1995), Sustainable agriculture in the semi-arid tropics through biological nitrogen fixation in grain legumes, *Plant Soil*, 174: 29–49.

Wani, S. P., Sarvesh, K. V., Dharmarajan, K. B. K. and Deepaja, S. M. (Eds) (2012), *Bhoochetana: Mission to Boost Productivity of Rainfed Agriculture Through Science-led Interventions in Karnataka.* International Crops Research Institute for the Semi-Arid Tropics, Patencheru 502 324, Andhra Pradesh, India, p.84. ISBN 978-92-9066-548-9.BOE 058.

Willey, R. W., Rao, M. R. and Natarajan, M. (1981), Traditional cropping systems with pigeonpeas and their improvement, pp. 11–25. In *Proceedings of the International Workshop on Pigeonpeas*, 15–19 December 1980, ICRISAT Center, India. Vol. 1. International Crops Research Institute for the Semi-Arid Tropics, Patencheru, A. P. 502 324, India.

Yadav, J. S. P. (1975), Annual Progress Report, All India Co-ordinated Scheme for Research on water management and Salinity, CSSRI, Karnal.

Yadav, R. P., Sharma, R. K. and Shrivastava, U. K. (1997), Fertility management in pigeonpea (*Cajanus cajan*) based intercropping system under rainfed conditions, *Indian J. Agron.*, 42(1): 46–9.

Index

Abiotic stress
common bean production 24
cowpea 217
groundnut 134–136
lentil 57–58
pigeonpea susceptibility 303
Acceso Peanut Enterprise Corporation 201–202
Acquired thermotolerance (ATT) 136
Aflatoxin contamination, in groundnut 131–132, 168–169
biological control 190
case study 195–203
controlling at village level 195–201
cropping systems 185–186
drying, storage and processing 190–192
factors impact 182–184
in Ghana 193–195
irrigation 189–190
minimizing 185–190
in Mozambique 193–195
nongovernmental organizations and private sector 201–202
pod and kernel damage 190
prediction models 184–185
resistant cultivars 186–187
soil fertility and plant nutrition 187
soil organic matter and texture 187–188
sowing and harvest date 188–189
in Uganda 193–195
value chain projects 193–203
in Zambia 189
African Agricultural Technology Foundation (AATF) 228
Afri-Nut Ltd. 202–203
Agricultural extension and advisory services 119
Agricultural Production Systems Simulator (APSIM) 184
Agro-ecology
lentil 53–54
sub-Saharan Africa
groundnut 125, 129, 137, 140
soybean 110, 113
Agronomic management, pigeonpea hybrids 349–350
Alectra vogelii 223–224
All India Co ordinated Pulses Improvement Project (AICPIP) 331, 332, 336, 349
All India Co ordinated Research Project (AICRP) 334, 348, 359
AMF. *see* Arbuscular mycorrhizal fungi (AMF)
Andean bean
genome-wide association studies 11, 13
symbiotic nitrogen fixation 8
yields 5, 6
Angular leaf spot (ALS) 25, 31
Anthracnose
bean 25, 32
lentil 63
Arachis hypogaea L. *see* Groundnut
Arbuscular mycorrhizal fungi (AMF)
common beans 37–38
faba bean 285–286
Arbuscules 162
Arid environments 311
Ascochyta blight (AB) 53–54
Aspergillus flavus 131–132, 137
aflatoxin biosynthesis pathway in 186
groundnut (*see* Groundnut)

Bacillus spp.
B. subtilis 163
B. thuringiensis 228–229
Bacterial artificial chromosome (BAC) 229
BBF. *see* Broad bed furrows (BBF)
Bean. *see also specific beans*
anthracnose 25, 32
common mosaic necrotic virus (BCMNV) 25
common mosaic virus (BCMV) 25, 32
golden mosaic virus (BGMV) 25
golden yellow mosaic virus (BGYMV) 25
root rots 33
Bean pod weevil 31
Bhoochetana project 339
Biofertilizers, groundnuts 161–162
Biological nitrogen fixation (BNF)
common bean 27, 28
cowpea 245
faba bean 258–259, 282–283
soybean 97–98, 108–109
Biomass
accumulation of soybean variety 115
pigeonpea 299–300
Biotechnology, for insect resistance 228–229
Biotic stress
common bean production 24
cowpea 217
groundnut 127–134
aflatoxin 131–132
foliar diseases 132–134
viruses 128–131
lentil 57–58
pigeonpea susceptibility 303
Blanket fertilizer application 157
BNF. *see* Biological nitrogen fixation (BNF)
Boron deficiency, groundnuts 161
Breeding method
for broomrape resistance 258
common beans 6
cowpea
drought and heat tolerance 224–225
germplasm resources 219–220

improved nutritional quality 226–227
International Institute of Tropical Agriculture 218–220, 227
limited programmes and past challenges 217–218
nitrogen fixation 225–226
objectives 219, 221
phosphorus 225–226
resistance to major diseases 223
resistance to major insects 224
variety development and testing 220
faba bean
heat, drought and waterlogging stresses 278
Orobanche resistance 277
resistance to foliar/soil-borne diseases 276–277
fragile ecosystems 310–311
groundnut 126–136, 138–139
abiotic stress resistance 134–136
biotic stress resistance 127–134
yields 126–127
lentil 58–59
pigeonpea
disease resistance 312
high-protein content 314–316
insect resistance 312–313
resistance to waterlogging 313–314
resistance to foliar diseases 257–258
Broad bed furrows (BBF) 332
Broomrape (*Orobanche* spp.) 53, 57, 58, 63, 275
breeding for resistance 258, 277
herbicide tolerance 256
resistance varieties 259–260
Bruchus rufimanus 275
Bulked segregant analysis (BSA) 84

Cajanus cajan L. Millspaugh. *see* Pigeonpea
Calcium deficiency, groundnut 160–161, 187–188
CCRP. *see* Collaborative Crop Research Program (CCRP)
Central and West Asia, North Africa (CWANA) regions, faba bean 274–276, 278
Centro Internacional de Agricultura Tropical (CIAT) 7, 37
Cercospora leaf spot *(Cercospora zonata)* 277
CGIAR. *see* Consultative Group for International Agricultural Research (CGIAR)
Chandra Shekhar Azad University of Agriculture and Technology (CSUAT) 331
China, faba bean in 270–274
area and production 270–272
cropping systems 272–273
export and import 274
soil fertility 273
Chitedze Agricultural Research Station 202–203
CIAT. *see* Centro Internacional de Agricultura Tropical (CIAT)
Climate change
extra-short-duration pigeonpea varieties 351–352
impact on soybean 118
lentil cultivation 74
Clinton Guistra Enterprise Partnership 201
Coat protein (CP) 128
Collaborative Crop Research Program (CCRP) 26
Common bacterial blight 32
Common bean 3
adaptation 3
in Africa 21, 23–24
biological nitrogen fixation 27, 28
breeding methods 6
cropping system 27–30
diseases management 31–33
angular leaf spot 31
anthracnose 32
bean common mosaic viruses 32
bean root rots 33
common bacterial blight 32
halo blight 32
economic significance 20
in Europe 22
future trends 13–14
genome-wide association studies 12
genomic analysis 12
ideotype 5
integrated pest management 30–31
in Latin America 21, 23–24
in Mexico 4
North America and Caribbean 22
nutrient management 30, 36–43
foliar fertilization 42–43
integrated soil fertility management 40–42
liming for soil acidity management 42
nitrogen 36, 38–39
phosphorus 36–38
origin and domestication 19–20
partitioning 5–6
production constraints 23–27
moisture stress 25
pests and diseases 25
seed quality considerations 26–27
soil fertility and acidity 23–25
quantitative trait loci analysis 8–11
seed yield QTL 9–11
spatial distribution 28
sub-Saharan Africa 20, 22–23
water management 33–36
drip irrigation 34–36
furrow irrigation 34, 35
wild germplasm 7
world production trends 20–23

yield gap 4
yields 4–5
component compensation 5
factors affecting 5–8
pyramid 6–7
under stress 7–8
Conservation tillage. *see* Tillage system
Consultative Group for International Agricultural Research (CGIAR) 358
Conventional intercropping system 28–29
Cotyledons 73, 131, 216
Cowpea 215
biotechnology for insect resistance 228–229
breeding
drought and heat tolerance 224–225
germplasm resources 219–220
improved nutritional quality 226–227
International Institute of Tropical Agriculture 218–220, 227
limited programmes and past challenges 217–218
nitrogen fixation 225–226
objectives 219, 221
phosphorus 225–226
resistance to major diseases 223
resistance to major insects 224
variety development and testing 220
future trends 227–230
global production 228
improved varieties
impact 228
performance 223
protein, iron and zinc contents 227
pyramiding genes for resistance 224
insect pests 217
international trials (CITs) 227
medium maturing semi-erect type varieties 222–223
origin 216
production constraints 217
regional preferences 216
Striga and *Alectra* resistance 223–224
versatility 216
West Africa 235–236
future trends 247
integrated pest management 245–247
intercropping systems 237–241
nutrient management 244–245
optimal population 236–237
rainfall distribution and planting dates 241–244
CP. *see* Coat protein (CP)
Crop geometry 332–333
Crop modelling, pigeonpea enhancing sustainability through 306–307
Cropping system
common bean 27–30
faba bean 272–273
lentil 77–79
pigeonpea 302, 339–341
Crop rotation 165
CSUAT. *see* Chandra Shekhar Azad University of Agriculture and Technology (CSUAT)
Cytoplasmic nuclear male sterility 316

Dehydration-responsive element binding (DREB) transcription factors 135
Democratic Republic of the Congo (DRC)
groundnut leaf miner 142
soybean 113, 114
Disease resistance, breeding pigeonpea for 312
DNA markers, soybean 84, 88–89
Double cropping
lentil 52, 78
soybean-cowpea 228
Double-up legumes technology 165
DRC. *see* Democratic Republic of the Congo (DRC)
DREB transcription factors. *see* Dehydration-responsive element binding (DREB) transcription factors
Drip irrigation
common bean 34–36
pigeonpea 344–346
Drought tolerance
common bean 25, 33
cowpea breeding 224–225
faba bean 255–256
groundnut 134–136, 182–183
DSSAT-CROPGRO-Peanut 184

Early leaf spot (ELS) groundnut 127, 132–134, 168
Economic significance, common bean 20
Effector-triggered immunity (ETI) 131
ESDP varieties. *see* Extra-short-duration pigeonpea (ESDP) varieties
Ethiopian Institute of Agricultural Research (EIAR) 257–258, 277
Etiella zinckenella (Treitschke) 30
Exagris Africa Ltd., 202–203
Expressed sequence tag (EST) 59, 98, 132, 229
Extra-short-duration pigeonpea (ESDP) varieties 329
climate and soil 351–352
harvesting 354
intercropping and rotations 353
nutrient management 353
plant population density 352–353
seed rate and treatment 352
sowing 352
water management 354
weed management 353–354
yields 354–355

Faba bean 269–270
achievements 259–262
biological nitrogen fixation 258–259, 282–283
breeding
broomrape resistance 258
heat, drought and waterlogging stresses 278
Orobanche resistance 277
resistance to foliar diseases 257–258
resistance to foliar/soil-borne diseases 276–277
in China 270–274
area and production 270–272
cropping systems 272–273
export and import 274
soil fertility 273
CWANA, 274–276
drought tolerance 255–256
foliar diseases 275–276
future trends 262–263
gene banks 278–279
genome mapping and markers 279–280
heat tolerance 255, 260–262
herbicide tolerance 256–257
insect pests 275–276
intercropping 282
marker-assisted selection 280
mineral nutrient requirements 283–287
Orobanche resistance varieties 259–260
pollen viability 260
production and productivity 253–254
relay cropping 282
status 274
water resources management 280–282
weed management 276
Faba bean necrotic yellow virus (FBNYV) 275, 276
Farming systems, groundnut 156
Farmyard manure (FYM) 161
Fertigation technology, pigeonpea 344–346
Fertilizer efficiency, pigeonpea 345–346
Flowering
cowpea 245–247
faba bean 255, 260, 281
pigeonpea 301–302, 342
Fluidigm approach 84
Focused Identification of Germplasm Strategy (FIGS) 58–59
Foliar diseases
breeding resistance to 257–258
faba bean 275–276
groundnut 132–134, 168
lentil 57
Foliar fertilization, common bean 42–43
Frankliniella fusca Hinds (tobacco thrips) 130
Frankliniella occidentalis (western flower thrips) 130
Furrow irrigation 34, 35
Fusarium wilt 53, 312, 313
Garbage-in, garbage-out (GIGO) QTL, 93
Genetic contamination, pigeonpea seed 305
Genetic resources
common beans 7
cowpea breeding 219–220, 229–230
faba bean 278–280
lentil 55–56
for trait mining 126
Genome by sequencing (GBS) analyses 13
Genome-wide association studies (GWAS) 8, 11, 12
Genomic selection (GS) 143–144
Genotype by environment (GxE) stability 26
Ghana, aflatoxin in groundnut 193–195
Global positioning systems (GPS) 172–173
Glycine max L Merr. *see* Soybean
GRD. *see* Groundnut rosette disease (GRD)
Groundnut 155
aflatoxin contamination 131–132, 168–169
biological control 190
case study 195–203
controlling at village level 195–201
cropping systems 185–186
drying, storage and processing 190–192
factors impact 182–184
in Ghana 193–195
irrigation 189–190
minimizing 185–190
in Mozambique 193–195
nongovernmental organizations and private sector 201–202
pod and kernel damage 190
prediction models 184–185
resistant cultivars 186–187
soil fertility and plant nutrition 187
soil organic matter and texture 187–188
sowing and harvest date 188–189
in Uganda 193–195
value chain projects 193–203
in Zambia 189
agronomic recommendation limitation 157–158
biofertilizers 161–162
boron deficiency 161
breeding method 126–136, 138–139
abiotic stress resistance 134–136
biotic stress resistance 127–134
yields 126–127
calcium deficiency in 160–161, 187, 188
case study 137–143
choice of variety/cultivar 158
confectionery 142–143
conservation tillage 159
cultivation under polythene mulch 166–167
drying 167–171
early leaf spot 127, 132–134, 168
ecology 156
farming systems 156

fertilizers and amendments 159–160
field insect pests 169–170
field selection 159
foliar diseases 132–134, 168
future trends 143
growth regulators 167
harvesting 123–125, 170–171
health benefits 123
improvement in Uganda 137–143
insect pests, storage of 170
intercropping/crop rotation 165
land preparation 159
landrace improvements 143
late leaf spot 127, 132–134, 137, 142
leaf spot disease 172
macronutrient deficiency 160–161
manganese deficiency 161
marker-assisted selection 137, 143–144
micronutrient deficiency 161, 164
nitrogen fixation 124
oil quality and content 136–137
pathogens 127
phosphorus deficiency in 160
plant protection practices 167–170
precision cultivation 172–173
regional improvements 142
response surface methodology 185
seed systems 173
seed treatment and priming 162–164
Serenut 1–4 series 141
soilborne fungal diseases 168
soil insect pests control 170
soil resources management 159–162
sowing 162, 164–165
specific leaf area 135
storage 170–171
varieties and replacements 141
varieties improvement 140–143
viruses 128–131, 169
water management 166
weed management 165
worldwide production trends 124
yield gap 156–157
zinc deficiency 161
Groundnut leaf miner *(Aproaerema modicella)* 142
Groundnut rosette disease (GRD) 169
biotic stress resistance 127–129
improvements for resistance to 140–141
regional improvements for resistance 142
symptoms 127
virus (GRAV) 128–129
Growth regulators, groundnuts 167, 172, 173

Halo blight 32
Harvest index
common beans 7
pigeonpea 300–301
Harvesting
early-maturing pigeonpea varieties 354
groundnut 123–125, 170–171, 188–189
lentil 79
pigeonpea 356
Heat stress
common bean 25
cowpea breeding 224–225
faba bean 255, 260–262
groundnut 134–136
Helicoverpa-tolerant cv. Abhaya 313
Herbicide tolerance, faba bean 256–257
Heterodera glycines 94
High-performance liquid chromatography (HPLC) 193
Human nutrition, pigeonpea 298
Hybrid technology, pigeonpea
adoption 317–319
evolution 316–317
high yields 318
promotion and marketing 319
seed production 318–319
two-parent hybrid breeding technology 319–320

ICPL 87119 (Asha) pigeonpea variety 312
Ideotype, common beans 5
IFLRC. *see* International Food Legumes Research Conferences (IFLRC)
IITA. *see* International Institute of Tropical Agriculture (IITA)
Indian Institute of Pulses Research (IIPR) 332, 348, 359
Insect pests
cowpea 217, 245–246
faba bean 275–276
groundnuts 169–170
pigeonpea 312–313
soybean 115–116
Integrated approaches, soybean yield 116–118
Integrated pest management (IPM) 185
common bean 30–31
cowpea 245–247
groundnuts 169
pigeonpea 313
Integrated soil fertility management (ISFM) 40–42
Intellectual property (IP) 84–85
Intercropping
common bean 27
conventional 28–29
cowpea 237–241
early-maturing pigeonpea varieties 353
faba bean 282
groundnuts 165
lentil 78
MBILI 28–29
pigeonpea 334–336

International Conference on Legume Genetics and Genomics 63
International Crops Research Institute for the Semi-Arid Tropics (ICRISAT) 126, 133, 187, 312–314
on Alfisols 159
in Asia and Africa 174
broad bed furrows 332
germplasm collection 129, 130
high-protein pigeonpea 298
hybrids of pigeonpea 350
pigeonpea varieties 329–330
sorghum intercrop 334
weeds control in pigeonpea 347
International Food Legumes Research Conferences (IFLRC) 63, 263
International Institute of Tropical Agriculture (IITA) 218–220, 227
IPM. *see* Integrated pest management (IPM)
Irrigation
common bean 34–36
groundnut 173, 189–190
pigeonpea 342–346
ISFM. *see* Integrated soil fertility management (ISFM)
Isopentenyltransferase (IPT) 135

Kano Agricultural and Rural Development Authority (KNARDA) 226
Kharif crops 78

Landrace, groundnut 143
Late leaf spot (LLS) groundnut 127, 132–134, 137, 142
Leafhoppers 30–31
Leaf spot disease, groundnut 127, 132–134, 137, 168, 172
Lens culinaris Medikus spp., 54. *see also* Lentil
from wild species 62–63
Lentil 51–52, 71
agro-ecologies 53–54
breeding methods 58–59
cropping systems 77–79
cultivation
climate requirements 74
land requirements 74
nutrient requirements 74–75
description 73
development 62–63
domestication and spread 55
drying and storage 79
fallow replacement with annual cropping 77
future trends 63–64, 80
genetic base 62–63
genetic resources 55–56
global production 52–53
harvesting 79
land preparation 75–77
marker-trait association studies 60–61
molecular markers 59
new technologies 59–61
phylogeny 54
planting methods 76–77
plant populations and spacing 77
production 53–54, 71–73
seed quality and treatments 75
seeds classification 51
storage of seed 79
sulphur 73, 75
weed control 79
yield enhancement 56
yield stability 57–58
Linkage disequilibrium (LD) analysis 8
Livestock nutrition 298–299
Lotus japonicus 263

Macronutrient deficiency, groundnuts 160–161
Manganese deficiency, groundnuts 161
Marker-assisted selection (MAS)
faba bean 280
groundnut 137, 143–144
lentil 59, 61
soybean 84
development 85–87
future trends 98
genetic and association map development 91–92
identification of polymorphism 89–91
methods for 85
QTL, 91–92, 97–98
recurrent cross populations 92–93
recurrent parent genotype 85, 92
scoring of phenotypes 93
selection programme 88–89
soybean cyst nematode 94–95, 97
steps involved in 86
sudden death syndrome 94–97
Marker-trait association studies, lentil 60–61
Maruca vitrata 313, 314
Maturity, pigeonpea 302, 305
MBILI intercropping system 28–29
The McKnight Foundation 26
Medicago truncatula 263, 280
Meds & Food for Kids (MFK) 201
Micro irrigation 343
Micronutrients
groundnuts 161, 164
pigeonpea 339
Microsatellites 84, 85, 88–89, 92, 95, 98
Mixed cropping
groundnuts 165
lentil 78
pigeonpea 334–336

Moisture stress
common bean 25
pigeonpea 341
Molecular markers
groundnut 136, 143–144
lentil 57, 59, 61
pigeonpea 304
soybean 93–98
Monocropping 78, 186
Monsanto Corporation 228
Mozambique, aflatoxin in groundnut 193–195
Multiple cropping 27, 158, 221
Mycotoxins 181–182, 192–193

National Semi-Arid Resources Research Institute (NaSARRI) 137, 140
Network for Genetic Improvement of Cowpea for Africa (NGICA) 228
Nitrogen
common bean 30, 36, 38–39
cowpea 225–226, 244–245
groundnuts 124
lentil 52
pigeonpea 336–337
soybean variety 112–114
Nitrogen fixation
biological
common bean 27, 28
cowpea 245
faba bean 258–259, 282–283
soybean 97–98, 108–109
symbiotic 8, 11, 25, 38
Nutrient management
common bean 30, 36–43
biological nitrogen fixation 28
foliar fertilization 42–43
integrated soil fertility management 40–42
liming for soil acidity management 42
nitrogen 30, 36, 38–39
phosphorus 36–38
cowpea 244–245
faba bean 283–287
lentil cultivation 74–75
pigeonpea 336–341
cropping systems 339–341
early-maturing pigeonpea varieties 353
micronutrients 339
nitrogen 336–337
phosphorus 337–338
potassium 338
security 298–299
sulphur 339
soybean 112–114

Orobanche spp., 53, 57, 58, 63, 275
breeding for resistance 258, 277
herbicide tolerance 256
resistance varieties 259–260

PAMP-triggered immunity (PTI) 131
Parasitic weed 53, 57, 58, 63, 275
breeding for broomrape resistance 258, 277
faba bean 256
herbicide tolerance 256
partial *Orobanche* resistance varieties 259–260
Partitioning, common beans 5–6
Peanut. *see* Groundnut
Peanut and Mycotoxin Innovation Laboratory (PMIL) 140
Peanut bud necrosis disease (PBND) 130, 169
Peanut bud necrosis virus (PBNV) 130, 169
Peanut stem necrosis diseases (PSND) 169
Peanut stripe disease (PStV) 169
PGPR. *see* Plant growth promoting rhizobacteria (PGPR)
Phaseolus vulgaris L. *see* Common bean
Phosphorus
common bean 36–38
cowpea 225–226, 244–245
deficiency in groundnut 160
faba bean 285–287
pigeonpea 337–338
Phosphorus solubilizing bacteria (PSB) 162, 330, 337, 338
Phosphorus-solubilizing microorganisms (PSM) 162
Photoperiod sensitivity 301–302
Phytoalexins 183–184, 186
Pigeonpea 297–298, 327–328
agronomic factors affecting 302–303
adaptation 302–303
maturity and cropping systems 302, 305
breeding
disease resistance 312
high-protein content 314–316
insect resistance 312–313
resistance to waterlogging 313–314
crop geometry 332–333
dibbled 333–334
drilling 333
drip fertigation 344–346
early-maturing varieties 351–355
climate and soil 351–352
extra-short-duration varieties 351
harvesting 354
intercropping and rotations 353
nutrient management 353
plant population density 352–353
seed rate and treatment 352
short-duration varieties 351
sowing 352
water management 354
weed management 353–354
yield 354–355

enhancing sustainability
through crop modelling 306–307
through efficient seed system 307–308
through plant breeding 308–316
fertilizer efficiency 345–346
future trends 320
generation turnover in 309
genetic factors influencing 303–306
guidelines for future research 357–358
harvesting 356
human nutrition 298
hybrid technology
adoption 317–319
agronomic management 349–350
evolution 316–317
high yields 318
promotion and marketing 319
seed production 318–319
two-parent hybrid breeding technology 319–320
intercropping 334–336
irrigation scheduling 344–345
livestock nutrition 298–299
micro irrigation 343
mixed cropping 334–336
nutrient management 336–341
cropping systems 339–341
micronutrients 339
nitrogen 336–337
phosphorus 337–338
potassium 338
sulphur 339
nutritional security 298–299
pan and crop coefficient 345
physiological factors affecting 299–302
biomass 299–300
flowering 301–302
harvest index 300–301
photoperiod sensitivity 301–302
rabi 355
seed treatment 330
selection of variety 329–330
sowing 330–333
depth 333
method 332–333
time 330–332
storage 356
summer 355–356
threshing 356
variability 304
wastelands 311
waterlogging 313–314, 328, 333, 351
water management 341–343
weed management 347–349
Pink-pigmented facultative methylotrophic (PPFMs) bacteria 259
Plant growth promoting rhizobacteria (PGPR) 162, 330
Polymorphism, MAS 89–91
Polythene mulch, groundnuts cultivation under 166–167
Potassium
groundnut 160
pigeonpea 338
Punjab Agricultural University (PAU) 331
Purdue Improved Crop Storage (PICS) bags 169
Quality Declared Seed (QDS) system 26–27

Quantitative trait loci (QTL) analysis
common beans 8–11, 13
groundnut 134
lentil 57
soybean 87, 91–92
biological nitrogen fixation 97–98
garbage-in, garbage-out 93
soybean cyst nematode 94–95, 97
sudden death syndrome 94–97

Rabi pigeonpea 331, 355
Rain-fed low and mid-hills 310–311
Ready-to-use therapeutic food (RUTF) 201
Recombinant inbred lines (RILs) 87
Recurrent cross populations 92–93
Recurrent parent genotype (RPG) 85, 92
Relay cropping 78, 282
Restriction fragment length polymorphism (RFLP) 88
Rhizobia, inoculation with 108–109
Rhizobium leguminosarum bv. *viciae* 52

Satellite RNA (satRNA) 128–129
SDS. *see* Sudden death syndrome (SDS)
Seed infection, symptoms of 26
Seed priming techniques 163–164
Seed quality
common bean 26–27
lentil 75
Seed treatment
early-maturing pigeonpea varieties 352
groundnuts 162–164, 173
pigeonpea 307–308, 330
Seed yield
extra-short-duration pigeonpea lines 308
Phaseolus vulgaris 9–10
and protein harvest 315
quantitative trait loci 11
Selectively amplified microsatellite polymorphic loci (SAMPL) 88
Serenut 1–4 groundnut series 141
Short-duration (SD) varieties
cowpea 221–222
pigeonpea 329, 331, 351–355
Single nucleotide polymorphisms (SNPs) 88–89
BARCBean3_6K Beadchip 8, 11, 13
common beans 7–11

lentil 56
soybean 84–86
Single seed descent (SSD) method 6
SNF. *see* Symbiotic nitrogen fixation (SNF)
Soil acidity
common bean 23–25, 42
faba bean 254, 259
Soilborne fungal diseases 168
Soilborne plant pathogens 163
Soil fertility
common bean 23–25, 40–42
faba bean in China 273
groundnut 187
Soil organic matter (SOM) 187–188
Soil plant analysis development (SPAD) 135
Soil resources management 159
Soil texture 187–188
Sowing
early-maturing pigeonpea varieties 352
groundnut 162, 164–165, 188–189
pigeonpea 330–333
Soybean
in Brazil 109–110
in Butere, west Kenya 117
choice of marker 88–89
Democratic Republic of the Congo 113, 114
diseases 115–116
DNA markers 84, 88
geneticists and breeders 83
intellectual property 84–85
marker-assisted selection 84
development 85–87
future trends 98
genetic and association map development 91–92
identification of polymorphism 89–91
methods for 85
QTL 91–92, 97–98
recurrent cross populations 92–93
recurrent parent genotype 85, 92
scoring of phenotypes 93
selection programme 88–89
soybean cyst nematode 94–95, 97
steps involved in 86
sudden death syndrome 94–97
in Nigeria 110–112
single-nucleotide polymorphisms 84–86
sub-Saharan Africa 105–107
access to output market 119
climate change impact 118
consumption 107–108
extension and advisory services 119
improved access to inputs 118
inoculation with rhizobia 108–109
input financing 118–119
integrated approaches 116–118
managing diseases and insect pests 115–116
nutrient management 112–114
optimizing plant population 114–115
production 107–108
suitability map 106
varieties 111–112
yields 108–112, 115
Soybean cyst nematode (SCN) 94–95, 97
Soybean rust 115–116
SPAD chlorophyll meter reading (SCMR) 135
Specific leaf area (SLA) groundnut 135
Stemphylium blight (SB) 53–54, 63
Storage
groundnuts 170–171
aflatoxin contamination 190–192
insect pests 170
lentil seed 79
pigeonpea 356
Stress. *see specific stress*
Striga gesnerioides 223–224
Strip intercropping system 238, 239, 241
Sub-Saharan Africa (SSA)
common bean 20, 22–23
soybean 105–107
access to output market 119
agro-ecology 110, 113
climate change impact 118
consumption 107–108
extension and advisory services 119
improved access to inputs 118
inoculation with rhizobia 108–109
input financing 118–119
integrated approaches 116–118
managing diseases and insect pests 115–116
nutrient management 112–114
optimizing plant population 114–115
production 107–108
suitability map 106
varieties 111–112
yields 108–112, 115
Sudden death syndrome (SDS) 94–97
Sulphur (S)
deficiency 161
lentil 73, 75
pigeonpea 339
Summer pigeonpea 355–356
Symbiotic nitrogen fixation (SNF) 8, 11, 25, 38

Tamil Nadu Agricultural University (TNAU) 344, 346
TaqMan 84–86, 89, 95
Tillage system
faba bean 281–282
groundnut 158, 159, 172, 175
lentil 74–76
Tomato spotted wilt virus (TSWV) 129–130, 169
Transpiration efficiency (TE) 135
2,3,6-Trichlorobenzoic acid (TCBA) 167

Triticum aestivum 11
TSWV. *see* Tomato spotted wilt virus (TSWV)
Two-parent hybrid breeding technology 319–320

Uganda, groundnut in
- aflatoxin contamination 193–195
- improvement 137–143

Unpredictable legume. *see* Groundnut

Vesicles 162
Vicia faba L. *see* Faba bean
Vigna unguiculata (L.) Walp. *see* Cowpea
Vigna vexillata 224
Viruses. *see also specific viruses*
- cowpea 222
- groundnut 128–131, 169

Wastelands, pigeonpea 311
Waterlogging, pigeonpea 313–314, 328, 333, 351
Water management
- common bean 33–36
  - drip irrigation 34–36
  - furrow irrigation 34, 35
- early-maturing pigeonpea varieties 354
- faba bean 280–282
- groundnuts 166
- pigeonpea 341–343

Water use efficiency (WUE)
- common bean 35–36
- groundnut 135
- pigeonpea 334, 341–343, 346

Weed management
- early-maturing pigeonpea varieties 353–354
- faba bean 276
- groundnuts 165
- lentil 79
- pigeonpea 347–349

West Africa, cowpea 235–236
- future trends 247
- integrated pest management 245–247
- intercropping systems 237–241
- manipulating planting dates to improve productivity 241–244
- nutrient management 244–245
- optimal population 236–237

Wheat-based cropping systems 310
Whole-genome shotgun (WGS) assembly 229
Winter lentil 78

Yield gap
- common beans 4
- groundnuts 156–157

Yields
- Andean bean 5, 6
- common beans 4–5
  - component compensation 5
  - factors affect 5–8
  - pyramid 6–7
  - under stress 7–8
- early-maturing pigeonpea varieties 354–355
- groundnut 126–127
- lentil 56–58
- pigeonpea hybrids 318
- soybean 108–112, 115–118

Zinc
- cowpea varieties 227
- faba bean 286–287
- groundnuts 161